Lecture Notes in Computer Science 3381

Commenced Publication in 1973
Founding and Former Series Editors:
Gerhard Goos, Juris Hartmanis, and Jan van Le

Peter Vojtáš
Mária Bieliková
Bernadette Charron-Bost
Ondrej Sýkora (Eds.)

SOFSEM 2005:
Theory and Practice
of Computer Science

31st Conference on Current Trends
in Theory and Practice of Computer Science
Liptovský Ján, Slovakia, January 22-28, 2005, Proceedings

 Springer

Volume Editors

Peter Vojtáš
P.J. Šafárik University, Department of Computer Science, Faculty of Science
Jesenná 5, 04154 Košice, Slovak Republic
E-mail: vojtas@upjs.sk

Mária Bieliková
Slovak University of Technology
Faculty of Informatics and Information Technologies
Ilkovičova 3, 812 19 Bratislava, Slovak Republic
E-mail: bielik@fiit.stuba.sk

Bernadette Charron-Bost
LIX, École Polytechnique
91128 Palaiseau Cedex, France
E-mail: charron@lix.polytechnique.fr

Ondrej Sýkora
Loughborough University, Department of Computer Science
Loughborough, Leicestershire LE11 3TU, UK
E-mail: o.sykora@lboro.ac.uk

Library of Congress Control Number: 2004117173

CR Subject Classification (1998): F.2, F.1, D.2, G.2, H.3, H.2.8, H.4, F.3

ISSN 0302-9743
ISBN 3-540-24302-X Springer Berlin Heidelberg New York

Springer is a part of Springer Science+Business Media

springeronline.com

© Springer-Verlag Berlin Heidelberg 2005
Printed in Germany

Typesetting: Camera-ready by author, data conversion by Scientific Publishing Services, Chennai, India
Printed on acid-free paper   SPIN: 11375982   06/3142   5 4 3 2 1 0

Preface

This volume contains papers selected for presentation at the 31st Annual Conference on Current Trends in Theory and Practice of Informatics – SOFSEM 2005, held on January 22–28, 2005 in Liptovský Ján, Slovakia.

The series of SOFSEM conferences, organized alternately in the Czech Republic and Slovakia since 1974, has a well-established tradition. The SOFSEM conferences were originally intended to break the Iron Curtain in scientific exchange. After the velvet revolution SOFSEM changed to a regular broad-scope international conference. Nowadays, SOFSEM is focused each year on selected aspects of informatics. This year the conference was organized into four tracks, each of them complemented by two invited talks:

– *Foundations of Computer Science* (Track Chair: Bernadette Charron-Bost)
– *Modeling and Searching Data in the Web-Era* (Track Chair: Peter Vojtáš)
– *Software Engineering* (Track Chair: Mária Bieliková)
– *Graph Drawing* (Track Chair: Ondrej Sýkora)

The aim of SOFSEM 2005 was, as always, to promote cooperation among professionals from academia and industry working in various areas of informatics. Each track was complemented by two invited talks.

The SOFSEM 2005 Program Committee members coming from 13 countries evaluated 144 submissions (128 contributed papers and 16 student research forum papers). After a careful review process (counting at least 3 reviews per paper), followed by detailed discussions in the PC, and a co-chairs meeting held on October 8, 2005 in Bratislava, Slovakia, 44 papers (overall acceptance rate 34.38%) were selected for presentation at SOFSEM 2005: 28 full contributed papers (acceptance rate 21.88%) and 16 short contributed papers selected by the SOFSEM 2005 PC for publication in the Springer LNCS proceedings volume.

An integral part of SOFSEM is the Student Research Forum. The forum offers the opportunity to publish, present and discuss student projects, to receive feedback on both the original results of scientific work as well as on work in progress. The Program Committee selected 7 papers for publication (from 16 submitted) for presentation at the Student Research Forum session. The best student paper by Martin Senft was selected and included in these proceedings.

We would like to thank all Program Committee members for their meritorious work in evaluating the submitted papers, as well as numerous additional referees who assisted the Program Committee members.

As editors of these proceedings, we are much indebted to all the contributors to the scientific program of the symposium, especially to the authors of the papers. Special thanks go to those authors who prepared the manuscripts according to the instructions and made life easier for us. We would also like to thank those who responded promptly to our requests for minor modifications and corrections in their manuscripts. The database and electronic support system for the Program Committee was designed by Rastislav Královič. Our special thanks go to Richard Královič for most of the hard technical work in preparing this volume. We are also thankful to the members of the Organizing Committee led by Dana Pardubská who made sure that the conference ran smoothly in a pleasant environment. Last, but not least, we want to thank Springer for excellent cooperation in the publication of this volume.

January, 2005   Mária Bieliková
                                                 Bernadette Charron-Bost
                                                 Ondrej Sýkora
                                                 Peter Vojtáš

Organization

Steering Committee

Branislav Rovan, *Chair* — Comenius University, Bratislava, Slovakia
Miroslav Bartošek, *Secretary* Masaryk University, Brno, Czech Republic
Mária Bieliková — Slovak University of Technology in Bratislava, Slovakia
Peter van Emde Boas — University of Amsterdam, The Netherlands
Keith G. Jefferey — CLRC RAL, Chilton, Didcot, Oxon, UK
Antonín Kučera — Masaryk University, Brno, Czech Republic
Július Štuller — Institute of Computer Science, Prague, Czech Republic
Gerard Tel — Utrecht University, The Netherlands
Petr Tůma — Charles University in Prague, Czech Republic

Program Committee

Mária Bieliková, *Co-chair* — Bratislava
Bernadette Charron-Bost, *Co-chair* — Palaiseau
Ondrej Sýkora, *Co-chair* — Loughborough
Peter Vojtáš, *Chair* — Košice

Claude Crepeau (Montréal)
Ulrich Eisenecker (Zweibrücken)
Ghica van Emde Boas (Heemstede)
Viliam Geffert (Košice)
Sergio Greco (Cosenza)
Patrick Healy (Limerick)
Juraj Hromkovič (Zürich)
Michael Kaufmann (Tübingen)
Rastislav Královič (Bratislava)
Jan Kratochvíl (Prague)
Giuseppe Liotta (Perugia)
Frederic Magniez (Orsay)
Peter Mellor (London)
Pavol Návrat (Bratislava)
Jerzy Nawrocki (Poznań)
Patrice Ossona de Mendez (Paris)

Catuscia Palamidessi (Palaiseau)
Dana Pardubská (Bratislava)
Dimitris Plexousakis (Heraklion)
Jaroslav Pokorný (Prague)
Karel Richta (Prague)
Mark Roantree (Dublin)
Klaus-Dieter Schewe (Palmerston North)
Václav Snášel (Ostrava)
Gerard Tel (Utrecht)
Ioannis Tollis (Dallas)
Petr Tůma (Prague)
Imrich Vrťo (Bratislava)
Krzysztof Wecel (Poznań)
Pavel Zezula (Brno)

Additional Referees

G. Andrejková
S. Barton
M. Batko
T. Biedl
H.-J. Boeckenhauer
D. Bongartz
S. Brlek
L. Bulej
N. Busi
T. Chothia
I. Černá
E. Danna
E. Di Giacomo
W. Didimo
S. Dobrev
V. Dohnal
V. Dujmovic
C. Durr
P. Eades
P. van Emde Boas
C. Endler
H. Fernau
S. Flesca
M. Forišek
F. Furfaro
J.M. Gabbay
J. Gajdošíková
M. Galbavý
D. Galmiche
A. Garg
L. Giordano
D. Gottesman
D. Gross
D. Gruska
P. Gurský
P. Hájek
P. Harrenstein

D. Hirschkoff
P. Hliněný
T. Horváth
L. Ištoňová
J. Ivančo
M. Jiřina Jr.
C. Johnen
G. Juhás
T. Kalibera
B. Katreniak
J. Kempe
M. Klazar
S. Kobourov
I. Koffina
G. Kokkinidis
J.-C. Konig
V. Koutný
S. Krajči
K. Kritikos
Ri. Královic
J. Kupke
S. Laplante
A. Laurent
F. Le Fessant
P. Lennartz
K. Lynch
E. Masciari
V. Mencl
D. Miller
P. Mlynarčík
I. Mlynková
F. Morain
F. Mráz
P. Nadeau
F.Z. Nardelli
L. Nebeský
R. Neruda

R. Ostertág
Ch. Papamanthou
M. Patrignani
J.-F. Perrot
I. Phillips
T. Pitner
M. Pizzonia
A. Pugliese
M. de Rougemont
M. Santha
S. Seibert
L. Sidirourgos
T. Skopal
A. Slobodová
J. Sochor
L. Stacho
J. Stern
L. Sanselme
D. Swierstra
A. Szabari
A. Tagarelli
G. Tel
W. Unger
F.D. Valencia
W.F. de la Vega
P. Veltri
P. Verbaan
Y. Verhoeven
F. Vernadat
M.G. Vigliotti
J. Vinař
M. Vomlelová
V. Vranić
G. Woeginger
R. de Wolf
I. Žežula

Organized by

Slovak Society for Computer Science

Faculty of Mathematics, Physics and Informatics,
Comenius University, Bratislava

in cooperation with

Czech Society for Computer Science

Faculty of Informatics and Information Technologies,
Slovak University of Technology, Bratislava

Faculty of Science, Institute of Informatics, P.J. Šafárik University, Košice

Supported by

European Association for Theoretical Computer Science

Slovak Research Consortium for Informatics and Mathematics

Organizing Committee

Vanda Hambálková
Vladimír Koutný
Rastislav Královič
Richard Královič
Zuzana Kubincová

Edita Máčajová
Marek Nagy
Dana Pardubská (chair)
Viliam Solčány

Table of Contents

The Best Student Paper

Short Contributed Papers

Discovering Treewidth

Hans L. Bodlaender

Institute of Information and Computing Sciences, Utrecht University,
P.O. Box 80.089, 3508 TB Utrecht, the Netherlands
hansb@cs.uu.nl

Abstract. Treewidth is a graph parameter with several interesting theoretical and practical applications. This survey reviews algorithmic results on determining the treewidth of a given graph, and finding a tree decomposition of small width. Both theoretical results, establishing the asymptotic computational complexity of the problem, as experimental work on heuristics (both for upper bounds as for lower bounds), preprocessing, exact algorithms, and postprocessing are discussed.

1 Introduction

About a quarter of a century, the notion of treewidth has now played a role in many investigations in algorithmic graph theory. While for a long time, the use of treewidth was limited to theoretical investigations, and it sometimes was believed that it could not play a role in practical applications, nowadays there is a growing tendency to use it in an actual applied setting.

An interesting example of this practical use of treewidth can be found in the work by Koster, van Hoesel, and Kolen [63], where tree decompositions are used to solve frequency assignment instances from the CALMA project, and other partial constraint satisfaction problems. The most frequent used algorithm to solve the inference problem for probabilistic, or Bayesian belief networks (often used in decision support systems) uses tree decompositions [67]. See e.g., also [1, 39].

Graphs of bounded treewidth appear in many different contexts. For an overview of graph theoretic notions that are equivalent to treewidth, or from which bounded treewidth can be derived, see [12]. Many probabilistic networks appear to have small treewidth in practice. Yamagucki, Aoki, and Mamitsuka [91] have computed the treewidth of 9712 chemical compounds from the LIGAND database, and discovered that all but one had treewidth at most three; the one exception had treewidth four. Thorup [86] showed that the control flow graph of goto-free programs, written in one of a number of common imperative programming languages (like C, Pascal) have treewidth bounded by small constants. See also [45].

Many problems can be solved in linear or polynomial time when the treewidth of the input graph is bounded. Usually, the first step of such an algorithm is to find a tree decomposition of small width. In this paper, we give an overview of

M. Bieliková et al. (Eds.): SOFSEM 2005, LNCS 3381, pp. 1–16, 2005.

algorithms for finding such tree decompositions. Nowadays, much work has been done on this topic, and we now have a rich theory, and intriguing experimental approaches.

After some preliminaries in Section 2, we survey exact algorithms (Section 3), approximation algorithms and upper bound heuristics (Section 4), lower bound heuristics (Section 5), and preprocessing and postprocessing methods (Section 6).

2   Preliminaries

The notion of treewidth was introduced by Robertson and Seymour in their work on graph minors [77]. Equivalent notions were invented independently, e.g., a graph has treewidth at most k, if and only if it is a partial k-tree. See [12] for an overview of notions equivalent to or related to treewidth. In this paper, we assume graphs to be undirected and simple. Many results also hold for directed graphs, and often they can be generalised to hypergraphs. $n = |V|$ denotes the number of vertices of graph $G = (V, E)$, $m = |E|$ its number of edges.

Definition 1. *A* tree decomposition *of a graph* $G = (V, E)$ *is a pair* $(\{X_i, i \in I\}, T = (I, F))$, *with* $\{X_i, i \in I\}$ *a collection of subsets of* V *(called* bags*), and* $T = (I, F)$ *a tree, such that*

1. $\bigcup_{i \in I} X_i = V$.
2. *For all* $\{v, w\} \in E$, *there is an* $i \in I$ *with* $v, w \in X_i$.
3. *For all* $v \in V$, $T_v = \{i \in I \mid v \in X_i\}$ *forms a connected subtree of* T.

The width *of a tree decomposition* $(\{X_i, i \in I\}, T = (I, F))$ *is* $\max_{i \in I} |X_i| - 1$. *The* treewidth *of* G *is the minimum width over all tree decompositions of* G.

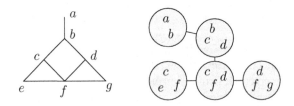

Fig. 1. A graph and a tree decomposition of width 2

Having a tree decomposition of small width in general implies that the graph has many separators of small size. E.g., consider a tree decomposition $(\{X_i, i \in I\}, T = (I, F))$, and choose a node $r \in I$ as root of T. Consider some node $i \in I$, and let G_i be the subgraph of G induced by the set V_i of vertices in sets X_j with $j = i$ or j is a descendant of i. Then, the definition of tree decomposition implies that all vertices in G_i that have a neighbour in G that does not belong to G_i belong to X_i. Hence, X_i separates all vertices in $V_i - X_i$ from all vertices in $V - V_i$. This property amongst others enables many dynamic programming algorithms on graphs of bounded treewidth. A useful lemma on tree decompositions (which can be seen as a rephrasing of the Helly property for trees, see [44, 22] is the following.

Lemma 1. *Let* $(\{X_i, i \in I\}, T = (I, F))$ *be a tree decomposition of* G. *Let* W *be a clique in* G. *Then there is an* $i \in I$ *with* $W \subseteq X_i$.

There are several equivalent definitions of the notion of treewidth. The various algorithms for determining the treewidth and finding tree decompositions are based upon different such notions. We review here those that we use later in this paper.

A graph $G = (V, E)$ is *chordal*, if and only if each cycle in G of length at least four has a chord, i.e., an edge between non-successive vertices in the cycle. There are two equivalent definitions of chordality that we will use. A *perfect elimination scheme* of a graph $G = (V, E)$ is an ordering of the vertices $v_1, \ldots, v_n$, such that for all $v_i \in V$, its higher numbered neighbours form a clique, i.e., for $j_1 > i$, $j_2 > i$, if $\{v_i, v_{j_1}\} \in E$ and $\{v_i, v_{j_2}\} \in E$, then $\{v_{j_1}, v_{j_2}\} \in E$. A graph $G = (V, E)$ is the *intersection graph of subtrees of a tree*, if and only if there is a tree $T = (I, F)$ and for each vertex $v \in V$ a subtree T_v of T, such that for all $v, w \in V$, $v \neq w$: $\{v, w\} \in E$, if and only if the trees T_v and T_w have at least one vertex in common.

Theorem 1 (See [42, 44]). *Let* $G = (V, E)$ *be a graph, The following statements are equivalent.*

1. *G is a chordal graph.*
2. *G has a perfect elimination scheme.*
3. *G is the intersection graph of subtrees of a tree.*

A *triangulation* of a graph $G = (V, E)$ is a chordal graph $H = (V, F)$ that contains G as a subgraph: $E \subseteq F$. A triangulation $H = (V, F)$ is *a minimal triangulation*, when there does not exist a triangulation $H' = (V, F')$ with $E \subseteq F' \subset F$ ($F' \neq F$).

Given a tree decomposition $(\{X_i, i \in I\}, T = (I, F))$ of G, we can build corresponding triangulation $H = (V, F)$: add to G an edge between each non-adjacent pair of vertices $\{v, w\}$ such that there is an $i \in I$ with $v, w \in X_i$. I.e., each bag X_i is turned into a clique. The graph H thus obtained is the intersection graph of subtrees $T_v = T[\{i \in I \mid v \in X_i\}]$ of T, thus chordal. The maximum cliquesize of H is exactly one larger than the width of the tree decomposition (compare with Lemma 1.)

Lemma 2. *The treewidth of a graph G equals the minimum over all triangulations H of G of the maximum clique size of H minus one.*

We can also build a tree decomposition from an ordering $v_1, \ldots, v_n$ of the vertices of the graph G. We first construct a triangulation H of G by the following *fill-in* procedure: initially, $H = G$, and then for $i = 1$ to n, we add to H, edges between yet non-adjacent higher numbered neighbours of v_i. After having done this, $v_1, \ldots, v_n$ is a perfect elimination scheme of H; the model of H as intersection graph of subtrees of a tree can be easily transformed to a tree decomposition of H and of G. Its width equals the maximum over all vertices of its higher numbered neighbours in the ordering in H.

3   Exact Algorithms

The TREEWIDTH problem: given a graph G, and an integer k, decide if the treewidth of G is at most k, is NP-complete [3]. This unsettling fact does not prevent us from wanting to compute the treewidth of graphs, and fortunately, in many cases, there are methods to effectively obtain the treewidth of given graphs.

Special Graph Classes. There are many results on the complexity of treewidth when restricted to special graph classes. We mention here a few of these. A highly interesting recent result was obtained by Bouchitté and Todinca, who found an algorithm to determine the treewidth of a graph in time, polynomial in the number of its minimal separators [29, 28]. Many graph classes have the property that each graph in the class has a polynomial number of minimal separators, e.g., permutation graphs, weakly chordal graphs.

Other polynomial time algorithms for treewidth for special graph classes can be found in e.g., [16, 23, 31, 32, 38, 59, 58, 60]. NP-completeness results appear amongst others in [24, 46]. See also [72]. Other older results are surveyed in [9].

Exponential Time Algorithms. Based upon the results from Bouchitté and Todinca [29, 28], Fomin et al. [41] found an exact algorithm for treewidth that runs in time $O^*(1.9601^n)$ time. (See [90] for the O^* notation and an introduction to exponential time algorithms.)

Algorithms with a running time of $O^*(2^n)$ are easier to obtain: one can show that the algorithm of [3] has this time, or build a dynamic programming algorithm following a technique first established for TSP by Held and Karp [49].

For small graphs, the treewidth can be computed in practice using *branch and bound*. Experiments have been published by Gogate and Dechter [43]. The algorithm searches for an ordering of the vertices that corresponds to a tree decomposition of small width, see Section 2, i.e., at each step, we select the next vertex in the ordering. Gogate and Dechter establish several rules to cut off branches during branch and bound. The algorithm can also be used as a heuristic, by stopping the branch and bound algorithm at a specific time and reporting the best solution found so far.

Fixed Parameter Cases. As we often want to use a tree decomposition for running a dynamic programming algorithm that is exponential in the width, we often want to test if the treewidth is smaller than some given constant k. Much work has been done of this fixed parameter case of treewidth. Here we let k denote the constant for which we want to test if the treewidth is at most k.

The first polynomial time algorithm for the problem was given by Arnborg, Corneil, and Proskurowski [3]. Their algorithm runs in $O(n^{k+2})$ time. A modification of this algorithm has been proposed and successfully experimentally evaluated by Shoikhet and Geiger [83].

Downey and Fellows introduced the theory of *fixed parameter tractability*. A problem with input parameter k and input size n is fixed parameter tractable, when there is a function f and a constant c, such that there is an algorithm that solves the problem in $f(k) \cdot n^c$ time, (in contrast to algorithms using $\Omega(n^{g(k)})$ time for some increasing function g). See [40]. The first result that showed that treewidth is fixed parameter tractable, i.e., solvable in $O(n^c)$ time for some constant c, for fixed treewidth k, was obtained by Robertson and Seymour [77, 78]. This result was fully non-constructive: from the deep results of their graph minor theory, one gets a non-constructive proof that there exists a characterisation that can be tested in $O(n^2)$ time. Later results, by Lagergren [64], Reed [76], Lagergren and Arnborg [65], Bodlaender and Kloks [15], and Bodlaender [10] improved upon either the constructivity or the running time. Finally, in [11], a linear time algorithm was given that checks if the treewidth is at most k, and if so, outputs the corresponding tree decomposition. That algorithm uses about $O(k^3)$ calls to the dynamic programming algorithm from [15], but the hidden constant in the 'O'-notation of this algorithm is horrendous, even for small values of k. Röhrig [79] has experimentally evaluated the linear time algorithm from [11]. Unfortunately, this evaluation shows that the algorithm uses too much time even for very small values of k (e.g., when $k = 4$.) A parallel variant of the algorithm from [11] was given by Bodlaender and Hagerup [14]. A variant with $O(k^2)$ calls to the algorithm of [15] was given by Perković and Reed [73].

The linear time algorithm for fixed k is attractive from a theoretical point of view: in many cases, an algorithm exploiting small treewidth would use the algorithm as a first step. From a practical point of view, the algorithm is useless however, and the quest remains for algorithms that are efficient from the implementation viewpoint. Fortunately, several heuristics appear to perform well in practice, as we will see in the next section.

Also, for very small values of k, there are special algorithms. Testing if the treewidth is one is trivially linear (the graph must be a forest), a graph has treewidth at most two, if and only if each biconnected component is a series parallel graph (see e.g., [25]), and testing if a graph is series parallel can be done in linear time by the algorithm of Valdes, Tarjan, and Lawler [88]. Arnborg and Proskurowski [4] give a set of six reduction rules, such that a graph has treewidth at most three, if and only if it can be reduced to the empty graph by means of these rules. These rules can be implemented such that the algorithm runs in linear time, see also [70]. Experiments show that these algorithms run very fast in practice. A more complicated linear time algorithm for testing if the treewidth of a graph is at most 4 has been given by Sanders [81]. As far as I know, this algorithm has not yet been tried out in practice.

4   Approximation Algorithms and Upper Bound Heuristics

There are many algorithms that approximate the treewidth. We can distinguish a number of different types, depending on whether the algorithm is polynomial for all values of k, and whether the algorithm has a guaranteed performance.

Polynomial Time Approximation Algorithms with a Performance Ratio. We first look at algorithms that are polynomial, even when k is not bounded, and that give a guarantee on the quality of the output. The first such approximation algorithm for treewidth was given in [13]. This algorithm gives tree decompositions with width at most $O(\log n)$ times the optimal treewidth. (See also [57].) It builds a tree decomposition by repeatedly finding balanced separators in the graph and subgraphs of it. Bouchitté et al. [27] and Amir [2] recently improved upon this result, giving polynomial time approximation algorithms with ratio $O(\log k)$, i.e., the algorithms output a tree decomposition of width $O(k \log k)$ when the treewidth of the input graph is k. It is a long standing and apparently very hard open problem whether there exist a polynomial time approximation algorithm for treewidth with a constant performance ratio.

Fixed Parameter Approximation Algorithms. There are also several approximation algorithms for treewidth that run in time, exponential in k. They either give a tree decomposition of width at most ck (for some constant c), or tell that the treewidth is more than k. See [2, 6, 64, 76, 78].

Upper Bound Heuristics. Many of the heuristics that have been proposed and are used to find tree decompositions of small width do not have a guarantee on their performance. However, amongst these, there are many that appear to perform very well in many cases.

A large class of these heuristics is based upon the same principle. As discussed in Section 2, a tree decomposition can be build from a linear ordering of the vertices. Thus, we can build in some way a linear ordering of the vertices, run the fill-in procedure, and turn the triangulation into a tree decomposition. Often, one already adds fill-in edges during the construction of the linear order.

A very simple heuristic of this type is the Minimum Degree heuristic: we repeatedly select the vertex v with the minimum number of unselected neighbours as the next vertex in the ordering, and turn the set of its unselected neighbours into a clique, then temporarily remove v. The Minimum Fill-in heuristic is similar, but now we select a vertex which gives the minimum number of added fill-in edges for the current step. More complicated rules for selecting next vertices have been proposed by Bachoore and Bodlaender [5], and by Clautiaux et al. [34, 33]. In some cases, improvements are thus made upon the simpler Minimum Degree or Minimum Fill-in heuristics. Also, sometimes orderings generated by algorithms originally invented for chordal graph recognition ([80, 85, 7] have been used as linear ordering to generate the tree decomposition from. These tend to give tree decompositions with larger width. See [62] for an experimental evaluation of several of these heuristics.

These heuristics can also be described using tree decompositions only: Select some vertex v (according to the criteria at hand, e.g., the vertex of minimum degree). Build the graph G', by turning the set of neighbours $N(v)$ of v into a clique, and then removing v. Recursively, compute a tree decomposition of G'. By Lemma 1, there must be a bag i^* with $N(v) \subseteq X_i$. Now, add a new node i_v to G, with $X_{i_v} = \{v\} \cup N(v)$, and make i_v adjacent to i^* in the tree. One can check that this gives a tree decomposition of G.

A different type of heuristic was proposed by Koster [61]. The main idea of the heuristic is to start with any tree decomposition, e.g., the trivial one where all vertices belong to the same bag, and then stepwise refine the heuristics, i.e., the heuristic selects a bag and splits it into smaller bags, maintaining the properties of tree decomposition.

There are several algorithms that, given a graph G, make a minimal triangulation of G. While not targeted at treewidth, such algorithms can be used as treewidth heuristic. Recently, Heggernes, Telle, and Villanger [47] found an algorithm for this problem that uses $o(n^{2.376})$ time; many other algorithms use $O(nm)$ time.

See [87] for an online database with some experimental results.

Heuristics with Local Search Methods. Some work has been done on using stochastic local search methods to solve the treewidth problem or related problems. Kjærulff [56] uses simulated annealing to solve a problem related to treewidth. Genetic algorithms have been used by Larrañaga et al. [66]. Clautiaux et al. [33] use tabu search for the treewidth problem. The running times of these meta heuristics is significantly higher, but good results are often obtained.

Approximation Algorithms for Special Graph Classes. Also, approximation algorithms have been invented with a guarantee on the performance for special graph classes, e.g., a ratio of 2 can be obtained for AT-free graphs [30], and a constant ratio can be obtained for graphs with bounded asteroidal number [27].

5   Lower Bound Heuristics

It is for several reasons interesting to have good lower bound heuristics for treewidth. They can be used in a subroutines in a branch and bound algorithm (as, e.g., is done in [43]), or in an upper bound heuristic (e.g., as part of the rule to select the next vertex of the vertex ordering [33]), and inform us about the quality of upper bound heuristics. Also, when a lower bound for the treewidth is too high, it may tells us that it is not useful to aim at a dynamic programming algorithm solving a problem with tree decompositions.

It is easy to see that the minimum degree of a vertex in G, and the maximum clique size of G are lower bounds for the treewidth. These bounds are often not very good, and the maximum clique size is NP-hard to compute. An improvement to these bounds is made with the *degeneracy*: the maximum over all subgraphs G' of G of the minimum vertex degree of G' [84, 68]. The degeneracy can be easily computed: repeatedly remove a vertex of minimum degree from the graph, and then report the maximum over the degrees of the vertices when they were removed.

An improvement to the degeneracy can be obtained by instead of removing a vertex, contracting it to one of its neighbours. This idea was found independently by Bodlaender, Koster, and Wolle [20], and by Gogate and Dechter [43]. The MMD+ heuristic thus works as follows: set $\ell = 0$, then repeat until G is empty: find a vertex v of minimum degree d in G; set $\ell = \max(\ell, d)$; contract v to a neighbour (or remove v if v is isolated). In [20], different rules to select the vertex to contract to are explored. The heuristic to select the neighbour of v of smallest degree performs reasonably well, but the heuristic to select the neighbour w of v such that v and w have the smallest number of common neighbours usually gives better lower bounds. (When v and w have a common neighbour x, then contracting v and w causes the two edges $\{v, x\}$ and $\{w, x\}$ to become the same edge. The rule thus tries to keep the graph as dense as possible.)

In [20], the related graph parameter of *contraction degeneracy*: the maximum over all minors G' of G of the minimum vertex degree of G' is introduced and studied. Computing the contraction degeneracy is NP-hard [20], but it can be computed in polynomial time on cographs [26].

A different lower bound rule, based on the Maximum Cardinality Search algorithm has been invented by Lucena [69]. Maximum Cardinality Search (originally invented as a chordal graph recognition algorithm by Tarjan and Yannakakis [85]) works as follows. The vertices of the graph are visited one by one. MCS starts at an arbitrary vertex, and then repeatedly visits an unvisited vertex which has the maximum number of visited neighbours. Lucena showed that when MCS visits a vertex that has at that point k visited neighbours, then the treewidth is at least k.

So, we can get a treewidth lower bound by constructing an MCS ordering of the vertices of G, and then reporting the maximum over all vertices of the number of its visited neighbours when it was visited. This bound is always at least the degeneracy (if G has a subgraph G' with minimum vertex degree k, then when the last vertex from G' is visited, it has at least k visited neighbours). A theoretical and experimental analysis of this lower bound was made by Bodlaender and Koster [17]. E.g., it is NP-hard to find an MCS ordering that maximises the yielded lower bound.

Other lower bounds based on the degree are also possible. Ramachandra-murthi [74, 75] showed that for all graphs that are not complete, the minimum over all non-adjacent pairs of vertex v and w of the maximum of the degree of v and w is a lower bound for the treewidth of G. (This bound can be shown as follows. Consider a tree decomposition of G, and repeatedly remove leaf nodes i from T with neighbour j in T with $X_i \subseteq X_j$. If we remain with a tree decomposition with one bag, the claim clearly holds. Otherwise, T has at least two leaf nodes, and each bag of a leaf node contains a vertex whose neighbours are all in the same leaf bag.) This lower bound usually is not very high, but when combined with contraction, it can give small improvements to the MMD+ lower bound. An investigation of this method, and other methods combining degree based lower bounds with contraction is made in [21].

An interesting technique to obtain better lower bounds was introduced by Clautiaux et al. in [34]. It uses the following result.

Lemma 3. *Let v, w be two vertices in G, and let v and w have at least $k + 2$ disjoint neighbours (vertex disjoint paths between them). Then G has treewidth at most k, if and only if $G + \{v, w\}$ has treewidth at most k.*

The *neighbour* or *path improved graph* of G is the graph obtained by adding edges between all pairs of vertices with at least $k + 2$ common neighbours (or vertex disjoint paths). The method of [34] now can be described as follows. Set ℓ to some lower bound on the treewidth of input graph G. Compute the (neighbour or path) improved graph G' of G (with $k = \ell$). Run some treewidth lower bound algorithm on G'. If this algorithm gives a lower bound larger than ℓ, then the treewidth of G is at least $\ell + 1$, and we add one to ℓ, and repeat, until no increase to ℓ is obtained. In [34], the degeneracy is used as lower bound subroutine, but any other lower bound can be used. Experimental results of this type can be found in [20]. The method gives significant increases to the lower bounds for many graphs, but also costs much time; the version where we use the neighbour improved graph gives smaller bounds but uses also less time when compared to the path improved graph. In [20], a heuristic is proposed, where edge contraction steps are alternated with improvement steps. This algorithm works well for small instances, but appears to use (too) much time on larger instances.

6   Preprocessing and Postprocessing

6.1   Preprocessing

There are several methods for preprocessing a graph before running an algorithm for treewidth on it. With preprocessing, we hope to decrease the size of the input graph. The algorithm for treewidth thus often runs on a smaller instance, and hence can be much faster. E.g., we first preprocess the graph, and then run a slow exact algorithm on the reduced instance.

Reduction Rules. Bodlaender et al. [19] give several reduction rules that are *safe* for treewidth. Besides a graph (initially the input graph), we maintain an integer variable *low* that is a lower bound for the treewidth of the input graph. We have that initially *low* $\leq tw(G)$, (e.g., *low*= 0. Each reduction rule takes G and *low*, and rewrites this to a smaller graph G', with possibly an updated value of *low*. A rule is safe, if, whenever we can rewrite a graph G with variable *low* to G' and *low'*, we have $\max(tw(G), low) = \max(tw(G'), low')$. It follows that when G'' and *low''* are obtained from G with a series of applications of safe rules, then the treewidth of G equals $\max(tw(G''), low'')$. The rules in [19] are taken from the algorithm from [4] to recognise graphs of treewidth at most three, or generalisations of these. Two of these rules are the *simplicial rule*: remove a vertex of degree d whose neighbours form a clique, and set *low* to $\max(d, low)$, and the *almost simplicial rule*: when v is a vertex of degree $d \leq low$ whose neighbourhood contains a clique of size $d - 1$, then add edges between non-adjacent neighbours of v and remove v. Experiments show that in many instances from practical problems, significant reductions can be obtained with these reduction rules [19]. Generalisations of the rules were given by van den Eijkhof and Bodlaender [89].

Safe Separators. A set of vertices $S \subseteq V$ is a *separator* in a graph $G = (V, E)$, if $G[V - S]$ contains more than one connected component. A separator is *inclusion minimal*, when it does not contain another separator of G as proper subset. A separator S in G is *safe for treewidth*, when the treewidth of G equals the maximum over all connected components W of $G[V - S]$ of the treewidth of the graph $G[W \cup S] + clique(S)$ (i.e., the graph obtained with vertices $V \cup S$, and edges between adjacent vertices in G, and each pair of vertices in S).

Thus, when we have a safe (for treewidth) separator S in G, we can split G into the parts of the form $G[W \cup S] + clique(S)$ for all connected components W of $G[V - S]$, and compute for each such part the treewidth separately. Hence, safe separators can be used for preprocessing for treewidth.

There are several types of safe separators that can be found quickly. For instance, every separator that is a clique is safe (see [71]), and clique separators can be found in $O(nm)$ time. Other safe separators are given in [18], e.g., inclusion minimal separators of size r that contain a clique of size $r - 1$; all inclusion separators of size two; minimum size separators S of size three such that at least two connected components of $G[V - S]$ contain at least two vertices. See also [18] for an experimental evaluation of the use of safe separators.

6.2   Postprocessing

Once we have found a tree decomposition of G, it sometimes is possible to modify the tree decomposition slightly to obtain one with a smaller width. This can be best explained by looking at the triangulation of G that corresponds to the tree decomposition.

Many heuristics yield tree decompositions whose corresponding triangulations are not always minimal triangulations, e.g., the minimum degree heuristic. (A few heuristics guarantee that the triangulation is always minimal.)

There are several algorithms, that, given a graph $G = (V, E)$, and a triangulation $H = (V, F)$ of G, find a minimal triangulation $H' = (V, F')$ of G that is a subgraph of H: $E \subseteq F' \subseteq F$ [8, 37, 48]. So, we can use the following postprocessing step when given a tree decomposition: build the corresponding triangulation, find a minimal triangulation (e.g., with an algorithm from [8, 48]) and then turn this minimal triangulation back into a tree decomposition.

7   Conclusions

There are several interesting notions that are related to treewidth, and that obtained also much attention in the past years, e.g., pathwidth, cliquewidth (see e.g. [36]). Very closely related to treewidth is the notion of branchwidth (treewidth and branchwidth differ approximately by at most a factor of 1.5). The branchwidth of planar graphs can be computed in polynomial time [82], and thus it is intriguing that the corresponding problem for treewidth is still open. Interesting experimental work on branchwidth has been done by Hicks [53, 54, 55, 51, 52, 50]. Cook and Seymour [35] used branch decompositions for solving the travelling salesman problem.

The many theoretic and experimental results on the treewidth problem show that finding a tree decomposition of small width is far from hopeless, even while the problem itself is NP-hard. Upper and lower bound heuristics appear to give good results in many practical cases, which can be further improved by post-processing; preprocessing combined with cleverly designed exact algorithms can solve many small instances exactly. There still are several challenges. Two theoretical questions are open for a long time, and appear to be very hard: Is there an approximation algorithm for treewidth with a constant performance ratio (assuming $P \neq NP$)? Does there exist a polynomial time algorithm for computing the treewidth of planar graphs, or is this problem NP-hard? Also, the quest for better upper and lower bound heuristics, more effective preprocessing methods, etc. remains.

Acknowledgement

I want to express my gratitude to the many colleagues who collaborated with me on the research on treewidth and other topics, helped me with so many things, and from who I learned so much in the past years. I apologise to all whose work should have been included in this overview but was inadvertingly omitted by me.

References

1. J. Alber, F. Dorn, and R. Niedermeier. Experimental evaluation of a tree decomposition based algorithm for vertex cover on planar graphs. To appear in Discrete Applied Mathematics, 2004.
2. E. Amir. Efficient approximations for triangulation of minimum treewidth. In *Proc. 17th Conference on Uncertainty in Artificial Intelligence*, pages 7–15, 2001.
3. S. Arnborg, D. G. Corneil, and A. Proskurowski. Complexity of finding embeddings in a k-tree. *SIAM J. Alg. Disc. Meth.*, 8:277–284, 1987.
4. S. Arnborg and A. Proskurowski. Characterization and recognition of partial 3-trees. *SIAM J. Alg. Disc. Meth.*, 7:305–314, 1986.
5. E. Bachoore and H. L. Bodlaender. New upper bound heuristics for treewidth. Technical Report UU-CS-2004-036, Institute for Information and Computing Sciences, Utrecht University, Utrecht, the Netherlands, 2004.

6. A. Becker and D. Geiger. A sufficiently fast algorithm for finding close to optimal clique trees. *Artificial Intelligence*, 125:3–17, 2001.
7. A. Berry, J. Blair, P. Heggernes, and B. Peyton. Maximum cardinality search for computing minimal triangulations of graphs. *Algorithmica*, 39:287–298, 2004.
8. J. R. S. Blair, P. Heggernes, and J. Telle. A practical algorithm for making filled graphs minimal. *Theor. Comp. Sc.*, 250:125–141, 2001.
9. H. L. Bodlaender. A tourist guide through treewidth. *Acta Cybernetica*, 11:1–23, 1993.
10. H. L. Bodlaender. Improved self-reduction algorithms for graphs with bounded treewidth. *Disc. Appl. Math.*, 54:101–115, 1994.
11. H. L. Bodlaender. A linear time algorithm for finding tree-decompositions of small treewidth. *SIAM J. Comput.*, 25:1305–1317, 1996.
12. H. L. Bodlaender. A partial k-arboretum of graphs with bounded treewidth. *Theor. Comp. Sc.*, 209:1–45, 1998.
13. H. L. Bodlaender, J. R. Gilbert, H. Hafsteinsson, and T. Kloks. Approximating treewidth, pathwidth, frontsize, and minimum elimination tree height. *J. Algorithms*, 18:238–255, 1995.
14. H. L. Bodlaender and T. Hagerup. Parallel algorithms with optimal speedup for bounded treewidth. *SIAM J. Comput.*, 27:1725–1746, 1998.
15. H. L. Bodlaender and T. Kloks. Efficient and constructive algorithms for the pathwidth and treewidth of graphs. *J. Algorithms*, 21:358–402, 1996.
16. H. L. Bodlaender, T. Kloks, D. Kratsch, and H. Mueller. Treewidth and minimum fill-in on d-trapezoid graphs. *J. Graph Algorithms and Applications*, 2(5):1–23, 1998.
17. H. L. Bodlaender and A. M. C. A. Koster. On the maximum cardinality search lower bound for treewidth, 2004. Extended abstract to appear in proceedings WG 2004.
18. H. L. Bodlaender and A. M. C. A. Koster. Safe separators for treewidth. In *Proceedings 6th Workshop on Algorithm Engineering and Experiments ALENEX04*, pages 70–78, 2004.
19. H. L. Bodlaender, A. M. C. A. Koster, F. van den Eijkhof, and L. C. van der Gaag. Pre-processing for triangulation of probabilistic networks. In J. Breese and D. Koller, editors, *Proceedings of the 17th Conference on Uncertainty in Artificial Intelligence*, pages 32–39, San Francisco, 2001. Morgan Kaufmann.
20. H. L. Bodlaender, A. M. C. A. Koster, and T. Wolle. Contraction and treewidth lower bounds. In S. Albers and T. Radzik, editors, *Proceedings 12th Annual European Symposium on Algorithms, ESA2004*, pages 628–639. Springer, Lecture Notes in Computer Science, vol. 3221, 2004.
21. H. L. Bodlaender, A. M. C. A. Koster, and T. Wolle. Degree-based treewidth lower bounds. Paper in preparation, 2004.
22. H. L. Bodlaender and R. H. Möhring. The pathwidth and treewidth of cographs. *SIAM J. Disc. Math.*, 6:181–188, 1993.
23. H. L. Bodlaender and U. Rotics. Computing the treewidth and the minimum fill-in with the modular decomposition. *Algorithmica*, 36:375–408, 2003.
24. H. L. Bodlaender and D. M. Thilikos. Treewidth for graphs with small chordality. *Disc. Appl. Math.*, 79:45–61, 1997.
25. H. L. Bodlaender and B. van Antwerpen-de Fluiter. Parallel algorithms for series parallel graphs and graphs with treewidth two. *Algorithmica*, 29:543–559, 2001.
26. H. L. Bodlaender and T. Wolle. Contraction degeneracy on cographs. Technical Report UU-CS-2004-031, Institute for Information and Computing Sciences, Utrecht University, Utrecht, the Netherlands, 2004.

27. V. Bouchitté, D. Kratsch, H. Müller, and I. Todinca. On treewidth approximations. *Disc. Appl. Math.*, 136:183–196, 2004.

28. V. Bouchitté and I. Todinca. Treewidth and minimum fill-in: Grouping the minimal separators. *SIAM J. Comput.*, 31:212–232, 2001.

29. V. Bouchitté and I. Todinca. Listing all potential maximal cliques of a graph. *Theor. Comp. Sc.*, 276:17–32, 2002.

30. V. Bouchitté and I. Todinca. Approximating the treewidth of at-free graphs. *Disc. Appl. Math.*, 131:11–37, 2003.

31. H. Broersma, E. Dahlhaus, and T. Kloks. A linear time algorithm for minimum fill in and tree width for distance hereditary graphs. *Disc. Appl. Math.*, 99:367–400, 2000.

32. H. Broersma, T. Kloks, D. Kratsch, and H. Müller. A generalization of AT-free graphs and a generic algorithm for solving triangulation problems. *Algorithmica*, 32:594–610, 2002.

33. F. Clautiaux, S. N. A. Moukrim, and J. Carlier. Heuristic and meta-heuristic methods for computing graph treewidth. *RAIRO Oper. Res.*, 38:13–26, 2004.

34. F. Clautiaux, J. Carlier, A. Moukrim, and S. Négre. New lower and upper bounds for graph treewidth. In J. D. P. Rolim, editor, *Proceedings International Workshop on Experimental and Efficient Algorithms, WEA 2003*, pages 70–80. Springer Verlag, Lecture Notes in Computer Science, vol. 2647, 2003.

35. W. Cook and P. D. Seymour. Tour merging via branch-decomposition. *Informs J. on Computing*, 15(3):233–248, 2003.

36. B. Courcelle, J. A. Makowsky, and U. Rotics. Linear time solvable optimization problems on graphs of bounded clique width. *Theor. Comp. Sc.*, 33:125–150, 2000.

37. E. Dahlhaus. Minimal elimination ordering inside a given chordal graph. In *Proceedings 23rd International Workshop on Graph-Theoretic Concepts in Computer Science WG'97*, pages 132–143. Springer Verlag, Lecture Notes in Computer Science, vol. 1335, 1997.

38. E. Dahlhaus. Minimum fill-in and treewidth for graphs modularly decomposable into chordal graphs. In *Proceedings 24th International Workshop on Graph-Theoretic Concepts in Computer Science WG'98*, pages 351–358. Springer Verlag, Lecture Notes in Computer Science, vol. 1517, 1998.

39. R. Dechter. Bucket elimination: a unifying framework for reasoning. *Acta Informatica*, 113:41–85, 1999.

40. R. G. Downey and M. R. Fellows. *Parameterized Complexity.* Springer, 1998.

41. F. V. Fomin, D. Kratsch, and I. Todinca. Exact (exponential) algorithms for treewidth and minimum fill-in. In *Proceedings of the 31st International Colloquium on Automata, Languages and Programming*, pages 568–580, 2004.

42. F. Gavril. The intersection graphs of subtrees in trees are exactly the chordal graphs. *J. Comb. Theory Series B*, 16:47–56, 1974.

43. V. Gogate and R. Dechter. A complete anytime algorithm for treewidth. In proceedings UAI'04, Uncertainty in Artificial Intelligence, 2004.

44. M. C. Golumbic. *Algorithmic Graph Theory and Perfect Graphs.* Academic Press, New York, 1980.

45. J. Gustedt, O. A. Mæhle, and J. A. Telle. The treewidth of Java programs. In D. M. Mount and C. Stein, editors, *Proceedings 4th International Workshop on Algorithm Engineering and Experiments*, pages 86–97. Springer Verlag, Lecture Notes in Computer Science, vol. 2409, 2002.

46. M. Habib and R. H. Möhring. Treewidth of cocomparability graphs and a new order-theoretic parameter. *ORDER*, 1:47–60, 1994.

47. P. Heggernes, J. A. Telle, and Y. Villanger. Computing minimal triangulations in time $O(n^\alpha \log n) = o(n^{2.376})$. To appear in proceedings SODA'05, 2005.

48. P. Heggernes and Y. Villanger. Efficient implementation of a minimal triangulation algorithm. In R. Möhring and R. Raman, editors, *Proceedings of the 10th Annual European Symposium on Algorithms, ESA'2002*, pages 550–561. Springer Verlag, Lecture Notes in Computer Science, vol. 2461, 2002.

49. M. Held and R. Karp. A dynamic programming approach to sequencing problems. *J. SIAM*, 10:196–210, 1962.

50. I. V. Hicks. Graphs, branchwidth, and tangles! Oh my! Working paper. http://ie.tamu.edu/People/faculty/Hicks/default.htm.

51. I. V. Hicks. Planar branch decompositions I: The ratcatcher. INFORMS Journal on Computing (to appear).

52. I. V. Hicks. Planar branch decompositions II: The cycle method. INFORMS Journal on Computing (to appear).

53. I. V. Hicks. *Branch Decompositions and their Applications*. Ph. d. thesis, Rice University, Houston, Texas, 2000.

54. I. V. Hicks. Branchwidth heuristics. *Congressus Numerantium*, 159:31–50, 2002.

55. I. V. Hicks. Branch decompositions and minor containment. *Networks*, 43:1–9, 2004.

56. U. Kjærulff. Optimal decomposition of probabilistic networks by simulated annealing. *Statistics and Computing*, 2:2–17, 1992.

57. T. Kloks. *Treewidth. Computations and Approximations*. Lecture Notes in Computer Science, Vol. 842. Springer-Verlag, Berlin, 1994.

58. T. Kloks. Treewidth of circle graphs. *Int. J. Found. Computer Science*, 7:111–120, 1996.

59. T. Kloks and D. Kratsch. Treewidth of chordal bipartite graphs. *J. Algorithms*, 19:266–281, 1995.

60. T. Kloks, D. Kratsch, and J. Spinrad. On treewidth and minimum fill-in of asteroidal triple-free graphs. *Theor. Comp. Sc.*, 175:309–335, 1997.

61. A. M. C. A. Koster. *Frequency Assignment - Models and Algorithms*. PhD thesis, Univ. Maastricht, Maastricht, the Netherlands, 1999.

62. A. M. C. A. Koster, H. L. Bodlaender, and S. P. M. van Hoesel. Treewidth: Computational experiments. In H. Broersma, U. Faigle, J. Hurink, and S. Pickl, editors, *Electronic Notes in Discrete Mathematics*, volume 8. Elsevier Science Publishers, 2001.

63. A. M. C. A. Koster, S. P. M. van Hoesel, and A. W. J. Kolen. Solving partial constraint satisfaction problems with tree decomposition. *Networks*, 40:170–180, 2002.

64. J. Lagergren. Efficient parallel algorithms for graphs of bounded tree-width. *J. Algorithms*, 20:20–44, 1996.

65. J. Lagergren and S. Arnborg. Finding minimal forbidden minors using a finite congruence. In *Proceedings of the 18th International Colloquium on Automata, Languages and Programming*, pages 532–543. Springer Verlag, Lecture Notes in Computer Science, vol. 510, 1991.

66. P. Larrañaga, C. M. H. Kuijpers, M. Poza, and R. H. Murga. Decomposing Bayesian networks: triangulation of the moral graph with genetic algorithms. *Statistics and Computing (UK)*, 7(1):19–34, 1997.

67. S. J. Lauritzen and D. J. Spiegelhalter. Local computations with probabilities on graphical structures and their application to expert systems. *The Journal of the Royal Statistical Society. Series B (Methodological)*, 50:157–224, 1988.

68. D. R. Lick and A. T. White. k-degenerate graphs. *Canadian Journal of Mathematics*, 22:1082–1096, 1970.
69. B. Lucena. A new lower bound for tree-width using maximum cardinality search. *SIAM J. Disc. Math.*, 16:345–353, 2003.
70. J. Matoušek and R. Thomas. Algorithms for finding tree-decompositions of graphs. *J. Algorithms*, 12:1–22, 1991.
71. K. G. Olesen and A. L. Madsen. Maximal prime subgraph decomposition of Bayesian networks. *IEEE Trans. on Systems, Man, and Cybernetics, Part B*, 32:21–31, 2002.
72. A. Parra and P. Scheffler. Characterizations and algorithmic applications of chordal graph embeddings. *Disc. Appl. Math.*, 79:171–188, 1997.
73. L. Perković and B. Reed. An improved algorithm for finding tree decompositions of small width. In P. Widmayer, editor, *Proceedings 25th Int. Workshop on Graph Theoretic Concepts in Computer Science, WG'99*, pages 148–154. Springer Verlag, Lecture Notes in Computer Science, vol. 1665, 1999.
74. S. Ramachandramurthi. A lower bound for treewidth and its consequences. In E. W. Mayr, G. Schmidt, and G. Tinhofer, editors, *Proceedings 20th International Workshop on Graph Theoretic Concepts in Computer Science WG'94*, pages 14–25. Springer Verlag, Lecture Notes in Computer Science, vol. 903, 1995.
75. S. Ramachandramurthi. The structure and number of obstructions to treewidth. *SIAM J. Disc. Math.*, 10:146–157, 1997.
76. B. Reed. Finding approximate separators and computing tree-width quickly. In *Proceedings of the 24th Annual Symposium on Theory of Computing*, pages 221–228, New York, 1992. ACM Press.
77. N. Robertson and P. D. Seymour. Graph minors. II. Algorithmic aspects of tree-width. *J. Algorithms*, 7:309–322, 1986.
78. N. Robertson and P. D. Seymour. Graph minors. XIII. The disjoint paths problem. *J. Comb. Theory Series B*, 63:65–110, 1995.
79. H. Röhrig. Tree decomposition: A feasibility study. Master's thesis, Max-Planck-Institut für Informatik, Saarbrücken, Germany, 1998.
80. D. J. Rose, R. E. Tarjan, and G. S. Lueker. Algorithmic aspects of vertex elimination on graphs. *SIAM J. Comput.*, 5:266–283, 1976.
81. D. P. Sanders. On linear recognition of tree-width at most four. *SIAM J. Disc. Math.*, 9(1):101–117, 1996.
82. P. D. Seymour and R. Thomas. Call routing and the ratcatcher. *Combinatorica*, 14(2):217–241, 1994.
83. K. Shoikhet and D. Geiger. A practical algorithm for finding optimal triangulations. In *Proc. National Conference on Artificial Intelligence (AAAI '97)*, pages 185–190. Morgan Kaufmann, 1997.
84. G. Szekeres and H. S. Wilf. An inequality for the chromatic number of a graph. *J. Comb. Theory*, 4:1–3, 1968.
85. R. E. Tarjan and M. Yannakakis. Simple linear time algorithms to test chordiality of graphs, test acyclicity of graphs, and selectively reduce acyclic hypergraphs. *SIAM J. Comput.*, 13:566–579, 1984.
86. M. Thorup. Structured programs have small tree-width and good register allocation. *Information and Computation*, 142:159–181, 1998.
87. Treewidthlib. http://www.cs.uu.nl/people/hansb/treewidthlib, 2004-03-31.
88. J. Valdes, R. E. Tarjan, and E. L. Lawler. The recognition of series parallel digraphs. *SIAM J. Comput.*, 11:298–313, 1982.

89. F. van den Eijkhof and H. L. Bodlaender. Safe reduction rules for weighted treewidth. In L. Kučera, editor, *Proceedings 28th Int. Workshop on Graph Theoretic Concepts in Computer Science, WG'02*, pages 176–185. Springer Verlag, Lecture Notes in Computer Science, vol. 2573, 2002.

90. G. J. Woeginger. Exact algorithms for NP-hard problems: A survey. In *Combinatorial Optimization: "Eureka, you shrink"*, pages 185–207, Berlin, 2003. Springer Lecture Notes in Computer Science, vol. 2570.

91. A. Yamaguchi, K. F. Aoki, and H. Mamitsuka. Graph complexity of chemical compounds in biological pathways. *Genome Informatics*, 14:376–377, 2003.

From Research Prototypes to Industrial Strength Open Source Products - The ObjectWeb Experience

Emmanuel Cecchet

INRIA/ObjectWeb – 655, Avenue de l'Europe – 38330 Montbonnot – France
Emmanuel.Cecchet@inria.fr

Abstract. Open source software has become a common way of disseminating research results. In this talk, we first introduce the motivations and implications of releasing research prototypes as open source software (OSS).

ObjectWeb is an international consortium fostering the development of open source middleware. We give an overview of tools available for OSS development and management based on ObjectWeb experiences. The infrastructure required for hosting such developments is also described.

We report various experiences and practices of small and large ObjectWeb projects in their way to reach the quality of industrial strength products. Finally, we summarize the lessons learned from the success and failures of these projects.

Keywords: open source, community, technology transfer, software engineering, metrics.

1 Introduction

Academic research activities often lead to the implementation of prototype software by either individuals such as master or Ph.D. students, or entire research groups. Open source has become a common way to disseminate research results to the academic community but there is an increasing industrial interest in open source software. However, for a research prototype to become an industrial strength open source product, there is a gap that we highlight in this article.

ObjectWeb [3] is an international open source consortium founded in 2001 by INRIA, Bull and France Telecom R&D. It brings together various academic and industrial partners to foster the development of professional component-based open source middleware. We report various experiences and practices of small and large ObjectWeb projects resulting from research prototypes, in their way to reach the quality of industrial strength products.

The outline of this paper is as follows. Section 2 introduces the motivations and implications of releasing a software as open source with an emphasis on the licensing issues and the difficulties to build a community. The ObjectWeb consortium organization and infrastructure is introduced in section 3. Section 4 describes ObjectWeb projects lifecycle and the best practices to build successful open source software. Section 5 summarizes the lessons learned from the various project experiences and we conclude in section 6.

M. Bieliková et al. (Eds.): SOFSEM 2005, LNCS 3381, pp. 17–27, 2005.

2   Open Source Software

Open source has definitely changed the software landscape. Linux is a good example of an industrial strength open source software (OSS) that grows from both industrial and individual contributions. In academia, research institutions, especially in Europe where patents on software still do not exist, OSS is considered as a good mean for technology transfer. We outline next the motivations and implications to release a research prototype as OSS.

2.1   Motivations and Implications

From an academic point of view, releasing a research prototype as OSS allows the community to get familiar with the technology, contribute to it, eventually reproduce experimental results in other environments and so on. However, the authors will expose themselves to critics by showing their design and code. Everybody is not ready to get that exposure but it is necessary to go through this step to be also exposed to contributions. Besides there is no magic with OSS and a technology will not get wide acceptance or improve just because it is released as open source.

One of the most common mistake is releasing an end-of-life product or a no more maintained research prototype as an OSS. Building a community around an OSS takes a lot of time and contributors are usually very few. Even the largest OSS developments have few main contributors, most of the community contributions being bug reports or small patches. An OSS can only survive if it has a sustained user community. This requires significant resources and work in terms of user support to bootstrap the community. This effort is often underestimated and leads to the death of a majority of OSS projects like the one that are found on sourceforge.net. The same kind of mistake is made by small companies who think that they will get extra engineering manpower by open sourcing their product. This often leads to the exploitation of the code base by competitors without getting short term contributions back from the community. Before taking such decision, the license must be carefully chosen.

Attracting users and contributors to an OSS requires promotion not only in the form of research papers but also some lobbying in the potentially interested communities. This can take the form of non-scientific publication, participation to wide audience events, collaboration with high visibility groups, getting references from other projects and so on. Managing the community so that it feels involved in the project is also very important to make it grow. Not only the code should be open, the users who are potential contributors must feel that they could influence the direction the project is heading if they get involved. Researchers are often not familiar with these tasks and fail to build an open source community around a good technology by neglecting these social aspects.

Even though you could have a large user community, getting industrial acceptance requires extra effort. OSS still suffers from an image of low quality software that could not be used in production. This is often due to the large number of available open source projects from which it is hard to evaluate the maturity or quality. It took a long time for Linux to get into real world business and there is still a strong belief that you need to pay to obtain quality software. However, if someone still has to pay for the development of OSS, there are several sustainable business models around OSS based on services, support or dual licensing.

2.2   Licenses

Discussing the various flavors of open source licenses is definitely out of the scope of this article, however we would like to highlight the main options and their implications.

The GNU Public License (GPL) [6] is a widely used license in the open source community, however it is often considered as business hostile due to its viral nature. Indeed, every product that embeds or links GPL code must also be released under GPL. This is why most of the common Linux libraries such as glibc are released under a non-viral BSD license. This allows commercial closed source products to be linked to the Linux GPL code via these non-GPL libraries. GPL enforces code modifications to be contributed back to the community but it limits their usage in commercial products.

LGPL (Lesser GNU Public License) [7] relaxes the GPL constraints by allowing the software to be included in any product being it closed or open source. However, modifications to the original code should be released under LGPL. LGPL is often a good choice for libraries or middleware.

The BSD license [8] is also very popular since it does not impose anything. Anyone can use the code, modify it and redistribute it without any constraint. Moreover, there is no obligation to contribute modifications back to the community. BSD like licenses are appreciated from professionals because they can freely exploit the code base and build commercial products out of it.

The APL (Apache Public License) [9] is similar to BSD but it requires every contributor to give his copyright to the Apache Software Foundation (ASF) who is legally responsible for the code. This means that the ASF owns the code but if someone commits stolen code in an Apache project, the ASF can be sued.

Note that the license choice is very important since changing a license requires the agreement of all copyright holders. Finding all contributors of a long-lived project and getting their agreement might quickly get impossible. The dual licensing model also requires the agreement of all copyright holders.

A well-known example of OSS using dual licensing is the MySQL database. MySQL AB is the company holding the copyright of the whole MySQL code base. MySQL code is released under GPL and prevents its distribution in non-GPL products. MySQL AB also sells for a fee the same code base under a license that has no commercial usage restriction. This approach requires that any MySQL contributor give up (or sell) his copyright to MySQL AB so that MySQL AB remains the copyright holder of the whole code base in order to continue its dual licensing business.

2.3   Building a Community

To be used in industrial environments, an OSS must prove its long term viability by building a community of both developers and users. A truly successful open source projects should survive the leaving of its original authors. For this to happen, the project must have a significant user base from which a critical mass of contributors can make the project live. We report next some best practices and experiences from various ObjectWeb projects.

In the early stages of the public release, it is very important to provide timely support to users. Choosing the right moment to release a software is not easy. A too immature software will discourage users and it will be very hard to convince them to look again at your piece of software later in time. The first impression is very important. A too polished software is not desirable too since users will only download and use the software without having anything to contribute in return. A good open source project is a balance of innovative core technology, minor bugs and enviable missing features.

There is no ready-made recipe to build a community and get external contributions, however a good user documentation is definitely a mandatory starting point. Design documents or developer documentations are of great help to let potential contributors get familiar with the code. To reach broader audiences, it is often necessary to write white papers or release technical brochures so that users get a quick insight of the project and can decide if they want to invest further in the software. A book edited by a recognized editor will definitely widen the credibility of the project and attract new users.

Internationalization is also a good mean to get local communities getting involved in the project. ObjectWeb has a very active community in Asia leading a translation effort that promotes the technology in this area of the world. It is very important for OSS to open their communities worldwide.

Last but not least, use cases and success stories are very important. It is very hard to get feedback on how people are using an OSS, be it a positive or negative experience. However, an OSS with industrial references has more credibility to the eyes of new users and comfort existing users in their choice.

3 The ObjectWeb Consortium

3.1 Organization

ObjectWeb is an open, not-for-profit, international consortium hosted by INRIA. The main purpose of ObjectWeb is the development of industrial-strength, component-based, distributed middleware technology, to be used and leveraged in many different fields, including e-commerce, enterprise information systems, and telecommunication services. As an open consortium, ObjectWeb has a collegial structure which is organized as follows:

- the *Board*, comprised of representatives from the different members of the consortium (both individuals and organizations), is responsible for the overall direction, policies, and strategy of the consortium. The daily operations are nevertheless the responsibility of the *Executive Committee*.
- the *College of Architects*, comprised of appointed individuals chosen for their expertise and their responsibilities in leading the development the ObjectWeb code base, oversees the technical orientations within the consortium and is ultimately responsible for the evolution and architectural integrity of the overall code base and approval of new projects.

One of the main tenets of the ObjectWeb philosophy is that projects should deliver middleware components that can be reused across a wide range of software platforms (e.g., J2EE) and application domains. In order to attain this goal, a major long-term objective of the ObjectWeb consortium is to ensure a coherent evolution of its code base by building upon a component model and a set of associated common software frameworks for building component assemblies.

3.2   Infrastructure

If open source software by itself is free, there is a cost for producing and releasing it. ObjectWeb does not provide any resource to help in the development of the various projects but it provides the infrastructure to host the projects. ObjectWeb benefits from INRIA's network infrastructure and employs 5 full-time people in the executive committee to run the consortium and the infrastructure.

After 3 years of existence, we report some numbers to give an insight on the current scale of the consortium. ObjectWeb currently hosts more than 80 projects, 285 active developers and more than 3200 registered users. There are about 180 mailing lists counting about 8000 subscribers. The web site shows a steadily increasing number of visits as the consortium grows. A single server is still sufficient to handle the 3.5 millions of hits per month on the projects and consortium web pages that accounts for about 40GB of data.

The development support is hosted on a separate machine that runs the GForge software (see section 3.3). The number of hits reaches up to 4.6 millions per month but hosting the software code and the various releases requires about 400GB of storage. A full-time engineer deals with the administration of these 2 main servers. Scaling and replicating these dynamic content Web servers is a concern and should be addressed in the near future.

3.3   GForge

GForge is a platform for collaborative software development. It provides tools for project and software management. GForge decentralizes the administration tasks by delegating to each project leader the management of his team. Administrators of a project can add new members to the team and assign different roles such as developer, doc writer and so on. Each team member can have different rights for every GForge tool including source code management repositories such as CVS or Subversion.

The task management module allows to track all tasks of the project and assign them to team members. It is possible to define priorities, deadlines, dependencies between tasks and also attach discussion threads to any task. The developers can report their progress by updating the completion percentage of the tasks they have been assigned. Gantt diagrams can be dynamically generated to get a quick graphical overview of the workplan.

Exposing all tasks, even unassigned tasks, allows the community to get a comprehensive view of the current status and perspectives of the project. Project administrators can also post job offers on GForge to attract new contributors to their projects.

GForge also integrates various trackers for bugs, support requests, patches and feature requests. Tasks can be associated to bugs or feature requests to have an homogeneous way to manage the development activities. These various trackers can be configured to send updates to the project mailing lists. Finally, GForge provides a news subsystem that allows projects to post news about significant events such as software releases or significant decisions or contributions.

The ObjectWeb Forge has been extended to provide each project with a personal Wiki space that allows developers to share notes and handle dynamic content in a more convenient way. The Wiki complements the other GForge tools.

4   ObjectWeb Practices

Open source projects have basically two different communities to talk with: users and developers. In both cases, mailing lists or forums are a common way to communicate. However, the expectation of these communities are quite different and a contributor is often first a user of the project.

Users usually get their first impression of the project from the Web site. Having a clear message explaining the purpose of the project, its current maturity (alpha, beta, stable, ...), links to documentation, screenshots and so on definitely helps to make a good impression. If the software is not well packaged and documented, this will be a show stopper for professional users.

The developer's community will be more interested in design documents as well as API specifications or well documented code. Displaying a comprehensive workplan with currently assigned and opened tasks definitely help people getting involved in the project.

To get a project known by the open source community requires a significant amount of efforts. Registering the project on sites like freshmeat.net and posting on the relevant forums and web sites will let people know about the project and look at it. RSS feeds are also a good mean to automatically broadcast new release announcements or other significant events.

4.1   Project Organization

Each ObjectWeb project has a leader who is responsible for taking the design decisions and managing the development team. It is important that all decisions are made transparent so that both user and developer communities see the openness of the project not only with regard to the source code but also for its modus operandi. All design options should always be discussed on the mailing list.

An important role of the project leader is to prioritize and schedule tasks to organize the development team work and present a comprehensive workplan. This role can of course be distributed among several project administrators but there should be one person responsible for coordinating the whole project.

In the early stages of a project, it is often better to have a single mailing list for users and developers to foster the development of the community. If the split occurs too early, users tend to post questions both on user and developer lists to get support. You may miss contributions by posting design related discussions on a dedicated developer mailing list whereas users may have brought very interesting feedback. The various ObjectWeb experiences have shown that dedicated user and developer mailing lists is only necessary for projects with very large user communities. However, a dedicated mailing list for CVS commits or trackers report might be desirable.

The project should also be organized in such a way that transparency can be achieved through traceability. Traceability requires tools such as an history file tracking all developments, detailed CVS logs, list of libraries used by the project (name, version, description, license, ...), mailing list archives and so on.

Within ObjectWeb, each project leader has a complete autonomy over the direction he wants to give to the project and who can join his team. Most successful projects have used a very permissive policy in accepting new committers. Oftentimes, contributors prefer submitting patches and not having CVS access because they are worried about breaking anything in the code base. But in practice, if an error occurs, source management tools such as CVS or SVN offers the necessary support to rollback bogus commits. As long as coding rules are well defined and a good source code management system is used, permissive acceptation policy allow the team to attract new contributors and build an active developer community. Large projects with a critical mass of developers can have much more restrictive policies but very few open source projects ever reach that size.

4.1.1 Developer Community

The project leader can manage his developer community using the ObjectWeb Forge tools. He can decide who will get CVS access, assign different roles with regard to bug, task, patch, release management and so on.

Fostering contributions from external developers is very much encouraged, but do not be mistaken into thinking that external contributors will be numerous, this does not happen often. Simply open sourcing a project is not sufficient to magically acquire contributions. Significant efforts in terms of communication, documentation, code legibility, education, and support to newcomers must be made if one wants to build a self-sustainable developer community around a project.

There is no policy regarding copyright transfer, but it is a common rule that each contributor retains his copyright. ObjectWeb requires that a contribution can not be rejected because a contributor wants to keep his copyright. A contribution can only be rejected if does not match the architectural or technical vision that the project leader wants to take for the project.

4.1.2 User Community

Depending on the target audience of a project, users could be middleware developers or a very large audience of people with heterogeneous skills. In all cases, a project does not survive without users. Providing support to users is a real challenging and resource consuming task, but it is necessary to make a successful project.

You have to encourage your users to provide you with feedback since most users take the product, try it and do not say anything about whether it suits their needs or not. An active mailing list is often synonymous of an healthy user community and users feel more comfortable posting on an active mailing list.

4.2 Federating Development Efforts

ObjectWeb is an open community (not just an open source community) and there are no strict rules on how a project must or should be managed. However, we encourage projects to federate their efforts to achieve our vision of middleware made of reusable components. Every project leader is responsible for the management of his project and has complete control over it. The leader has the entire freedom on architectural and technical choices, such as on how he manages the communities around the project or he wants to communicate on the project (marketing, …). The only constraints is that the project remains in the ObjectWeb scope.

The ObjectWeb College of Architects can review projects to evaluate their status. Typically, they may ask the project team to report at a quarterly architecture meeting. The purpose of the review is to initiate discussions on how to move forward with possible synergies with other projects or working groups, how to integrate contributions to/from the ObjectWeb community and especially with regard to the component model and its frameworks. This allows the overall projects to make progress with the help of the community.

ObjectWeb projects are using various metrics to measure their progress. These metrics include: traffic and number of subscribers on the mailing list, number of contributors and contributions (feedback, bug reports, lines of code contributes, patches, …), web site statistics (hits, unique visitors, downloads, …) and number of users (success stories, usage in production, …).

As the area of middleware is constantly evolving, it is natural that parts of the overall code base may become deprecated or inactive. Different projects might also combine their effort into a new common code base. It is not possible, and it is not the intent, to drop projects and delete their code base. However, deprecated or inactive projects should be clearly advertised so that users are properly informed when they choose a technology. An old code base is often very useful and we think that it is good to keep past projects in an "attic" space where it is always possible to consult them.

4.3   Software Development

Open source software development has a short development cycle. New releases can be carried out very quickly with an almost immediate feedback from the community that helps testing and fixing bugs.

4.3.1   Documentation

Documentation is probably one the most important but often overlooked part of an open source project. Introductory documents such as white papers, brochures or research articles help newcomers apprehending the concepts of the project and getting a first impression on the scope and quality of the software.

A well documented code allows the automatic generation of API documentations using tools such as Javadoc or Doxygen [1]. This does not replace specification or design documents but they are complementary and are often a useful reference for developers.

4.3.2   Coding

It is important to clearly expose coding style and conventions in a document such as a developer guide so that all developers can adopt the project rules. A good way to enforce these rules is to share IDE parameters in the CVS repository. For example, Eclipse [4] settings as well as .project and .classpath files can be shared in CVS. The settings will provide all developers with the same formatting rules, Checkstyle [2] settings, code templates including copyright notice and standard tags such as TODO and FIXME which will relate to specific tasks.

The project should include all libraries necessary to rebuild the whole software and possibly include the source code of third-party libraries so that it can be easily introspected using tools like Eclipse. The compilation chain must also be documented and automated as much as possible. Including descriptions in ant targets or Makefiles will help the users to understand and rebuild the software.

One of the common issue is having cross-platform starting scripts. The usage of an automated installer tool such as IzPack [5] can accommodate the platform specific discrepancies and prevent the usage of environment variables or manual user configuration which is often error prone and leads to increased support effort.

4.3.3 Testing

To assess the robustness of an open source software, an extensive regression test suite is needed. Unit testing is now widely available and many tools exist [10] to automate the execution of these tests and generate reports. It is a good practice to make the test execution reports available online for each release or nightly build.

Unit testing might not be enough for certain pieces of software, especially those dealing with distributed system issues where real deployment must be performed. Specific load injection tools might be required to test these configurations. These tools can also be used to perform performance measurements or evaluate the variations of various quality of services provided by the software.

4.3.4 Software Releases

Packaging and distributing the software is a key aspect of the success of an open source software. It is necessary to have at least a source and binary distribution that includes a comprehensive set of documentation. Additional automated distributions such as nightly build releases or CVS snapshots are also common. It is a good practice to include test suite reports in the source and automated distributions.

It is very important to keep an history of all releases and log all fixes and feature adding between releases so that it is always possible to diagnose problems related to a specific release of the software. Best practices include using CVS tags for every release and branches for major releases.

Packaging such as compressed archives (.tar.gz or .zip) is common but more advanced installer tools such as IzPack can automate the installation process and provide the user with a friendly graphical interface. For an industrial strength open source product, it is necessary to integrate smoothly in the targeted installation environment such as having icons in the menu bars or automatic integration as a service when applicable.

Linux specific packaging such as rpm or dbm allows for a quick adoption by these communities. This packaging can often be handled by the communities belonging to the specific Linux distributions. Moreover, it is possible to define software dependencies and have an integrated support for uninstallation or upgrades. The JPackage project [11] provides a common packaging of Java related open source software for all Linux distributions.

4.4 Contributions

Contributions are necessary for an open source project to make progress. All Object-Web projects experience difficulties in getting feedback from the community. Most common forms of contributions are feature request, bug reports, patches and document translations. However use case and success stories are hard to gather and unsatisfied users usually don't contribute back negative experiences on the mailing list which prevent oftentimes the teams to address properly the related issues.

Any form of feedback can be used as an input for the project. Bug reports for example should lead to the creation of new test cases. Getting users to contribute on the mailing list to provide support to new users can spare resources of the core team. But for users to be aware of this, there is a significant amount of education that must be proactively made by the project team.

Examples or demos are also very important to show various use cases scenarios that can address various user needs. This complements documentation, tutorial, courses or training materials. Having translation for the various documents can help reaching new communities like Asia.

Code contributions are less frequent and most open source projects relies on few core developers. Even if coding conventions have been made explicit, it is safe for the development team to integrate manually the first contributions of an external contributors before granting CVS write access.

5 Lessons Learned

It is hard if even possible to define guidelines for making a successful open source project. However, we learned a few lessons from the success and failures experienced by the 80 ObjectWeb projects.

First of all, a clear definition of the project and its goal is mandatory. If the project implements a recognized standard or addresses a specific and identified need, it makes the task much more easier for users to understand the purpose of the project. But the most critical moment is probably choosing when to publicly release the software.

A project should be mature enough so that it is usable but it shouldn't be too polished else it will not attract any external contribution since there will be few opportunities for bug reports or patches. A successful open source community arises from project that have a good balance of maturity, usability, bugs and missing but desirable features. It is very important for the project team to be always honest and to never oversell the software. It will take a long time for a disappointed user to come back to a project if he ever comes back.

Sharing all design or configuration choices on the project *mailing list* helps to involve the community and attract new contributors. Oftentimes users try the software right away without reading the documentation and therefore, examples, demos and default configuration files handling the most common setups are very important. Moreover, it is common that a piece of software is used for something it was not designed for.

Promoting the projects can only be done properly by the project teams but it is a time consuming task. Having an exhaustive web site that chooses relevant keywords to be properly indexed by search engines can enhance the visibility of a project. Presence on Web forums such as slashdot.org, TheServerSide.com or freshmeat.net is free but covering wide audience events such as LinuxWorld or even more academic events has a greater impact. It is often hard to get *success stories* because many corporate users are still reluctant to communicate about their usage of open source software.

The legibility of the source code can also have a significant impact on the project adoption. One ObjectWeb project was implemented using all the subtleties of a programming language making the code very efficient but hard to understand and to re-use. The project ended up with no contributor and died when its original contributor died.

Similar problems arise from research prototypes developed by a single student or for the length of a financed project. At the end of the project or when the student leaves, if nothing has been done to promote the project and build a community around it fro the start of the project, it is more likely that the project will not survive. The long-term viability is often a concern for industrial users because they want to use the technology without being involved in its development.

6   Conclusion

ObjectWeb is an international consortium that is fostering the development of open source middleware by academics and industrials since 2001. We have reported on the ObjectWeb consortium experience in fostering the technology transfer of research prototypes as open source software. From the motivation and implication of releasing software as open source, we have presented various practices and lessons learned to build successful open source software and communities.

Establishing such a consortium requires significant resources but once the process has been bootstrapped, we strongly believe that open source is probably one of the best technology transfer mean for research prototypes that provides software commons.

References

[1] The Doxygen documentation system – http://www.doxygen.org.
[2] Checkstyle.- http://checkstyle.sourceforge.net/.
[3] The ObjectWeb consortium – http://www.objectweb.org.
[4] The Eclipse project – http://eclipse.org.
[5] The IzPack installer - http://www.izforge.com/izpack/.
[6] The GNU General Public License- http://www.gnu.org/copyleft/gpl.html.
[7] The GNU Lesser General Public License - http://www.gnu.org/ copyleft/ lesser.html.
[8] The BSD License - http://www.opensource.org/licenses/bsd-license.php.
[9] Apache Software License - http://www.opensource.org/licenses/apachepl.php.
[10] Unit, Testing Resources for Extreme Programming - http://www.junit.org/.
[11] The JPackage project - http://jpackage.org/.

How Hard Is It to Take a Snapshot?

Faith Ellen Fich

Department of Computer Science,
University of Toronto,
Toronto, Canada
fich@cs.utoronto.ca

Abstract. The snapshot object is an important and well-studied primitive in distributed computing. This paper will present some implementations of snapshots from registers, in both asycnhronous and synchronous systems, and discuss known lower bounds on the time and space complexity of this problem.

1   Introduction

An important problem in shared memory distributed systems is to obtain a consistent view of the contents of the shared memory while updates to the memory are happening concurrently. For example, in a sensor network, simultaneous readings of all the sensors may be needed, even when the measured readings change frequently. In a distributed computation, periodically recording the global state can facilitate error recovery, since the computation can be restarted from the most recent backup, rather than from the beginning of the computation. Ideally, one wants to collect these checkpoints without interrupting the computation. Checkpoints can also be useful for debugging distributed programs, allowing invariants that have been violated to be found more easily.

To obtain a consistent view, it is not enough to simply collect the entire contents of memory by reading one value at a time. The difficulty is that the first values that have been read may be out of date by the time that later values are read. The problem can be formalized as the implementation of a snapshot object that can be accessed concurrently by different processes. A snapshot object consists of a set of $m > 1$ components, each capable of storing a value. Processes can perform two different types of operations: UPDATE any individual component or atomically SCAN the entire collection to obtain the values of all the components. A single-writer snapshot object is a restricted version in which there are the same number of processes as components and each process can UPDATE only one component.

It is often much easier to design fault-tolerant algorithms for asynchronous systems and prove them correct if one can think of the shared memory as a snapshot object, rather than as a collection of individual registers. Snapshot objects have been used to solve randomized consensus [3, 4], approximate agreement [8], and to implement bounded concurrent time stamps [15] and other types of objects [5, 16].

M. Bieliková et al. (Eds.): SOFSEM 2005, LNCS 3381, pp. 28–37, 2005.

Unbounded time stamps are particularly easy to generate using a single-writer snapshot object. Each process stores the time stamp from its last request (or 0, if has not yet requested one) in its component. To obtain a new time stamp, a process performs a SCAN and adds 1 to the maximum time stamp contained in the result. Then the process updates its component with its new time stamp. This ensures that a request that is started after another request has completed will receive a larger time stamp. Note that concurrent requests can receive the same or different time stamps.

Because snapshot objects are much easier for programmers to use, researchers have spent a great deal of effort on finding efficient implementations of them from registers, which, unlike snapshot objects, are provided in real systems.

Our asynchronous shared memory system consists of n processes, $p_1, \ldots, p_n$ that communicate through shared registers. At each step of a computation, one process can either read from one register or write a value to one register. In addition, it can perform a bounded number of local operations that do not access shared registers. The order in which processes take steps is determined by an adversarial scheduler. In synchronous shared memory, at each step, every active process either reads from one register or writes a value to one register, in addition to performing a bounded amount of local work. An active process is a process that has neither completed its task nor crashed. The order in which the shared memory operations occur during a step is determined by an adversarial scheduler.

We consider linearizable (atomic) implementations [17], where each SCAN and UPDATE operation appears to take effect at some instant during the time interval in which it is executed. Implementations should also be wait-free. This means that there is an upper bound on the number of steps needed for any process to perform any operation, regardless of the behaviour of other processes. In particular, processes do not have to wait for other processes to finish. Thus wait-free implementations tolerate crash failures.

The rest of this paper will present an overview of some different implementations of snapshot objects from registers and discuss known lower bounds. Both the time complexity and the number of registers are considered, but not the size of the registers. Various techniques are known to reduce register size [9, 11, 12, 15, 19], but these are beyond the scope of this paper.

2   An Asynchronous Implementation Using m Registers

Our first implementation uses one shared register for each component of the snapshot object. Although simply collecting the values of the registers may give an invalid view, a valid view *is* guaranteed if two collects are performed by some process and, meanwhile, the values of none of the registers are changed. Note that the value of a register may change and then change back again. To distinguish between this possibly bad situation and the situation when no changes have occurred, each process performing an UPDATE to a component writes its identifier and a sequence number to the corresponding register together with the new value.

However, a process can repeatedly collect the values without ever getting two consecutive identical results. To overcome this problem, when a process performs UPDATE, it first performs SCAN and writes the result, together with the new value, its identifier, and a sequence number to the appropriate register. Now, when a process performs two collects that have different results, it could use the result of a SCAN that is stored in one of the registers.

Unfortunately, the stored result may be out of date: It may be a memory state that only occurred before the process began its SCAN. To ensure that a process p performing a SCAN gets a memory state that occurs while it is executing, it repeatedly performs collects until either two consecutive collects are the same (in which case, the result of this collect is its result) or it sees that some other process has performed at least two different UPDATES since p performed its first collect. In the second case, p returns the last result written by this other process, which is from a SCAN that is part of an UPDATE that began after p began its SCAN.

One of these two cases will occur by the time a process has performed $n + 2$ collects. Each collect takes takes $\Theta(m)$ steps. Thus the time complexity of a SCAN is $O(mn)$. Since an UPDATE contains an embedded SCAN, its time complexity is also $O(mn)$. This algorithm, although presented in a slightly different form, is from Afek, Attiya, Dolev, Gafni, Merritt, and Shavit's paper [1]. It also appears in [13].

3   An Implementation of a Snapshot Object from a Single-Writer Snapshot Object

Anderson [2] shows how to construct a snapshot object with m components shared by n processes using a single-writer snapshot object shared by n processes. The idea is that each process stores information about every component of the snapshot object in its component of the single-writer snapshot object. Specifically, it stores the value of the last UPDATE operation it performed to each component, together with a time stamp for each of those UPDATES.

To perform a SCAN of the snapshot object, a process performs a SCAN of the single-writer snapshot object and, for each component, returns the value with the latest time stamp. Process identifiers are used to break ties.

To perform an UPDATE to component i of the snapshot object, a process performs a SCAN of the single-writer snapshot object. The result of the SCAN gives it the time stamp of the last UPDATE to component i performed by each process. It generates a new time stamp that is later than all of these. Then the process UPDATES its single-writer component with the new value and time stamp for component i and the old values and time stamps for all other components.

Thus, if there is an implementation of a single-writer snapshot object in which SCAN takes time $S(n)$ and UPDATE takes time $U(n)$ in the worst case, then there is an implementation of a snapshot object in which SCAN takes $O(S(n))$ time and UPDATE takes $O(S(n) + U(n))$ time.

4   Asynchronous Single-Writer Snapshot Implementations Using Lattice Agreement

Lattice agreement is a decision problem that is closely related to the single-writer snapshot object. In this problem, each process gets an element of some lattice as input and must output an element that is at least as large as its input, but no larger than the least upper bound of all the inputs. Furthermore, all outputs must be comparable with one another.

It can easily be solved in $O(1)$ time using a single-writer snapshot object: Each process UPDATES its component with its input, performs a SCAN, and takes the least upper bound of all the lattice elements in the result as its output. Conversely, a single-writer snapshot object in which each process performs at most one operation can be implemented using single-writer registers and an instance of the lattice agreement problem [7]. To perform an UPDATE, a process writes the new value to its single-writer register and then performs a SCAN, throwing away the result. To perform a SCAN, a process collects the values in all the single-writer registers and uses this view as an input to lattice agreement, where one view is less than or equal to another view in the lattice if the set of components that have new values in the first view is a subset of those in the other view. A process uses its output from the lattice agreement problem as the result of its SCAN. The reason a process performing an UPDATE with some value v performs an embedded SCAN is to ensure that there isn't a later UPDATE by another process whose new value is collected by some SCAN that doesn't collect v.

Attiya, Herlihy, and Rachman [7] extend this idea to allow processes to perform an unbounded number of SCAN and UPDATE operations using an unbounded number of instances of lattice agreement. First, each process writes a sequence number to its single-writer register together with its new value. In the lattice, one view is less than or equal to another view if each component of the first view has a sequence number that is less than or equal to the sequence number in the same component of the other view. A process p that wants to perform a SCAN (or an embedded SCAN as part of an UPDATE) uses the view it collects as input to an instance of lattice agreement that it has not already solved. There may be multiple instances of lattice agreement that are active simultaneously. Process p either chooses the latest of these, if it has not already solved it, or starts a new instance. After getting an output from this instance, p checks whether a later instance of lattice agreement is active. If not, it can use this view as its result. Otherwise, it performs a second collect and uses the view from it as input to the latest instance that is now active. If, after getting its output to this instance, there is an even later active instance, there is some other process that participated in the same instance and got a valid view. Then p uses this view as its result.

Inoue, Chen, Masuzawa, and Tokura [18] show how to solve agreement in this lattice in $O(n)$ time. From the discussion above, this implies the existence of an n-component single-writer snapshot implementation from an unbounded number of registers in which SCAN and UPDATE take $O(n)$ time. They consider a complete binary tree of height $\lceil \log_2 n \rceil$, with each process assigned to a different leaf. At its leaf, a process has a view consisting of only one component, which contains its new value. The process proceeds up the tree, at each successive level doubling the length of its view, until, at the root, it has a view that includes all n components. At each internal node of height k, a process solves an instance of lattice agreement among the 2^k processes that are associated with leaves in its subtree, using an input formed from its resulting view at the previous level. This takes $O(2^k)$ time. The fact that the inputs of processes arriving from the same child are comparable with one another ensures that all the outputs are comparable with one another.

Attiya and Rachman [9] also use lattice agreement to show that an n-process single-writer snapshot object can be implemented from single-writer registers so that the time complexities of both SCAN and UPDATE are in $O(n \log n)$. Like the previous implementation, they use a complete binary tree of height $\lceil \log_2 n \rceil$ that contains a single-writer register for each process at every node. However, their approach to solving lattice agreement is quite different. First, they consider the case when at most n SCAN and UPDATE operations are performed. To perform an UPDATE, a process writes its new value, together with a sequence number, to its single-writer register at the root of the tree and then performs an embedded SCAN. To perform a SCAN, a process traverses the tree from the root down to a leaf. All the processes that end at the same leaf will return the same view. A process collects the views written to the single-writer registers at each internal node it reaches. It decides whether to go left or right depending on how the sum of the sequence numbers it sees compares to some appropriately chosen threshold. A process with a sum that is at or below the threshold goes left and writes the same view at this child as it wrote at the node from which it came. A process that goes right computes a new view by taking the union of the views that it collected and writes this view at this child. In either case, the process repeats the same procedure at the child, until it reaches a leaf. An important observation is that the sequence number in a component of a view written to a right child is at least as large as the sequence number in the same component of a view written to its left sibling. This ensures that the views that processes write at different leaves are comparable with one another. To handle the general case, the operations are divided into virtual rounds, each containing n operations. They use a variety of mechanisms, including bounded counters and handshakes, to separate the operations from different rounds so that they do not interfere with one another.

5   Faster Asynchronous Snapshot Implementations

Israeli, Shaham, and Shirazi [20] present an implementation of an n-component single-writer snapshot object $\mathcal{A}$ with $O(n)$ SCAN time and $O(S(n) + U(n))$

UPDATE time from a single n-component single-writer snapshot object $\mathcal{B}$ with SCAN time $S(n)$ and UPDATE time $U(n)$ and n single-writer registers. Each process uses its component in the base object $\mathcal{B}$ to store both the current value of its component and a sequence number that it increments each time it performs an UPDATE. After performing an UPDATE, a process then SCANS $\mathcal{B}$ and writes the resulting view to its register. Since $\mathcal{B}$ is linearizable, all the views that are written can be totally ordered, using the sequence numbers contained in the views. To perform a SCAN of $\mathcal{A}$, a process simply reads the views written in the registers and chooses the most recent one. Applying the transformation in Section 3 yields a snapshot object with $O(n)$ SCAN time and $O(S(n)+U(n))$ UPDATE time. In particular, using Attiya and Rachman's implementation in Section 4 yields a snapshot object with $O(n)$ SCAN time and $O(n \log n)$ UPDATE time.

The same paper also gives a similar implementation of an n-component single-writer snapshot object $\mathcal{A}'$ with $O(n)$ UPDATE time and $O(n + S(n) + U(n))$ SCAN time. Here, each process stores a possibily inconsistent view in its register. To UPDATE $\mathcal{A}'$, a process reads the views of all processes and, from them, constructs a new view whose i'th component contains the maximum of the sequence numbers in component i together with its associated value. It increments the sequence number in its own component of this view and replaces its value with the new value. It then writes this view to its register. To SCAN $\mathcal{A}'$, a process first reads the registers of all processes and constructs a new view, as above. It UPDATES its component of the base object $\mathcal{B}$ with this view and then SCANS $\mathcal{B}$. Finally, it constructs a new view, as above, from the views returned by the SCAN, which it returns as its result. Again, using Attiya and Rachman's implementation in Section 4 yields a single-writer snapshot object with $O(n)$ UPDATE time and $O(n \log n)$ SCAN time.

6   Lower Bounds for Asynchronous Snapshots

Jayanti, Tan, and Toueg [23] use a carefully constructed covering argument to prove that, in any implementation of an n-component single-writer snapshot object, there is an execution of a SCAN that must read from at least $n - 1$ different registers. Since a $\min\{m, n\}$-component single-writer snapshot object is a special case of an m-component snapshot object shared by n processes, this implies a lower bound of $\min\{m, n\} - 1$ on the space and SCAN time of any implementation of the latter.

Fatourou, Fich, and Ruppert [14, 13], using a different covering argument, prove that any implementation of an m-component snapshot object shared by $n \geq m$ processes requires at least m registers, matching the upper bound in Section 2. They introduce the concept of a *fatal configuration*, which has a set of processes each of which is about to write to a different register and are in the midst of performing UPDATES to a smaller number of components. No correct implementation using m or fewer registers can reach a fatal configuration; otherwise, there is an insufficient number of registers to record UPDATES to the other components. This fact implies a significant amount of information about such implementations: SCANS never write to registers, each UPDATE operation writes to only one register, UPDATE operations to the same component write to the same register, and UPDATE operations to different components write to different registers. Hence, at least m registers are needed.

Using these results together with an intricate covering argument, they prove that, for $m < n$, any space-optimal implementation takes $\Omega(mn)$ time to perform a SCAN in the worst case. This also matches the upper bound in Section 2. For $m = n - 1$, the lower bound is $\Omega(n^2)$, in contrast to the implementations in Section 4 that use single-writer registers and run significantly faster. Note that any implementation using single-writer registers only needs to use one single-writer register per process: All of the single-writer registers into which a process can write can be combined into one single-writer register having many fields.

Israeli and Shirazi [IS98] proved an $\Omega(n)$ lower bound on UPDATE time for implementations of an n-component single-writer snapshot object using only single-writer registers. Their proof uses an elegant combinatorial argument. This can be extended to an $\Omega(m)$ lower bound on UPDATE time for implementations of an m-component snapshot object from m (multi-writer) registers.

7   Synchronous Snapshot Implementations and Lower Bounds

There is a simple implementation of a synchronous m-component snapshot object due to Neiger and Singh [24] in which SCAN and UPDATE each take $m + 1$ steps in the worst case. It uses m registers, each of which holds the current value of one component. Their idea is to divide time into blocks of $m + 1$ slots, one for reading each component and one for writing. To perform an UPDATE, a process waits until the next write slot and then writes the new value into the appropriate register. To perform a SCAN, a process just reads the values from the registers in the order specified by the next m read slots. It is possible to linearize every operation that includes a write slot in some order during that write slot so that all the views that result are consistent. They also prove that in any implementation using m registers, each of which can hold the value of only one component, $S(n) + U(n) \in \Omega(m)$, where $S(n)$ is the worst case SCAN time and $U(n)$ is the worst case UPDATE time.

A small variant of this idea allows SCANS to be performed in one step, while UPDATES take at most $2m + 1$ steps. In this case, one (larger) register holds the value of the entire snapshot object, which can be read at any time. Time is divided into blocks of $2m$ slots. A process that wants to UPDATE component i reads the register during slot $2i - 1$ and then write backs its contents in the next step, with its new value replacing the old value of component i. Brodsky and Fich [10] show that, with $2m - 1$ registers, it is possible to improve the UPDATE time to $O(\log m)$. The registers are conceptually organized as a strict balanced binary tree, where each of the m leaves contains the value of a different component and, provided no UPDATES are in progress, each internal node contains the concatenation of the values of its children. A process performs a SCAN by reading the register at the root. To perform an UPDATE, a process writes the new value to the corresponding leaf and then propagates this information up the tree to the root. Information from its siblings and from the siblings of each of its ancestors is also propogated to the root at the same time.

Brodsky and Fich present another implementation in which UPDATE takes at most 3 steps and SCAN takes at most $3m - 1$ steps. It uses m registers, one for each component. Each register has two fields, a top-field and a bottom-field, each of which holds a component value. Time is divided into two alternating phases, each consisting of m consecutive steps. During top-write phases, a process performing an UPDATE writes its new value to the top-field of the specified component and processes performing SCANS read from the bottom-fields. Similarly, during bottom-write phases, a process performing an UPDATE writes its new value to the bottom-field of the specified component and processes performing SCANS read from the top-fields. This ensures that writes to top-fields don't interfere with reads from top-fields and writes to bottom-fields don't interfere with reads from bottom-fields. In addition, each register contains a third field to which processes write when they UPDATE the corresponding component. The information in this field allows a process performing a SCAN to determine whether the top-field or the bottom-field contains the value of the most recent UPDATE to that component.

They also show how to combine their two implementations to perform UPDATE in $O(\log(m/c))$ time and SCAN in $O(c)$ time, for any positive integer $c \leq m$. For $n < m$, they get the same results, but with m replaced by n.

Finally, using an information theoretic argument, they prove a general trade-off $U(n) \in \Omega(\log(\min\{m, n\}/S(n)))$ between the worst case UPDATE time $U(n)$ and the worst case SCAN time $S(n)$, matching their upper bounds.

8   Conclusions

This paper presents a brief survey of different implementations of snapshot objects from registers, together with some complexity lower bounds. It is not intended to be comprehensive. There are a number of interesting aspects this paper does not cover, for example, randomized implementations [7], implementations that are adaptive to contention among processes [6], and implementations from more powerful objects [7, 25, 22].

Acknowledgements

I am grateful to Hagit Attiya for many interesting discussions about snapshot objects and to Alex Brodsky, Danny Hendler, and Kleoni Ioannidou for helpful comments about this paper. This work was supported by the Natural Sciences and Engineering Research Council of Canada.

References

1. Y. Afek, H. Attiya, D. Dolev, E. Gafni, M. Merritt, and N. Shavit, *Atomic Snapshots of Shared Memory*, JACM, volume 40, number 4, 1993, pages 873–890.
2. J. Anderson, *Multi-Writer Composite Registers*, Distributed Computing, volume 7, number 4, 1994, pages 175–195.
3. James Aspnes, *Time- and Space-Efficient Randomized Consensus*, Journal of Algorithms, volume 14, number 3, 1993, pages 414–431.
4. James Aspnes and Maurice Herlihy, *Fast, Randomized Consensus Using Shared Memory*, Journal of Algorithms, volume 11, number 2, 1990, pages 441–461.
5. James Aspnes and Maurice Herlihy, *Wait-Free Data Structures in the Asynchronous PRAM Model*, 2nd ACM Symposium on Parallel Algorithms and Architectures, 1990, pages 340–349.
6. Hagit Attiya, Arie Fouren, and Eli Gafni, *An Adaptive Collect Algorithm with Applications*, Distributed Computing, volume 15, 2002, pages 87–96.
7. Hagit Attiya, Maurice Herlihy, and Ophir Rachman, *Atomic Snapshots Using Lattice Agreement*, Distributed Computing, volume 8, 1995, pages 121–132.
8. Hagit Attiya, Nancy Lynch, and Nir Shavit, *Are Wait-free Algorithms Fast?* JACM, volume 41, number 4, 1994, pages 725–763.
9. H. Attiya and O. Rachman, *Atomic Snapshots in O(n log n) Operations*, SIAM Journal on Computing, volume 27, number 2, 1998, pages 319–340.
10. Alex Brodsky and Faith Ellen Fich, *Efficient Synchronous Snapshots*, 23rd Annual ACM Symposium on Principles of Distributed Computing, 2004, pages 70-79.
11. C. Dwork, M. Herlihy, and O. Waarts, *Bounded Round Numbers*, 12th Annual ACM Symposium on Principles of Distributed Computing, 1993, pages 53–64.
12. C. Dwork and O. Waarts, *Simple and Efficient Concurrent Timestamping or Bounded Concurrent Timestamps are Comprehensible*, 24th Annual ACM Symposium on Theory of Computing, 1992, pages 655–666.
13. Panagiota Fatourou, Faith E. Fich, and Eric Ruppert, *A Tight Time Lower Bound for Space-Optimal Implementations of Multi-Writer Snapshots*, 35th Annual ACM Symposium on Theory of Computing, 2003, pages 259–268.
14. Panagiota Fatourou, Faith E. Fich, and Eric Ruppert, *Space-Optimal Multi-Writer Snapshot Objects Are Slow*, 21st Annual ACM Symposium on Principles of Distributed Computing, 2002, pages 13–20.
15. Rainer Gawlick, Nancy Lynch, and Nir Shavit, *Concurrent Timestamping Made Simple*, Israel Symposium on the Theory of Computing and Systems, LNCS volume 601, 1992, pages 171–183.
16. Maurice Herlihy, *Wait-Free Synchronization*, ACM Transactions on Programming Languages and Systems, volume 13, number 1, 1991, pages 124–149.
17. Maurice Herlihy and Jeannette Wing, *Linearizability: A Correctness Condition for Concurrent Objects*, ACM Transactions on Programming Languages and Systems, volume 12, number 3, 1990, pages 463–492.
18. Michiko Inoue, Wei Chen, Toshimitsu Masuzawa, and Nobuki Tokura. Linear Time Snapshots Using Multi-writer Multi-reader Registers. Proceedings of the 8th International Workshop on Distributed Algorithms, LNCS volume 857, 1994, pages 130–140.
19. A. Israeli and M. Li, *Bounded Time Stamps*, Distributed Computing, volume 6, number 4, 1993, pages 205–209.
20. A. Israeli, A. Shaham, and A. Shirazi, *Linear-Time Snapshot Implementations in Unbalanced Systems*, Mathematical Systems Theory, volume 28, number 5, 1995, pages 469–486.

21. A. Israeli and A. Shirazi, *The Time Complexity of Updating Snapshot Memories*, Information Processing Letters, volume 65, number 1, 1998, pages 33–40.
22. Prasad Jayanti, *f-Arrays: Implementation and Applications*, 21st Annual ACM Symposium on Principles of Distributed Computing, 2002, pages 270–279.
23. P. Jayanti, K. Tan, and S. Toueg, *Time and Space Lower Bounds for Nonblocking Implementations*, SIAM Journal on Coomputing, volume 30, 2000, pages 438–456.
24. G. Neiger and R. Singh, *Space Efficient Atomic Snapshots in Synchronous Systems*, Technical Report GIT-CC-93-46, Georgia Institute of Technology, College of Computing, 1993.
25. Yaron Riany, Nir Shavit, and Dan Touitou, *Towards a Practical Snapshot Algorithm*, Theoretical Computer Science, volume 269, 2001, pages 163–201.

Logical Foundations for Data Integration

Maurizio Lenzerini

Dipartimento di Informatica e Sistemistica "Antonio Runerti",
Università di Roma "La Sapienza",
Via Salaria 113, I-00198 Roma, Italy
http://www.dis.uniroma1.it/~lenzerini

Integrating heterogeneous data sources, which are distributed over highly dynamic computer networks, is one of the crucial challenges at the current evolutionary stage of Information Technology infrastructures. Large enterprises, business organizations, e-government systems, and, in short, any kind of internetworking community, need today an integrated and virtualized access to distributed information resources, which grow in number, kind, and complexity.

Several papers published in the last decades point out the need for a formal approach to data integration. Most of them, however, refer to an architecture based on a global schema and a set of sources. The sources contain the real data, while the global schema provides a reconciled, integrated, and virtual view of the underlying sources. As observed in several contexts, this centralized achitecture is not the best choice for supporting data integration, cooperation and coordination in highly dynamic computer networks. A more appealing architecture is the one based on peer-to-peer systems. In these systems every peer acts as both client and server, and provides part of the overall information available from a distributed environment, without relying on a single global view.

In this talk, we review the work done for rigorously defining centralized data integration systems, and then we focus on peer-to-peer data integration, with the aim of singling out the principles that should form the basis for data integration in this architecture. Particular emphasis is given to the problem of assigning formal semantics to peer-to-peer data integration systems. We discuss two different methods for defining such a semantics, and we compare them with respect to the above mentioned principles.

References

1. K. Aberer, M. Punceva, M. Hauswirth, and R. Schmidt. Improving data access in P2P systems. *IEEE Internet Computing*, 2002.
2. P. A. Bernstein, F. Giunchiglia, A. Kementsietsidis, J. Mylopoulos, L. Serafini, and I. Zaihrayeu. Data management for peer-to-peer computing: A vision. In *Proc. of the 5th Int. Workshop on the Web and Databases (WebDB 2002)*, 2002.
3. A. Calì, D. Lembo, and R. Rosati. Query rewriting and answering under constraints in data integration systems. In *Proc. of the 18th Int. Joint Conf. on Artificial Intelligence (IJCAI 2003)*, 2003. To appear.
4. D. Calvanese, G. De Giacomo, M. Lenzerini, and R. Rosati. Data integration in p2p systems. In *Databases, Information Systems, and Peer-to-Peer Computing*, pages 77–90. Springer, LNCS 2944, 2004.

M. Bieliková et al. (Eds.): SOFSEM 2005, LNCS 3381, pp. 38–40, 2005.

5. D. Calvanese, G. De Giacomo, M. Lenzerini, and R. Rosati. Logical foundations of peer-to-peer data integration. 2004. To appear.

6. L. Camarinha-Matos, H. Afsarmanesh, C. Garita, and C. Lima. Towards an architecture for virtual enterprises. *J. Intelligent Manufacturing*, 9(2), 1998.

7. T. Catarci and M. Lenzerini. Representing and using interschema knowledge in cooperative information systems. *J. of Intelligent and Cooperative Information Systems*, 2(4):375–398, 1993.

8. A. Doan, P. Domingos, and A. Halevy. Learning to match the schemas of data sources: a multistrategy approach. *Machine Learning Journal*, 2003.

9. R. Fagin, P. G. Kolaitis, R. J. Miller, and L. Popa. Data exchange: Semantics and query answering. In *Proc. of the 9th Int. Conf. on Database Theory (ICDT 2003)*, pages 207–224, 2003.

10. M. Fitting. Basic modal logic. In *Handbook of Logic in Artificial Intelligence and Logic Programming*, volume 1, pages 365–448. Oxford Science Publications, 1993.

11. E. Franconi, G. M. Kuper, A. Lopatenko, and L. Serafini. A robust logical and computational characterisation of peer-to-peer database systems. In *Databases, Information Systems, and Peer-to-Peer Computing*, pages 64–76. Springer, LNCS 2944, 2004.

12. M. Friedman, A. Levy, and T. Millstein. Navigational plans for data integration. In *Proc. of the 16th Nat. Conf. on Artificial Intelligence (AAAI'99)*, pages 67–73. AAAI Press/The MIT Press, 1999.

13. S. Gribble, A. Halevy, Z. Ives, M. Rodrig, and D. Suciu. What can databases do for peer-to-peer? In *Proc. of the 4th Int. Workshop on the Web and Databases (WebDB 2001)*, 2001.

14. A. Halevy, Z. Ives, D. Suciu, and I. Tatarinov. Schema mediation in peer data management systems. In *Proc. of the 19th IEEE Int. Conf. on Data Engineering (ICDE 2003)*, 2003.

15. A. Y. Halevy. Answering queries using views: A survey. *Very Large Database J.*, 10(4):270–294, 2001.

16. J. Heflin and J. Hendler. A portrait of the semantic web in action. *IEEE Intelligent Systems*, 16(2):54–59, 2001.

17. R. Hull. Managing semantic heterogeneity in databases: A theoretical perspective. In *Proc. of the 16th ACM SIGACT SIGMOD SIGART Symp. on Principles of Database Systems (PODS'97)*, pages 51–61, 1997.

18. R. Hull, M. Benedikt, V. Christophides, and J. Su. E-services: a look behind the curtain. In *Proc. of the 22nd ACM SIGACT SIGMOD SIGART Symp. on Principles of Database Systems (PODS 2003)*, pages 1–14. ACM Press and Addison Wesley, 2003.

19. A. Kementsietsidis, M. Arenas, and R. J. Miller. Mapping data in peer-to-peer systems: Semantics and algorithmic issues. pages 325–336. ACM Press and Addison Wesley, 2003.

20. C. Koch. Query rewriting with symmetric constraints. In *Proc. of the 2nd Int. Symp. on Foundations of Information and Knowledge Systems (FoIKS 2002)*, volume 2284 of *Lecture Notes in Computer Science*, pages 130–147. Springer, 2002.

21. M. Lenzerini. Data integration: A theoretical perspective. In *Proc. of the 21st ACM SIGACT SIGMOD SIGART Symp. on Principles of Database Systems (PODS 2002)*, pages 233–246, 2002.

22. H. J. Levesque and G. Lakemeyer. *The Logic of Knowledge Bases*. The MIT Press, 2001.

23. J. Madhavan and A. Y. Halevy. Composing mappings among data sources. In *Proc. of the 29th Int. Conf. on Very Large Data Bases (VLDB 2003)*, pages 572–583, 2003.

24. P. McBrien and A. Poulovassilis. Defining peer-to-peer data integration using both as view rules. In *Databases, Information Systems, and Peer-to-Peer Computing*, pages 91–107. Springer, LNCS 2944, 2004.

25. M. P. Papazoglou, B. J. Kramer, and J. Yang. Leveraging Web-services and peer-to-peer networks. In *Proc. of the 15th Int. Conf. on Advanced Information Systems Engineering (CAiSE 2003)*, pages 485–501, 2003.

26. J. D. Ullman. Information integration using logical views. In *Proc. of the 6th Int. Conf. on Database Theory (ICDT'97)*, volume 1186 of *Lecture Notes in Computer Science*, pages 19–40. Springer, 1997.

Recent Advances in Graph Drawing

Petra Mutzel

Vienna University of Technology,
Favoritenstr. 9–11 E186, A-1040 Vienna, Austria
mutzel@ads.tuwien.ac.at
http://www.ads.tuwien.ac.at

Abstract. Graph drawing is a very active field and deals with the visualization of discrete structures in such a way that they are easy to read and understand. Apart from the classical applications of graph drawing, such as VLSI design and data base visualization, recently, new emerging application areas have arisen. E.g., in bioinformatics it is of growing interest to visualize biochemical networks such as protein interaction networks, regulatory and signaling pathways. In software engineering and for process modelling the visualization of UML class diagrams is getting increasingly important. For the analysis of the world wide web it is important to visualize internet connections.

The new application areas demand new challenges to the field of graph drawing. One of the main current challenges is to deal effectively with very large graphs. Recent technological advances have brought increased data volumes and data complexity. Moreover, graph drawing has to deal with dynamic graphs that evolve over time. In this context, interactive visualization is getting increasingly important.

Recently, in graph drawing a lot of effort has been spent to attack these problems. But also in classical topics like crossing minimization a lot of progress has been made.

We give a survey on the recent trends and advances in graph drawing. This includes challenging new and classical problems as well as theoretical and practical algorithmic solutions and software tools.

M. Bieliková et al. (Eds.): SOFSEM 2005, LNCS 3381, p. 41, 2005.
© Springer-Verlag Berlin Heidelberg 2005

The Hyperdatabase Network – New Middleware for Searching and Maintaining the Information Space

Hans-Jörg Schek

Swiss Federal Institute of Technology, Zurich (ETH) and
University for Health Sciences,
Medical Informatics, and Technology Tyrol (UMIT)

.

Abstract. The hyperdatabase network, a project at ETH Zurich since 1998, is a new middleware for the information space and includes new multi-object multi-feature search facilities for multimedia objects. It applies database technology at the level of services and combines peer-to-peer, grid, and service-oriented architectures.

1 Introduction

In the past we talked about single database systems or federated systems consisting of a few participating databases. In the future, we expect an ever increasing number of data sources, reaching from traditional databases and large document and web page collections, down to embedded information sources in mobile "smart" objects as they will occur in a pervasive computing environment. Therefore, not only the immense amount of information demands new thoughts but also the number of different information sources and their coordination poses a great challenge for the development of the appropriate information infrastructure. We talk about the continuous, "infinite" information, shortly called the "information space". Information in this space is distributed, heterogeneous, and undergoes continuous changes. So, the infrastructure for the information space must provide convenient tools for accessing information via sophisticated search facilities and for combining or integrating search results from different sources (1), for developing distributed applications for analyzing and processing information (2), and for transactional (workflow) processes that ensure consistent propagation of information changes and simultaneous invocations of several (web) services (3). For the implementation of such an infrastructure we must strive for functions including recoverability, scalability, and availability. As far as possible such an infrastructure should avoid global components. Rather a peer-to-peer middleware must be provided that has some self-configuration and adaptation features to various applications and their load characteristics. A synthesis of service orientation, peer-to-peer, and grid middleware is the target of a new middleware starting from the basis of extended technology in databases and in information retrieval.

M. Bieliková et al. (Eds.): SOFSEM 2005, LNCS 3381, pp. 42–46, 2005.

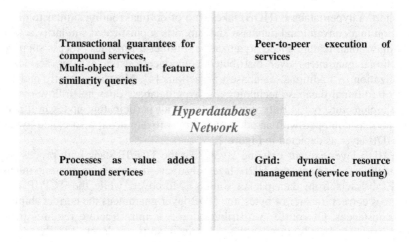

Fig. 1. Hyperdatabase network as synthesis of DB technology, SoA, Grid, and P2P

2 The Hyperdatabase Network – Vision and Prototype

In the presentation we will elaborate some of the aspects in these areas and report on our hyperdatabase network research at ETH and UMIT. The hyperdatabase vision [1,2,3] was established at ETH Zurich several years ago with the objective to identify a new middleware infrastructure based on well-understood concepts evolving from database technology. While a database system handles data records, a hyperdatabase system deals with services and service invocations. Services in turn may be using a database system. Therefore, we gave our vision the name hyperdatabase, i.e., a software layer for services on top of databases.

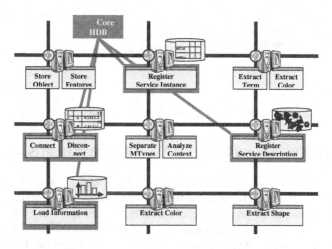

Fig. 2. Each node is equipped with the hyperdatabase layer

In short, a hyperdatabase (HDB) takes care of optimal routing similar to query optimization in a conventional database and supports sophisticated similarity search and high-dimensional feature space organization. Further it provides process support with transactional guarantees over distributed components using existing services as a generalization of traditional database transactions [4]. Most importantly and in contrast to traditional database technology, a hyperdatabase does not follow monolithic system architecture but is fully distributed over all participating nodes in a network. Every node is equipped with an additional thin software layer, a so-called hyperdatabase (HDB layer as depicted in Figure 2.

The HDB layer extends existing layers like the TCP/IP stack with process related functionalities. As such, the HDB layer abstracts from service routing much like TCP/IP abstracts from data packet routing. Moreover, while the TCP/IP protocol guarantees correct transfer of bytes, the HDB layer guarantees the correct shipment of process instances. Of course, a distributed process infrastructure requires that each service provider locally installs this additional software layer. Ideally, this layer comes together with the operating system much like the TCP/IP stack does. We have implemented a prototype system called OSIRIS (short for Open Service Infrastructure for Reliable and Integrated Process Support) following these principles [5]. The main functions of the OSIRIS layer are shown in Figure 3.

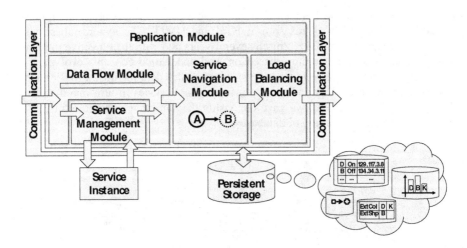

Fig. 3. The hyperdatabase layer in the OSIRIS implementation

3   Search Engine and the ETH World Information Space

A good example for an information space is the virtual campus of ETH within the ETHWorld project [6]. The virtual campus is a meeting place, the information space for students, assistants, professors, researchers, alumni and other interested people. The participants of the information space occur in two roles: as information providers and as information searchers. Search engines are mediators between them. ETHWorld not only supports textual searches but also visual ones. Participants are able to search with photos, images, or drawings.

In our part of the ETHWorld project we study and contribute similarity searches in multimedia documents as they are found in various web sites of ETH. The ISIS system (Interactive **SI**milarity Search) [7], part of OSIRIS, offers effective descriptors for the different kinds of document components and efficient search infrastructure for complex similarity searches. [8,9,10,11]. Complex similarity search means that a user may refer to several feature types and he/she may use several example objects as a query. We call this a multi-feature, multi-object query. In addition, user-friendly relevance feedback mechanisms, supported by information visualization, are built in.

In contrast to traditional web search engines that keep their data periodically up-to-date with the consequence of outdated information and dangling links, the ISIS search engine is maintained by the OSIRIS middleware. Changes in the information space are automatically and guaranteed propagated to all nodes as soon as possible. OSIRIS monitors local changes and triggers related maintenance processes that keep search indexes up-to-date.

Fig. 4 shows an insertion of a new molecule image and the related service invocations of this maintenance process. The features and context of the image are analyzed (e.g. texture, color, shape, surrounding text, links) and special search structures are updated accordingly. The ETHWorld case study also exemplifies that source information collections must be connected to computation intensive services for text classification, term extraction, image processing. Any time new services can be added to the system.

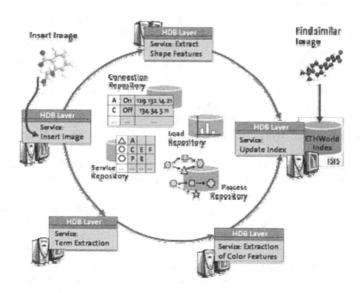

Fig. 4. OSIRIS/ISIS in the ETHWorld application

The hyperdatabase infrastructure takes care of the correct execution of processes. We do not apply a 2PC protocol because the subsystems are autonomous. Instead, we make sure that a process, once started will come to a well-defined termination, even in case of concurrent processes or in case of failures. We call this generalized atomicity and isolation. Beyond maintenance processes we have various user processes. They are handled with the same mechanisms. Examples are search processes over many search engines or electronic registration, similar to administrative workflow, or the production of reports. We use a graphical interface for process definition and verification [12].

Acknowledgement

Many individuals, students and members of the DBS Group, have contributed. Special thanks go to Klemens Böhm, Gert Brettlecker, Michael Mlivoncic, Heiko Schuldt, Christoph Schuler, Can Türker and Roger Weber.

References

1. H.-J. Schek, H. Schuldt, R. Weber. Hyperdatabases – Infrastructure for the Information Space. In: Proc. 6th IFIP Working Conf. on Visual Database Systems, S. 1–15, 2002.
2. H.-J. Schek, H. Schuldt, C. Schuler, C. Türker, R. Weber. Hyperdatenbanken zur Verwaltung von Informationsräumen. it-Information Technology, 2004.
3. H.-J. Schek and H. Schuldt and C. Schuler and R. Weber: Infrastructure for Information Spaces. In: Proceedings of the 6. East-European Conference on Advances in Databases and Information Systems (ADBIS'2002), Bratislava, Slovakia, September 2002.
4. H. Schuldt, G. Alonso, C. Beeri, H.-J. Schek. Atomicity and Isolation for Transactional Processes. ACM Transactions on Database Systems, 27(1):63–116, March 2002.
5. C. Schuler, R. Weber, H. Schuldt, H.J. Schek: Scalable Peer-to-Peer Process Management The OSIRIS Approach. In: Proc. of IEEE International Conference on Web Services (ICWS), San Diego, California, USA, July 2004.
6. ETHWorld – The Virtual Campus of ETH Zürich. http://www.ethworld.ethz.ch.
7. ISIS – Interactive Similarity Search. http://www.isis.ethz.ch.
8. R. Weber, H.-J. Schek, S. Blott. A Quantitative Analysis and Performance Study for Similarity Search Methods in High-Dimensional Spaces. In: Proc.24th Int. Conf. on Very Large Data Bases, New York, Sept. 1998.
9. Weber, M.Mlivoncic: Efficient Region-Based Image Retrieval. 12th International Conference on Information and Knowledge Management (CIKM'03), New Orleans, LA, USA, November 2003.
10. K. Böhm, M. Mlivoncic, H.-J. Schek, R. Weber: Fast Evaluation Techniques for Complex Similarity Queries. 27th Int. Conf. on Very Large Databases (VLDB), Roma, Italy, September 2001.
11. M. Mlivoncic, C. Schuler, C. Türker: Hyperdatabase Infrastructure for Management and Search of Multimedia Collections. Digital Library Architectures: Peer-to-Peer, Grid, and Service-Orientation, Proc. of the Sixth Thematic Workshop of the EU Network of Excellence DELOS on Digital Library Architectures, S. Margherita di Pula (Cagliari), Italy, June 2004.
12. R. Weber, C. Schuler, H. Schuldt, H.-J. Schek, P. Neukomm. Web Service Composition with O'GRAPE and OSIRIS. In: Proc. 29th Int. Conf. on Very Large Data Bases, Berlin, Sept. 2003.

Architecture of a Business Framework for the .NET Platform and Open Source Environments

Thomas Seidmann

Cdot AG, Wilen SZ, Switzerland
Thomas.Seidmann@cdot.ch
http://www.cdot.ch/thomas

Abstract. This paper contains a description of the architecture and components of a software framework for building enterprise style applications based on the .NET platform. The process and achieved results of porting this business framework to an open source platform based on GNU/Linux/Mono is described as well.

1 Introduction

Many software companies focused on building potentially large distributed data-driven applications are faced similar problems during the life cycle of their projects: the technology they use as a base for their solutions does not provide all the necessary abstractions or their abstraction level is too low. Usually they end up with a framework, in many cases even more of them in various projects, often depending on the particular project members involved in them. This fact leads to software maintenance problems as well as internal incompatibilities. To fill this gap we've developed a business framework (in further text referred to as 'framework') for so called enterprise application as a part of a broader product called Cdot-InSource, which comprises in addition to the business framework also services like education, project analysis, design and jump starting the implementation and consulting. The business framework part of Cdot-InSource is based on the .NET Framework platform authored by Microsoft and can be used as a foundation for a wide variety of applications, which might (but don't necessarily have to) employ Web services.

The outline of the rest of this paper is as follows. In section 2 we'll give an overview of the overall architecture of the framework contained in Cdot-InSource. Section 3 contains the description of the building blocks contained in the framework. Section 4 explains the blueprints for building and testing real-life applications using the framework. In section 5 we'll explain extending possibilities of the framework with respect to platform changes and additional requirements. Finally, in section 6 we'll described the objectives and effort of porting the framework to an open source platform as well as the current state and lessons learned so far. We conclude the paper with some thoughts of working in a "virtual company" and its social consequences.

M. Bieliková et al. (Eds.): SOFSEM 2005, LNCS 3381, pp. 47–52, 2005.

2   Architecture of the Business Framework

The architecture of the framework can best be explained based on applications it is intended for. We are looking at typical multi-tier applications with a relational DBMS used as a data store, a hierarchy of business services built on stateless components, a web service tier used as an entry point to the service layer and a set of client (front-end) applications. Figure 1 illustrates this architecture and also gives some hints about the platform support and contains some forward references to the next section.

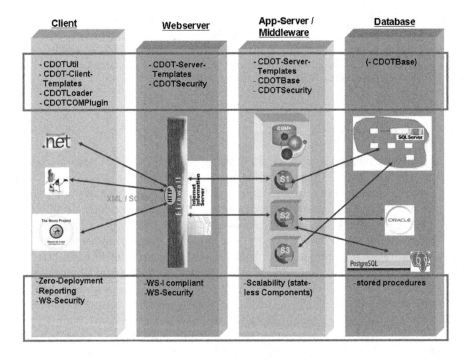

Fig. 1. Cdot-InSource architecture

The **web service tier** exhibits high adherence to current standards as formulated by the WS-I consortium. For example instead of inventing own authentication mechanisms, WS-Security is used for this purpose. When using Microsoft's ASP.NET implementation, this tier is normally hosted in the Internet Information Server (IIS), although other alternatives exist as well (see section 6). In addition to authentication, web services are usually responsible for authorization, although this task may be also delegated to the business services. It is of crucial importance that no business logic is implemented in the web service code itself, but delegated to the business services instead. The web service tier may form a hierarchy of layers (by aggregating simpler services into complex ones), but normally this won't be the case.

The **business service tier** can and usually will be structured into some hierarchy of layers with data access components (normally using DBMS-native data providers and protocols for accessing the data store) at the bottom and components performing complex business logic at the top. As already mentioned before, components of this tier are supposed to be stateless in order to achieve scalability of the whole system. There may be exceptions to this rule, for example in case of some special singleton objects, but generally state should be kept either in the database or in the client tier. To allow for higher scalability and/or integration with external systems the business service tier may (even partially) utilize middleware services like COM+ for functionality like distributed transactions or message queuing. Hosting of the business service components can be accomplished in a variety of ways starting from in-process with respect to the web service code, through an operating system service up to hosting in the middleware services. The same goes for the communication between these two layers, which can be done with .NET remoting or middleware-provided protocols (for example DCOM).

The primary **client tier** platform is, as in the case of the previous two tiers, the .NET framework, thus effectively forming a thin .NET client (also known as Windows Forms client). .NET offers advanced techniques for distributing portion of client code over a network, which can be used for zero (sometimes called 'no touch') deployment of front-end programs. As shall be seen in section 6, efforts are being undertaken for enabling platforms other than .NET to be able to execute .NET managed code.

Passing data between the various tiers is based on the notion of ADO.NET datasets, which represent an in-memory database with excellent serialization mechanisms for communication as well, even via web services. Thus instead of inventing own value object types, datasets (both strongly typed and untyped) are used for this purpose. The support for Java clients, although depicted on figure 1, is currently limited, due to the lack of dataset functionality. This problem of toolkit compatibility is tackled as well.

3   Components of the Framework

Flowing from the previous section, all program code contained in the business framework consists of MSIL, so called managed code. The basic components of the framework in the context of Cdot-InSource are depicted on figure 2, forming the central part of the picture. These components can be described as follows:

– Cdot.Security contains code augmenting the WS-Security implementation as well as an authorization sub-framework for performing RBAC (Role Based Access Control) independently from the one contained in .NET. This component is used by the client, web service and business service tiers.
– Cdot.Base represents an abstraction layer above ADO.NET for database access offering data provider independence to some extent, thus making the task of porting to a different DBMS platform easier. This component should be used only by program code of the business service tier.

- **Cdot.Office** contains glue code and abstraction shims for interaction with the operating system shell (Windows, X11) and some office productivity software systems (MS Office, OpenOffice). Both the use by the client tier as well as the business services is thinkable.
- **Cdot.Util** is the biggest component in terms of class count and is currently dedicated to .NET (Windows Forms) clients. This reaches from zero deployment support code through specialized user input widgets (controls) to a user interface configuration and internationalization sub-framework.

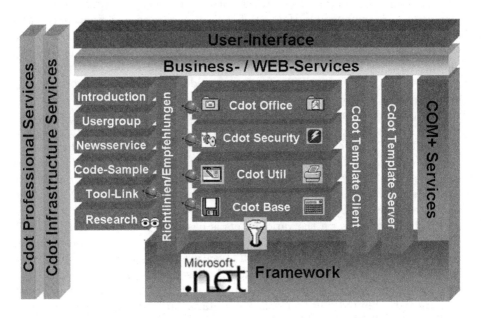

Fig. 2. Cdot-InSource components

Not shown on the figure are various third party components the framework relies in addition to the .NET platform. A complete enumeration of such components is clearly outside of the scope of this paper. Let us just mention on noticeable component: a grid library. Data driven enterprise applications often need a way to represent data in a tabular form using some kind of a data grid. Cdot.Util contains a shim layer offering virtually the same programming interface on top of various underlying grid implementations, which simplifies their replacement in client code. One of those grid implementations (actually the preferred one) resides in a third party component library. Another example of a third party component is a report generator, which we integrated into the framework instead of writing our own one.

4 Building and Testing Enterprise Applications Based on the Framework

In the previous section, figure 2 contains an important architectural component in the right central part of the picture – templates for service- (server-) and client-side code which serve as a blueprint for building applications. Although not covering every last detail of the framework, these templates represent a major help in starting a new project from the ground. Besides showing the way to structure the code in the various layers of an application they also stress out the importance of including testing code from the very beginning of the coding phase based on the principle of unit testing. A third-party (open source) unit testing tool called NUnit [1] is used for this purpose. Including testing code is deemed important particularly in the service component code and the templates show the way of accomplishing this task.

Unit testing is employed also inside the framework itself for the purpose of regression tests. Next section contains some notes on this.

In addition to templates and (simple) code samples for the various components of the business framework a complex sample is available, which should serve as a demonstration of the use of the whole framework. This sample contains the implementation of all tiers as described in section 2.

5 Extensibility of the Framework

The business framework of Cdot-InSource shouldn't be viewed as a static piece of code. Instead, it is a dynamical system evolving in time; additions and modifications are performed on a daily basis. Of course, since users of the framework are depending on the stability of their platform, some stringent rules have to be followed by the framework developers: incompatible changes should be avoided as far as possible and regression tests should be run after every significant change of the framework code to insure that previous functionality is still guaranteed. New functionality must come hand in hand with corresponding new testing code.

6 Porting the Framework to the MONO Platform

While .NET is widely accepted as an application platform, there are exceptions to this. The call for an alternative to platforms authored by Microsoft can sometimes and someplace be clearly heard. Until these days the only possible usable answer was the switch to the Java platform. The state of the MONO project [2] changes the view of things slightly. This project is aimed at offering an open source alternative (binary compatible) to .NET, developed mainly on top of GNU/Linux, running on a wide range of operating system platforms including even Windows.

---

[1] http://www.nunit.org
[2] http://www.go-mono.com

The promising state of the MONO project lead directly to the idea of using the very same business framework on this platform in addition to .NET. We will denote the adaptation process as porting even though the aim is to use exactly the same MSIL files (assebmlies) on both platforms.

The porting process comprises following problems to be solved:

1. Enabling the Web service and application server components to execute on MONO.
2. Enabling the client components, which normally are Windows Forms application components, to be able to execute in the Mono environment.

The first item turned out to be pretty straightforward with MONO version 1.0. We have managed to get the web service and business service tiers of an application to run on GNU/Linux, utilizing the Apache web server for the former. The business service tier in this particular application was originally hosted in a Windows service, so we've adapted it to execute as a background process (daemon) employing .NET remoting for the communication with the web server. The only real problem turned out to be the WS-Security implementation contained within MONO, which we had to modify and augment with missing functionality.

The second item is much more complicated and tricky, since the state of Windows Forms emulation in MONO by means of the Windows emulator Wine [3] is very immature and also due to the fact, that there is a number of third party components used with Cdot-InSource which must be taken as-is without the possibility to adapt them to MONO. This item is subject to work-in-progress right now.

7   Conclusions and Future Work

The Cdot-InSource business framework is evolving in time, but the platform is evolving and changing as well. Already today the extent of the next major overhaul of the .NET platform in Windows Longhorn can be glimpsed. It is the job of a good business framework to cope with such platform changes and offer the access to technologies as WinFX or Indigo with little or no impact on client code to the framework.

An interesting aspect of the business framework apart of its actual functionality and contents is the relationship to customers. It is not meant as a shrink-wrapped off-the shelf product, but instead it is based on a partnership model with customers. That way, with a relatively small number of customers (a maximum of several tens), it is possible to react on their needs and let flow the fulfillment of theses needs into the framework. It seems, that this customer relationship model is quite attractive and appreciated, since we've encountered positive feedback on this in many places.

---

[3] http://www.winehq.com

Progress on Crossing Number Problems*

László A. Székely

University of South Carolina, Columbia, SC 29208, USA
`szekely@math.sc.edu`

Abstract. Crossing numbers have drawn much attention in the last couple of years and several surveys [22], [28], [33], problem collections [26], [27], and bibliographies [40] have been published. The present survey tries to give pointers to some of the most significant recent developments and identifies computational challenges.

1 Computational Complexity

To avoid repetitions, we assume that the Reader has already some familiarity with crossing numbers. Earlier surveys [22], [27], [28] or [33] contain lots of important information not shown here.

The *crossing number* of a graph, $CR(G)$, is the minimum number of edge crossings, in any drawing of a graph in the plane, where every crossing point is counted for every pair of edges crossing there. Crossing numbers, introduced by Paul Turán 60 years ago [38], studied in graph theory, and then in the theory of VLSI, have become the most important graph parameter measuring the deviation from planarity. Pach and Tóth [24] (and independently Mohar) introduced two new variants of the crossing number problem:

the pairwise crossing number $CR\text{-}PAIR(G)$ is equal to the minimum number of unordered pairs of edges that cross each other at least once (i.e. they are counted once instead of as many times they cross), over all drawings of G; and

the odd crossing number $CR\text{-}ODD(G)$ is equal to the minimum number of unordered pairs of edges that cross each other odd times, over all drawings of G where edges touching each other in a common point is forbidden.

In Tutte's work [39] another kind of crossing number is implicit:

the independent-odd crossing number $CR\text{-}IODD(G)$ is equal to the minimum number of unordered pairs of non-adjacent edges that cross each other odd times, over all touching-free drawings of G. The inequalities immediately follow from the definitions:

$$CR\text{-}IODD(G) \leq CR\text{-}ODD(G) \leq CR\text{-}PAIR(G) \leq CR(G). \qquad (1)$$

No example of strict inequality is known. Pach [22] considers the problem of equality in (1) to be the most important open problem on crossing numbers.

---

* This research was supported in part by the NSF grant DMS 0302307.

M. Bieliková et al. (Eds.): SOFSEM 2005, LNCS 3381, pp. 53–61, 2005.

Open Problem 1. *Use computational tools to find strict inequalities in (1).*

The difficulty of Problem 1 lies in the fact that the problem $CR(G) \leq k$ is NP-complete if k is part of the input, see Garey and Johnson [14]. (Pach and Tóth [24] extended this result to CR-ODD, and also proved that CR-$PAIR$ is NP-hard.) It has been known that testing planarity, and therefore testing $CR(G) \leq k$ for any fixed k can be done in polynomial time. A recent algorithm of Grohe [15] tests $CR(G) \leq k$ for any fixed k in quadratic time.

Leighton and Rao [18] designed the first provably good approximation algorithm for crossing numbers. This algorithm approximates $n + CR(G)$, where n is the number of vertices, within a factor of $\log^4 n$ for degree bounded graphs. A recent paper of Even, Guha, and Schieber [11] reduced the factor to $\log^3 n$.

Open Problem 2. *Is there an approximation algorithm for $n + CR(G)$ within constant multiplicative factor, for degree bounded graphs?*

Note that the term n dominates $n + CR(G)$ if the crossing number is small, and therefore the approximation provides nothing in this case.

Open Problem 3. *Is there an approximation algorithm for $CR(G)$, if $CR(G)$ is small?*

A negative error term of sum of degree squares is present in the bisection width lower bound, for the crossing number—this is the source of n in $n + CR(G)$ above. Pach and Tóth [26] asked if this sum of degree squares can be reduced, but noted that it cannot go below n. Mysteriously, the sum of degree squares also comes in as negative error term in the embedding method [28], although we did not write it there in this way and just used an upper bound for it. In a third way, the sum of degree squares is a negative error term in the lower bound for the convex (alternatively, outerplanar or one-page) crossing number [29]. (The convex crossing number problem requires the placement of the vertices on a circle and edges are drawn in straight line segments.)

In view of the NP-completeness of testing $CR(G) \leq k$, it is no surprise that we have in general no short proofs to tight lower bounds for crossing numbers. However, hard problems often have interesting and efficiently tractable special cases. I suggest further analysis of the following optimality criterion [35]:

Let us be given a touching-free drawing D of the simple graph G, in which any two edges cross at most once, and adjacent edges (edges with an endpoint in common) do not cross. Let us associate with every edge $e = \{x, y\} \in E(G)$ an arbitrary vertex set $A_e \subseteq V(G) \setminus \{x, y\}$. If the edges e and f are non-adjacent, then we define the *parity* of this edge pair as 0 or 1 according to

$$par(e, f) = |e \cap A_f| + |f \cap A_e| \quad \text{modulo} \quad 2. \tag{2}$$

If non-adjacent edges e, f cross in D, then we write $e \times_D f$, otherwise write $e \|_D f$.

Theorem 1. *Using the notation above, the condition that for all choices of the sets A_e the inequality*

$$\sum_{\substack{par(e,f)=1 \\ e \times_D f}} 1 \leq \sum_{\substack{par(e,f)=1 \\ e \|_D f}} 1 \tag{3}$$

holds, implies that D realizes $CR(G)$.

It is hard to verify whether this criterion holds, since checking the condition for all possible sets A_e requires exponential time. However, inequality (3), which looks like a correlation inequality formulated about random edge pairs, may have a less exhaustive and more theoretical proof for some graphs and their drawings, if they show a high degree of structure. A natural candidate would be Zarankiewicz' drawing [32] of the complete bipartite graph, which is conjectured to be optimal.

Open Problem 4. *Find graph and drawing classes for which (3) can be verified in polynomial time.*

(Note that if Theorem 1 applies, then $CR(G) = CR\text{-}IODD(G)$.) Sergiu Norine has found an optimality criterion that seems to be different from (3) for complete bipartite graphs.

2   Origin of Lower Bounds

It is easy to prove by induction that a graph on $n \geq 3$ vertices and m edges has crossing number $\geq m - 3n + 6$. The base case is that a graph with $3n - 5$ edges, which must be non-planar, and must have crossing number ≥ 1. Ajtai *et al.* [3] and Leighton [17] independently discovered that for graphs with $m \geq cn$ edges, the crossing number is at least

$$CR(G) \geq \frac{c-3}{c^3}\frac{m^3}{n^2}. \tag{4}$$

A folklore probabilistic proof for (4) can be found in [28] and in [2]. The best current constant in (4) is $\frac{1}{33.75}$ for $m \geq 7.5n$, due to Pach and Tóth [25].

Once I had a discussion with János Pach on what is the origin of lower bounds for crossing numbers. He thought bisection width, I thought Leighton's bound (4). (Since then, I proved using Leighton's bound (4) the following theorem of Pach, Spencer and Tóth [23] that was first proved by the bisection width bound: if G has girth $> 2r$ and $m \geq 4n$, then $CR(G) = \Omega(m^{r+2}/n^{r+1})$ [34].) However, there is no doubt that Euler's formula is behind every lower bound. Therefore, I would find instructive

Open Problem 5. *Prove from first principles that a graph on n vertices, drawn in the plane in straight line segments without crossings, has at most $O(n)$ edges.*

Among "first principles" I disallow the theory of planar graphs, Euler's formula, etc. What I allow is using coordinates of points, pigeonhole principle, and the observation that two line segments cross if it follows from calculations with the coordinates of their endpoints.

3   The Rectilinear Revolution

A well-known theorem of Fáry [13] asserts that a graph is planar if and only if it admits a planar drawing in straight line segments. Bienstock and Dean [7] showed that if $CR(G) \leq 3$, then G admits a straight line drawing realizing its crossing number, but there exists graphs with crossing number 4 that require any prescribed number of crossings in any straight line drawing. Therefore, it is necessary to introduce the *rectilinear crossing number* $CR\text{-}LIN(G)$, which is the minimum number of crossings of G in any straight line drawing. Clearly $CR(G) \leq CR\text{-}LIN(G)$.

Ábrego and Fernández-Merchant [1] made a breakthrough by showing $CR\text{-}LIN(K_n) \geq \frac{1+o(1)}{64}n^4$, the conjectured lower bound for $CR(K_n)$. Lovász, Veszter-gombi, Wagner and Welzl [19] improved the lower bound by showing $CR\text{-}LIN(K_n) \geq (\frac{1}{64} + \frac{10^{-5}}{24} + o(1))n^4$, separating $CR(K_n)$ from $CR\text{-}LIN(K_n)$. Balogh and Salazar [6] increased the tiny separation of Lovász *et al.* manifold.

These new lower bounds are based on lower bounds on the number of *planar k-sets*. The k-set problem asks what is the largest possible number of k-element subsets of a set of n points in general position in the plane, which can be separated from the remaining $n - k$ by a straight line. Crossing numbers are also useful for k-sets, since the best upper bound for the number of k-sets, Dey [9] uses the crossing number method discussed in Section 4.

Aichholzer, Aurenhammer, and Krasser [4] developed a description of relative point locations in the plane, called *order types*, which allowed them to evaluate $CR\text{-}LIN(K_{11}) = 102$ and $CR\text{-}LIN(K_{12}) = 153$. Aurenhammer maintains the rectilinear crossing number webpage [5], which gives exact values of or lower and upper bounds to $CR\text{-}LIN(K_n)$, up to $n = 45$. Optimal drawing(s) also can be downloaded from [5].

Open Problem 6. *Make a conjecture for the rectilinear crossing number of* K_n.

Shahrokhi, Sýkora, Székely, and Vrťo [29] showed that for the convex crossing number $CR^*(G)$ (defined in Section 1),

$$CR^*(G) = O([CR(G) + \sum_v d_v^2] \log n). \tag{5}$$

Since we have $CR(G) \leq CR\text{-}LIN(G) \leq CR^*(G)$, we have that notwithstanding the examples of Bienstock and Dean [7] cited above, in "non-degenerate" cases there is at most a $\log n$ times multiplicative gap between the crossing number and the rectilinear crossing number. It is worth noting that the estimate in (5) is tight for an $n \times n$ grid [29]. This example is kind of degenerate since the error term dominates the RHS. However, several examples have been found where $CR\text{-}LIN(G) = \Theta(CR(G) \log n)$ [8].

4   Szemerédi-Trotter Type Theorems

I introduced the *crossing number method* [31] to prove Szemerédi–Trotter type theorems. The method requires setting lower and upper bounds for the crossing number of some graph, in order to get a bound for extremal problem. The lower bound is usually (4). For example, the original Szemerédi-Trotter theorem [36] claims that

Theorem 2. *For n points and l straight lines in the Euclidean plane, the number of incidences among the points and lines is $O(\{nl\}^{2/3} + n + l)$.*

Proof. The proof is quoted from [31]. We may assume without loss of generality that all lines are incident to at least one point. Let $\#i$ denote the number of incidences among the points and the lines. Define a graph G drawn in the plane such that the vertex set of G is the set of the given n points, and join two points with an edge drawn as a straight line segment, if the two points are two consecutive points on one of the lines. This drawing shows that $cr(G) \leq \binom{l}{2}$. The number of points on any of the lines is one plus the number of edges drawn along that line. Therefore the total number of incidences minus l is a lower bound for the number of edges in G. Formula (4) finishes the proof: either $4n \geq \#i - l$ or $CR(G) \geq \frac{1}{64}(\#i - l)^3/n^2$.   □

Theorem 2 clearly applies in every situation where we substitute the lines with curves, such that any two curves intersect in at most α points, and any two points are contained at most β curves, where α and β are fixed constants. For example, $\alpha = \beta = 2$, if we deal with translates of a strictly convex closed curve. This is the case if we consider incidences of unit circles and points.

A famous problem of Erdős [12] in geometry asks how many unit distances can be among n points in the plane. Given a point set with u unit distances, draw unit circles around the points. Clearly the number of incidences among the n points and the n unit circles is $2u$. However, the extension of the Szemerédi-Trotter theorem discussed above implies that $2u = \#i = O(n^{4/3})$, the best known estimate [30], which falls short of the conjectured $n^{1+\epsilon}$ for every $\epsilon > 0$.

Iosevich [16] gives an excellent survey of the many applications of the Szemerédi-Trotter theorem to analysis, number theory, and geometry. Here we cite one beautiful application to number theory that is relevant for our further discussion. Elekes [10] proved that

Theorem 3. *Any n distinct real numbers have $\Omega(n^{1.25})$ distinct sums or products.*

Proof. Let A denote a set of n real numbers. Consider the point set $P = \{(ab, c + d) : a, b, c, d \in A\}$ and the line set $L = \{\{(et, f + t) : t \in \mathbb{R}\} : e, f \in A\}$, where the inner set is a straight line given in parametric form. Clearly every line passes through n points from P, so $n^3 \leq \#i$. On the other hand, $\#i = O(\{|P||L|\}^{2/3} + |P| + |L|)$. Since $|L| = n^2$, simple algebra yields that $|P| = \Omega(n^{2.5})$. However, P is the Cartesian product of the sets $A \cdot A$ and $A + A$, and therefore one of these sets is as large as required.   □

5   Complex Szemerédi-Trotter Theorem

Is Theorem 3 valid for complex numbers? Elekes' proof would immediately give a positive answer if Theorem 4 is still valid for points and lines in a plane above the *complex* field. Note that a *stronger* bound cannot hold in the complex case, since Theorem 4 is tight within a multiplicative constant over the whole range of n and l [21]. Any sets of real points and real lines are also sets of complex points and complex lines, where the imaginery parts happen to be zero.

Csaba Tóth announced a proof to the complex Szemerédi-Trotter theorem a few years ago but the result has not been published yet [37]. The proof mimicks the original hard proof of Szemerédi and Trotter from [36], but every geometric step is more involved.

It is a natural question whether the crossing number method admits a complex analogue. The crucial point is that we miss a concept corresponding to graph drawing.

Perhaps an analogue to the planar straight line drawing could be the following. We draw a triplet system in $\mathbb{C}^2$ instead of a graph. The vertices of the triplet system are represented by points in $\mathbb{C}^2$. All 3 vertices of every triplet must be on a complex line. As the complex plane is isomorphic to $\mathbb{R}^4$, we represent the triplet by the convex hull of its vertices in $\mathbb{R}^4$. We require that triangles representing triplets share only a common vertex or a common edge.

Open Problem 7. *Show that there exists a constant K, such that a triplet system on n vertices, drawn as above, has at most Kn triplets.*

This could be the analogue of the fact that a planar graph can have only linear number of edges. This seems to be a more formidable problem. Solution to Problem 5 can give hint on how to proceed with Problem 7. The analogue of the straight line drawing allowing crossings could be the following, that I call a *linear triplet drawing*: triplets drawn as above may share a vertex, an edge, or may have a single crossing point which is not a vertex of either of them. Clearly a linear triplet drawing with m triplets and n vertices have at least $m - Kn$ crossings (if Problem 7 is answered affirmatively). We are going to make an analogue to the probabilistic folklore proof of (4). Assume $m \geq 2Kn$ for a linear a triplet drawing with c crossings. Pick vertices independently with probability $p = \sqrt{2Kn/m}$. By our choice of p, we have $m/p^3 = 2Kn/p^5$. Consider a triplet picked, if all 3 vertices are picked. The picked vertices in this way define an induced triplet system and an induced drawing of it. The number of vertices n', the number of edges m', and the number of crossings c' in the induced subdrawing are random variables. For any case, we have the following inequality between our random variables: $c' \geq m' - Kn'$. Taking expectation, we obtain $p^6 c \geq p^3 m - Kpn$. Doing the algebra, we obtain a counterpart of (4): $c \geq m/(2p^3) = \frac{m^{5/2}}{2(2K)^{3/2}n^{3/2}}$.

Let us see how to prove a counterpart of Theorem 4 modulo the claim above. Consider n points and l lines in the complex plane. We may assume without loss of generality that all lines are incident to at least one point. Within every complex line containing at least 3 points, build a triangulation on the points as vertices. The number of triangles is between linear functions of the number of points on that complex line, if there are at least 3 points on the complex line. Hence $\#i = \Theta(m) - 2l$, where m is the total number of triangles (or triplets) created, and $\#i$ is the total number of incidences. l complex lines have at most $\binom{l}{2}$ crossings, but the number of crossings is at least

$$\Omega\left(\frac{(\Theta(\#i) - 2l)^{5/2}}{n^{3/2}}\right)$$

or $m \leq 2Kn$. In the second case $\#i = O(n + l)$, and in the first case $\#i = O(n + l + l^{4/5}n^{3/5})$. Unfortunately, the last formula is not symmetric in n and l. However, using the point-line duality, we also have $\#i = O(n + l + n^{4/5}l^{3/5})$. Combining them $\#i = O(n + l + (nl)^{7/10})$, which is somewhat weaker than Theorem 4.

Acknowledgment. I thank Éva Czabarka for a careful reading of the manuscript.

References

1. Ábrego, B. M., and Fernández-Merchant, S., A lower bound for the rectilinear crossing number, *manuscript*.
2. Aigner, M., Ziegler, G. M., *Proofs from the Book*, Springer-Verlag, Berlin, 1998.
3. Ajtai, N., Chvátal, V., Newborn M., and E. Szemerédi, E., Crossing-free subgraphs, *Annals of Discrete Mathematics* **12** (1982) 9–12.
4. Aichholzer, O., Aurenhammer, F., and Krasser, H., On the crossing number of complete graphs, in: *Proc. Ann. ACM Symp. Computational Geometry*, 19–24, Barcelona, Spain, 2002.
5. Aurenhammer, F., On the Rectilinear Crossing Number, http://www.igi.tugraz.at/auren/
6. Balogh, J., and Salazar, G., On k-sets, convex quadrilaterals, and the rectilinear crossing number of K_n, *submitted*.
7. Bienstock, D., Dean, N., Bounds for rectilinear crossing numbers, *J. Graph Theory* **17** (1991), 333–348.
8. Czabarka, É., Sýkora, O., Székely L. A., and Vrťo, I., Convex crossing Numbers, circular arrangement problem, and isoperimetric functions, *submitted*.
9. Dey, T., Improved bounds for planar k-sets and related problems, *Discrete Comput. Geom.* **19** (1998) 373–382.
10. Elekes, G., On the number of sums and products, *Acta Arith.* **81** (1997)(4) 365–367.
11. Even, G., Guha, S., Schieber, B., Improved approximations of crossings in graph drawings and VLSI layout areas, in: *Proc. 32nd Annual Symposium on Theory of Computing, STOC'00* (ACM Press, 2000) 296–305.
12. Erdős, P., On sets of distances of n points, *Amer. Math. Monthly* **53**(1946), 248–250.
13. Fáry, I., On straight line representations of graphs, *Acta Univ. Szeged Sect. Sci. Math.* **11** (1948) 229–233.
14. Garey, M. R., and Johnson, D. S., Crossing number is NP-complete, *SIAM J. Alg. Discrete Methods* **4** (1983) 312–316.

15. Grohe, M., Computing crossing numbers in quadratic time, *Proc. 32nd Annual ACM Symposium on the Theory of Computing, STOC'01*, 231–236.
16. Iosevich, A., Fourier analysis and geometric combinatorics, *to appear*.
17. Leighton, F. T., *Complexity Issues in VLSI*, (MIT Press, Cambridge, 1983).
18. Leighton, F. T., and Rao, S., An approximate max flow min cut theorem for multicommodity flow problem with applications to approximation algorithm, in: *Proc. 29th Annual IEEE Symposium on Foundations of Computer Science* (IEEE Computer Society Press, Washington, DC, 1988) 422–431, and *J. ACM* **46** (1999) 787–832.
19. Lovász, L., Vesztergombi, K., Wagner, U., and Welzl, E., Convex quadrilaterals and k-sets, in: *Towards a Theory of Geometric Graphs*, ed. J. Pach, *Contemporary Mathematics* **342**, Amer. Math. Soc. 2004, 139–148..
20. A. Owens, On the biplanar crossing number, IEEE Transactions on Circuit Theory CT-18 (1971) 277–280.
21. Pach, J., and Agarwal, P. K., *Combinatorial Geometry*, Wiley and Sons, New York, 1995.
22. Pach, J., Crossing numbers, in: *Discrete and Computational Geometry* Japanese Conference JCDCG'98, Tokyo, Japan, December 1998, eds. J. Akiyama, M. Kano, M. Urabe, Lecture Notes in Computer Science Vol. 1763 (Springer Verlag, Berlin, 2000) 267–273.
23. Pach, J., Spencer, J., and Tóth, G., New bounds on crossing numbers, *Proc. 15th ACM Symposium on Computational Geometry* (ACM, 1999) 124–133; and *Discrete Comp. Geom.* **24** (2000) 623–644.
24. Pach, J., and Tóth, G., Which crossing number is it anyway? in: *Proc. 39th Annual Symposium on Foundation of Computer Science*, (IEEE Press, Baltimore, 1998) 617–626; and *J. Comb. Theory* Ser B **80** (2000) 225–246.
25. Pach, J., and Tóth, G., Graphs drawn with few crossings per edge, *Combinatorica* **17** (1997) 427–439.
26. Pach, J., Tóth, G., Thirteen problems on crossing numbers, *Geombinatorics* **9** (2000) 194–207.
27. Richter, R. B., and Salazar, G., A survey of good crossing number theorems and questions, *manuscript*.
28. Shahrokhi, F., Sýkora, O., Székely, L. A. and Vrťo, I., Crossing numbers: bounds and applications, in: *Intuitive Geometry*, eds. I. Bárány and K. Böröczky, Bolyai Society Mathematical Studies **6** (János Bolyai Mathematical Society, Budapest, 1997) 179–206.
29. Shahrokhi, F., Sýkora, Székely, L. A., and Vrťo, I., The gap between the crossing number and the convex crossing number, in: *Towards a Theory of Geometric Graphs*, ed. J. Pach, *Contemporary Mathematics* **342**, Amer. Math. Soc. 2004, 249–258.
30. Spencer, J., Szemerédi, E., Trotter, W. T., Unit distances in the Euclidean plane, in: *Graph Theory and Combinatorics* (B. Bollobás, ed.), Academic Press, London, 1984, 293–308.
31. Székely, L. A., Crossing numbers and hard Erdős problems in discrete geometry, *Combinatorics, Probability and Computing* **6** (1997) 353–358.
32. Székely, L. A., Zarankiewicz crossing number conjecture, in: Kluwer Encyclopaedia of Mathematics, Supplement III Managing Editor: M. Hazewinkel Kluwer Academic Publishers, 2002, 451–452.
33. Székely, L. A., A successful concept for measuring non-planarity of graphs: the crossing number, *Discrete Math.* **276** (2003), 1–3, 331–352.

34. Székely, L. A., Short proof for a theorem of Pach, Spencer, and Tóth, in: *Towards a Theory of Geometric Graphs*, ed. J. Pach, *Contemporary Mathematics* **342**, Amer. Math. Soc. 2004, 281–283.

35. Székely, L. A., An optimality criterion for the crossing number, *submitted*.

36. Szemerédi, E., Trotter, W. T., Extremal problems in discrete geometry, *Combinatorica* **3** (1983) 381–392.

37. Tóth, C. D., The Szemerédi-Trotter theorem in the complex plane, *manuscript*.

38. Turán, P., A note of welcome, *J. Graph Theory* **1** (1977) 7–9.

39. Tutte, W. T., Toward a theory of crossing numbers, *J. Combinatorial Theory* **8** (1970) 45–53.

40. Vrťo, I., Crossing Numbers of Graphs: A Bibliography
    http:/sun.ifi.savba.sk/~imrich/

Greedy Differential Approximations for Min Set Cover

C. Bazgan, J. Monnot, V. Th. Paschos, and F. Serrière

LAMSADE, Université Paris-Dauphine,
Place du Maréchal De Lattre de Tassigny, 75775 Paris Cedex 16, France
{bazgan, monnot, paschos, serriere}@lamsade.dauphine.fr

Abstract. We present in this paper differential approximation results for MIN SET COVER and MIN WEIGHTED SET COVER. We first show that the differential approximation ratio of the natural greedy algorithm for MIN SET COVER is bounded below by $1.365/\Delta$ and above by $4/(\Delta + 1)$, where Δ is the maximum set-cardinality in the MIN SET COVER-instance. Next, we study an approximation algorithm for MIN WEIGHTED SET COVER and provide a tight lower bound of $1/\Delta$.

1   Introduction

Given a family $\mathcal{S} = \{S_1, \ldots, S_m\}$ of subsets of a ground set $C = \{c_1, \ldots, c_n\}$ (we assume that $\cup_{S_i \in \mathcal{S}} S_i = C$), a set-cover of C is a sub-family $\mathcal{S}' \subseteq \mathcal{S}$ such that $\cup_{S_i \in \mathcal{S}'} S_i = C$; MIN SET COVER is the problem of determining a minimum-size set-cover of C. MIN WEIGHTED SET COVER consists of considering that sets of $\mathcal{S}$ are weighted by positive weights; the objective becomes then to determine a minimum total-weight cover of C.

Given $I = (\mathcal{S}, C)$ and a cover $\hat{\mathcal{S}}$, the sub-instance $\hat{I}$ of I induced by $\hat{\mathcal{S}}$ is the instance $(\hat{\mathcal{S}}, C)$. For simplicity, we identify in what follows a feasible (resp., optimal) cover $\mathcal{S}'$ (resp., $\mathcal{S}^*$) by the set of indices N' (resp., N^*) of the sets of the cover, i.e., $\mathcal{S}' = \{S_i : i \in N'\}$ (resp., $\mathcal{S}^* = \{S_i : i \in N^*\}$).

For an instance $(\mathcal{S}, C)$ of MIN SET COVER, its *characteristic graph* $B = (L, R; E)$ is a bipartite graph B with color-classes $L = \{1, \ldots, m\}$, corresponding to the members of the family $\mathcal{S}$ and $R = \{c_1, \ldots, c_n\}$, corresponding to the elements of the ground set C; the edge-set E of B is defined as $E = \{(i, c_j) : c_j \in S_i\}$.

A cover $\mathcal{S}'$ of C is said to be *minimal* (or *minimal for the inclusion*) if removal of any set $S \in \mathcal{S}'$ results in a family that is not anymore a cover for C.

Consider an instance $(\mathcal{S}, C)$ of MIN SET COVER and a minimal set-cover $\mathcal{S}'$ for it. Then, for any $S_i \in \mathcal{S}'$, there exists $c_j \in C$ such that S_i is the only set in $\mathcal{S}'$ covering c_j. Such a c_j will be called *non-redundant with respect to* $S_i \in \mathcal{S}'$; furthermore, S_i itself will be called *non-redundant for* $\mathcal{S}'$. With respect to the characteristic bipartite graph B' corresponding to the sub-instance I' of I induced by $\mathcal{S}'$ (it is the subgraph B' of B induced by $L' \cup R$ where $L = N'$), for any $i \in L'$, there exists a $c \in R$ such that $d(c) = 1$, where, for a vertex v of

M. Bieliková et al. (Eds.): SOFSEM 2005, LNCS 3381, pp. 62–71, 2005.

a graph G, $d(v)$ denotes the degree of v. In particular, there exists at least $|N'|$ non-redundant elements, one for each set. For simplicity, we will consider only one non-redundant element with respect to $S_i \in \mathcal{S}'$. Moreover, we assume that this element is c_i for the set $i \in N'$. Thus, the set of non-redundant elements with respect to $\mathcal{S}'$ considered here is $C_1 = \{c_i, \ i \in N'\}$.

In this paper we study differential approximability for MIN SET COVER in both unweighted and weighted versions. Differential approximability is analyzed using the so-called differential approximation ratio defined, for an instance I of an **NPO** problem Π (an optimization problem is in **NPO** if its decision version is in **NP**) and an approximation algorithm A computing a solution S for Π in I, as $\delta_A(I) = |\omega(I) - m_A(I, S)|/|\omega(I) - \mathrm{opt}(I)|$ where $\omega(I)$ is the value of the worst Π-solution for I, $m_A(I, S)$ is the value of S and $\mathrm{opt}(I)$ is the value of an optimal Π-solution for I. For an instance $I = (\mathcal{S}, C)$ of MIN SET COVER, $\omega(I) = m$, the size of the family $\mathcal{S}$. Obviously, this is the maximum-size cover of I. Finally, standard approximability is analyzed using the standard approximation ratio defined as $m_A(I, S)/\mathrm{opt}(I)$.

Surprisingly enough, differential approximation, although introduced in [1] since 1977, has not been systematically used until the 90's ([2,3] are, to our knowledge, the most notable uses of it) when a formal framework for it and a more systematic use started to be drawn ([4]). In general, no apparent links exist between standard and differential approximations in the case of minimization problems, in the sense that there is no evident transfer of a positive, or negative, result from one paradigm to the other. Hence a "good" differential approximation result does not signify anything for the behavior of the approximation algorithm studied when dealing with the standard framework and vice-versa. As already mentioned, the differential approximation ratio measures the quality of the computed feasible solution according to both optimal value and the value of a worst feasible solution. The motivation for this measure is to look for the placement of the computed feasible solution in the interval between an optimal solution and a worst-case one. Even if differential approximation ratio is not as popular as the standard one, it is interesting enough to be investigated for some fundamental problems as MIN SET COVER, in order to observe how they behave under several approximation criteria. Such joint investigations can significantly contribute to a deeper apprehension of the approximation mechanisms for the problems dealt. A further motivation for the study of differential approximation is the stability of the differential approximation ratio under affine transformations of the objective function. This stability often serves in order to derive differential approximation results for minimization (resp., maximization) problems by analyzing approximability of their maximization (resp., minimization) equivalents under affine transformations. We will apply such transformation in Section 4.

We study in this paper the performance of two approximation algorithms. The first one is the classical greedy algorithm studied, for the unweighted case and for the standard approximation ratio, in [5, 6] and, more recently, in [7]. For this algorithm, we provide a differential approximation ratio bounded below

by $1.365/\Delta$ when $\Delta = \max_{S_i \in \mathcal{S}}\{|S_i|\}$ is sufficiently large. We next deal with MIN
WEIGHTED SET COVER and analyze the differential approximation performance
of a simple greedy algorithm that starts from the whole $\mathcal{S}$ considering it as
solution for MIN WEIGHTED SET COVER and then it reduces it by removing the
heaviest of the remaining sets of $\mathcal{S}$ per time until the cover becomes minimal.
We show that this algorithm achieves differential approximation ratio $1/\Delta$.

Differential approximability for both MIN SET COVER and MIN WEIGHTED
SET COVER have already been studied in [8] and discussed in [9]. The differential
approximation ratios provided there are $1/\Delta$, for the former, and $1/(\Delta + 1)$,
for the latter. Our current work improves (quite significantly for the unweighted
case), these old results. Note also that an approximation algorithm for MIN SET
COVER has been analyzed also in [4] under the assumption $m \geqslant n$, the size of the
ground set C. It has been shown that, under this assumption, MIN SET COVER is
approximable within differential approximation ratio $1/2$. More recently, in [9],
under the same assumption, MIN SET COVER has been proved approximable
within differential approximation ratio $289/360$.

It is proved in [4] that *if $\boldsymbol{P} \neq \boldsymbol{NP}$, then inapproximability bounds for standard
(and differential) approximation for* MAX INDEPENDENT SET *hold as differential
inapproximability bounds for* MIN SET COVER. *Consequently, unless $\boldsymbol{P} = \boldsymbol{NP}$,*
MIN SET COVER *is not differentially approximable within $O(n^{\epsilon-(1/2)})$, for any
$\epsilon > 0$.* This result implies that approximation ratios of the same type as in
standard approximation (for example, $O(1/\ln \Delta)$, or $O(1/\log n)$) are extremely
unlikely for MIN SET COVER in differential approximation. Consequently, differential approximation results for MIN SET COVER cannot be trivially achieved by
simply transposing the existing standard approximation results to the differential framework. This is a further motivation of our work.

In what follows, we deal with non-trivial instances of (unweighted) MIN SET
COVER. An instance I is non-trivial for unweighted MIN SET COVER if the two
following conditions hold simultaneously: (i) no set $S_i \in \mathcal{S}$ is a proper subset of
a set $S_j \in \mathcal{S}$, and (ii) no element in C is contained in I by only one subset of $\mathcal{S}$
(i.e., there is no non-redundant set for $\mathcal{S}$).

2   The Natural Greedy Algorithm for Min Set Cover

Let us first note that a lower bound of $1/\Delta$ can be easily proved for the differential
ratio of any algorithm computing a minimal set cover. We analyze in this section
the differential approximation performance of the following very classical greedy
algorithm for MIN SET COVER, called SCGREEDY in the sequel:

1. compute $S_i \in \operatorname{argmax}_{S \in \mathcal{S}}\{|S|\}$; set $N'' = N'' \cup \{i\}$ (N'' is initialized to $\emptyset$);
2. update I setting: $\mathcal{S} = \mathcal{S} \setminus \{S_i\}$, $C = C \setminus S_i$ and, for any $S_j \in \mathcal{S}$, $S_j = S_j \setminus S_i$;
3. repeat Steps 1 to 2 until $C = \emptyset$;
4. range N'' in the order sets have been chosen and assume $N'' = \{i_1, \ldots, i_k\}$;
5. Set $N' = N''$; for $j = k$ downto 1: if $N' \setminus \{i_j\}$ is a cover then $N' = N' \setminus \{i_j\}$;
6. output N' the minimal cover computed in Step 5.

Theorem 1. *For Δ sufficiently large, algorithm SCGREEDY achieves differential approximation ratio $1.365/\Delta$.*

Proof. Consider N'' and the sets $\mathcal{S}'' = \{S'_{i_1}, S'_{i_1} \ldots, S'_{i_k}\}$, computed in Step 4 with their residual cardinalities, i.e., as they have been chosen during Steps 1 and 2; remark that, so-considered, the set $\mathcal{S}''$ forms a partition on C. On the other hand, consider solution N' output by the algorithm SCGREEDY and remark that family $\{S'_i : i \in N'\}$ does not necessarily cover C.

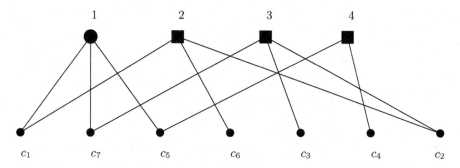

Fig. 1. An example of application of Step 5 of SCGREEDY

As an example, assume some MIN SET COVER-instance $(\mathcal{S}, C)$ with $C = \{c_1, \ldots, c_7\}$ and suppose that execution of Steps 1 to 4 has produced a cover $N'' = \{1, 2, 3, 4\}$ (given by the sets $\{S_1, S_2, S_3, S_4\}$). Figure 1 illustrates characteristic graph B', i.e., the subgraph of $B = (L, R; E)$ induced by $L' \cup R$ where L' and R correspond to the sets N'' and C respectively. It is easy to see that N'' is not minimal and application of Step 5 of SCGREEDY drops the set S_1 out of N''; hence, $N' = \{2, 3, 4\}$. The residual parts of S_2, S_3 and S_4 are $S'_2 = \{c_2, c_6\}$, $S'_3 = \{c_3\}$ and $S'_4 = \{c_4\}$, respectively. Note that Step 5 of SCGREEDY is important for the solution returned in Step 6, since solution N'' computed in Step 4 may be a worst solution (see the previous example) and then, $\delta(I, \mathcal{S}') = 0$.

Consider an optimal solution N^* given by the sets S_i, $i \in N^*$ and denote by $\{S^*_i\}$, $S^*_i \subseteq S_i$, $i \in N^*$, an arbitrary partition of C (if an element c is covered by more than one sets S_i, $i \in N^*$, then c is randomly assigned to one of them). Let $C'_1 = \{c_i : i \in N' \setminus N^*\}$ be a set of non-redundant elements with respect to N'; obviously, by construction $|C'_1| = |N' \setminus N^*|$. Finally, set $N^*_1 = \{j \in N^* : \exists c \in C'_1, c \in S^*_j\}$. We deduce $N^*_1 \subseteq N^* \setminus N'$, since any element $c \in C'_1$ is non-redundant for N' (otherwise, there would exist at least a $c \in C'_1$ covered twice: one time by a set in $N' \setminus N^*$ and one time by a set in $N' \cap N^*$, absurd by the construction of C'_1). Finally, set $\bar{N} = \{1, \ldots, m\} \setminus (N' \cup N^*)$. Observe that, using the notations just introduced, we have:

$$\delta(I, \mathcal{S}') = \frac{|N^*_1| + |N^* \setminus (N' \cup N^*_1)| + |\bar{N}|}{|N' \setminus N^*| + |\bar{N}|} \tag{1}$$

Consider the bipartite graph $B'' = (L'', R''; E'')$ with $L'' = N_1^* \cup (N' \setminus N^*)$, $R'' = C_1'$ and $(i, c_j) \in E''$ iff $i \in S_j^*$ or $i = j$. This graph is a partial graph of the characteristic bipartite graph B' induced by $L' = N_1^* \cup (N' \setminus N^*)$ and $R' = C_1'$. By construction, B'' is not connected and, furthermore, any of its connected components is of the form of Figure 2.

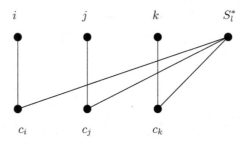

Fig. 2. A connected component of B''

For $i = 1, \ldots, \Delta$, denote by x_i the number of connected components of B'' corresponding to sets S_l^* of cardinality i. Then, by construction of this sub-instance, we have:

$$|N_1^*| = \sum_{i=1}^{\Delta} x_i \tag{2}$$

$$|N' \setminus N^*| = |C_1'| = \sum_{i=1}^{\Delta} i \cdot x_i \tag{3}$$

Consider $z \in [1, \Delta]$ such that $|C_1'| = i_0 |N_1^*|$ where $i_0 = \Delta/z$. One can easily see that i_0 is the average cardinality of sets in N_1^* (when we consider the sets S_i^*, $i \in N_1^*$, that form, by construction, a partition on C_1'). Indeed,

$$i_0 = \frac{1}{|N_1^*|} \sum_{i \in N_1^*} |S_i^*| = \frac{\sum_{i=1}^{\Delta} i \cdot x_i}{\sum_{i=1}^{\Delta} x_i} \tag{4}$$

We have immediately from (1), (2) and (3):

$$\delta(I, S') \geqslant \frac{|N_1^*|}{|N' \setminus N^*|} = \frac{|N_1^*|}{|C_1'|} = \frac{1}{i_0} = \frac{z}{\Delta} \tag{5}$$

Consider once more the component of Figure 2, suppose that set S_ℓ^* has cardinality i and denote it by $S_\ell^* = \{c_{\ell_1}, \ldots, c_{\ell_i}\}$ with $\ell_1 < \ldots < \ell_i$. By greedy rule of SCGREEDY, we deduce that the sets $S_{\ell_1}', \ldots, S_{\ell_i}'$ (recall that we only consider the residual part of the set) have been chosen in this order (cf. Steps 4 and 5 of

SCGREEDY) and verify $|S'_{\ell_p}| \geqslant i+1-p$ for $p = 1, \ldots, i$. Consequently, there exist $(i-1)+(i-2)+\ldots+1 = i(i-1)/2$ elements of C not included in C'_1. Iterating this observation for any connected component of B'' we can conclude that there exists a set $C_2 \subseteq C$, outside set C_1, of size at least $|C_2| \geqslant \sum_{i=1}^{\Delta} i(i-1)x_i/2$. Elements of C_2 are obviously covered, with respect to N^*, by sets either from N_1^*, or from $N^* \setminus N_1^*$. Suppose that sets of N_1^* of cardinality i (there exist x_i such sets), $i = 1, \ldots, \Delta$, cover a total of $k_i x_i$ elements of C_2. Therefore, there exists a subset $C'_2 \subseteq C_2$ of size at least: $|C'_2| \geqslant \sum_{i=1}^{\Delta}((i(i-1)/2) - k_i)x_i$. The elements of C'_2 are covered in N^* by sets in $N^* \setminus N_1^*$. In order that C'_2 is covered, a family $N_2^* \subseteq N^* \setminus N_1^*$ of size

$$|N_2^*| \geqslant \frac{1}{\Delta} \cdot \sum_{i=1}^{\Delta} \left(\frac{i(i-1)}{2} - k_i \right) x_i \tag{6}$$

is needed. Dealing with N_2^*, suppose that for a $y \in [0,1]$: (i) $(1-y)|N_2^*|$ sets of N_2^* belong to $N^* \setminus N'$ (indeed, they belong to $N^* \setminus (N' \cup N_1^*)$) and (ii) $y|N_2^*|$ sets of N_2^* belong to $N^* \cap N'$.

We study the two following cases: $y \leqslant (\Delta - 1)/\Delta$ and $y \geqslant (\Delta - 1)/\Delta$.

The first case is equivalent to $(1-y) \geqslant 1/\Delta$ and then, taking into account that $k_i \leqslant \Delta - i$, we obtain: $(1-y)|N_2^*| \geqslant |N_2^*|/\Delta \geqslant \sum_{i=1}^{\Delta}((i(i-1)/(2\Delta^2)) + (i/\Delta) - 1)x_i$. Using (1), (2), (3) and (6), we deduce:

$$\delta(I, S') \geqslant \frac{|N_1^*| + |N_2^*|/\Delta}{|N' \setminus N^*|} \geqslant \frac{\sum_{i=1}^{\Delta} \left(\frac{i(i-1)}{2\Delta^2} + \frac{i}{\Delta} \right) x_i}{\sum_{i=1}^{\Delta} i \cdot x_i} = \frac{1}{\Delta} + \frac{\sum_{i=1}^{\Delta} f(i)x_i}{\sum_{i=1}^{\Delta} i \cdot x_i} \tag{7}$$

where $f(x) = x(x-1)/(2\Delta^2)$, with $1 \leqslant x \leqslant \Delta$. We will now show the following inequality (see also (4)) that $i_0 = (\sum_{i=1}^{\Delta} ix_i)/(\sum_{i=1}^{\Delta} x_i)$:

$$\frac{\sum_{i=1}^{\Delta} f(i) \cdot x_i}{\sum_{i=1}^{\Delta} i \cdot x_i} \geqslant \frac{f(i_0)}{i_0} \tag{8}$$

Remark that (8) is equivalent to $\sum_{i=1}^{\Delta} f(i) \cdot (x_i / \sum_{i=1}^{\Delta} x_i) \geqslant f(i_0)$. On the other hand, since f is convex, we have by Jensen's theorem $\sum_{i=1}^{\Delta} z_i f(i) \geqslant f(\sum_{i=1}^{\Delta} iz_i)$, where $z_i \in [0,1]$, $\sum_{i=1}^{\Delta} z_i = 1$. Setting $z_i = x_i / \sum_{i=1}^{\Delta} x_i$, (8) follows.

Thus, since $i_0 = \Delta/z$ and we study an asymptotic ratio in Δ, (7) becomes

$$\delta(I, S') \geqslant \frac{1}{\Delta} + \frac{1}{2\Delta^2} \left(\frac{\Delta}{z} - 1 \right) \approx \frac{1}{\Delta} + \frac{1}{2\Delta z} \tag{9}$$

Expression (9) is decreasing with z, while (5) is increasing with z. Equality of both ratios is reached when $2z^2 - 2z - 1 = 0$, i.e., for $z = (2 + \sqrt{12})/4 \approx 1.365$.

We now deal with case $y \geqslant (\Delta - 1)/\Delta$. Sub-family $N_2^* \cap N'$ (of size $y|N_2^*|$) is, by hypothesis, common to both N' (the cover computed by SCGREEDY) and N^*. Minimality of N' implies that, for any set $i \in N_2^* \cap N'$, there exists at least one element of C non-redundant with respect to S_i. So, there exist at least $|C_3| = |N_2^* \cap N'|$ elements of C outside C_1' and C_2.

Some elements of C_3 can be covered by sets in N_1^*. In any case, for the sets $\{j_1, \ldots, j_{x_i}\}$ of N_1^* of cardinality i with respect to the partition S_ℓ^*, there exist at most $(\Delta - (i + k_i))x_i$ elements of C_3 that can belong to them (so, these elements are covered by the residual set $S_{j_p} \setminus S_{j_p}^*$ for $p = 1, \ldots, x_i$). Thus, there exist at least

$$|C_3'| = |C_3| - \sum_{i=1}^{\Delta} (\Delta - (i + k_i))\, x_i = y\,|N_2^*| - \sum_{i=1}^{\Delta} (\Delta - (i + k_i))\, x_i \qquad (10)$$

elements of C_3 not covered by sets in N_1^*. Since initial instance $(\mathcal{S}, C)$ is non-trivial, elements of C_3' are also contained in sets N_3 either from $N^* \setminus N_1^*$, or from $\bar{N}$. So, the family N_3 has size at least $|C_3'|/\Delta$. Moreover, using (6), (10) and $y \leqslant 1$, we get:

$$|N_3| \geqslant \frac{y\,|N_2^*|}{\Delta} - \sum_{i=1}^{\Delta} \frac{(\Delta - (i + k_i))\, x_i}{\Delta} \geqslant y \sum_{i=1}^{\Delta} \frac{i(i-1)}{2\Delta^2} x_i + \sum_{i=1}^{\Delta} (\frac{i}{\Delta} - 1) x_i \qquad (11)$$

We so deduce: $\delta(I, \mathcal{S}') \geqslant (|N_1^*| + |N_3 \setminus \bar{N}| + |\bar{N}|)/(|N' \setminus N^*| + |\bar{N}|)$, which, taking into account that $|\bar{N}| \geqslant |\bar{N} \cap N_3|$, finally becomes:

$$\delta(I, \mathcal{S}') \geqslant \frac{|N_1^*| + |N_3|}{|N' \setminus N^*| + |N_3|} \qquad (12)$$

Note, furthermore, that function $(a + x)/(b + x)$ is increasing with x, for $a \leqslant b$ and $x > -b$. Therefore, using (2), (3), (11) and $y \geqslant (\Delta - 1)/\Delta$, (12) becomes:

$$\delta(I, \mathcal{S}') \geqslant \frac{\displaystyle\sum_{i=1}^{\Delta} \left(\frac{(\Delta-1)\cdot i(i-1)}{2\Delta^3} + \frac{i}{\Delta} \right) x_i}{\displaystyle\sum_{i=1}^{\Delta} \left(i + \frac{(\Delta-1)\cdot i(i-1)}{2\Delta^3} + \frac{i}{\Delta} - 1 \right) x_i} \qquad (13)$$

Set now $f(x) = (\Delta - 1) \cdot (x(x-1)/2\Delta^3) + (x/\Delta)$; (13) can now be expressed as:

$$\delta(I, \mathcal{S}') \geqslant \frac{\displaystyle\sum_{i=1}^{\Delta} f(i) x_i}{\displaystyle\sum_{i=1}^{\Delta} (f(i) + i - 1)\, x_i} \qquad (14)$$

With the same arguments, as for the convexity of f, we deduce from (14):

$$\delta(I, \mathcal{S}') \geqslant \frac{f(i_0)}{f(i_0) + i_0 - 1} = \frac{\frac{(\Delta-1)\cdot i_0(i_0-1)}{2\Delta^3} + \frac{i_0}{\Delta}}{i_0 + \frac{(\Delta-1)\cdot i_0(i_0-1)}{\Delta^3} + \frac{i_0}{\Delta} - 1} \qquad (15)$$

Recall that we have fixed $i_0 = \Delta/z$. If one assumes that Δ is arbitrarily large, one can simplify calculations by replacing $i_0 - 1$ by i_0. Then, (15) becomes:

$$\delta\left(I, S'\right) \geqslant \frac{\frac{i_0^2}{2\Delta^2} + \frac{i_0}{\Delta}}{\frac{i_0^2}{2\Delta^2} + \frac{i_0}{\Delta} + i_0} \geqslant \frac{\frac{1}{2z^2} + \frac{1}{z}}{\frac{1}{2z^2} + \frac{1}{z} + \frac{\Delta}{z}} \approx \frac{1}{2z\Delta} + \frac{1}{\Delta} \qquad (16)$$

Ratio given by (5) is increasing with z, while the one of (16) is decreasing with z. Equality of both ratios is reached when $2z^2 - 2z - 1 = 0$, i.e., for $z \approx 1.365$.

So, in any of the cases studied above, the differential approximation ratio achieved by SCGREEDY is greater than, or equal to, $1.365/\Delta$ and the proof of the theorem is now complete. $\qquad\square$

Proposition 1. *There exist* MIN SET COVER-*instances where the differential approximation ratio of* **SCGREEDY** *is* $4/(\Delta + 2)$ *for any* $\Delta \geqslant 3$.

Proof. Assume a fixed $t > 1$, a ground set $C = \{c_{ij} : i = 1, \ldots, t-1, j = 2, \ldots, t, j > i\}$ and a system $S = \{S_1, \ldots, S_t\}$, where $S_i = \{c_{ji} : j < i\} \cup \{c_{ij} : j > i\}$, for $i = 1, \ldots, t$. Denote by $I_t = (S, C)$ the instance of MIN SET COVER defined on C and S.

Remark that the smallest cover for C includes at least $t - 1$ sets of S. Indeed, consider a family $S' \subseteq S$ of size less than $t - 1$. Then, there exists $i_0 < j_0$ such that neither S_{i_0}, nor S_{j_0} belong to S'. In this case element $c_{i_0 j_0} \in C$ is not covered by S'. Note finally that, for I_t, the maximum size of the subsets of S is $\Delta = t - 1$. Indeed, for any $i = 1, \ldots, t$, $|\{c_{ji} : j < i\}| = i - 1$ and $|\{c_{ij} : j < i\}| = t - i$; so, $|S_i| = t - 1$.

Fix an even Δ and build the following instance (S, C) for MIN SET COVER:

$$C = \left\{ a_{ij}, a'_{ij} : i, j = 1, \ldots, \frac{\Delta}{2}, j > i \right\} \cup \left\{ b_{ij} : i = 1, \ldots, \frac{\Delta + 2}{2}, j = 1, \ldots, \Delta \right\}$$

$$S_i^1 = \{a_{ji} : j < i\} \cup \{a_{ij} : j > i\} \cup \left\{ b_{ji} : j = 1, \ldots, \frac{\Delta + 2}{2} \right\}, \ i = 1, \ldots, \frac{\Delta}{2}$$

$$S_i^2 = \{a'_{ij} : j < i\} \cup \{a'_{ij} : j > i\}$$

$$\cup \left\{ b_{jk} : j = 1, \ldots, \frac{\Delta + 2}{2}, k = i + \frac{\Delta}{2} \right\}, \ i = 1, \ldots, \frac{\Delta}{2}$$

$$S^j = \left\{ S_i^j : i = 1, \ldots, \frac{\Delta}{2} \right\}, \ j = 1, 2$$

$$S^3 = \left\{ \left\{ \{b_{ij} : j = 1, \ldots, \Delta\}, i = 1, \ldots, \frac{\Delta + 2}{2} \right\} i = 1, \ldots, \frac{\Delta + 2}{2} \right\}$$

Set $S = S^1 \cup S^2 \cup S^3$. Notice that, $\forall S_i \in S$, $|S_i| = \Delta$. Hence, during its first iteration, SCGREEDY can choose a set in S^3. Such a choice does not reduce cardinalities of the remaining sets in this sub-family; so, during its first $(\Delta + 2)/2$ iterations, SCGREEDY can exclusively choose all sets in S^3. Remark that such choices entail that the surviving instance is the union of two disjoint instances $I_{\Delta/2}$ (i.e., instances of type I_t, as the ones defined in the beginning of this section, with

$t = \Delta/2$), induced by the sub-systems $(\mathcal{S}^1, \{a_{ij}\})$ and $(\mathcal{S}^2, \{a'_{ij}\})$. According to what has been discussed at the beginning of the section, any cover for such instances uses at least $(\Delta/2) - 1$ sets. So, for a set-cover $\mathcal{S}'$ computed by SCGREEDY (remark that $\mathcal{S}'$ is minimal), we finally have: $m(\hat{I}, \mathcal{S}') \geqslant (3\Delta/2) - 1$. Furthermore, since $\mathcal{S}^1 \cup \mathcal{S}^2$ is a cover for C: $\mathrm{opt}(\hat{I}) = \Delta$; finally, $\omega(\hat{I}) = (3\Delta/2) + 1$. In all, $\delta(\hat{I}, \mathcal{S}') = 4/(\Delta + 2)$.

For the case of odd Δ (that is omitted here), it can be proved that there exist an instance of MIN SET COVER for which SGREEDY attains differential ratio at least $4/(\Delta + 1)$.   □

3   Differential Approximation for Min Weighted Set Cover

Consider an instance $I = (\mathcal{S}, C, \boldsymbol{w})$ of MIN WEIGHTED SET COVER, where $\boldsymbol{w}$ denotes the vector of the weights on the subsets of $\mathcal{S}$ and the following algorithm, denoted by WSC in what follows:

- order sets in $\mathcal{S}$ in decreasing weight-order (i.e., $w_1 \geqslant \ldots \geqslant w_m$); let $N = \{1, \ldots, m\}$ be the set of indices in the (so-ordered) $\mathcal{S}$; set $N' = N$;
- for $i = 1$ to m: if $N' \setminus \{i\}$ covers C, then set $N' = N' \setminus \{i\}$;
- output N'.

Proposition 2. *WSC achieves differential approximation ratio bounded below by $1/\Delta$. This bound is asymptotically tight.*

Proof. We use in what follows notations introduced in Section 2. Observe that $N \setminus N' = \bar{N} \cup (N^* \setminus N')$ and $N \setminus N^* = \bar{N} \cup (N' \setminus N^*)$ where we recall that $\bar{N} = N \setminus (N^* \cup N')$. Denoting, for any $i \in N$, by w_i the weight of S_i, and, for any subset $\mathcal{X} \subseteq N$, by $w_{\mathcal{X}}$, the total weight of the sets with indices in $\mathcal{X}$, i.e., the quantity $\sum_{i \in \mathcal{X}} w_i$, the differential approximation ratio of WSC becomes

$$\delta(I, N') = \frac{w_{N \setminus N'}}{w_{N \setminus N^*}} \qquad (17)$$

Let $C_c = \{c_j : \exists i \in N' \cap N^*, c_j \in S_i\}$ be the set of elements covered by $N' \cap N^*$ and let $\bar{C}_c = C \setminus C_c$ be the complement of C_c with respect to C. It is easy to see that both $N' \setminus N^*$ and $N^* \setminus N'$ cover $\bar{C}_c$. Obviously, $C'_1 \subseteq \bar{C}_c$ (recall that $C'_1 = \{c_i : i \in N' \setminus N^*\}$ is a set of non-redundant elements with respect to sets of $N' \setminus N^*$ and that any element of C'_1 is covered by sets in $N^* \setminus N'$).

Consider the sub-instance of I induced by $(N' \setminus N^* \cup N^* \setminus N', C'_1)$. Fix an index $i \in N^* \setminus N'$ and denote by $S_i^* = \{c_{i_1}, \ldots, c_{i_k}\}$ the restriction of S_i to C'_1, i.e., $S_i^* = S_i \cap C'_1$. Assume that $S_i^* \neq \emptyset$; as it will be understood just below if this is not the case, then the approximation ratio of WSC will be even better. Obviously, since sets $i_1, \ldots, i_k$ have been chosen by WSC (i.e., $\{i_1, \ldots, i_k\} \subseteq N'$), $w_{i_j} \leqslant w_i$ and, $k \leqslant \Delta$, we get: $\sum_{j=1}^{k} w_{i_j} \leqslant \Delta w_i$. Summing it for all $i \in N^* \setminus N'$, we obtain $w_{N' \setminus N^*} \leqslant \Delta w_{N^* \setminus N'}$ and then, $w_{N \setminus N^*} \leqslant \Delta w_{N^* \setminus N}$. Expression (17) suffices now to conclude the proof of the ratio. The proof of the tightness is omitted here. □

4   Max Hypergraph Independent Set

An instance $(\mathcal{S}, C)$ of MIN SET COVER can also be seen as a hypergraph H where C is the set of its vertices and $\mathcal{S}$ is the set of its hyper-edges. Then MIN SET COVER consists of determining the smallest set of hyper-edges covering C. The "dual" of this problem is the well-known MIN HITTING SET problem, where, on $(\mathcal{S}, C)$ one wishes to determine the smallest subset of C hitting any set in $\mathcal{S}$. MIN HITTING SET and MIN SET COVER are approximate equivalent in both standard and differential paradigms (see, for example, [10]; the former is the same as the latter modulo the inter-change of the roles of $\mathcal{S}$ and C). On the other hand another well-known combinatorial problem is MAX HYPERGRAPH INDEPENDENT SET where given $(\mathcal{S}, C)$, one wishes to determine the largest subset C' of C such that no $S_i \in \mathcal{S}$ is a proper subset of C'. It is easy to see that for MAX HYPERGRAPH INDEPENDENT SET and MIN HITTING SET, the objective function of the one is an affine transformation of the objective function of the other, since a hitting set is the complement with respect to C of a hypergraph independent set. Consequently, the differential approximation ratios of these two problems coincide, and coincide also (as we have seen just above) with the differential approximation ratio of MIN SET COVER. Hence, the results of the previous sections identically apply for MAX HYPERGRAPH INDEPENDENT SET also.

References

1. Ausiello, G., D'Atri, A., Protasi, M.: On the structure of combinatorial problems and structure preserving reductions. In: Proc. ICALP'77. Lecture Notes in Computer Science, Springer-Verlag (1977)
2. Ausiello, G., D'Atri, A., Protasi, M.: Structure preserving reductions among convex optimization problems. J. Comput. System Sci. **21** (1980) 136–153
3. Zemel, E.: Measuring the quality of approximate solutions to zero-one programming problems. Math. Oper. Res. **6** (1981) 319–332
4. Demange, M., Paschos, V. Th.: On an approximation measure founded on the links between optimization and polynomial approximation theory. Theoret. Comput. Sci. **158** (1996) 117–141
5. Johnson, D.S.: Approximation algorithms for combinatorial problems. J. Comput. System Sci. **9** (1974) 256–278
6. Lovász, L.: On the ratio of optimal integral and fractional covers. Discrete Math. **13** (1975) 383–390
7. Slavík, P.: A tight analysis of the greedy algorithm for set cover. In: Proc. STOC'96. (1996) 435–441
8. Demange, M., Grisoni, P., Paschos, V. Th.: Differential approximation algorithms for some combinatorial optimization problems. Theoret. Comput. Sci. **209** (1998) 107–122
9. Hassin, R., Khuller, S.: z-approximations. J. Algorithms **41** (2001) 429–442
10. Ausiello, G., Crescenzi, P., Gambosi, G., Kann, V., Marchetti-Spaccamela, A., Protasi, M.: Complexity and approximation. Combinatorial optimization problems and their approximability properties. Springer, Berlin (1999)

A Methodology of Visual Modeling Language Evaluation

Anna Bobkowska

Gdańsk University of Technology,
Narutowicza 11/12, 80-952 Gdańsk, Poland
annab@eti.pg.gda.pl

Abstract. In order to achieve more maturity in Visual Modeling Language (VML) engineering, effective methodologies of VML evaluation are necessary. This paper proposes a methodology for evaluating VML from a cognitive perspective. It discusses methodological problems of applying psychological results for VML evaluation. It then presents research on a CD-VML methodology that is based on cognitive dimensions (CD). This research covers analysis of the application of CDs in the context of VMLs, empirical studies of UML with an original CD questionnaire, customization of CDs for VML evaluation and empirical studies of use cases with a CD-VML-UC questionnaire - a product of the CD-VML methodology for use case diagram.

1 Introduction

Models play an important role in the activities of software development, integration and maintenance. As software systems and their models become increasingly complex, numerous interrelated diagrams are used for documentation. Sometimes a few interrelated representations of the same system are required and there is a need of management of not only relationships between the diagrams in the model but also many interrelated models. Characteristics of performing software engineering tasks, such as the level of their difficulty, the time of modeling, the time of searching for information on the diagrams, the number of mistakes and possibilities of automation depend on Visual Modeling Language (VML), which is used for modeling. The standard VML nowadays is Unified Modeling Language [14], but there are also many domain-specific modeling languages in use.

VML engineering is an emerging discipline which aims to provide methods for designing high quality VMLs. Important components of VML engineering are methodologies for evaluating several VML characteristics. VML is a complex phenomenon which integrates visual and linguistic features. Additionally, VMLs might be used different contexts. Therefore many methods of evaluation should be designed. We assume that a good methodology of VML evaluation should be:

- easy to use - not requiring knowledge of complex theories;
- flexible - allowing evaluation in different contexts of use, e.g. type of the system under development or type of model application;
- effective - allowing discovery of all important issues within a perspective;
- efficient – carried out in a short time.

The idea of this research was to propose a tool for VML evaluation from a cognitive perspective. It is believed that a cognitive fit of the technology to its users

M. Bieliková et al. (Eds.): SOFSEM 2005, LNCS 3381, pp. 72–81, 2005.

can increase efficiency, decrease cost, improve software quality and allow for easier learning, use and maintenance [11]. However, the use of cognitive and social sciences in software engineering is not an easy task. There are many loosely related psychological theories that answer only selected questions, and often they were verified with limited scope. Psychological research has also its own methodological problems, which result from the fact that none can really 'touch' entities inside the human mind. Thus, there is a need for checking the usefulness of a particular theory in the software engineering context, and usually also for its customization. Solid empirical studies are then essential for verification. There have been attempts of such research, but these have encountered several problems [6], e.g. mismatch between software engineering and psychology in programming. Other problems are concerned with different values in research (explanation vs. practical use) and the level of required precision, which is much higher in software engineering.

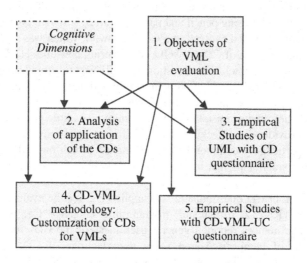

Fig. 1. Schema of research methodology: research areas and their relationships

Fig. 1 presents a schema of research methodology. In the first step we defined objectives for VML evaluation methodology. An analysis of application of cognitive dimensions (CD)[1,2,8] in the VML context was then made and empirical study with an original CD questionnaire was performed. These suggested a need for customization, so in the next step we developed a CD-VML methodology and verified it in another empirical study with a CD-VML-UC questionnaire - a product of the CD-VML methodology for the use case model.

Section 2 describes cognitive dimensions and the analysis of application of CDs in the VML context. Section 3 presents results of the empirical study of UML with an original CD questionnaire. Section 4 presents customization of CDs for VMLs and the resulting CD-VML methodology. Section 5 contains a description of the empirical study of evaluating use cases with CD-VML-UC questionnaire, and section 6 presents results of this evaluation. Section 7 summarizes the research and draws conclusions.

2   Analysis of Application of CDs for VML Evaluation

The quality of application of a theory or a method for a given purpose depends on the quality of the theory or method and appropriateness of using it for this purpose. Theoretical analysis enables better their understanding, their characteristics and background. With this analysis, the decision about application is based on more than just belief that a method does what it promises to do. The analysis of the application of CDs for VML evaluation included acquiring knowledge about CDs and CD framework, its history of evolution, background theories, known applications of the CD questionnaire and reasoning about possibilities of applying it in VML context.

Cognitive Dimensions [8] were designed for evaluation of usability of 'notational systems'. Notation is 'how users communicate with the system; notations can include text, pictures, diagrams, tables, special symbols or various combinations of these. Notational systems include many different ways of storing and using information – books, different ways of using pencil and paper, libraries, software programs, and small electronic devices.' The idea evolved to a CD framework which was a kind of research tool about different aspects of CDs as well as discovering new ones [1]. The intention of providing a Cognitive Dimensions Questionnaire [2] was to provide a discussion tool that allows for the evaluation and improvement of notational systems in an effective and cheap way. For example, the cognitive dimension 'error-proneness' is described as follows 'the notation invites mistakes and the system gives little protection' and the CD questionnaire provides for checking it the following questions: 'Do some kinds of mistake seem particularly common or easy to make? Which ones? Do you often find yourself making small slips that irritate you or make you feel stupid? What are some examples?'

CD framework or CD questionnaire was applied e.g. for the evaluation of real-time temporal logic TRIO, Z formalism, programming language C#, and spreadsheets [4,10,12,13]. The authors reported general usefulness of the CDs for the evaluation. However, some of them discovered some imperfections of CDs. They included: working with cognitive dimensions requires hard mental operations; some cognitive dimensions overlap; there are no satisfaction criteria for the cognitive dimensions; some questions were difficult to interpret in a given context and customization for specific applications was suggested. A software engineer while reading the term of 'dimensions' might expect to cover essential parts of cognitive psychology with characteristics known from our experience with three-dimensional space or algebra. However, the CD framework is not complete, but still under development and the cognitive dimensions are more alike some criteria of evaluation.

Whilst making a decision about the potential usefulness of CDs for VML evaluation one should notice that the cognitive perspective is important for VML evaluation. Additionally, VMLs with CASE tools can be classified as notational systems. The problems may result from the fact that contemporary models and CASE tools are more complex than notational systems and, in consequence, terminology might be misleading, e.g. the concept of 'notation element' in CDs is different from the 'model element' in UML - one 'model element' can be related to many 'notation elements'.

3   Empirical Study of UML Evaluation with a CD Questionnaire

The goal of the study was to evaluate the application of a CD questionnaire as a method of VML evaluation by trying it in practice. Seven persons took part in the study: three academic staff (one of them with experience of VML design), one software developer from industry and three students.

The participants were given the following tasks:

- to evaluate UML and the selected CASE tool by filling in the CD questionnaire;
- to reflect and comment on the results achieved with the method and to express opinions about its usefulness for evaluation.

The CD questionnaire is nine pages long and participants filled it in about an hour. Questionnaires were completed and analyzed against criteria of a good methodology. The answers were not precise and had little practical value for improvement. Some examples were written down rather than all elements carefully examined. The answers were not always clear and quite often contained such expressions as 'it depends...', 'in general...' or 'it is possible...'. The analysis of comments and opinions indicated that participants were aware of these problems. They estimated the level of details in the gathered data as medium or low. None of the responders discovered anything new. They did not find the questionnaire very useful for UML evaluation. Only the VML designer appreciated the value of having any methodology of evaluation. Final satisfaction of using the CD questionnaire as a method of VML evaluation was medium or low. A detailed description of this study can be found in [3].

The main reported problems about the CD questionnaire as a methodology of VML evaluation and lessons learned for further research included:

- The method should enable consideration of all VML elements separately. The CD questionnaire contains questions which do not distinguish between notation (UML) and a (CASE) tool which all participants needed to consider separately. This deficiency did not allow the capture of contradictory or complementary influence of several aspects of the 'notational system'.
- A higher level of precision is required. The CD questionnaire was considered too general and imprecise. Responders could not answer some questions because they were vague and responders did not know what they exactly meant. Additionally some measures of success were suggested.
- Preparation of the participants is needed. The vocabulary of the questionnaire should fit to the mental representations of their users. All responders had some problems with understanding the meaning of some questions in the context of visual modeling languages. They would like to know the theory beyond these questions.

This stage of the research was concluded with the need of customization of the CDs for VMLs to help eliminate the discovered problems.

4   Customization: CD-VML Methodology

The approach to customization of the original CD questionnaire for VML evaluation is presented in Fig. 2. In the first stage, we redefined cognitive dimensions for VML evaluation and we generalized knowledge about contemporary VML in a definition of VML. A model of mental representations was then made and all elements were integrated into a CD-VML template.

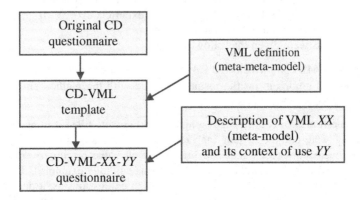

Fig. 2. Approach to customization of the CD questionnaire for VML

The basic assumption underlying the redefinition of CDs was to focus on *mechanisms* delivered by the language for creating models. Modeling, visual elements and mechanisms, at this level of abstraction, were basic language elements as well as methods for building more complex constructions. The list and names of the CDs were intact, but their definitions were adapted to VML, e.g. new meaning of 'error-proneness' is 'mechanisms which invite mistakes' and its motivation (for users who don't know CDs in detail) is 'to find typical errors and their reasons in the VML.' The definition of Visual Modeling Language assumes it is an internally integrated:

- set of *model elements* and *modeling mechanisms* used to relate *model elements* and operate on them resulting in *constructions* (relationships, groupings, etc.); together with *constraints* that describe possible relationships between them – defined by ABSTRACT SYNTAX;
- set of designations (basic *meanings*) for the *model elements* and *modeling mechanisms* and *constructions* defined in *ABSTRACT SYNTAX* together with hints on how to produce meaning of the *model* – described by SEMANTICS;
- set of possible *applications* (contexts of use for language *SEMANTICS*) and their conceptual models represented by *goals and tasks* – described by PRAGMATICS;
- set of *notation elements* and *visual mechanisms* (e.g. link, co-location) related to *model elements* and *modeling mechanisms* that allow for building *diagrams* and, thus, visualizing the model – described by NOTATION.

It would not be possible to achieve the precision of the questions without description of the basic concepts of the mental representations. They included: concept, mental construction, conceptual model, perspective and mental activity. The result of this integration was a universal CD-VML template, which shifted the CD questionnaire from the general area of the notational systems to the area of VML.

In the second stage, questionnaires for several VMLs in their specific context of use can be developed. Each CD-VML-*XX*-*YY* questionnaire can be generated on the basis of the CD-VML template, description of VML - *XX* and its context of use - *YY*. In order to facilitate focusing responders' attention they contain tables with relevant elements of concrete visual modeling language elements to be considered. The resulting questionnaire should be an efficient tool for assessing all combinations of the evaluation aspects and VML elements. In the case of the CD-VML-UC questionnaire, part of the modeling language under evaluation *XX* were use cases according to the UML standard (abbreviated as UC) and *YY* was its default context of use for creating models, using them and changing them (thus, the last part was omitted).

The flexibility of the methodology is manifested by the observation, that the CD-VML template can be customized for many VMLs for use in different contexts. The criteria of effectiveness, efficiency and ease of use were also considered. The experience with questionnaires applied in software engineering suggest that people work in a hurry and need support of detailed questions that combine pieces of information from different conceptual models. Otherwise, it is very likely that useful information will not be discovered. The CD-VML-UC questionnaire had several features which made it easy to use, effective and efficient:

- It grouped questions according to cognitive dimensions and provided their descriptions and motivations;
- It reduced the number of open questions about VML elements and mechanisms (in order to reduce time needed to remind model elements), and contained the space for describing reasons, inserting comments, suggestions for improvements and presenting examples with the possibility to comment on them;
- It used different colors assigned to concepts related to syntax, semantics, pragmatics, notation and mental representations.

In order to give an example, we present questions from the section of the questionnaire related to error-proneness. It starts with questions to find common mistakes whilst modeling and using diagrams: 'Whilst modeling, what kinds of mistakes are particularly common or easy to make? Which kinds of mistakes you must be careful not to make? Whilst using diagrams, which misunderstandings are very likely to happen?' Then, there is a table with listing of all model elements and notation elements and space for problem descriptions. It is headed with the questions: 'Which model elements or constructions or notation element or visual mechanisms are these mistakes concerned with? What are the problems?' Below this table there is a space for explanations, examples, comments and suggestions for improvement of the VML or suggestions for special features of the CASE tool.

5   Empirical Study with CD-VML-UC Questionnaire

The goal of this study was to verify the CD-VML methodology by trying out an example: evaluation of use case diagrams with CD-VML-UC questionnaire in practice. Use case diagram [14] is a popular and quite simple UML diagram, but some controversies around its details still exist [5,7,9,15]. The participants of this study were forty-five of the best students (out of about one hundred forty in total). The study took place at the end of the term, in which they had a lecture and practical on Analysis and Design of Computer Systems. Thus, they had knowledge about software development lifecycle and application of object-oriented methods in its early phases. The only possible problem of having students as experimental participants was their small experience – they modeled only one (but usually quite complex) system during the practical. However their strength was that they could remember very well what was easy to do whilst modeling, what were their problems and questions, what mistakes they were making and which other mechanisms they would need to specify a particular view of their system.

The participants were given the tasks like in the previous study:

- to evaluate use cases by filling in the CD-VML-UC questionnaire;
- to reflect and comment on the results achieved with the method and to express opinions about its usefulness for evaluation.

The CD-VML-UC questionnaire was eighteen pages long and the study took about four hours. Before they started filling the questionnaire in they were given a tutorial, which aimed at delivering additional method-specific knowledge. The tutorial included information about the approach and students' role in the study, the theory of VML, information about cognitive dimensions and structure of the questionnaire. During the study they could use UML standard documentation to check things they were not sure about.

The results of the study are described in section 6. Here, we summarize conclusions from this study. Students confirmed simplicity and usefulness of use cases, but they also discovered a large number of problems, gave reasonable explanations to them and quite often made suggestions for improvements. Their discoveries covered all the problems reported in the related work. In comparison to the previous study (described in section 3), in which no problems related to use cases were discovered, this is a great improvement in the methodology. This brings more evidence to the hypothesis, that more detailed questionnaires help in acquiring more detailed data. The level of detail of the individual answers was satisfactory: problems usually were described with details and examples and even simple diagrams for illustration of the problem were added. In the section of comments they stated the following strengths of the methodology: the precision of questions, large area of covered issues and ease of use. Among the weaknesses they mentioned: too large length of the questionnaire and redundancy in the sense of similar questions in different parts of the questionnaire or at least similar answers to them. Most participants discovered something new about different aspects of use cases or modeling. While describing the problems many students had awareness that many aspects depend on the developers' decisions and size of the system. Usefulness of the method was evaluated as high and results were considered as important for use case model improvement.

6   Reported Problems for Use Case Diagram

Most elements of the use case diagram were evaluated as easy to understand and use. The lowest score had UseCaseInstance and ExtensionPoint. Students often considered ExtensionPoint as unclear, useless, redundant with extend relationship and making problems in modeling in the case of several extensions. The problems with UseCaseInstance were concerned with lack of precision, lack of a notational symbol and unclear purpose. For other elements reported by participants deficiencies can be classified to the following groups: a 'stick man' as actor's notational element; problems with appropriate level of abstraction of use cases; difficult decision of choice between include, extend or generalize relationships; readability of the include and extend stereotype labels; need to see business perspective; need of mechanisms of use case decomposition and potential problems with generalizations. The methodology at this point does not deliver rules necessary to judge which of them are important.   The purpose of this section is to present empirical data which were achieved with this method. On the other hand, we are aware that other perspectives, e.g. consistency with other diagrams, might provide other arguments.

A 'stick man' as notational element of the actor is very intuitive when actor represents human users of the system. It is not intuitive at all when it represents another system, which cooperates with the system under development. Students suggested other icons, e.g. computer icon for another system. The UML standard gives such a possibility, but most CASE tools provide only default notation of a 'stick man'.

The participants reported problems with appropriate level of abstraction of use cases:

- Use cases, which are included or extend another use case are not strictly units of functionality visible from the outside of the system;
- It is difficult to keep the same level of abstraction on the diagram – especially using inclusions and extensions requires exceptions from this; when we assume that criteria for decomposing into included and extending use cases are shared functionality and conditional performance, it is still confusing with regards to other functionality at the same level of detail;
- Roles of actors in interaction with a use case are not clear, for example when there are two or more actors associated with the same use case, the diagram does not show if all of them can initialize this use case or some of them only interact with the system.

The confusions in deciding which of the modeling mechanisms to choose from possible 'include', 'extend', and 'generalization' result from the fact that they represent very similar concepts. In   real systems there is not such a clear-cut distinction about 'always-sometimes' and quite often there are situations in which something that is usually included can be omitted. The difference may depend on the actor when at least two of them are associated with a use case.

The problems with readability of the visual constructions with use of include and extend stereotypes labels were motivated as follows:

- The stereotype labels take a lot of space, they are difficult to distinguish and it takes more time to read them in comparison to possible presentations of them as different kinds of arrows or lines;
- It is difficult to see proper associations in the complex diagrams, e.g. when there are crossings of relationships or when there is a large distance between related use cases or there are chains of relationships;

- Direction of the 'extend' arrow is confusing – there is a lexical consistency, but not a visual one, which is related to how people usually read the diagram.

Many students suggested other kinds of arrows or lines to depict them.

The lacking aspects of 'business perspective' were order of performing use cases in time; spatial or organizational relationships between actors; correlations with organizational business activity; the roles of actors in use cases and relationships between actors who are associated with separate use cases and at the same time their interaction with the system is associated with the same goal.

The concepts related to mechanisms of use case decomposition were: a possibility to indicate levels of abstraction and navigate between them; a possibility to extend a use case to a use case diagram and possibility to group use cases. Generalization of both use cases and actors was considered as a concept that sometimes might be useful and consistent with an object-oriented paradigm, but its use may cause more problems than benefits. It is a difficult concept, it extends the time of reading the models and will not be easy to understand for potential customers.

These observations are consistent with those reported in related work which suggest problems with use case relationships [5,7], included and extending use cases [7], or limitations of use cases in modeling context of the system, relating systems involved in the same business and specifying constraints between use case occurrences [15].

7   Conclusions and Future Work

Evaluating a methodology of VML evaluation is a tricky task, because there are many parameters of the study. An empirical approach, in which individuals perform some activities and then reflect and comment on the results, seems to be the only way to capture ideas with a practical value. However, the role of the theoretical analysis of usefulness of a given theory as a basis for a method of VML evaluation should not be dismissed. In this research, the main problems of overlapping, complexity and precision were already discovered during the analysis of the CD framework.

This research delivered a CD-VML methodology for evaluating any contemporary VML from a cognitive perspective. It satisfies the criteria of ease of use, effectiveness and flexibility and only partially the criterion of efficiency. It is a result of the integration of the VML definition with the cognitive dimensions. The VML definition delivered a precise model of the VML and the CDs delivered a set of criteria related to the cognitive aspects. A challenge for future work on evaluating VMLs from a cognitive perspective is to focus on confirmation of positive features of the VMLs and to find ways of estimating the significance of discovered problems.

Acknowledgements

A part of this research was performed at University of Kent, UK and was supported by SegraVis Research Training Network. I would like to thank Stuart Kent for many useful discussions about visual modeling languages and Thomas Green for replying to my questions about cognitive dimensions. I am also grateful to participants of my empirical studies for their nice attitude and the results they delivered.

References

1. Blackwell, A.F., Britton, C., Cox, A. Green, T.R.G., Gurr, C.A., Kadoda, G.F., Kutar, M., Loomes, M., Nehaniv, C.L., Petre, M., Roast, C., Roes, C., Wong, A. And Young, R.M. Cognitive Dimensions of Notations: Design tools for cognitive technology. In M. Beynon, C.L. Nehaniv, and K. Dautenhahn (Eds.) Cognitive Technology 2001 (LNAI 2117). Springer-Verlag, (2001).pp. 325-341.
2. Blackwell, A.F., Green, T.R.G.: A Cognitive Dimensions questionnaire optimised for users. In: Proceedings of the Twelfth Annual Meeting of the Psychology of Programming Interest (PPIG) Group. (2000) 137-152.
3. Bobkowska A.E.: Cognitive Dimensions Questionnaire Applied to Visual Modelling Language Evaluation - a Case Study. In: Proceedings of the Fifteenth Annual Workshop of the PPIG (2003)
4. Clarke, S.: Evaluating a new programming language. In G. Kadoda (Ed.) Proceedings of the Thirteenth Annual Meeting of the PPIG. (2001) pp. 275-289.
5. Cox K.: Cognitive Dimensions of use cases: feedback from a student questionnaire. In: Proceedings of the Twelfth Annual Meeting of the PPIG (2000)
6. Detienne F.: Software Design – Cognitive Aspects, Springer-Verlag (2002)
7. Genova G., Llorens J., Quintana V.: Digging into Use Case Relationships. In: Proceedings of <UML>2002, Lecture Notes on Computer Science 2460 (2002)
8. Green, T. R. G.: Cognitive dimensions of notations. In: People and Computers . Cambridge University Press. Cambridge (1989) 443-460
9. Isoda S.: A Critique of UML's Definition of the Use-Case Class. In: Proceedings of <UML>2003, Lecture Notes on Computer Science 2863 (2003)
10. Kutar, M., Britton, C. And Wilson, J.: Cognitive Dimensions: An experience report. In A.F. Blackwell & E. Bilotta (Eds.) Proceedings of the Twelfth Annual Meeting of the Psychology of Programming Interest Group (2000) pp. 81-98.
11. Shneiderman B.: Software Psychology. Human Factors in Computer and Information Systems. Winthrop Publishers Inc. (1980)
12. Triffitt, E., Khazaei, B.: A Study of Usability of Z Formalism Based on Cognitive Dimensions. In J. Kuljis, L. Baldwin and R. Scoble (Eds), Proceedings of the Fourteenth Annual Meeting of the PPIG (2002) pp. 15-28
13. Tukiainen, M.: Evaluation of the Cognitive Dimensions Questionnaire and Some Thoughts about the Cognitive Dimensions of Spreadsheet Calculation. In G. Kadoda (Ed.) Proceedings of the Thirteenth Annual Meeting of the PPIG. (2001) pp 291-301.
14. Unified Modeling Language v. 1.4 ., available at http://www.omg.org/uml/
15. Wegmann A., Genilloud G.: "The Role of "Roles" in Use Case Diagrams", Proceedings of 3rd International Conference on the Unified Modeling Language York, UK (2000).

Local Computations on Closed Unlabelled Edges: The Election Problem and the Naming Problem (Extended Abstract)

Jérémie Chalopin

LaBRI, Université Bordeaux I, 351 cours de la Libération, 33405 Talence, France
chalopin@labri.fr

1 Introduction

The different local computations mechanisms are very useful for delimiting the borderline between positive and negative results in distributed computations. Indeed, they enable to study the importance of the synchronization level and to understand how important is the initial knowledge. A high level of synchronization involved in one atomic computation step makes a model powerful but reduces the degree of parallelism. Charron-Bost et al. [1] study the difference between synchronous and asynchronous message passing models. The model studied in this paper involves more synchronization than the message passing model: an elementary computation step modifies the states of two neighbours in the network, depending only on their current states. The information the processors initially have can be global information about the network, such as the size, the diameter or the topology of the network. The initial knowledge can also be local: each node can initially know its own degree for example. Another example of local knowledge is the existence of a port numbering: each processor locally gives numbers to its incident edges and in this way, it can consistently distinguish its neighbours. In Angluin's model [2], it is assumed that a port numbering exists, whereas it is not the case in our model. In fact, we obtain a model with a strictly lower power of computation by relaxing the hypothesis on the existence of a port numbering.

The Model. A network of processors will be represented as a simple connected undirected graph. As usual the vertices represent processors and edges direct communication links. The state of each processor is represented by the label $\lambda(v)$ of the corresponding vertex.

An elementary computation step will be represented by relabelling rules of the form given schematically in Figure 1. If, in a graph G, there are two neighbours labelled X and Y then applying this rule we replace X (resp. Y) by a new label X' (resp. Y'). The labels of all other graph vertices are irrelevant for such a computation step and remain unchanged. The computations using uniquely this type of relabelling rules are called in this paper *local computations on closed unlabelled edges*. Thus an algorithm in our model is simply given by some (possibly infinite but always recursive) set of rules of the type presented in Figure 1.

M. Bieliková et al. (Eds.): SOFSEM 2005, LNCS 3381, pp. 82–91, 2005.

A run of the algorithm consists in applying the relabelling rules specified by the algorithm until no rule is applicable, which terminates the execution. The relabelling rules are applied asynchronously and in any order, which means that given the initial labelling usually many different runs are possible.

$$X \quad Y \quad \longrightarrow \quad X' \quad Y'$$

Fig. 1. Graphical form of a rule for local computations on closed unlabelled edges

Election, Naming and Enumeration. The election problem is one of the paradigms of the theory of distributed computing. It was first posed by LeLann [3]. A distributed algorithm solves the election problem if it always terminates and in the final configuration exactly one processor is marked as *elected* and all the other processors are *non-elected*. Moreover, it is supposed that once a processor becomes *elected* or *non-elected* then it remains in such a state until the end of the algorithm. The naming problem is another important problem in the theory of distributed computing. The aim of a naming algorithm is to arrive at a final configuration where all processors have unique identities. The enumeration problem is a variant of the naming problem whose aim is to give to each node a unique number between 1 and the size of the graph. These problems are important since they constitute basic initial steps of many other distributed algorithms.

Related Works. Graphs where election is possible were already studied for different types of basic computation steps and particular types of network topology (tree, grid, torus, ring with a known prime number of vertices, etc.), see [4].

Yamashita and Kameda [5] characterize the graphs for which there exists an election algorithm in the message passing model and they study the importance of the port numbering in [6].

Mazurkiewicz [7] considers an asynchronous computation model where in one computation step, labels are modified on a subgraph consisting of a node and its neighbours, according to certain rules depending on this subgraph only. His characterization of the graphs where enumeration and election are possible can be formulated using coverings [8]. In this model, the port numbering does not give a more powerful model, since in each computation step, a node can always distinguish its neighbours.

Chalopin and Métivier [9] consider three different asynchronous models that are defined by the rules presented in Figure 2. Note that, contrary to the model we examine in the present paper, all these models allow edge labelling. In fact, allowing to label the edges is equivalent to the existence of a port numbering, since in these models, it is always possible for a processor to consistently identify its neighbours. Consequently, the first model studied in [9] is equivalent to the model of Angluin [2]. It turns out that for all models of Figure 2 the election and naming problems can be solved on a graph G if and only if G is not a covering of any graph H not isomorphic to G, where H can have multiple edges but no self-loop. Mazurkiewicz [10] has also studied the first model described in Figure 2 and he gives an equivalent characterization thanks to equivalence relations over the vertices and the edges of the graph.

In the conclusion of [6], Yamashita and Kameda underline the importance of the edge labelling and it is a natural question to wonder if the edge labelling, or equivalently the existence of a port numbering, modify the power of the different models of Figure 2. Boldi et al. [11] consider a model where the network is a directed multigraph. When a processor is activated, it changes its state depending on its previous state and on the states of its ingoing neighbours. They characterize the graphs that admits an election algorithm using fibrations, that are generalization of coverings to directed graphs. Chalopin et al. [12] consider a model where an elementary computation step modifies the state of one vertex depending on its current state and the state of one of its neighbours; as in the model studied here, the edges are not labelled. In this model, naming and election are not equivalent and characterizations are given using submersions that are locally surjective morphisms. The comparison between the characterization given in [11], in [12] and in [9] shows that for the second and the third model of Figure 2, it gives strictly more powerful models to allow the edges to be labelled. In this paper, we complete the study of the importance of the port numbering: the characterization we give of the graphs for which the naming and the election problems can be solved for the model of Figure 1 is very different of the characterization given in [9]. Moreover, we can remark that the three models of Figure 2 that are equivalent when the edges can be labelled are no longer equivalent when this hypothesis is relaxed.

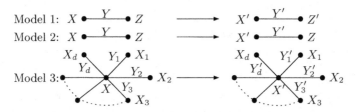

Fig. 2. Elementary relabelling steps for the models studied in [9]

Main Results. We introduce in Section 2 the notion of pseudo-covering, that is a generalization of coverings. We prove in Section 3 that naming and election can be solved on a graph G if and only if G is minimal for the pseudo-covering relation (Theorem 1).

The problems are solved constructively: we encode an enumeration algorithm with explicit termination by local computations on closed unlabelled edges that work correctly for all graphs where these problems are solvable. This algorithm uses some ideas from Mazurkiewicz'algorithm [7]. However, in the models considered in [11, 9, 7, 10, 6], a node can consistently distinguish its neighbours whereas it is impossible in the model studied here. Each execution of our algorithm on a graph G computes a labelling that induces a graph H such that G is a pseudo-covering of H. Consequently, there exists an integer k such that each label of the final labelling of G appears exactly k times in the graph; it is not the case for the model studied in [12]. In our solution, stamps are associated to synchronizations between neighbours. These associated stamps solve the problem, but they introduce a non-trivial difficulty in the proof of the termination of the algorithm: we must prove that stamps are bounded.

Imposed space limitations do not allow to present all the proofs in the paper.

2   Preliminaries

Graphs. The notations used here are essentially standard [13]. We consider finite, undirected, connected graphs $G = (V(G), E(G))$ with vertices $V(G)$ and edges $E(G)$ without multiple edges or self-loop. Two vertices u and v are said to be adjacent or neighbours if $\{u, v\}$ is an edge of G and $N_G(v)$ will stand for the set of neighbours of v. An edge e is incident to a vertex v if $v \in e$ and $I_G(v)$ will stand for the set of all the edges of G incident to v.

A homomorphism between graphs G and H is a mapping $\gamma \colon V(G) \to V(H)$ such that if $\{u, v\} \in E(G)$ then $\{\gamma(u), \gamma(v)\} \in E(H)$. We say that γ is an isomorphism if γ is bijective and γ^{-1} is a homomorphism.

A graph H is a subgraph of G, noted $H \subseteq G$, if $V(H) \subseteq V(G)$ and $E(H) \subseteq E(G)$. A subgraph H of G is called a partial graph of G if G and H have the same set of vertices.

A matching F of a graph G is a subset of $E(G)$ such that for every $e, e' \in F$, $e \cap e' = \emptyset$: F is a set of disjoint edges of G. A matching F of G is perfect if every vertex $v \in V(G)$ is the endvertex of exactly one edge $e \in F$.

Throughout the paper we will consider graphs where vertices are labelled with labels from a recursive label set L. A graph labelled over L is a couple $\mathbf{G} = (G, \lambda)$, where G is the underlying non labelled graph and $\lambda \colon V(G) \to L$ is the (vertex) labelling function. Let H be a subgraph of G and λ_H the restriction of the labelling $\lambda \colon V(G) \to L$ to $V(H)$. Then the labelled graph $\mathbf{H} = (H, \lambda_H)$ is called a subgraph of $\mathbf{G} = (G, \lambda)$; we note this fact by $\mathbf{H} \subseteq \mathbf{G}$. A homomorphism of labelled graphs is just a homomorphism that preserves the labelling.

For any set S, $|S|$ denotes the cardinality of S. For any integer q, we denote by $[1, q]$ the set of integers $\{1, 2, \ldots, q\}$.

Coverings and Pseudo-Coverings. A graph G is a *covering* of a graph H via γ if γ is a surjective homomorphism from G onto H such that for every vertex v of $V(G)$ the restriction of γ to $I_G(v)$ is a bijection onto $I_H(\gamma(v))$. The covering is proper if G and H are not isomorphic. A graph G is called *covering-minimal* if every covering from G to some H is a bijection.

A graph G is a *pseudo-covering* of H via a morphism φ modulo a graph G' if G' is a partial graph of G that is a covering of H via the restriction $\varphi_{|G'}$ of φ to G'. The pseudo-covering is proper if G and H are not isomorphic. A graph G is said *pseudo-covering-minimal* if there does not exist a graph H such that G is a proper pseudo-covering of H. An example of pseudo-covering is given in Figure 3. Naturally, coverings and pseudo-coverings of labelled graphs are just coverings and pseudo-coverings of underlying graphs such that the associated morphisms preserve the labelling.

If $\mathbf{G}$ is a pseudo-covering of a graph $\mathbf{H}$ via φ modulo $\mathbf{G}'$, then for every edge $f = \{w_1, w_2\} \in E(H)$, $\varphi_{|G'}^{-1}(f)$ is a perfect matching of $\varphi^{-1}(\{w_1, w_2\})$. Consequently, there exists an integer q such that for every vertex $v \in V(H)$, $|\varphi^{-1}(v)| = q$.

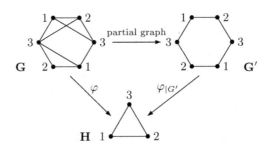

Fig. 3. The graph $\mathbf{G}$ is a pseudo-covering of $\mathbf{H}$ via the mapping φ modulo $\mathbf{G}'$ where φ maps each vertex of $\mathbf{G}$ labelled i to the unique vertex of $\mathbf{H}$ with the same label i. This pseudo-covering is proper and the graph $\mathbf{H}$ is pseudo-covering-minimal

Local Computations on Closed Unlabelled Edges. For any set $\mathcal{R}$ of edge local relabelling rules of the type described in Figure 1 we shall write $\mathbf{G} \,\mathcal{R}\, \mathbf{G}'$ if $\mathbf{G}'$ can be obtained from $\mathbf{G}$ by applying a rule of $\mathcal{R}$ on some edge of $\mathbf{G}$. Obviously, $\mathbf{G}$ and $\mathbf{G}'$ have the same underlying graph G, only the labelling changes for the endvertices of exactly one (active) edge. Thus, slightly abusing the notation, $\mathcal{R}$ will stand both for a set of rules and the induced relabelling relation over labelled graphs. The reflexive transitive closure of such a relabelling relation is noted $\mathcal{R}^*$. The relation $\mathcal{R}$ is called *noetherian* on a graph $\mathbf{G}$ if there is no infinite relabelling sequence $\mathbf{G}_0 \,\mathcal{R}\, \mathbf{G}_1 \,\mathcal{R}\, \ldots$, with $\mathbf{G}_0 = \mathbf{G}$. The relation $\mathcal{R}$ is noetherian on a set of graphs if it is noetherian on each graph of the set. Finally, the relation $\mathcal{R}$ is called noetherian if it is noetherian on each graph. Clearly noetherian relations code always terminating algorithms.

The following lemma is a counterpart of the lifting lemma of Angluin [2] adapted to pseudo-coverings; it exhibits a strong link between pseudo-coverings and local computations on closed unlabelled edges. An immediate corollary is that there does not exist any algorithm using local computations on closed unlabelled edges that solves the election problem or the naming problem on a graph $\mathbf{G}$ that is not pseudo-covering-minimal.

Lemma 1 (Lifting Lemma). *Let $\mathcal{R}$ be a relabelling relation encoding an algorithm using local computations on closed unlabelled edges and let $\mathbf{G}_0$ be a pseudo-covering of $\mathbf{H}_0$. If $\mathbf{H}_0 \,\mathcal{R}^*\, \mathbf{H}_1$ then there exists $\mathbf{G}_1$ such that $\mathbf{G}_0 \,\mathcal{R}^*\, \mathbf{G}_1$ and $\mathbf{G}_1$ is a pseudo-covering of $\mathbf{H}_1$.*

3   An Enumeration Algorithm

In this section, we describe a Mazurkiewicz-like algorithm $\mathcal{M}$ using local computations on closed unlabelled edges that solves the enumeration problem on a pseudo-covering-minimal graph **G**.

Each vertex v attempts to get its own number between 1 and $|V(G)|$. A vertex chooses a number and exchanges its number with its neighbours. If during a computation step, two neighbours exchange their numbers, a stamp o is given to the operation such that two operations involving the same vertex have different stamps. Each node broadcasts its number, its label and its *local view* (the numbers of its neighbours and the stamps of the operations of exchange associated to each neighbour) all over the network. If a vertex u discovers the existence of another vertex v with the same number, then it compares its local view with the local view of v. If the label of u or the local view of u is "weaker", then u chooses another number and broadcasts it again with its local view. At the end of the computation, every vertex will have a unique number if the graph is pseudo-covering-minimal.

The main difference with Mazurkiewicz'algorithm is the existence of the stamps o. The algorithm we will describe below computes a graph **H** such that **G** is a pseudo-covering of **H**. To define a pseudo-covering, we need to define a morphism and a subset of $E(G)$. As in Mazurkiewicz'algorithm, the numbers of the nodes will be used to define the morphism φ whereas the stamps o will be used to select the edges of G.

Labels. We consider a labelled graph $\mathbf{G} = (G, \lambda)$. For each vertex $v \in V(G)$, the label of v is the pair $(\lambda(v), c(v))$ where $\lambda(v)$ is the initial label of v whereas $c(v)$ is a triple $(n(v), N(v), M(v))$ representing the following information obtained during the computation:

- $n(v) \in \mathbb{N}$ is the *number* of the vertex v computed by the algorithm;
- $N(v) \in \mathcal{N}$ is the *local view* of v. If the node v has a neighbour v', some relabelling rules will allow v and v' to add $n(v')$ in $N(v)$ and $n(v)$ in $N(v')$. Each time this operation is done between two neighbours a stamp o is given to the operation and $(n(v'), o)$ is added to $N(v)$ (resp. $(n(v), o)$ is added to $N(v')$). Consequently, $N(v)$ is a finite set of pairs (n, o);
- $M(v) \subseteq \mathbb{N} \times L \times \mathcal{N}$ is the *mailbox* of v and contains the whole information received by v at any step of the computation, i.e., the numbers, the labels and the local views of the nodes of the network.

The fundamental property of the algorithm is based on a total order on local views, as defined in [7], such that the local view of any vertex cannot decrease during the computation. Consider a vertex v such that the local view $N(v) \in \mathcal{N}$ is the set $\{(n_1, o_1), (n_2, o_2), \ldots, (n_d, o_d)\}$, we assume that $n_1 > n_2 > \ldots > n_d$ and we say that the d-tuple $((n_1, o_1), (n_2, o_2), \ldots, (n_d, o_d))$ is the ordered representation of $N(v)$. We define a total order $\prec$ on such ordered tuples using the alphabetical order; it induces naturally a total order on $\mathcal{N}$. We assume that the set of labels L is totally ordered by $<_L$ and we extend $\prec$ on $L \times \mathcal{N}$.

The Relabeling Rules. We now describe the relabelling rules of the algorithm; the first rule $\mathcal{M}_0$ is a special rule that extends the initial label $\lambda(v)$ of each vertex to $(\lambda(v),(0,\emptyset,\emptyset))$. The rules $\mathcal{M}_1$ and $\mathcal{M}_2$ are very close to the rules of Mazurkiewicz's algorithm.

The first rule enables two neighbours v and v' having different mailboxes to share the information they have about the labels present in the graphs.

$$(l_1,(n_1,N_1,M_1)) \quad (l_2,(n_2,N_2,M_2)) \qquad (l_1,(n_1,N_1,M')) \quad (l_2,(n_2,N_2,M'))$$

$\mathcal{M}_1: \quad \bullet \! \bullet \qquad \longrightarrow \qquad \bullet \! \bullet$

If $M_1 \neq M_2$ then $M' := M_1 \cup M_2$.

The second rule enables a vertex v to change its number if $n(v) = 0$ or if there exists a vertex v' such that $n(v) = n(v')$ and v has a weaker label or a weaker local view than v'.

$$(l,(n,N,M)) \qquad (l,(k,N,M'))$$

$\mathcal{M}_2: \qquad\qquad \bullet \qquad \longrightarrow \qquad \bullet$

If $n = 0$ or $\exists(n,l',N') \in M$ such that $(l,N) \prec (l',N')$
then $k := 1 + \max\{n' \mid \exists(n',l',N') \in M\}$ and $M' := M \cup \{(k,l,N)\}$.

The third rule enables a node having a neighbour with exactly the same label to change its number. If this rule is applied, the number of each node is inserted in the local view of the other with a stamp o associated to the operation that is different of the other stamps associated to operations involving one of the two nodes. Moreover, when the number $n(v')$ of a neighbour v' of v is inserted in $N(v)$, all the elements (m,o) belonging to $N(v)$ such that $m \leq n(v')$ are deleted from the local view. The rationale behind this deletion step is explained in the next rule $\mathcal{M}_4$ below.

$$(l,(n,N,M)) \qquad (l,(n,N,M)) \qquad (l,(k,N_1,M')) \qquad (l,(n,N_2,M'))$$

$\mathcal{M}_3: \quad \bullet \! \bullet \qquad \longrightarrow \qquad \bullet \! \bullet$

If $\quad n > 0$ and $\forall(n,l',N') \in M, (l',N') \preceq (l,N)$
then $k := 1 + \max\{n' \mid \exists(n',l',N') \in M\}$;
$\qquad o := 1 + \max\{o' \mid \exists(n',o') \in N\}$;
$\qquad N_1 := N \setminus \{(n',o') \in N \mid n' \leq n\} \cup \{(n,o)\}$;
$\qquad N_2 := \{(k,o)\}$ and $M' := M \cup \{(k,l,N_1),(n,l,N_2)\}$.

The fourth rule enables two neighbours v and v' to exchange their numbers if an update is needed, i.e., if there does not exist o such that $(n_2,o) \in N_1$ and $(n_1,o) \in N_2$. As for the precedent rule, if the number $n(v')$ of a neighbour v' of v is inserted in $N(v)$, all the elements (m,o) belonging to $N(v)$ such that $m \leq n(v')$ are deleted.

The role of the stamp o associated to the operation is to ensure that at the end of the computation, if the local view of a vertex v_0 contains (n,o), it means that it has a neighbour v_0' such that $n(v_0') = n$, $(n(v_0),o) \in N(v_0')$ and such that the rule $\mathcal{M}_3$ or $\mathcal{M}_4$ was applied to these two vertices; an interesting consequence is that in the final labelling, $|\{v \mid n(v) = n(v_0)\}| = |\{v \mid n(v) = n(v_0')\}|$.

$$(l_1, (n_1, N_1, M)) \quad (l_2, (n_2, N_2, M)) \quad (l_1, (n_1, N_1', M')) \quad (l_2, (n_2, N_2', M'))$$

$\mathcal{M}_4:$ •————————•   ⟶   •————————•

If $n_1, n_2 > 0, n_1 \neq n_2,$
     $\forall (n_1, l_1', N_1') \in M, (l_1', N_1') \preceq (l_1, N_1),$
     $\forall (n_2, l_2', N_2') \in M, (l_2', N_2') \preceq (l_2, N_2),$
     and $\nexists o \mid (n_2, o) \in N_1$ and $(n_1, o) \in N_2$
then $o := 1 + \max\{o' \mid \exists (n', o') \in N_1 \cup N_2\};$
     $N_1' := N_1 \setminus \{(n', o') \in N_1 \mid n' \leq n_2\} \cup \{(n_2, o)\};$
     $N_2' := N_2 \setminus \{(n', o') \in N_2 \mid n' \leq n_1\} \cup \{(n_1, o)\};$
     $M' := M \cup \{(n_1, l_1, N_1'), (n_2, l_2, N_2')\}.$

The intuitive justification for the deletion of all the (m, o) is the following. If there is a synchronization between two neighbours v and v', then they should agree on an integer o_0 and add $(n(v), o_0)$ to $N(v')$ and $(n(v'), o_0)$ to $N(v)$. But, it is possible that v synchronized with v' in the past and in the meantime v' has changed its identity number or has synchronized with another vertex w such that $n(w) = n(v)$. In this case, to remain in a consistent state, the vertex v should modify its local view to remove the old identity number of v' and the o associated to this precedent synchronization. The trouble is that v has no means to know which of the pairs (m, o) belonging to its view $N(v)$ should be deleted. However, since our algorithm assures the monotonicity of subsequent identity numbers of each vertex and monotonicity of subsequent o involving the node v', we know that the couple (m, o) to remove is such that $(m, o) <_{Lex} (n(v'), o_0)$ Therefore, by deleting all such (m, o) from the local view $N(v)$, we are sure to delete all invalid information. Of course, in this way we risk to delete also the legitimate current informations about other neighbours of v from its view $N(v)$. However, v can recover this information just by (re)synchronizing with all such neighbours.

Properties. In the following, we consider an execution of the algorithm on a graph $\mathbf{G}$. We will denote by $(\lambda(v), (n_i(v), N_i(v), M_i(v))$ the label of the vertex v after the ith computation step.

We can see that the label of each node can only "increase" during the computation. Indeed, for each step i, for each vertex v, $n_i(v) \leq n_{i+1}(v)$, $N_i(v) \preceq N_{i+1}(v)$ and $M_i(v) \subseteq M_{i+1}(v)$. Moreover, if a vertex v knows the existence of a node with the number m (i.e., $\exists (m, l, N) \in M_i(v)$), then for each $m' \leq m$, there exists a node w such that $n_i(w) = m'$. An immediate corollary of this property is that after each computation step the numbers of the nodes is a set $[1, k]$ with $k \leq |V(G)|$.

We will now prove that each execution of $\mathcal{M}$ on a graph $\mathbf{G}$ is finite. In fact, we just have to prove that the values of $n(v)$, $N(v)$ and $M(v)$ are bounded for each vertex v . Since we already know that $n(v) \leq |V(G)|$, we just have to prove that the stamps o are also bounded. It will imply that $N(v)$ and $M(v)$ can only take a finite number of values. From the properties described above, there exists a step i_0 such that $\forall i \geq i_0, \forall v \in V(G), n_i(v) = n_{i_0}(v)$ and therefore the rules $\mathcal{M}_2$ and $\mathcal{M}_3$ cannot be applied after the step i_0. Consider two neighbours v and w such that $n_{i_0}(v) > n_{i_0}(w)$ and two steps $j_2 > j_1 > i_0$ where the rule $\mathcal{M}_4$ is applied to the edge $\{v, w\}$. Then, there must exist an edge $\{v', w'\}$ with $(n_{i_0}(v), n_{i_0}(w)) <_{Lex} (n_{i_0}(v'), n_{i_0}(w'))$ and a step $j \in [j_1, j_2]$ where the rule $\mathcal{M}_4$ is applied to $\{v', w'\}$. Consequently, the rule $\mathcal{M}_4$ can only be applied a finite number of time over each edge and we can ensure the termination of the algorithm.

For each execution of the algorithm over $\mathbf{G}$, a graph $\mathbf{H}$ is associated to the final labelling with $V(H) = \{n(v) \mid v \in V(G)\}$ such that $\mathbf{G}$ is a pseudo-covering of $\mathbf{H}$. If $\mathbf{G}$ is pseudo-covering-minimal, then $\mathbf{G} \simeq \mathbf{H}$. Consequently, for every run of the enumeration algorithm, the graph associated to the final labelling is isomorphic to $\mathbf{G}$ and therefore the set of numbers the vertices have is exactly $[1, |V(G)|]$. Moreover, once a vertex gets the number $|V(G)|$, it knows that all the vertices have a different number that will not change any more and therefore it can detect the termination. We can therefore transform the enumeration algorithm into an election algorithm, by choosing to elect the node that gets the number $|V(G)|$. From these results and Lemma 1, we have the following theorem.

Theorem 1. *For every graph* $\mathbf{G}$, *it is equivalent to solve the following problems on* $\mathbf{G}$ *with local computations on closed unlabelled edges: naming, naming with explicit termination and election. These problems can be solved on* $\mathbf{G}$ *if and only if* $\mathbf{G}$ *is a pseudo-covering-minimal graph.*

4   Comparison with Other Models

It is easy to see that each algorithm encoded by local computations on closed unlabelled edges can be translated in the models of Mazurkiewicz [7], Angluin [2] and Chalopin and Métivier [9]. And the algorithms encoded in the model of [12] can be encoded by local computations on closed unlabelled edges.

In the models of Mazurkiewicz [7], Angluin [2], Chalopin and Métivier [9], Yamashita and Kameda [5] and Boldi et al. [11], the election and the naming problems can be solved in the graph G_1 of Figure 4. Nevertheless, this graph G_1 is a pseudo-covering of the graph H_1. Therefore, it is not possible to solve the election problem by using local computations on closed unlabelled edges.

If we consider the pseudo-covering-minimal graphs G_2 and G_3 of Figure 4, we can solve the election problem over these graphs with local computations on closed unlabelled edges. But the election problem cannot be solved over G_3 in the models studied in [11] and in [12]. Moreover there does not exist any election algorithm for the graph G_2 in the model studied in [5].

Consequently, our model is strictly less powerful than the models studied by Mazurkiewicz [7], by Angluin [2] and by Chalopin and Métivier [9], but strictly more powerful than the model studied by Chalopin et al. [12]. And the power

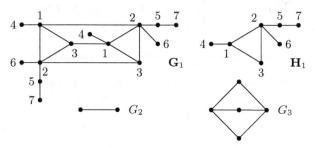

Fig. 4. Different graphs that show the differences between the different models

of computation of our model is not comparable to the power of the models of Yamashita and Kameda [5, 6] and Boldi et al. [11].

References

1. Charron-Bost, B., Mattern, F., Tel, G.: Synchronous, asynchronous and causally ordered communication. Distributed Computing **9** (1996) 173–191
2. Angluin, D.: Local and global properties in networks of processors. In: Proc. of the 12th Symposium on Theory of Computing. (1980) 82–93
3. LeLann, G.: Distributed systems: Towards a formal approach. In Gilchrist, B., ed.: Information processing'77, North-Holland (1977) 155–160
4. Tel, G.: Introduction to distributed algorithms. Cambridge University Press (2000)
5. Yamashita, M., Kameda, T.: Computing on anonymous networks: Part i - characterizing the solvable cases. IEEE Transactions on parallel and distributed systems **7** (1996) 69–89
6. Yamashita, M., Kameda, T.: Leader election problem on networks in which processor identity numbers are not distinct. IEEE Transactions on parallel and distributed systems **10** (1999) 878–887
7. Mazurkiewicz, A.: Distributed enumeration. Inf. Processing Letters **61** (1997) 233–239
8. Godard, E., Métivier, Y., Muscholl, A.: Characterization of classes of graphs recognizable by local computations. Theory of Computing Systems **37** (2004) 249–293
9. Chalopin, J., Métivier, Y.: Election and local computations on edges (*extended abstract*). In: Proc. of FOSSACS'04. Number 2987 in LNCS (2004) 90–104
10. Mazurkiewicz, A.: Bilateral ranking negotiations. Fundamenta Informaticae **60** (2004) 1–16
11. Boldi, P., Codenotti, B., Gemmell, P., Shammah, S., Simon, J., Vigna, S.: Symmetry breaking in anonymous networks: Characterizations. In: Proc. 4th Israeli Symposium on Theory of Computing and Systems, IEEE Press (1996) 16–26
12. Chalopin, J., Métivier, Y., Zielonka, W.: Election, naming and cellular edge local computations (*extended abstract*). In: Proc. of ICGT'04. Number 3256 in LNCS (2004) 242–256
13. Rosen, K.H., ed.: Handbook of discrete and combinatorial mathematics. CRC Press (2000)

A Hierarchical Markovian Mining Approach for Favorite Navigation Patterns

Jiu Jun Chen, Ji Gao, Jun Hu, and Bei Shui Liao

College of Computer Science, Zhejiang University,
Hangzhou 310027, Zhejiang, China
rackycjj@zju.edu.cn

Abstract. This paper presents a new approach based on Hierarchical Markov user model to mine the favorite navigation patterns. First some new notions in the model, such as state, navigation behavior, mean staying time and favorite, are defined. And the user navigation behavior hierarchy is then constructed. Based on the hierarchical Markov user model, an algorithm is designed to mine the user favorite navigation paths, and the results are good.

1 Introduction

User navigation behavior patterns are useful knowledge in practice, which can help create a more robust web service. Some methods, such as OLAP, DM, fuzzy theory, stochastic process, and support vector machines are used for user pattern mining [1, 2]. Agrawal and Srikant [3] adopted sequential mining techniques to discover web access patterns and trends. In [4], the authors proposed the maximal forward references to break down user sessions into transactions for mining access patterns.

Most solutions discover patterns simply according to user's access frequency in web logs. It is inaccurate. As we all know, pages, which visited frequently, may not show that users have more interest in them, such as page that is only to be utilized the links of a page to another page. At the same time, web user interests are changeable, and it is difficult to track the exact pattern of web users. In order to solve this issue, we first introduce the hierarchical Markov user model to study the navigation characters. Based on the hierarchical model, an algorithm is proposed to extract the user favorite navigation paths, and experimental example results are discussed.

2 Hierarchical Markovian Mining for Navigation Patterns

We model the navigation activities as a Markov process for the following reasons: Firstly, the information about the user navigation pattern is changeable, and Markov model can extract the changes dynamically. Secondly, the web user is

M. Bieliková et al. (Eds.): SOFSEM 2005, LNCS 3381, pp. 92–95, 2005.

largely unknown from the start, and may change during the exploration, and Markov process is suited to resolve those time series problem. Some definitions in the model are used.

Definition 1. State is a collection of one or more pages of the website with similar functions. Besides n functional states, the model contains other two special states, Entry and Exit. We assume that web user always enters the Entry state before making any request, and resides in Exit state after exiting the process;

Definition 2. In a limited time, the sequence of user's relative web requests in a website is defined as user navigation path. It can be viewed as a sequence set of states. It represents the navigation behavior of web user while exploring. Web user may remain in one state, which can be viewed as reading the contents of web pages, and also make transitions among states, which represents the requests of pages.

Definition 3. Transition probability defines as the probability of transition from state i to state j. We suppose that if there are n kinds of different transitions to leave one state, the state that has higher transition probability reveals user interest.

Definition 4. Mean staying time is the mean time which the process remains in one state before making a transition to another state. The longer staying time, the more interested visiting. We suppose that if there are n kinds of different translations to leave one state, those states that have long staying time reveal user interest. The pages that are only for a user to pass have limited staying time. Although this page have many visited times, it can lower the interest level according to the staying time.

Definition 5. Favorite level integrates the weight of transition probability and mean staying time while evaluating the interest level of the visited state. It can prevent from only mining visited states with high probability and low staying time. In formula(1), p_{ij} is transition probability, t_{ij} is mean staying time .

$$\begin{cases} f_{ij} = \dfrac{p_{ij} \times t_{ij}}{(\sum\limits_{j=2}^{n+1} p_{ij})(\sum\limits_{j=2}^{n+1} t_{ij})/(n)^2} & \begin{array}{l} i \in (2, n+1) \\ j \in (2, n+2) \end{array} \\ f_{1j} = Fthreshold; & j \in (2, n+1) \\ f_{(n+2)(n+2)} = Fthreshold \\ f_{i1} = 0 \quad i \in (1, n+2) \\ f_{(n+2)j} = 0; & j \in (2, n+1) \\ f_{1(n+2)} = 0 \end{cases} \tag{1}$$

Based on those definitions, hierarchical Markov user models are constructed. It assumes that, (a) Any transition can be made between functional states; (b) No transition can be made to the Entry state; (c) No transition can be made from Exit state to any state other itself; (d) No transition can be made from Entry state to Exit state. As shown in figure (1), each node is a visited states or pages, which aggregated the statistical information of web user navigational behavior, such as Transition Probability, Favorite Level and so on. The edges are used to represent the transition between states. For the Hierarchical Markov user Model, the edges that both Transition Probability and Favorite Level are higher than the threshold are the favorite naviga-tion paths.

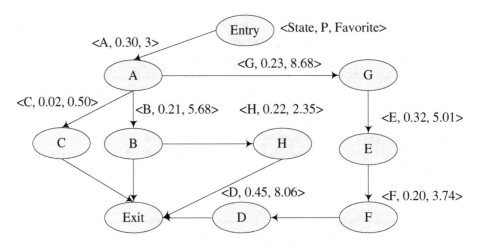

Fig. 1. Hierarchical Markov User Model

The following algorithm is used to mine the user favorite navigational path from Hierarchical Markov User Model.

```
Algorithm HMFNPM(Output){ Favorite navigation paths}
Initialize:TEMP_STACK=NULL
FindFavoritePath(HMUM,Fthreshold,Pthreshold )
    repeat
        If S is Exit state, then ReturnPreviousNode(Node S);
        Else VisitNode(Node S);
    until TEMP_STACK IS NULL.
    Return the favorite paths set;
End
VisitNode(Node S)
    If S.F>=Fthreshold and S.P>=Pthreshold then
        Node S is marked as a favorited node;
    Else Node S is not a favorite node;
    Push S into TEMP_STACK
    S<-Child(S);
End
ReturnPreviousNode(Node S, TEMP_STACK)
```

```
If TEMP_STACK is not NULL then
   Pop NODE Q from TEMP_STACK
   If Q is a favorite node, then
      Create new path in favorite navigation pathe set
         Output node Q into path
         repeat
            Next node R in TEMP_STACK
            output R into path
         until R is not a favorite node or R is a Exit state
   S<-Q;
End
```

An example is shown in figure 1, transition probability threshold and favorite threshold are set to be 0.2 and 3 respectively. First, a transition from Entry to A state is made automatically. State A is visited, its transition probability and favorite are higher than predefined threshold, and it is marked as a favorite node, put it to the temp stack and get its child B. Because its transition probability and favorite level are higher, state B is view as one favorite node. Then we get the child node of B, because it is Exit state. We get the previous node B, and return one candidate favorite path from temp stack, (Entry, A, B, Exit). The steps are executed and finally, we can get the favorite paths sets, (Entry, A, B, Exit), (Entry, A, B, H, Exit), (Entry, A, G, E, F, D, Exit).

3   Conclusions

In this paper, we proposed a hierarchical Markovian mining method to learn the navigation patterns of web user. We constructed a hierarchical Markov model to track and represent the user behaviors dynamically. Some concepts in the model are defined, which reflect user navigation behavior accurately. Based on the model, we designed an algorithm to mine user's favorite path, and it show that it is effective.

References

1. Nasraoui, O., Petenes, C.: An intelligent web recommendation engine based on fuzzy approximate reasoning. IEEE International Conference on Fuzzy Systems, (2003) 1116-1121.
2. Rohwer, Judd, A.: Least squares support vector machines for direction of arrival estima-tion. IEEE Antennas and Propagation Society, AP-S International Symposium (Digest), 1 (2003) 57-60.
3. Agrawal, R., Srikant, R.: Mining sequential patterns. Proceedings of the 11th Interna-tional Conference on Data Engineering, Taipei, Taiwan, (1995) 3-14.
4. Chen, M., S., Park, J., S., Yu, P., S.: Efficient data mining for path traversal patterns. IEEE Transaction, Knowledge Data Engineering, 10(2) (1998) 209-221.

Non-planar Orthogonal Drawings
with Fixed Topology
Extended Abstract

Markus Chimani, Gunnar W. Klau, and René Weiskircher

Institute of Computer Graphics and Algorithms,
Vienna University of Technology, Austria
{mch, gunnar, weiskircher}@ads.tuwien.ac.at

Abstract. This paper discusses the calculation of bend minimal shapes
for non-planar graphs with given topology. Based on the Simple-Kandinsky drawing standard – a simplification of the more complex Kandinsky
standard – we show the disadvantage of using standard models for this
task: We show that the minimal bend count is suboptimal, when these
models are applied to non-planar graphs; it is therefore beneficial to
extend these standards.

We define such an extension for Simple-Kandinsky called SKANPAG
(Simple-Kandinsky for Non-Planar Graphs). It treats edge crossings in a
special way by letting them share identical grid points where appropriate.
Hence it allows crossings of whole bundles of edges instead of single edges
only. Besides having a reduced number of bends, drawings following this
standard are easier to read and consume less area than those produced
by the traditional approaches.

In this paper, we show a sharp upper bound of the bend count, if
the standard Simple-Kandinsky model is used to calculate shapes for
non-planar graphs. Furthermore, we present an algorithm that computes
provably bend-minimal drawings in the SKANPAG standard.

1   Introduction

We consider the problem of producing an ortohognal drawing of a graph in
the plane. The three-phased *topology-shape-metrics* approach [1, 2, 3] breaks the
drawing problem into three subproblems:

1. *Topology/Planarization:* The first phase calculates a topology of the given
   graph. If the graph is non-planar, this includes augmenting the graph with
   dummy-nodes which represent edge crossings. The main objective is to produce as few crossings as possible.
2. *Shape/Orthogonalization:* The second phase calculates an orthogonal shape
   for the given topology. Such a shape defines the bends on the edges, and
   the angles between adjacent edges. The main objective is to produce as few
   bends as possible.

M. Bieliková et al. (Eds.): SOFSEM 2005, LNCS 3381, pp. 96–105, 2005.

3. *Metrics/Compaction:* The third phase calculates the final dimensions for the given shape, by assigning lengths to the edge segments. The main objective is to minimize the size of the resulting drawing.

In this paper we focus on the second phase of the approach, and propose an extension of the well-known Simple-Kandinsky drawing model, which is also known as Simple-Podevsnef [4]. Note that Simple-Kandinsky is a simplification of the more complex Kandinsky/Podevsnef standard (**P**lanar **o**rthogonal **d**rawing with **e**qual **v**ertex **s**izes and **n**on-**e**mpty **f**aces) [5]. All of the above models define orthogonal drawings with equal node size where multiple edges can be attached to a single side of a node. In contrast to Kandinsky, Simple-Kandinsky has certain restrictions on how these bundles split up. We give a brief introduction to the Simple-Kandinsky model and its corresponding algorithm, which is based on a min-cost-flow network, in Section 2.

Our extension, which is called SKANPAG (**S**imple-**Kan**dinsky for **N**on-**P**lanar **G**raphs), is discussed in Section 3. It allows the dummy-nodes introduced in the planarization phase to share positions on the drawing grid under certain conditions. In general, this leads to a reduction of the number of bends, as well as to a smaller area required by the drawing. Furthermore, it increases the overall readability of the resulting drawing.

We show an upper bound of the bend count for the use of classic Simple-Kandinsky on non-planar graphs in Section 4. In Section 5 we present an algorithm that generates a bend-minimal SKANPAG-compliant shape (for a given topology). To our knowledge, this algorithm is the first that can draw non-planar graphs with the minimum number of bends in the Simple-Kandinsky model. Although our method is based on an integer linear program (ILP), its running time is very low in practice, even for large and dense graphs – drawing, e. g, the K_{25} takes under 50 seconds. This is discussed in detail in Section 6. We conclude with Section 7 where we also present our ideas for further research in this field.

Since most of the proofs are quite technical and long, we only outline them in this extended abstract. Details can be found in [6, 7].

A related approach has been followed by Fößmeier and Kaufmann in [8] for the Kandinsky drawing standard. However, their method relies on the incorrect assumption that all LPs with integer coefficients would result in integer solution vectors and is beyond remedy [6].

2   The Simple-Kandinsky Model

The first polynomial approach to generate bend-minimal shapes was given by Tamassia [9], and is based on transforming the problem into a min-cost-flow network. This original method is restricted to graphs with maximum degree four.

Several attempts have been made to extend the method to the larger class of planar graphs with arbitrary node degrees. We focus on a model where nodes are drawn with uniform size. Fößmeier and Kaufmann have introduced the *Kandinsky* drawing standard [5]. By using a fine grid for the edges and a coarser grid for the nodes, it is possible that several edges emanate from the same side of a node, forming 0° angles between them. Furthermore, the standard does not allow faces with angle sum 0°. The authors present a minimum-cost flow approach with additional constraints on the flow that can be realized, for instance, using an ILP approach as in, e. g., [10] or in the AGD library [11].

Recently, Bertolazzi, Di Battista, and Didimo have proposed a simplification of this standard – which we will refer to as the *Simple-Kandinsky standard* – by adding the following restrictions [4]:

(S1) Each node with degree ≥ 4 has at least one edge emanating from each of its four sides. No node with degree ≤ 4 is allowed to have 0° angles.

(S2) For each two neighboring edges that leave a node on the same side, the first bend on the rightmost edge is a right bend.

The authors present a polynomial-time algorithm based on a network flow model that computes bend-minimal drawings of planar graphs with fixed topology in this standard. Here, we present a slightly modified but equivalent flow model for this task. Although larger by a constant factor, it is favorable for our purposes due to its simplicity, consistency, and straightforward extendibility.

Let $G = (V, E)$ be a planar graph with given topology, characterized by the face set F with outer face f_o. We create an underlying min-cost-flow network with two different types of directed edges: (a) arcs from the nodes $v \in V$ to incident faces $f \in F$ with zero cost, and (b) arcs with unit cost between adjacent faces.

Each unit transported in this network corresponds to a 90° angle: A flow of x units on an arc between a node $v' \in V$ and a face $f' \in F$ implies an angle of $x \cdot 90°$ on f' at v'. Each flow unit on an arc from the face f' to f'' ($\in F$) introduces a 90° bend on the edge $e' \in E$ that is on the border of both faces.

The capacities of the arcs as well as the supplies and demands of the nodes in the network are straightforward, and similar to the original Tamassia model: Each node $v \in V$ has a supply of four units which corresponds to 360°; each face $f \in F - \{f_o\}$ has a demand of $2 \deg(f) - 4$, while the outer face f_o has a demand of $2 \deg(f_o) + 4$, where $\deg(.)$ denotes the number of edges bounding a face.

To satisfy constraints (S1) and (S2) – and therefore guarantee valid drawings – we have to apply an augmentation to the network for each high-degree node in

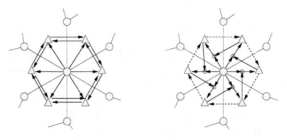

Fig. 1. Network construction (left) and correctly augmented network (right); (circles and triangles represent nodes and faces, resp., dashed arcs are unchanged by the augmentation, irrelevant arcs are not shown)

V (see Fig. 1): We add a cyclic substructure of demand-free nodes. Each of these nodes is the target of an arc of type (b) which causes right bends and an arc of type (a). Furthermore, we insert arcs between the new nodes and the faces surrounding the high-degree node; these arcs have a lower flow bound of one unit. The construction guarantees that if two edges leave a node on the same side, forming a 0° angle, the right edge of this bundle has to have a right bend.

3   Theory of Hyper-Faces – The Skanpag Model

The key to bend-minimality for non-planar graphs is to allow certain dummy-nodes to share a grid point, see Fig. 2(a). Nodes that share a grid point are said to be *merged*. Faces that become empty by such a merge, are said to be *collapsed*.

If not stated otherwise, let $G_o = (V_o, E_o)$ be a non-planar, simple, and self-loop-free graph. The graph $G = (V, E)$ denotes the planarization of G_o (based on a topology T_o). Let T be the planar topology of G that the drawing should be based on, and F the face set implied by T. We have $V = V_o \cup D$, where D is the set of the dummy-nodes introduced by the planarization.

Definition 1. *If an edge $e_o \in E_o$ is split up into several sub-edges $e_i \in E$ during the planarization, we call the set $\{e_i\}$ the* meta-edge *of any of these e_i.*

To extend Simple-Kandinsky for non-planar graphs in the most general way, we have to demand its properties only for the underlying non-planar graph G_o, not for the planarized graph G. Hence we demand a right bend on bundles for the meta-edges instead of only for their first sub-edges.

This leads to the analysis of bundles of meta-edges, and when their dummy-nodes are allowed to merge. Two meta-edges can only (partially) merge, if they emit from a common source node $s \in V_o$ with $\deg(s) > 4$ (see property (S1)).

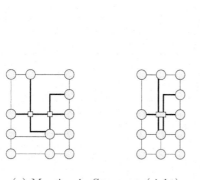

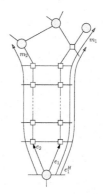

(a) Merging in SKANPAG (right) compared to Simple-Kandinsky (left)

(b)   Meta-edges (m_1, m_2), hyper-face, and hyper-edge (e_1^H)

Fig. 2. Definitions. Original nodes are circles, dummy-nodes are squares

Therefore we can deduce the two different structures of faces in F which might collapse:

- *collapsible triangle ("coltri"):* A face that has exactly one incident node with a degree greater than four (the source node s of a bundle), and two incident dummy nodes.
- *collapsible quad ("colquad"):* A face that has exactly four incident dummy nodes.

This leads to the concept of *hyper-faces* (see Fig. 2(b)):

Definition 2. *Let e_1 and e_2 be the edges incident to a coltri f_0 and its high degree node s. A hyper-face f_0^H is a sequence of faces that starts with f_0, contains any number of colquads, and ends with a face other than a colquad. All its faces have to be incident to a sub-edge of the meta-edges of e_1 and to a sub-edge of the meta-edge of e_2.*

Note that we only consider simple graphs; hence a hyper-face end-face can never be a coltri.

We can merge a pair of dummy-nodes that lie on the border of such a hyper-face, see Fig. 2(a), as long as there are no bends which cause a split up (*"de-merge"*).

If not stated otherwise, we will give directions on the hyper-face in its *natural orientation*, where the coltri is at the bottom (as in Fig. 2(b)). We assume that m_1 (the meta-edge of e_1) is on the right side of the hyper-face. To satisfy the Simple-Kandinsky constraints, we have to assure that m_1 has at least one right bend before any left bend, and before m_2 (the meta-edge of e_2) has a right bend.

It is clear that such a right bend has to happen before we would be forced to merge a dummy-node with an original node. Hence we can specify the constraint that such a right bend has to happen on the *hyper-edge*:

Definition 3. *The hyper-edge e_1^H is an ordered subset of the meta-edge m_1 and contains all the edges of the meta-edge that are on the boundary of sub-faces of the according hyper-face (see Fig. 2(b)). The analogously defined subset of m_2 is the* partner edge *of the hyper-edge.*

We can summarize these observations: SKANPAG has to force a right bend on each hyper-edge, if its respective coltri has an opening angle of 0°. We can merge dummy-nodes if and only if their two meta-edges leave a node on the same side, and do not have any bends that would cause a split up of the bundle. Note that edges incident to a high degree node that have not been split up by the planarization step still have to satisfy property (S2) of standard Simple-Podevsnef instead.

4   Quality Guarantee of Simple-Kandinsky

Prior to SKANPAG, the only way to draw non-planar graphs was to use a planar drawing standard as a heuristics. We look at the special case of using Simple-

Kandinsky, because of its strong relationship to SKANPAG . Furthermore, Simple-Kandinsky seems to be a good compromise between simplicity, execution time, and quality. Hence it is quite interesting to assess the quality difference between Simple-Kandinsky used as a heuristics and SKANPAG on non-planar graphs.

Note that for planar graphs, the minimum number of bends is equal for Simple-Kandinsky and for SKANPAG. Since there are no hyper-faces in planar graphs, there are no additional constraints, and hence no difference in the solution.

Also note that the minimal bend count in Simple-Kandinsky can never be lower than SKANPAG's: Each valid Simple-Kandinsky solution defines a valid SKANPAG solution with the same number of bends because Simple-Kandinsky's right-bend constraint is a specialization of the corresponding constraint for hyper-faces in SKANPAG.

Theorem 1. *For any given planarized graph $G = (V, E)$ and its planar topology T with h hyper-faces, Simple-Kandinsky requires at most h more bends than* SKANPAG.

Proof. We assume that a SKANPAG-algorithm has calculated a valid shape. The following observation holds true for every hyper-face h_i: If the opening angle of h_i is 90° or if there is a right bend on the first sub-edge of h_i, this shape is valid for Simple-Kandinsky, too.

If the coltri has an opening angle of 0° and there is no right bend on the first sub-edge of the hyper-edge of h_i, it is not a valid Simple-Kandinsky shape (Fig. 3(a), left). But we can achieve a related Simple-Kandinsky shape by increasing the opening angle from 0° to 90° and adding one right bend on the bundle partner of h_i (Fig. 3(a), right).

This transformation is not influenced by collapsed coltris to the left of h_i, since these have to be extended by analogous bends themselves to become Simple-Kandinsky compliant. Neither does the transformation influence its surrounding area, since its overall shape remains the same.

Hence we need at most one more bend for each hyper-face to transform a SKANPAG shape into a valid Simple-Kandinsky drawing.   □

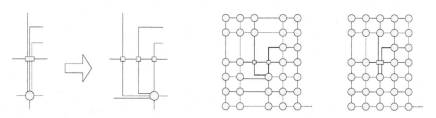

(a) A valid hyper-face in SKANPAG may need one more bend in Simple-Kandinsky

(b) Example where the heuristic solution is as bad as it gets

Fig. 3. Simple-Kandinsky as a heuristics

Corollary 1. *The bound given in Theorem 1 is tight.*

Proof. Fig. 3(b) shows the graph used as a building block, containing exactly one hyper-face. We can put an arbitrary number of these blocks next to each other (joined by a simple edge). For every block, Simple-Kandinsky needs at least two bends, but SKANPAG requires only one. Hence Simple-Kandinsky requires exactly h bends more than SKANPAG. Note that each block is optimally planarized, since there exists no other planarization generating less than two dummy-nodes per block. □

5   An Algorithm for Bend Minimal Skanpag-Drawings

Due to lack of space, we only outline the following algorithms. We implemented them as part of the AGD library [11] which is available freely for research purposes. Details, as well as the corresponding proofs, can be found in our technical paper [7] or in [6].

To solve the bend minimization problem, we use an ILP that models the underlying min-cost-flow network introduced in Section 2 and contains the right-bend-on-hyper-edge constraint.

While in Simple-Kandinsky the simple right-bend rule is enough to guarantee valid drawings, this is no longer true for SKANPAG. Hence we need an in-depth analysis of the situations that could render a correct drawing of hyper-faces impossible. Note that if all hyper-faces are drawn correctly, the complete graph will be drawn correctly (see the proof of Simple-Kandinsky's validity in [4]).

If the coltri of a hyper-face has an opening angle of 90°, the drawing of the hyper-face is always valid (this follows from the correctness of Simple-Kandinsky). Hence we only have to analyze the case of a 0° opening angle.

By careful enumeration we can deduce two different types of situations that inhibit valid drawings of a hyper-face:

1. A flow unit that leaves a collapsed sub-face to its left side, and reenters over the same side. (Although this situation seems like a contradiction to the objective function, it occurs in bend-minimal drawings – both in SKANPAG and in Simple-Kandinsky – to generate necessary right bends)
2. A flow unit sent from a sub-face f_1 into its neighboring sub-face f_2 below it.
   (a) The flow unit is not sent directly backwards from f_2 to f_1
   (b) The flow unit is sent directly backwards from f_2 to f_1

We can show that we can prohibit all type 1 and type 2a errors by simple changes of the network structure, without cutting off the optimal solution. The errors of type 2b, however, are more complicated. They could, e. g., be avoided by additional constraints in the ILP, but these would introduce new 0/1-variables. Therefore we propose a different approach using a repair-function:

We solve the ILP without handling type 2b errors and thus generate an *almost valid* solution. This solution can then be repaired by a polynomial helper function. The principle of this repair function is to analyze the remaining errors and move the corresponding flows to other positions. It ensures that all bend properties and constraints are still satisfied afterwards, but the formerly invalid flows do not generate errors anymore.

Our repair-function has an upper time bound of $O(h^2 l_{max}^2)$, where h is the number of hyper-faces and l_{max} the cardinality of the longest hyper-face. This bound can be estimated more generously as $O(|E|^2)$.

6   Computational Results

The *Rome graphs* [12] are a well established collection of 11529 graphs based on 112 graphs taken from real-world applications. They have between 10 and 100 nodes each, and 3280 graphs are planar. But even the non-planar graphs are quite sparse and have therefore very few dummy-nodes after the planarization step. Nearly 7000 graphs do not have any hyper-face at all and 1500 others contain only one. Thus we know from Theorem 1 that we cannot expect big differences between Simple-Kandinsky and SKANPAG . The implementation of our new method solves all these graphs without any need for the repair-function, nor does the LP-relaxation ever produce non-integer solutions.

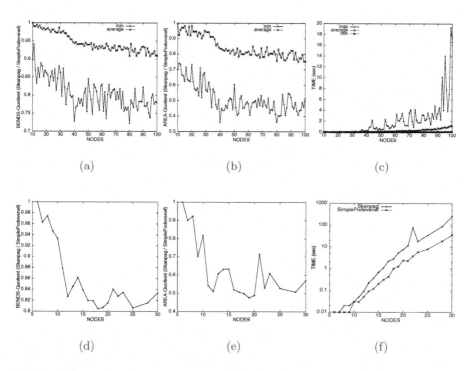

Fig. 4. Performance of SKANPAG, relative to Simple-Kandinsky. Top row: squared Rome graphs; bottom row: complete graphs. The peak at (f) is because the ILP-solver had to do several branches

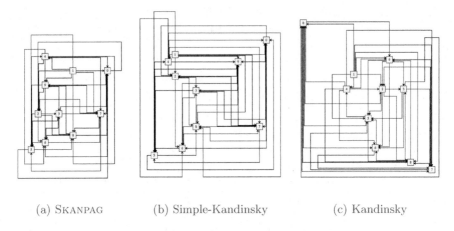

(a) SKANPAG   (b) Simple-Kandinsky   (c) Kandinsky

Fig. 5. Graph with 10 nodes and 42 edges, drawn using three different standards

Since this test suite is too sparse to show the difference between SKANPAG and Simple-Kandinsky, we *squared* each of the Rome graphs. The resulting graphs are still not extremely dense, but have more hyper-faces: only about 1200 still have none.

If not otherwise stated, the statistics compare SKANPAG to Simple-Kandinsky. As Fig 4(a) shows, we need nearly 10% less bends in the average case for big graphs, with peak values of up to 25%. The size of the drawing (Fig 4(b))is reduced by over 20% on average. In some cases, we reduce the area consumption by 60%. As Fig 4(c) shows, the runtime performance of SKANPAG is acceptable even for large and dense graphs.

We also tested SKANPAG with complete graphs (up to K_{30}), to demonstrate the quality advantage for dense graphs. Note that the planarization of K_{30} has nearly 11000 dummy nodes. Fig 4(d) shows that we can save 15%-20% of bends for all such graphs with over 11 nodes; the area savings are even higher (up to 50%, see Fig. 4(e)). Figure 4(f) shows the runtime performance of SKANPAG and Simple-Kandinsky.

Figure 5 shows an example of a quite dense graph, drawn by Kandinsky, Simple-Kandinsky, and SKANPAG (equally scaled). More examples and details on the statistics can be found in [6].

7   Conclusion and Further Work

We have presented a new approach for drawing non-planar graphs that takes the special properties of dummy nodes into account. Our algorithm guarantees the minimum number of bends for any given topology following the Simple-Kandinsky properties, by the use of an integer linear program. It is the first approach to solve this problem and due to our polynomial time repair function, the runtime is acceptable even for large and dense graphs.

Note that our algorithm can also be used for drawing clustered graphs orthogonally. These are usually drawn by modeling the cluster boundaries as circles consisting of dummy-nodes and dummy-edges [13, 14]. By treating the dummy-nodes on the cluster boundaries just like any other dummy-node, we can achieve savings in bends and area compared to previously known algorithms.

The complexity of producing bend-minimal drawings in the SKANPAG model is still unknown. This – as well as the complexity proof for the Kandinsky model itself – is an interesting field of future study.

References

1. Batini, C., Nardelli, E., Tamassia, R.: A layout algorithm for data flow diagrams. IEEE Trans. Softw. Eng. **12** (1986) 538–546
2. Battista, G.D., Eades, P., Tamassia, R., Tollis, I.: Graph Drawing. Prentice Hall (1998)
3. Tamassia, R., Battista, G.D., Batini, C.: Automatic graph drawing and readability of diagrams. IEEE Trans. Syst. Man. Cybern., SMC-18(1) (1988)
4. Bertolazzi, P., Di Battista, G., Didimo, W.: Computing orthogonal drawings with the minimum number of bends. IEEE Transactions on Computers **49** (2000) 826–840
5. Fößmeier, U., Kaufmann, M.: Drawing high degree graphs with low bend numbers. In Brandenburg, F.J., ed.: Proc. of the 3rd Int. Symp. on Graph Drawing (GD 1995). Volume 1027 of Lecture Notes in Computer Science., Passau, Germany, Springer (1996) 254–266
6. Chimani, M.: Bend-minimal orthogonal drawing of non-planar graphs. Master's thesis, Vienna University of Technology, Department of Computer Science, Austria (2004)
7. Chimani, M., Klau, G., Weiskircher, R.: Non-planar orthogonal drawings with fixed topology. Technical Report TR 186 1 04 03, Institute of Computer Graphics and Algorithms, Vienna University of Technology (2004)
8. Fößmeier, U., Kaufmann, M.: Algorithms and area bounds for nonplanar orthogonal drawings. In: Proc. 6th Symposium on Graph Drawing (GD'97). Volume 1353 of LNCS. (1997) 134–145
9. Tamassia, R.: On embedding a graph in the grid with the minimum number of bends. SIAM J. Comput. **16** (1987) 421–444
10. Eiglsperger, M., Fößmeier, U., Kaufmann, M.: Orthogonal graph drawing with constraints. In: 10th Annual ACM-SIAM Symposium on Discrete Algorithms (SODA 1999). (1999) 3–11
11. Jünger, M., Klau, G.W., Mutzel, P., Weiskircher, R.: AGD: A Library of Algorithms for Graph Drawing. Mathematics and Visualization. In: Graph Drawing Software. Springer (2003) 149–172
12. Di Battista, G., Garg, A., Liotta, G.: An experimental comparison of three graph drawing algorithms (extended abstract). In: Proceedings of the eleventh annual symposium on Computational geometry, ACM Press (1995) 306–315
13. Battista, G.D., Didimo, W., Marcandalli, A.: Planarization of clustered graphs. In: Graph Drawing (Proc. GD 2001). Volume 2265 of LNCS. (2001) 60–74
14. Lütke-Hüttmann, D.: Knickminimales Zeichnen 4–planarer Clustergraphen. Master's thesis, Saarland University, Department of Computer Science, Saarbrücken, Germany (2000)

A Topology-Driven Approach to the Design of Web Meta-search Clustering Engines*

Emilio Di Giacomo, Walter Didimo, Luca Grilli, and Giuseppe Liotta

Dipartimento di Ingegneria Elettronica e dell'Informazione,
Università degli Studi di Perugia, Perugia, Italy
{digiacomo, didimo, grilli,liotta}@diei.unipg.it

Abstract. In order to overcome the limits of classical Web search engines, a lot of attention has been recently devoted to the design of Web meta-search clustering engines. These systems support the user to browse into the URLs returned by a search engine by grouping them into distinct semantic categories, which are organized in a hierarchy. In this paper we describe a novel topology-driven approach to the design of a Web meta-search clustering engine. By this approach the set of URLs is modeled as a suitable graph and the hierarchy of categories is obtained by variants of classical graph-clustering algorithms. In addition, we use visualization techniques to support the user in browsing the categories hierarchy.

1   Introduction

As the amount of information accessible on the Web super-exponentially increases, there is a growing consensus that the paradigm adopted by most popular search engines to output the results of a query is becoming inadequate (see e.g. [6, 20, 21]). Indeed, such output typically consists of a ranked list of URLs, which may be very long and difficult to browse for the interested user. Especially in cases of polisemy (one word may have several different meanings) it is useful that the output returned by a search engine be organized into categories grouping documents strongly related from a semantic view point.

Indeed, a lot of attention has been recently devoted to the design of *Web meta-search clustering engines* (see e.g. [6, 8, 14, 16, 17, 20, 21]). A Web meta-search clustering engine is a system that behaves as follows: (i) It receives a query from the user, and forwards it to one or more Web search engines that can be automatically accessed over the Internet; (ii) It selects a subset of the URLs returned by the search engines and groups these URLs into clusters, which are presented to the users as a hierarchy of categories. In order to make its output expressive and to allow a reasonable on-line interaction, a Web meta-search clustering engine should satisfy the following basic requirements:

---

* Research supported in part by "Progetto ALINWEB: Algoritmica per Internet e per il Web", MIUR Programmi di Ricerca Scientifica di Rilevante Interesse Nazionale.

M. Bieliková et al. (Eds.): SOFSEM 2005, LNCS 3381, pp. 106–116, 2005.

– *Efficiency.* The system should be sufficiently fast in reacting to a user query.
– *Effective Clustering.* The URLs in each cluster should be strongly related from a semantic view point.
– *Expressive Labeling.* Each category should be labeled with a word or a sentence that clearly reflects the topic shared by its URLs.

As for the first requirement, it is clear that efficiency cannot be achieved if the meta-search clustering engine relies on downloading and parsing all documents that are returned by the search engine. Therefore, those techniques that compute communities from the Web graph by assuming that the pages can be downloaded and parsed (for example for extracting hyperlinks) cannot immediately be used for the design of a meta-search clustering engine. A very limited list of such clustering techniques includes [1, 7, 9, 11, 16].

Concerning the other two requirements, there is no general consensus about the best design approach to achieve them. Evaluating the effectiveness of a cluster and the expressiveness of its label strongly depends on the semantic context of the query and on the user judgment. As a matter of fact, even the most effective Web meta-search clustering engines usually end-up by presenting many "meaningful" categories together with a few "inexpressive" categories on some specific queries. This phenomenon is particularly uncomfortable if the categories whose labels do not seem to make much of sense actually contain documents that are semantically related to meaningful categories and contain strategic information for the user.

In this paper we describe a novel topology-driven approach to the design of a Web meta-search clustering engine. By this approach the set of URLs is modeled as a suitable graph and the computation of the categories and of their associated labels is obtained by variants of classical graph-clustering algorithms. The topology-driven approach turns out to be comparable with traditional text-based strategies for the definition of the clusters hierarchy. In addition, our approach makes it natural to use graph visualization techniques to support the user in handling inexpressive labels. Namely, categories with inexpressive labels can be visually related to more meaningful ones. To the best of our knowledge, there is only one Web meta-search engine (Kartoo[1]) that supports the user with graph visualization. However, Kartoo only shows the inclusion relationships between documents and categories, and it does not provide sufficient support to visually comprehend the semantic connections among different clusters.

Our main results can be listed as follows:

– We introduce and study the concept of *snippet graph* that describes semantic relations between the snippets of the URLs returned by a search engine.
– We engineer and modify known clustering techniques to identify categories by a topological analysis of the snippet graph.
– A new visualization paradigm is introduced to display the clusters hierarchy. It uses advanced graph drawing methods to visualize semantic relationships between apparently unrelated categories.

---

[1] http://www.kartoo.com/

– We present a prototype of WhatsOnWeb, a Web meta-search clustering engine
that has been designed according to the topology-driven approach.
– Finally, the results of an experimental analysis that show the behavior of
WhatsOnWeb and that compares it with state-of-the-art systems are discussed.

2   Web Meta-search Clustering Systems: State of the Art

In the field of Web meta-search clustering engines, there are two different types
of systems: commercial and academic systems. As also pointed out by other
authors [6], commercial systems currently offer the best performances, both in
terms of quality of the results and in terms of efficiency. Unfortunately, there
is no scientific publication that describes the technology of commercial tools.
Among them, Vivisimo [2] is considered from the majority of people as the most
effective Web meta-search clustering engine. A limited list of other commercial
meta-search clustering engines includes Groxis, iBoogie, Kartoo, Mooter.

In the academic systems scenario, an important contribution comes from
Grouper [19, 20], that uses a clustering approach based on the analysis of the
so called *Web snippets*. A Web snippet (or simply snippet) is a small portion of
text that a Web search engine returns associated with each URL in response to
a user query; it represents a brief summary of the document addressed by the
URL. A Web snippet approach consists in computing clusters and their labels by
analyzing text similarity between snippets. In [19] the authors describe a linear
time algorithm (called *STC algorithm*) based on a suffix tree data structure,
and provide experiments that show the effectiveness and the efficiency of their
approach with respect to previous clustering techniques. They also performed experiments that compare the precision of several document clustering algorithms
(included STC) when applied on snippets instead of on whole documents. Their
experiments show that this precision does not decrease significantly, i.e., snippet
clustering is in most cases a "good" approximation of document clustering. Some
recent works related to the Web snippet approach are [6, 14, 17, 21]. In particular, SnakeT[3] [6] is a well documented academic system that appears promising
especially for what concern the labeling of categories.

Although the current systems based on the Web snippet approach use different specific algorithms to determine clusters, most of them adopt a common
general strategy. They first compute a set of so called *base clusters*, possibly overlapping, and then recursively merge those clusters that have large overlapping
and organize them in a hierarchy, from bottom to top. Both the computation
of base clusters and their organization in a hierarchy are mainly based on text
analysis techniques, while clustering algorithms that rely on the topology of a
computed graph have in our opinion not yet received enough attention in the
specific context of designing Web meta-search clustering engines.

---

[2] http://www.vivisimo.com/
[3] http://roquefort.di.unipi.it/

3   A Topology-Driven Approach

The topology-driven approach is simple and intuitive. It is based on the new concept of snippet graph of a set of URLs, which implicitly represents a global view of the semantic relationships between these URLs; for efficiency reasons, these relationships are constructed analyzing the title and snippet text associated with each URL. A high-level description of the approach and of its benefits is as follows. Let U be a set of URLs returned by a Web search engine in response to a user query. (i) We define the *snippet graph* of U as a weighted graph whose vertices represent the elements of U and whose edges represent relationships between the snippets of the elements of U. The weight of each edge informs about the strength of the relationship between the end-vertices of the edge. Intuitively, the snippet graph summarizes a self-contained set of semantic relationships between elements of U. (ii) Once a snippet graph G of U have been computed, we apply on G an algorithm that extracts clusters in G, and that organizes them into a hierarchy, which will be presented to the user. The clustering algorithm works within the assumption that groups of strongly connected vertices are semantically related. (iii) Clusters are labeled by using some text information associated with the vertices and the edges of the snippet graph. These information are definitively stored during the snippet graph construction.

The proposed topology-driven approach has two main advantages with respect to previous snippet clustering strategies: (a) The snippet graph gives a global view of the relationships between a set of URLs. Computing clusters by considering the structure of this graph appears more reasonable and, intuitively, more powerful than independently evaluating text similarity between pairs of snippets. Also, there are several effective algorithms that can be used to construct a clusters hierarchy in a general graph (see e.g. [2, 10, 13, 18]). (b) Constructing a clusters hierarchy on the snippet graph makes it natural to handle the relationships between distinct clusters. These relationships can be visualized in order to support the user when the labeling of some categories is not sufficiently expressive.

3.1   The Snippet Graph

Let U be a set of URLs returned by one or more search-engines in response to a user query. In the following, if $u \in U$, we call *text of* u the concatenation of the title and the snippet associated with u. We analyze the text of all elements of U and, as usual, we clean them by removing both stop-words and words that appear in the user query. We then apply a stemming algorithm on the cleaned texts; with every computed stem s we associate a bag B_s of words whose stem is equal to s. Sets B_s will be used later for labeling the clusters. Also, we assign a score f_s to each stem s; f_s measures the importance of a semantic relationship between URLs whose texts contain s. Several criteria can be used to determine function f_s. The simplest is to assign a constant value to each stem, for example $f_s = 1$. Another possibility is to adopt a function based on the frequency of s in all the texts of the elements of U.

The *snippet graph* G of U is a labeled weighted graph defined as follows: (i) G has a vertex v_u associated with each element $u \in U$. The label of v_u can be either the title of u or its description as URL. (ii) G has an edge $e = (v_{u_1}, v_{u_2})$ (where $u_1, u_2 \in U$) iff the texts of u_1 and u_2 share a non-empty set S of (stemmed) words. The weight w_e of e is the sum of all scores of the elements of S, i.e. $w_e = \sum_{s \in S} f_s$. The label of e is a list of all words of S.

3.2   Computing Clusters in the Snippet Graph

In order to determine clusters in the snippet graph, we look for *communities of vertices*, i.e. set of vertices that are strongly related from a topological view point. Among the wide range of graph clustering approaches proposed in the literature we point our attention on those adopting a recursive decomposition strategy based on edge connectivity (see, e.g. [2, 10, 13]). Namely, the clusters hierarchy is determined by recursively cutting some edges that disconnect the graph; they mainly differ for the criteria used to choose the next edge to be removed. We apply the definition of community recently given by Brinkmeier [2], which is strongly related to the definition of k-component introduced by Matula [12]. Let G be a graph and let S be a subset of vertices of G. According to the definition in [2], the *community of S in G* is the largest subgraph of G of maximum edge connectivity among all subgraphs containing S. This definition appears quite natural and is particularly effective for two main reasons: (i) The community of every subset of vertices of G is uniquely determined. This is not true for the other definitions based on edge connectivity. (ii) It naturally induces a clusters hierarchy on G that implicitly describes all the communities of subsets of vertices of G. Each cluster of the hierarchy represents a different community viewed at a specific level of abstraction, and the vertices of G are atomic communities. Graph G together with its clusters hierarchy can be described as a clustered graph according to the definition given by Feng et al. [5]. The clusters hierarchy is described as a tree, where the leaves are associated with the vertices of G and the internal nodes are the non-atomic clusters of G. We call this tree the *community tree* of G.

Let G be the snippet graph of a set U of URLs. We use the decomposition strategy in [2] as a basic tool for constructing the *cluster hierarchy of the snippet graph* G. This cluster hierarchy, which we denote as H_G, is a tree whose internal nodes represent the different semantic categories. The root of H_G is a node representing the whole set of URLs. The nodes of level i ($i > 0$) are referred to as *level-i categories*. In particular, the level-1 categories partition the set of all URLs and represent the macro categories initially shown to the user. Each of these categories can be itself partitioned into disjoint sub-categories, that are level-2 categories, and so on. The leaves of H_G represent URLs. Notice that, H_G does not allow repetition of URLs, i.e. all its leaves correspond to distinct URLs. This implies that every URL is put in exactly one deepest category, and therefore deepest clusters never overlap. We now describe how to compute H_G. The set of level-i categories of H_G is denoted by L_i.

Computing and Labeling Level-1 Categories. The level-1 categories are computed as follows:

(1) Initially set L_1 as empty; (2) Compute the community tree T of G and consider the nodes of T that have only leaf-children. Add these nodes to L_1; (3) Let G' be the graph obtained by deleting from G all vertices corresponding to the children of nodes in L_1, and let T' be the community tree of G'. Again, add to L_1 those nodes that have only leaf-children; (4) Iterate Step 2 until a desired set of level-1 categories has been determined. In practice, $7 - 8$ iterations are enough to compute the most meaningful categories.

To label the level-1 category represented by a node μ of H_G, we apply the following procedure. Let G_μ be the subgraph of G induced by the children of μ, and let S be the set of stemmed words associated with the edges of G_μ. If $s \in S$ is associated with k $(k > 0)$ edges, then kf_s is called the *total score of s in G_μ*. Denote by $S_{max} \subseteq S$ the subset of stemmed words of S that have maximum total score in G_μ. For each stemmed word $s \in S_{max}$, we select one representative word $b_s \in B_s$; then, we label μ with the set of words in $\mathcal{B}_\mu = \cup_{s \in S_{max}} b_s$. In practice, $\mathcal{B}_\mu$ often consists of a single word and rarely consists of more than two or three words. Keeping category labels short, but still expressive, simplifies the amount of information that the user must deal with.

Computing and Labeling Level-i Categories. The computation of the level-i categories $(i > 1)$ is done similarly to the level-1 categories by using a recursive approach that progressively simplifies the snippet graph. Namely, suppose that all the level-i categories have been computed, for some $i \geq 1$. We compute the level-$(i+1)$ categories as follows: (1) For each node μ of H_G representing a level-i category, let G_μ be the subgraph of G induced by the vertices in μ. Compute the graph G_μ^{-}, obtained from G_μ by subtracting to the edge weights and labels the contribution of the words in $\mathcal{B}_\mu$. (2) Compute L_{i+1} by applying on every G_μ^{-} the same procedure as the one applied on G for computing L_1. The nodes in L_{i+1} obtained from G_μ^{-} are made children of μ in H_G.

When the decomposition algorithm of a level-i category μ does not produce sub-categories for μ, we stop the decomposition process for the subtree rooted at μ. Concerning the labeling procedure of level-i categories, it is the same as the one for level-1 categories.

Computing the Leaves. Let L_i $(i \geq 1)$ be the set of the deepest categories of H_G. For each $\mu \in L_i$, we make all the vertices of G in μ as children of μ. They represent the leaves of H_G. Each leaf of H_G is labeled with the title of its corresponding URL.

Concerning the theoretical time complexity of constructing H_G, it mainly depends on the complexity of computing a community tree. Indeed, both the number of levels of H_G and the number of iterations applied to determine the nodes of each level can be set to constant numbers, using an empirical evaluation. To compute the community tree, Brinkmeier [2] describes an $O(n^2m + n^3 \log n)$ technique based on recursively applying a minimum cut algorithm several times (n and m are the number of vertices and the number of edges of the input graph, respectively). However, as also discussed in [2], the computation of a minimum cut can be avoided in many cases during the whole decomposition, using some simple considerations on the degree of the vertices. This dramatically reduces the running time in practice (see Section 4).

3.3   A New Visualization Paradigm

Most of the current Web meta-search engines adopt a simple visualization paradigm. The user can browse the categories through the same interface used to explore the content of a file system. The main limit of this approach is that no information are provided to the user about the relationships between distinct categories. These information can be of help, especially when the labeling of the categories is not sufficiently expressive. Also, recursively exploring nested nodes in the categories tree can rapidly lead to a very "high" drawing, which cannot be effectively displayed on a computer screen and hence difficult to manage for the user. Further, re-drawing the categories tree from scratch during the browsing may significantly change the user mental map. For the above motivations, beside to the standard tree of categories, we propose a new visualization paradigm, which has the following characteristics: (i) A clustered drawing of the snippet graph G is presented to the user; the clusters correspond to the nodes of H_G. (ii) In each cluster we do not show the relationships induced by words in the label of the cluster, because their existence can be implicitly assumed due to the nature of the cluster itself. We show all the other semantic connections; they mainly give information about the relationships between distinct categories. (iii) Clusters can be expanded or contracted at each time preserving "as much as possible" the user mental map. To this aim we adopt an engineered version of orthogonal graph drawing techniques described in [3, 15]. Expanding or contracting clusters make it possible to explore the categories at different abstraction levels.

For reasons of space we do not describe in detail our visualization paradigm (for the details see [4]). Figure 1 shows an example of drawing for the query "Armstrong". The category "Lance" and its subcategory "Story" are expanded.

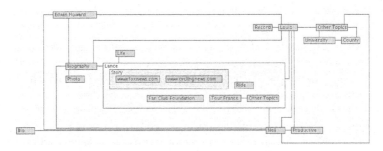

Fig. 1. Map of the level-1 categories for the query "Armstrong". The category "Lance" and its subcategory "Story" are expanded

4   Experimental Analysis

We designed a system, called `WhatsOnWeb`, according to the principles of the topology-driven approach presented in Section 3. A first prototype implementation of `WhatsOnWeb` is available on-line [4]. In this implementation we retrieve data from `Google` and correctly handle the English language. We used `WhatsOnWeb` to perform a preliminary experimental analysis that evaluates the effectiveness of our topology-driven approach from a user perspective. In our experiments we did not involve human judgments, since it can strongly vary with the chosen subjects and it does not allow us to replicate the experiments in future works and comparisons. Instead, we aim at detecting a number of objective parameters that can be automatically measured and that reflect some important aspects for the browsing of the user. We take into account three main aspects:

Efficiency. We measured the time required by our system to generate a response to a user query.

Structure of the Hierarchy. In order to evaluate the quality of the hierarchy from the structural side, we measured the following parameters and compare them with those obtained by the results of `Vivisimo`. *Depth of the hierarchy*: Maximum number of levels in the whole hierarchy and average number of levels in the subtrees rooted at level-1 categories. This gives an idea of the number of steps that the user may should perform to reach a page of interest, both in the average and in the worst case. *Width of the hierarchy*: Maximum number and average number of children of each category. This gives an idea of the number of elements that the user should may scan when expands a category.

Coherence of Categories. Structural properties do not suffice to evaluate the quality of a hierarchy. Each category should group documents sharing a common topic and the label of the category should clearly reflect this topic. We define simple but reasonable parameters to partially estimate the coherence and the labeling correctness of a clusters hierarchy H_G. Namely, for each category C, let D_C be the union of all documents in C and let L_C be the set of words in the label of C. Also, let W_C be the number of occurrences of words of L_C in D_C, and let W be the number of occurrences of words of L_C in the documents contained in the siblings of C plus C itself. The *coherence percentage of C* is defined as $coher(C) = 100\frac{W_C}{W}$. The *coherence percentage of H_G* is defined as

$$coher(H_G) = \sum_{C \in H_G}(coher(C) * |D_C|)/\sum_{C \in H_G}|D_C|.$$

We also define the *incoherence percentage of C* as the percentage of documents of C that do not contain words of L_C. The *incoherence percentage of H_G* is the average of all cluster incoherence percentages.

For the experiments we used a set Q of 9 query strings. The strings in Q define (i) *Ambiguous queries*: `Armstrong`, `Jaguar`, `Mandrake`. (ii) *Generic queries*:

---

[4] `http://zico.diei.unipg.it/~whatsonweb`

Health, Language, Machine. (iii) *Specific queries*: Mickey Mouse, Olympic Games, Steven Spielberg.

All the experiments were performed on a PC Pentium IV, 2.7 GHz, and 512 MB RAM. Figure 2 shows the data about the depth and the width of the cluster hierarchies computed by WhatsOnWeb against those computed by Vivisimo. For every query in Q, WhatsOnWeb analyzed the same number of snippets considered by Vivisimo, which usually ranges in the interval $[160, 250]$. As shown in the figure, the structure of the hierarchies constructed by the two systems are very similar in many cases.

Concerning efficiency, Figure 2 also shows the CPU time spent by WhatsOnWeb on each query: beside the overall time required by the whole computation, we also reported the times required by the two main phases of our technique: construct the snippet graph and compute the labeled clusters hierarchy. The number of the processed snippets (pages) is also shown. The results show that our clustering algorithm is quite fast in practice.

Query string	Max Depth WW	Max Depth VS	Avg Depth WW	Avg Depth VS	Max Width WW	Max Width VS	Avg Width WW	Avg Width VS	Pages	Snip. Graph	Cluster Hier.	Total
Armstrong	4	5	2,13	2,38	30	40	3,93	3,62	222	2,12	3,09	5,21
Jaguar	4	5	2,18	2,53	22	32	3,71	3,26	187	1,88	1,84	3,72
Mandrake	4	5	2,13	2,39	24	47	4,35	3,71	219	2,25	6,96	9,21
Health	4	5	2,25	2,56	16	39	3,87	3,22	181	1,97	7,36	9,32
Language	4	5	2,17	2,64	18	44	4,24	3,30	169	1,78	4,32	6,10
Machine	4	5	2,12	2,29	25	45	4,71	3,61	235	2,35	2,87	5,22
Mickey Mouse	4	5	2,06	2,38	34	47	4,78	3,52	240	2,60	4,57	7,17
Olympic Games	4	5	2,14	2,51	29	41	3,98	3,58	228	2,50	9,26	11,77
Steven Spielberg	4	5	2,06	2,69	31	36	5,00	3,26	211	2,24	4,39	6,63

Fig. 2. Structure of the cluster hierarchy (WW= WhatsOnWeb, VS=ViviSimo) and on the CPU time of WhatsonWeb

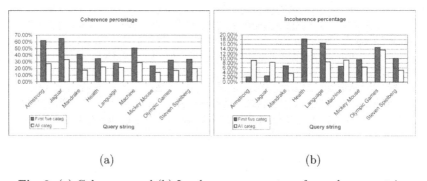

|(a)|(b)|

Fig. 3. (a) Coherence and (b) Incoherence percentage for each query string

Figure 3 shows some results about the coherence and the incoherence percentages of the hierarchies computed by WhatsOnWeb. Namely, WhatsOnWeb ranks the category in the computed hierarchy using a *cluster relevance score* that combines the Google rank for the documents of a cluster with the number of documents in the cluster itself. The user is typically interested in the top categories of the hierarchy. Thus we measured the cluster coherence/incoherence of the first five categories that WhatsOnWeb presents to the user. Also, we measured the cluster coherence/incoherence of the whole hierarchy. The experiments put in evidence two main aspects: (i) The first five categories usually have a cluster coherence that is significantly greater than for the rest of the categories (it rises up to 60% in the average), especially for the ambiguous queries. This result is positive if we consider that the union of the first five categories typically groups more than 50% of the documents. (ii) The cluster incoherence is lower than 10% for most of the instances, although for some queries it increases up to 16-18%.

References

1. Y. Asano, H. Imai, M. Toyoda, and M. Kitsuregawa. Finding neighbor communities in the web using inter-site graph. In *DEXA'03*, pages 558–568, 2003.
2. M. Brinkmeier. Communities in graphs. In *IICS'03*, volume 2877 of *LNCS*, pages 20–35, 2003.
3. G. Di Battista, W. Didimo, M. Patrignani, and M. Pizzonia. Orthogonal and quasi-upward drawings with vertices of prescribed sizes. In *GD '99*, volume 1731 of *LNCS.*, pages 297–310, 1999.
4. E. Di Giacomo, W. Didimo, L. Grilli, and G. Liotta. A topology-driven approach to the design of web meta-search clustering engines. Technical Report RT-004-04, Dip. Ing. Elettr. e dell'Informaz., Univ. Perugia, 2004.
5. Q. Feng, R. F. Choen, and P. Eades. How to draw a planar clustered graph. In *COCOON'95*, volume 959 of *LNCS*, pages 21–31, 1995.
6. P. Ferragina and A. Gulli. The anatomy of a clustering engine for web-page snippet. In *ICDM'04*, 2004.
7. G. Flake, S. Lawrence, and C. L. Giles. Efficient identification of web communities. In *ACM SIGKDD*, pages 150–160, 2000.
8. B. Fung, K. Wang, and M. Ester. Large hierarchical document clustering using frequent itemsets. In *SDM*, 2003.
9. G. Greco, S. Greco, and E. Zumpano. Web communities: Models and algorithms. *World Wide Web*, 7:58–82, 2004.
10. E. Hartuv and R. Shamir. A clustering algorithm based on graph connectivity. *Information Processing Letters*, 76:175–181, 2000.
11. N. Imafuji and M. Kitsuregawa. Finding a web community by maximum flow algorithm with hits score based capacity. In *DASFAA'03*, pages 101–106, 2003.
12. D. W. Matula. k-components, clusters, and slicing in graphs. *SIAM J. Appl. Math.*, 22(3):459–480, 1972.
13. M. E. J. Newman and M. Girvan. Finding and evaluating community structure in networks. *Phys. Rev. E 69*, 2004.
14. S. Osinski, J. Stefanowski, and D. Weiss. Lingo: Search results clustering algorithm based on singular value decomposition. *IIS 2004*, pages 359–368, 2004.
15. R. Tamassia. On embedding a graph in the grid with the minimum number of bends. *SIAM J. on Comput.*, 16(3):421–444, 1987.
16. Y. Wang and M. Kitsuregawa. Link based clustering of web search results. In *WAIM 2001*, volume 2118 of *LNCS*, pages 225–236, 2001.
17. D. Weiss and J. Stefanowski. Web search results clustering in polish: Experimental evaluation of carrot. In *IIS 2003*, 2003.

Computing Upward Planar Drawings Using Switch-Regularity Heuristics⋆

Walter Didimo

Dipartimento di Ingegneria Elettronica e dell'Informazione,
Università degli Studi di Perugia, Via G. Duranti,
06125 Perugia, Italy
Università di Perugia
didimo@diei.unipg.it

Abstract. Let G be an upward planar embedded digraph. The classical approach used to compute an upward drawing of G consists of two steps: (i) A planar st-digraph including G is constructed adding a suitable set of dummy edges; (ii) A polyline drawing of the st-digraph is computed using standard techniques, and dummy edges are then removed. For computational reasons, the number of dummy edges added in the first step should be kept as small as possible. However, as far as we know, there is only one algorithm known in the literature to compute an st-digraph including an upward planar embedded digraph. In this paper we describe an alternative heuristic, which is based on the concept of *switch-regularity* introduced by Di Battista and Liotta (1998). We experimentally prove that the new heuristic significantly reduces the number of dummy edges added to determine the including st-digraph. For digraphs with low density, such a reduction has a positive impact on the quality of the final drawing and on the overall running time required by the drawing process.

1 Introduction

The *upward drawing* convention is commonly used to display acyclic digraphs representing hierarchical structures, like for example PERT diagrams and class inheritance diagrams. In an upward drawing each vertex is represented as a point of the plane and all edges are drawn as curves monotonically increasing in a common direction (for example the vertical one). An *upward planar drawing* is an upward drawing with no edge crossing (see, e.g. Figure 1(c)). It is known that not all planar digraphs admit an upward planar drawing, and the upward planarity testing problem is in general NP-complete [1]. Bertolazzi et. al [2] proved that the upward planarity testing problem can be solved in polynomial time when the planar embedding of the digraph is assigned. Namely, given a planar embedded digraph G, they introduce the concept of *upward planar embedding* of G, which is a labeled embedding that specifies the type of angles at source- and sink-vertices

---

⋆ This work is partially supported by the MIUR Project ALINWEB.

M. Bieliková et al. (Eds.): SOFSEM 2005, LNCS 3381, pp. 117–126, 2005.

inside each face of G; the label of an angle informs if that angle will be "large" (greater than π) or "small" (less than π) in the final drawing. The authors prove that an upward planar drawing of G exists if and only if there exists an upward planar embedding of G with certain properties; such an upward embedding can be computed in polynomial time if it exists. An algorithm is described in [2] that computes a drawing of an upward planar embedded digraph G in two steps: **Step 1**: Construct an including planar st-digraph of G adding a suitable set of dummy edges; **Step 2**: Compute a polyline drawing of the st-digraph using standard techniques (see e.g. [3, 4]), and then remove dummy edges.

It is important that the number of dummy edges added in **Step 1** be kept as small as possible. Indeed, dummy edges restrict the choices that can be performed in **Step 2** to determine a good layout of the digraph, and they also influence the running time of **Step 2**. However, as far as we know, the augmentation technique described in [2] is the only one existing in the literature.

In this paper we investigate a new approach for augmenting an upward planar embedded digraph to an st-digraph. The main results of this paper are listed below:

- We provide a new characterization for the class of *switch-regular* upward embeddings (Section 2). This class has been introduced by Di Battista and Liotta in [5]. Our characterization is similar to the one given in [6] for defining turn-regular orthogonal representations.
- We use the above characterization to design a new polynomial-time algorithm that computes an st-digraph including an upward planar embedded digraph (Section 3).
- We experimentally prove that the new algorithm significantly reduces the number of dummy edges added to determine the including st-digraph when compared with the technique in [2]. This reduction dramatically reduces the total edge length of the computed drawings for low-density digraphs and has a positive impact on the overall running time of the drawing process (Section 4).

2   Switch-Regular Upward Embeddings

2.1   Preliminaries

In this paper we concentrate on planar digraphs with a given planar embedding. We use a notation that is slightly revised with respect to the one in [2, 5]. Let G be an embedded planar digraph. An *upward planar drawing* Γ of G is a planar drawing of G such that: (i) Γ preserves the embedding of G. (ii) All the edges of G are drawn as curves monotonically increasing in the vertical direction. If G admits an upward planar drawing, it is called an *upward planar digraph*.

A vertex of G is *bimodal* if its incident list can be partitioned into two (possibly empty) lists, one consisting of incoming edges and the other consisting of outgoing edges. If all vertices of G are bimodal then G and its embedding are called *bimodal*.

Let f be a face of an embedded planar bimodal digraph G. Visit the boundary of f counterclockwise. Let $s = (e_1, v, e_2)$ be a triplet such that v is a vertex of the boundary of f and e_1, e_2 are incident edges of v that are consecutive on the boundary of f. Triplet s is called a *switch* of f if the direction of e_1 is opposite to the direction of e_2 (note that e_1 and e_2 may coincide if G is not biconnected). If e_1 and e_2 are both incoming, then s is a *sink-switch* of f. If they are both outgoing, then s is a *source-switch* of f. Let $2n_f$ be the number of switches of f. The *capacity* of f is defined as $c_f = n_f - 1$ if f is an internal face, and $c_f = n_f + 1$ if f is the external face.

An assignment of the sources and sinks of G to its faces such that the following properties hold is *upward consistent*: (i) A source (sink) is assigned to exactly one of its incident faces. (ii) For each face f, the number of sources and sinks assigned to f is equal to c_f.

The following theorem gives a characterization of the class of upward planar embedded digraphs.

Theorem 1. [2] *Let G be an embedded planar bimodal digraph. G is upward planar if and only if it admits an upward consistent assignment.*

If G has an upward consistent assignment then the upward planar embedding of G corresponding to that assignment is a labeled planar embedding of G, called *upward planar embedding* of G, that has the following properties: (i) For each face f the switches of f are labeled S or L. (ii) A source-switch or a sink-switch of f is labeled L if it is assigned to f, otherwise it is labeled S.

If f is a face of an upward planar embedding, the circular list of labels of f is denoted by σ_f. Also, S_{σ_f} and L_{σ_f} denote the number of S and L labels of f, respectively.

Property 1. [2] If f is a face of an upward planar embedding then $S_{\sigma_f} = L_{\sigma_f} + 2$ if f is internal, and $S_{\sigma_f} = L_{\sigma_f} - 2$ if f is external.

Given an upward planar embedding of a graph G, it is possible to construct an upward planar drawing of G such that every angle at a source- or a sink-switch of f is greater than π when the switch is labeled L and it is less than π when the switch is labeled S. Figures 1(a) and 1(c) show an upward planar embedded digraph and a corresponding upward planar drawing.

An internal face f of an upward planar embedding is *switch-regular* if σ_f does not contain two distinct maximal subsequences σ_1 and σ_2 of S-labels such that $S_{\sigma_1} > 1$ and $S_{\sigma_2} > 1$. The external face f is *switch-regular* if σ_f does not contain two consecutive S labels. An upward planar embedding is *switch-regular* if all its faces are switch-regular. For example, the upward planar embedding of Figure 1(a) is not switch-regular, since face f is not switch-regular. All the other faces are switch-regular.

As shown in [2], given an upward planar embedding of a digraph G it is possible to construct a planar st-digraph including G by adding a new source s^*, a new sink t^*, the edge (s^*, t^*), and a suitable set of dummy edges, called *saturating edges*. More formally, the saturating edges are iteratively added applying the following rules.

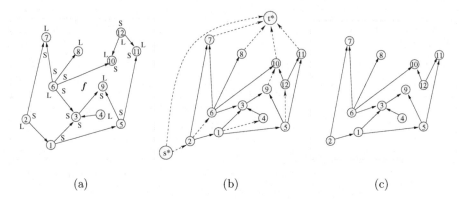

Fig. 1. (a) A bimodal digraph G with a given upward planar embedding. (b) A drawing of an st-digraph including G. The dashed edges represent a complete saturator of G. (c) An upward planar drawing of G corresponding to the upward embedding

- If $s = (e_1, v, e_2)$ and $s' = (e_1', v', e_2')$ are two source-switches of a face f such that s is labeled S and s' is labeled L (see Figure 2(a)), then we can add a saturating edge $e = (v, v')$ splitting f into two faces f' and f''. Face f' contains the new source-switch (e_1, v, e) labeled S, and f'' contains the new source-switch (e_2, v, e) labeled S. Also, v' does not belong to any switch labeled L in f' and f''. We say that s *saturates* s' and the new embedding is still upward planar.
- If $s = (e_1, v, e_2)$ and $s' = (e_1', v', e_2')$ are two sink-switches of a face f such that s is labeled L and s' is labeled S (see Figure 2(b)), then we can add a saturating edge $e = (v, v')$ splitting f into two faces f' and f''. Face f' contains the new sink-switch (e, v', e_2') labeled S, and f'' contains the new sink-switch (e_1', v', e) with labeled S. Also, v does not belong to any switch labeled L in f' and f''. We say that v' *saturates* v and the new embedding is still upward planar.
- Once all faces have been decomposed by using the two rules above, we can add saturation edges that either connect a sink-switch labeled L of the external face to t^*, or s^* to a source-switch labeled L of the external face. After the insertion of these edges the embedding is still upward planar.

A *saturator* of a face f of G is a set of saturating edges that connect vertices of f. Such a saturator is *complete* if no more saturating edges can be added to decompose f. A maximal set of saturating edges for all faces of G is called a *complete saturator* of G. Figure 1(b) shows a complete saturator of the upward planar embedded digraph in Figure 1(a).

Theorem 2. [5] *Let f be a face of an upward planar embedding of a digraph G. Face f has a unique complete saturator if and only if f is switch-regular.*

Theorem 2 implies that an upward planar embedded digraph has a unique complete saturator if and only if it is switch-regular.

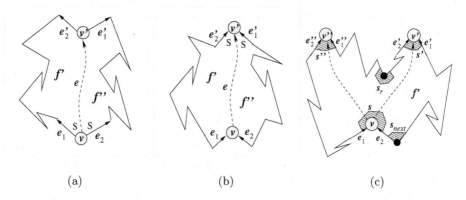

(a)   (b)   (c)

Fig. 2. (a) A saturating edge between two source-switches. (b) A saturating edge between two sink-switches. (c) Illustration of the proof of Theorem 3

2.2 Characterizing Switch-Regular Upward Embeddings

In this section we give a new characterization of switch-regular upward embeddings; it is strongly related to the definition of turn-regular orthogonal representation given in [6].

Let G be an embedded bimodal planar digraph with a given upward planar embedding. Let f be a face of G. A *reflex switch* of f is a switch of f that has label L. A *convex switch* of f is a switch of f that has label S. Denote by Σ_f the circular list of switches of f while visiting the boundary of f counterclockwise (clearly, $|\Sigma_f| = |\sigma_f|$). For any switch $s \in \Sigma_f$, we define $turn(s) = -1$ if s is reflex, and $turn(s) = 1$ if s is convex.

Let $s' = (e_1', v, e_2')$ and $s'' = (e_1'', v'', e_2'')$ be two switches in Σ_f. Denote by $\Sigma_f(s', s'')$ the subsequence of Σ_f from s' (included) to s'' (excluded). We define the following function:

$$rotation_f(s', s'') = \sum_{s \in \Sigma_f(s', s'')} turn(s).$$

In particular, by Property 1, $rotation_f(s, s) = +2$ for any switch s of an internal face f. If f is external, then $rotation_f(s, s) = -2$. Also $rotation_f(s', s'') = rotation_f(s', s) + rotation_f(s, s'')$, for each ordered sequence $s', s, s'' \in \Sigma_f$. Let $\{s', s''\}$ be an ordered pair of reflex switches of f. We say that $\{s', s''\}$ is a pair of *kitty corners* of f if one of the following holds:

- $rotation_f(s', s'') = +1$.
- $rotation_f(s', s'') = -3$ and f is external.

Note that by Property 1, if $\{s', s''\}$ is a pair of kitty corners of a face f (internal or external), then $\{s'', s'\}$ is a pair of kitty corners of f, too. Indeed, if f is internal and $rotation_f(s', s'') = +1$, then $rotation_f(s'', s') = +1$. Also if f is external and $rotation_f(s', s'') = +1$, then $rotation_f(s'', s') = -3$. In the upward planar embedding of Figure 1(a), denoted by $s' = ((3,9), 9, (5,9))$ and $s'' = ((12,11), 12, (12,10))$ in face f, we have $rotation_f(s', s'') = +1$, and therefore s' and s'' are kitty corners of f. The following theorem summarizes the main result of this section.

Theorem 3. *A face of an upward planar embedding is switch-regular if and only if it has not kitty corners.*

Proof. We prove the statement for an internal face f. A similar proof works for the external face. Let f be a switch-regular face. Suppose by contradiction that f contains a pair $\{s', s''\}$ of kitty corners and consider the subsequence $\Sigma_f(s', s'')$. Since $rotation_f(s', s'') = +1$, then in $\Sigma_f(s', s'')$ the number of convex switches is equal to the number of reflex switches plus one. Therefore, since s' is a reflex switches, in $\Sigma_f(s', s'')$ there is necessarily two consecutive switches labeled S. Applying the same reasoning, there must be two consecutive labels S in the subsequence $\Sigma_f(s'', s')$. Since s' and s'' are labeled L, we have found two maximal subsequences of S-labels both of size greater than one. Therefore, f is not switch-regular, a contradiction.

Conversely, let f be a face that does not contain kitty corners. Suppose by contradiction that f is not switch-regular. By Theorem 2 f has not a unique saturator. This implies that in f there is a reflex switch $s = (e_1, v, e_2)$ that can be saturated by at least two distinct convex switches, say $s' = (e_1', v', e_2')$ and $s'' = (e_1'', v'', e_2'')$. Assume, without loss of generality, that s, s', and s'' are source-switches. Each of the two saturating edges (v, v') and (v, v'') would split f, keeping the embedding upward planar. Refer to the notation of Figure 2(c): Denote by s_{next} the switch that follows s in Σ_f, and let f' be the face to the right of the saturating edge (v, v'). Since the complete rotation of an internal face is always 2, we have that $rotation_{f'}(s_{next}, s') = 2 - rotation_{f'}(s', s_{next}) = +1$. Also, since s is a reflex switch of f, we have that $rotation_f(s, s') = 0$. By a similar reasoning applied on the face to the right of the saturating edge (v, v''), we have that $rotation_f(s, s'') = 0$, and hence $rotation_f(s', s'') = 0$. This implies that in the subsequence $\Sigma_f(s', s'')$ there exists at least one reflex switch s_r such that $rotation_f(s, s_r) = +1$ (indeed s' is labeled S, and the number of L labels in $\Sigma_f(s', s'')$ is equal to the number of S labels). Therefore, $\{s, s_r\}$ is a pair of kitty corners, a contradiction.

3   A Switch-Regularity Heuristic

Let G be a planar bimodal digraph with a given upward planar embedding. A linear-time algorithm is described in [2] that constructs an st-digraph including G by computing a complete saturator of G. This algorithm recursively decomposes every face f, searching on the boundary of f subsequences of three consecutive switches s, s', s'' such that both s and s' are labeled S, while s'' is labeled L. Each time such a sequence is found, the algorithm splits f into two faces by adding a saturating edge connecting the vertices of s and s''. When there are no more sequences of labels SSL in a face, then the algorithm suitably connects the reflex switches of the external face with s^* and t^*. In the following we call SimpleSat the algorithm in [2].

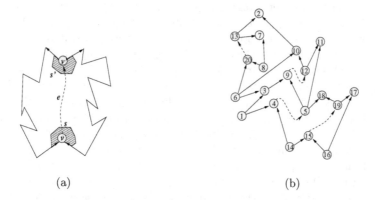

(a)   (b)

Fig. 3. (a) A face having kitty corners s, s' is split by edge $e = (v, v')$. (b) An upward planar embedding augmented to a switch-regular one; dummy edges are dashed

We use the characterization of Theorem 3 to design a new heuristic for computing a planar st-digraph including G. For each face f, we test if f is switch-regular on not. If f is not switch-regular, we apply an $O(deg(f))$ procedure that finds a pair $\{s, s'\}$ of kitty corners of f, where $deg(f)$ is the degree of f. Such a procedure uses the same technique described in [6] for detecting a pair of kitty corners in a face of an orthogonal representation. Once $\{s, s'\}$ is detected, we split f by adding to G a dummy edge connecting the vertices of s and s' (see Figure 3(a)). Observe that, after the insertion of this edge, the new embedding is still an upward planar embedding. We recursively decompose all faces that are not-switch regular by applying the above algorithm, so to make the upward planar embedding switch-regular. In Figure 3(b) it is shown an upward planar embedding that is augmented to a switch-regular one, by using the above strategy. When the upward planar embedding becomes switch-regular, we apply the SimpleSat algorithm so to add the edges that are still needed to construct an st-digraph including G. Note that, since we apply algorithm SimpleSat on a switch-regular upward embedding, the complete saturator determined by this algorithm is uniquely defined. We call our heuristic SrSat. From the time complexity view point, SrSat takes $O(n^2)$ time in the worst case, since there may be an $O(n)$ number of kitty corners, and the detection of each pair requires $O(n)$ time. In practice however, since SrSat adds less edges than SimpleSat, the overall running time of the drawing algorithm is reduced (see Section 4).

4   Experimental Results

We implemented and experimented heuristics SimpleSat and SrSat on a large test suite of graphs, in order to compare their performances. The test suite is

composed by two subsets, Small-graphs and Large-graphs, of upward planar embedded digraphs. Subset Small-graphs consists of 800 graphs having
number of vertices in $\{10, 20, \ldots, 100\}$ and density (edge-to-vertex ratio) in
$\{1.2, 1.4, 1.6, 1.8\}$. Subset Large-graphs consists of 120 graphs having number
of vertices in $\{500, 600, \ldots, 1500, 2000\}$ and density ranging from 1.2 to 1.3.

All graphs have been randomly generated. Since we were interested in upward
planar embedded digraphs, the design of the generation algorithm was a difficult
task. We used two different approaches to generate the graphs for the two subsets.
Namely, each graph in subset Small-graphs was generated by the following
procedure: (i) Generate a connected (possibly non-planar) graph with a uniform
probability distribution; (ii) Compute a spanning tree of this graph by randomly
choosing the root vertex; (iii) Randomly add edges to the spanning tree until
the desired density value is reached: Each time we choose a new edge, this edges
is added only if the planarity of the graph is not violated otherwise the edge is
discarded; (iv) Once a planar embedded graph has been generated, an upward
orientation is assigned to the graph by applying the network-flow algorithm
described in [7].

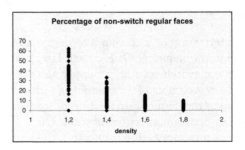

Fig. 4. Percentage of non-switch regular faces with respect to the graph density (edge-
to-vertex ratio)

The above algorithm requires the application of a planarity testing algorithm
each time a new edge is chosen for possible insertion. Therefore, it does not allow
the generation of large graphs in a reasonable time. In order to generate graphs
for the subset Large-graphs we used a different and faster algorithm. First we
generate a planar embedded biconnected graph by the technique described in [8],
and then we still assign an orientation to the graph by the algorithm in [7]. Thus,
the graphs in this subset are biconnected.

All experiments were performed on a PC Pentium M, 1.6 Ghz, 512 MB
RAM, and Linux OS. We first measured the percentage of non-switch regular
faces of the input graphs. As already found in other similar experiments on
orthogonal representations [6], this percentage decreases exponentially with the
increasing of the graph density (see Figure 4). For graphs of density 1.2 we
have up to 60% of faces that are not switch-regular. Indeed, the performances of
algorithms SimpleSat and SrSat differ especially on low density graphs, while
the two algorithms perform quite similarly on high density graphs. For the above

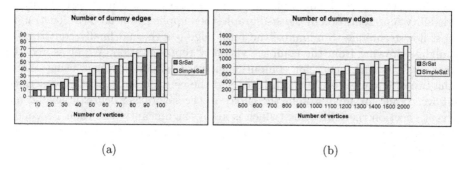

Fig. 5. Number of dummy edges added by the two heuristics with respect to the number of vertices: (a) Small graphs (10-100 vertices). (b) Large graphs (500-2000 vertices)

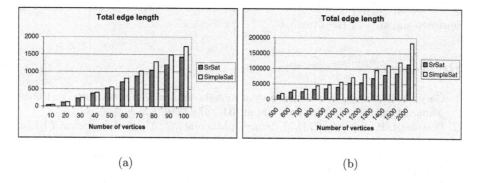

Fig. 6. Total edge length of the drawing with respect to the number of vertices: (a) Small graphs (10-100 vertices). (b) Large graphs (500-2000 vertices).

motivation, in the following we only discuss the results for graphs with low density.

Figure 5 depicts the number of dummy edges added by the two algorithms. Notice that, both for small and large graphs algorithm SrSat adds in the average the 16% of edges less than algorithm SimpleSat.

The reduction of the number of dummy edges added by SrSat has a positive impact on the drawing readability. Namely, while the area of the drawings computed by SrSat and SimpleSat is comparable in the average (we omit the charts on the area for space reasons), applying SrSat instead of SimpleSat dramatically reduces the total edge length of the final drawings (see Figure 6); for large graphs the improvement is about 25% in the average, and it increases up to 36% for graphs with 2000 vertices.

Finally, although the asintotically cost of SrSat is $O(n^2)$, this heuristic causes a reduction of the overall running time in practice, due to the "small" number of dummy edges added with respect to SimpleSat. In fact, on the st-digraphs computed by the two heuristics we applied the same algorithm to determine a polyline drawing of the digraph (see. e.g. [3]). This algorithm first computes a visibility representation of the st-digraph, then applies on it an $O(n^2 \log n)$ min-cost-flow technique to minimize the total edge length, and finally constructs a polyline drawing from the compact visibility representation. We measured the overall CPU time spent for computing upward planar drawings using one of the two different saturating heuristics and the compaction algorithm described above. While for small graphs the choice of the saturating heuristic has no relevant effect on the CPU time (which is always significantly less than 1 second), applying SrSat against SimpleSat on large graphs reduces the overall CPU time of about 8% in the average.

5   Open Problems

One of the open problems that naturally rises from the results of this paper is the design of novel saturating heuristics based on switch-regularity that also improve the area of the drawings.

References

1. Garg, A., Tamassia, R.: On the computational complexity of upward and rectilinear planarity testing. SIAM J. on Comput. **31** (2001) 601–625
2. Bertolazzi, P., Battista, G.D., Liotta, G., Mannino, C.: Upward drawings of triconnected digraphs. Algorithmica **6** (1994) 476–497
3. Di Battista, G., Eades, P., Tamassia, R., Tollis, I.G.: Graph Drawing. Prentice Hall, Upper Saddle River, NJ (1999)
4. Kaufmann, M., Wagner, D.: Drawing Graphs. Springer Verlag (2001)
5. Di Battista, G., Liotta, G.: Upward planarity checking: "faces are more than polygon". In: Graph Drawing (Proc. GD '98). Volume 1547 of Lecture Notes Comput. Sci. (1998) 72–86
6. Bridgeman, S., Di Battista, G., Didimo, W., Liotta, G., Tamassia, R., Vismara, L.: Turn-regularity and optimal area drawings of orthogonal representations. Computational Geometry: Theory and Applications **16** (2000) 53–93
7. Didimo, W., Pizzonia, M.: Upward embeddings and orientations of undirected planar graphs. Journal of Graph Algorithms and Applications **7** (2003) 221–241
8. Bertolazzi, P., Di Battista, G., Didimo, W.: Computing orthogonal drawings with the minimum number of bends. IEEE Trans. on Computers **49** (2000) 826–840

Serial and Parallel Multilevel Graph Partitioning Using Fixed Centers

Kayhan Erciyeş[1,2], Ali Alp[3], and Geoffrey Marshall[1]

[1] California State University San Marcos
333 S.Twin Oaks Valley Rd., San Marcos, CA 92096, U.S.A.
kerciyes@csusm.edu, marsh021@csusm.edu
[2] İzmir Institute of Technology, Urla, İzmir, TR-35340, Turkey
[3] Ege University International Computer Institute,
Bornova, İzmir, TR-35100, Turkey
alpali@bornova.ege.edu.tr

Abstract. We present new serial and parallel algorithms for multilevel graph partitioning. Our algorithm has coarsening, partitioning and uncoarsening phases like other multilevel partitioning methods. However, we choose fixed nodes which are at least a specified distance away from each other and coarsen them with their neighbor nodes in the coarsening phase using various heuristics. Using this algorithm, it is possible to obtain theoretically and experimentally much more balanced partitions with substantially decreased total edge costs between the partitions than other algorithms. We also developed a parallel method for the fixed centered partitioning algorithm. It is shown that parallel fixed centered partitioning obtains significant speedups compared to the serial case. ...

1 Introduction

The aim of a graph partitioning algorithm is to provide partitions such that the number of vertices in each partition is averaged and the number of edge-cuts between the partitions is minimum with a total minimum cost. Graph partitioning finds applications in many areas including parallel scientific computing, task scheduling, VLSI design ana operation research. One important area of research is on searching algorithms that find good partitions of irregular graphs to map computational meshes to the high performance parallel computer processors for load balancing such that amount of computation for each processor is roughly equal with minimum communication among them. Solution of sparse linear systems where the graph representing the coefficient matrix is partitioned for load balancing among processors is one area that research is directed [1][2]. Recently, graph partitioning algorithms are used in mobile ad-hoc networks to form clusters for dynamic routing purposes [3][4]. In *multi-constraint approach* [5], each vertex is assigned a vector of k weights that represent the work associated with that vertex in each of the k computational phases. The aim is to partition the graph so that each of the k weights is balanced as well as the sum of edge weights are minimized. In the related *multi-objective model* [6], the partition tries to minimize several cost functions at the same time. Each edge is given a vector of j weights where different cost functions are an element of this vector. The

M. Bieliková et al. (Eds.): SOFSEM 2005, LNCS 3381, pp. 127–136, 2005.

partitioning then tries to balance vertex weights by minimizing the cost functions. In *skewed partitioning* [7], each vertex can have k preference values showing its tendency to be in each one of k sets which is taken into consideration by the partitioning algorithm. Partition refinement is an important step to improve partition quality. Kernighan and Lin (KL) [8] provided a greedy method to swap a subset of vertices among partitions to reduce edge connectivity further. During each step of KL algorithm, a pair of vertices, one from each partition are selected and swapped to give a reduced total edge weight among the partitions if possible. *Multilevel graph partitioning* is a comparatively new paradigm for graph partitioning and consists of coarsening, initial partitioning and refinement steps [9][10]. In the coarsening phase, a set of vertices is selected and are collapsed to form a coarser graph. This step is performed sufficient times to get a simple graph which can be divided into the required partitions by a suitable algorithm. The obtained partitions are projected back by uncoarsening and refinement by algorithms such as KL along this process. Chaco[11], METIS[12] and SCOTCH[13] are popular multilevel partitioning tools used in diverse fields.

In this study, we propose new serial and parallel multilevel algorithms for graph partitioning. We compare these methods with the multilevel graph partitioning method of [10]. In Section 2, background including the multilevel graph partitioning method is reviewed. Section 3 describes our serial graph partitioning algorithm which is called *Fixed Centered Partitioning* (FCP). Parallel implementation of the FCP algorithm on a cluster of workstations is given in Section 4. Finally, experimental results obtained so far are presented and comparison of the algorithm with other algorithms are outlined in Section 5 and the conclusions are given in Section 6.

2 Background

The multilevel graph partitioning model has proven to be very robust, general and effective. The idea is simply approximation of a large graph by a sequence of smaller and smaller graphs after which the smallest graph can be partitioned into p partitions by a suitable algorithm. This partition is then brought back to the original graph by refinements. Consider a weighted graph $G_0=(V_0,E_0)$ with weights both on vertices and edges. A multilevel graph partitioning algorithm consists of the following phases.

Coarsening Phase : In the coarsening phase, the graph G_0 is transformed into a sequence of smaller graphs G_1, G_2 ..... , G_m such that $|V_0| > |V_1| > |V_2| > \cdots > |V_m|$. In most coarsening schemes, a set of vertices of G_i is combined to form a single vertex of the next level coarser graph G_{i+1}. Let $V=\{v_i\}$ be the set of vertices of G_i combined to form vertex v of G_{i+1}. The weight of vertex v is set equal to the sum of the weights of the vertices in $V=\{v_i\}$. Also, in order to preserve the connectivity information in the coarser graph, the edges of v are the union of the edges of the vertices in v_i. The *maximal matching* of the graph is a set of edges such that there is not a pair adjacent on the same vertex. Various approaches to find maximal matching exist. The *heavy edge matching* (HEM) computes the matching M_i such that the weight of the edges in M_i is high. The vertices are visited in random order, but the collapsing is performed with the vertex that has the heaviest weight edge with the chosen vertex. In *Random Matching* (RM), vertices are visited in random order and an adjacent vertex is chosen in random order as well [10]. During the successive coarsening phases, the weight of vertices and edges increase.

Partitioning Phase : The second phase of the multilevel algorithm computes a high-quality partition P_m of the coarse graph $G_m=(V_m, E_m)$ by a suitable algorithm such that each partition contains roughly equal vertex weights and the sum of the weights of the edge cuts between the partitions is minimum.

Uncoarsening/Refinement Phase : During this phase, the partition P_m of the coarser graph G_m is projected back to the original graph, by going through $G_{m-1}, G_{m-2}, ..., G_1$. Since each vertex of G_{i+1} contains a distinct subset of vertices of G_{m-1}, obtaining P_i from P_{i+1} is done by simply assigning the set of vertices collapsed in G_{i+1} to the partition $P_{i+1}[v]$. Algorithms such as KL are usually used to improve partition quality.

3   Serial Fixed Centered Partitioning

The method we propose has coarsening, partitioning and uncoarsening phases as in the other multilevel partitioning methods. We however choose fixed initial nodes called *centers* and collapse the vertices around these centers which must have at least a fixed distance to the other selected center nodes. The FCP algorithm is described in Section 3.1, the formal analysis of the algorithm is stated in Section 3.2. and an example partition is given in Section 3.3.

3.1   Serial FCP Algorithm

The Serial FCP algorithm can be formally described as in Fig.1. Inputs to the algorithm are the initial graph G_0, number of partitions required and the two heuristics, HC and HM. Since we have fixed centers that do not change as the graph gets coarsened, a way to allocate these centers initially is needed. The first approach we employed is to run *Breadth-First-Search* (BFS) algorithm for all the nodes in the graph and find p center nodes which have the maximum distance between them. BFS, however, is time consuming as it has a runtime of $O(n^3)$. Secondly, we may choose the centers randomly with the constraint that each center has at least some predetermined distance among them. The third approach chooses the centers randomly with no constraints. The minimum distance heuristic h_1 between any two centers may be associated to the diameter value of the graph and the number of partitions by $h_1 = 2d/p$ where d is the diameter of the graph and p is the number of partitions required. The possible heuristics used to locate the centers initially could be summarized as follows:

- HC_1 : Apply Breadth-First-Search (BFS) to G_0 and find p centers that are $2d/p$ distance from each other
- HC_2 : Choose centers randomly with the condition that they are at least $2d/p$ distance from each other
- HC_3 : Choose the centers at random with no constraints

Once the centers are chosen, FCP proceeds by collapsing a neighbor vertex at each iteration to the fixed centers as shown in Fig.1 using a second heuristic, HM. Two possible values for HM are the *Heaviest Edge Neighbor* (HEN) or *Random Neighbor* (RN). Based on the heuristic chosen, the *Collapse* function performs the collapse operation of a marked neighbor node with the center which simply merges the marked vertex to the center by adding its weight to the center, removing the edge between them and inserting any previously coexisting edges between them by adding the weights of the edges and representing them as a single edge with this weight.

Procedure *Serial_FCP*
        Input : G_0 : initial graph
                p : number of partitions
                HC : heuristic to allocate initial centers
                HM : heuristic to mark neighbor nodes
    1.   *Locate_centers*(G_0, HC);
    2.   for i=1 to $\lfloor n/p \rfloor$ do
    3.   for each center c do
    4.   *Collapse*(G_i, c, HM);

Fig. 1. Serial FCP Algorithm

3.2   Analysis of Serial FCP

The time complexity and the quality of the partitions of the Serial FCP can be stated in the following theorems:

Theorem 1: FCP performs partitioning of $G(V,E)$ in $O(\lfloor n/p \rfloor)$ steps where $|V| = n$ and p is the number of partitions required. The time complexity of the total collapsing of FCP is $O(n)$.

Proof : FCP simply collapses p nodes with its heaviest edges at each step resulting in $\lfloor n/p \rfloor$ steps. Since there are p collapsing at each step, total time complexity is $O(n)$.

Corollary 1: FCP performs partitioning of $G(V,E)$ such that the final partitions have $O(\lfloor n/p \rfloor + 1)$ vertices.

Proof: The FCP collapses one node to each partition and the total number of steps is $O(\lfloor n/p \rfloor)$ by Theorem 1. In the last step, there will be $O(p \bmod n)$ nodes to collapse which means that the final partitions will have a maximum of $n/p+1$ nodes.

3.3   FCP Examples

Let us illustrate RM, HEM and FCP by the example shown in Fig. 1 where (a) is RM, (b) is HEM, (c) is FCP and heavy lines are used to show matchings. The initial graphs and outputs of RM and HEM are reproduced from [14]. The output graphs are formed after $\Theta(5)$ collapses for RM and HEM but $\Theta(6)$ for FCP after two steps. For this particular example, we see that FCP performs much better with a total edge cost of 16 compared to RM (30) and HEM (24). We also get 3 vertices per partition with respect to 2 vertices in RM and HEM. If three partitions were required, we would have stopped for FCP but continue with matching for CM and HEM. Moreover, FCP does not have a matching phase, therefore it has much better runtimes than RM and HEM.

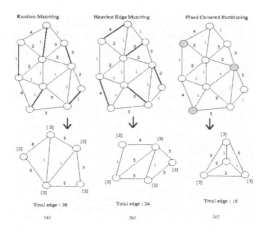

Fig. 2. Comparison of RM (a), HEM (b) and FCP (c) in an arbitrary network

4   Parallel Fixed Centered Partitioning

The proposed parallelization of FCP consists of three phases. In the first phase, the determination of the diameter of the network is done in parallel. The coordinatorsends the adjacency list of the graph to individual workers and each worker then estimates its local diameter of the graph by performing BFS on its local partition. The coordinator gathers all of the local diameters and estimates the total diameter. It then locates the centers based on this diameter and sends the identities of the centers to each processor.

Process *Parallel_FC_Coordinator*

Input : G_0 : initial graph
        p : number of partitions
        HC : heuristic to allocate initial centers
        HM : heuristic to mark neighbor nodes

  1.   /* Locate Centers */
  2.   *Send* adjacency list and their identities to slaves
  3.   *Receive* local diameters from all slaves
  4.   estimate diameter of the graph and determine center nodes
  5.   *Send* center nodes to the slaves.
  6.   /* Wait for local collapses */
  7.   *forall* workers
  8.   *Receive* the collapsed nodes from the worker
  9.   *while* there are nodes to be collapsed /* Check Overlaps */
10.   *Receive* node identities from slaves
11.   *Send* COLLAPSED or NOT_COLLAPSED to workers
12.   mark NOT_COLLAPSED nodes as collapsed

Fig. 3. Coordinator pseudocode for Parallel FCP

In the second phase, each processor collapses the graph around its designated center independently until a predetermined *h2* times such that no overlap would occur. The heuristic *h2* used is set as d/p^2 for the implementation. In the third phase, each processor attempts to collapse possibly overlapping regions with others. Therefore, every time a number of nodes are to be collapsed, acknowledgment from the coordinator is searched to check that these nodes have not been collapsed before. The coordinator and worker pseudocodes are shown in in Fig. 3 and Fig. 4.

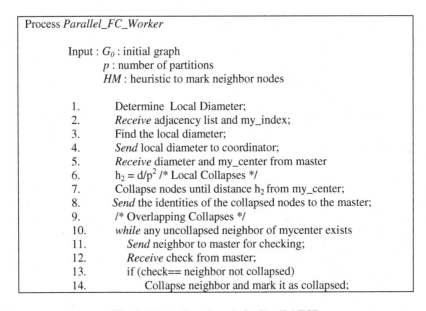

Process *Parallel_FC_Worker*

Input : G_0 : initial graph
         p : number of partitions
         HM : heuristic to mark neighbor nodes

1.   Determine Local Diameter;
2.   *Receive* adjacency list and my_index;
3.   Find the local diameter;
4.   *Send* local diameter to coordinator;
5.   *Receive* diameter and my_center from master
6. $h_2 = d/p^2$ /* Local Collapses */
7.   Collapse nodes until distance h_2 from my_center;
8.   *Send* the identities of the collapsed nodes to the master;
9.   /* Overlapping Collapses */
10.   *while* any uncollapsed neighbor of mycenter exists
11.   *Send* neighbor to master for checking;
12.   *Receive* check from master;
13.   if (check== neighbor not collapsed)
14.   Collapse neighbor and mark it as collapsed;

Fig. 4. Worker Pseudocode for Parallel FCP

5   Results

5.1   Results for Serial Centered Node Matching

We implemented the graph partitioning using HEM, RM and FCP (alternatively called Centered Matching - CM) for various randomly created matrix sizes (128*128, 256*256, 512*512, 1024*1024, 2048*2048). The graphs represented by the matrices are partitioned on Ultra 5 Sun Sparc servers which run Solaris7 as operating system and runtimes of partitioning algorithms are compared. Center nodes in FCP are found by two different heuristics as HC_1 and HC_2. by running the BFS algorithm for all nodes or randomly choosing and checking for a distance between as described in Section 3.1. As shown in Fig. 5, the first FCP (CM with random center) method is the fastest as expected since FCP does not have a matching phase. The second FCP method (CM with BFS) is the slowest because BFS is executed on all nodes of the graph to find center nodes. In Fig. 6, the total edge costs between the partitions are plotted for FCP (with HC2 and HC3) and HEM and RM. It may be seen that both FCP methods have a significant decreased total edge costs between the partitions.

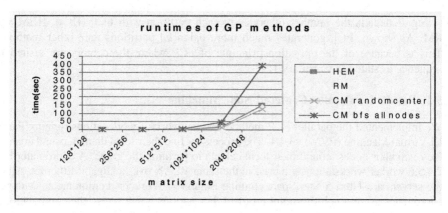

Fig. 5. Execution Times for theGraph partitioning methods

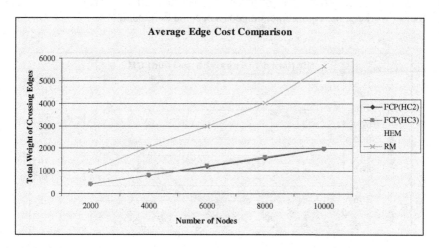

Fig. 6. Edge Cost Comparison of the Four Algorithms

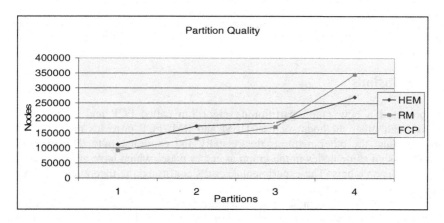

Fig. 7. Number of nodes in each part for all graph partitioning methods(2048*2048 matrix)

Fig. 7 depicts the number of nodes in each partition with FCP (HC$_2$), HEM and RM. As shown, FCP generates much more balanced partitions than other methods. This is because of the operation principle of FCP where the centers are visited in sequence as stated in Corollary 1.

5.2 Results for Parallel Centered Node Matching

We implemented the parallel FCP method on a network of workstations running Parallel Virtual Machine (PVM) v3.4.4. The processors find their local diameters and coarsen their neighbor nodes around their local centers to partition the graph. A coordinator and 2,4,5,6 worker workstations are used in the same VLAN over a Gigabit Ethernet. All of the servers are Ultra 5 Sun Sparc running Solaris 7. Workers communicate with the coordinator using a point-to-point protocol. For various graph sizes, the computational run times are recorded. Fig. 8 displays the results of the Parallel FCP for various number of processors ranging from 1 to 6. It may be seen that after a threshold value of the number of workers, the communication costs become dominant.

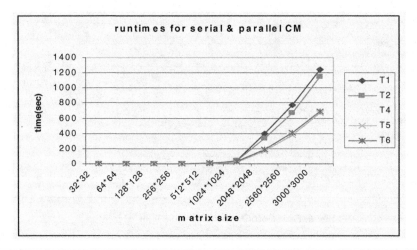

Fig. 8. Comparison of serial and parallel runtimes of FCP. T$_1$: serial runtime. T$_i$ represents the execution time with i workers

6   Conclusions

We proposed a new graph partitioning algorithm called FCP and compared this method with other methods such as HEM and RM. FCP provides more favorable partitions than the other methods theoretically and experimentally. However, FCP needs to assign some nodes in the graph as center nodes which have at least a certain distance to each others. The experimental results confirmed the theoretical FCP properties in terms of the runtime, total edge cost between the partitions and the partition quality. FCP using any heuristic performed much better than HEM and RM in terms of total edge cost and partition quality. For the runtime, FCP with HC2 and HC3 resulted in lower times than HEM and RM but FCP with HC1 proved to be the slowest as expected. FCP does not have a matching phase which results in a faster execution time and divides the graph into almost equal partitions as stated in Corollary 1. We also developed a parallel version of FCP. In this case, the diameter of the graph is estimated after each processor finds their local diameters and then the centers can be found by the coordinator. After finding centers where each of is assigned to different workers, each workstation continues collapsing the neighbor nodes of its center until there are no uncollapsed nodes left by getting acknowledgement from the coordinator after some predetermined value. The efficiency of the parallel algorithm rises for larger graphs until a threshold value where the communication costs start to dominate.

Our general conclusions can be summarized as follows. FCP provides improvement over the other graph matching algorithms such as RM and HEM in three aspects. Firstly, it does not have a matching phase, therefore it is faster. Secondly, it provides almost equal partitions with significantly lower total edge costs between the partitions than the other methods and thirdly it is suitable for parallel processing as each center collapses independently. One negative aspect of FCP is the initial marking of the fixed centers and random allocation could be a solution in this case. Also, the parallel algorithm employed requires a heuristic to be able to perform collapsing without any supervision initially. We are looking into more efficient ways of finding centers and performing FCP in parallel. Another interesting research direction would be modifying the FCP to perform multi-constraint, multi-objective graph partitioning. In this case, several cost functions would need to be minimized when choosing the vertices to collapse.

References

1. Hendrickson, B. and Kolda, T., G. : Partitioning Rectangular and Structurally Nonsymmetric Sparse Matrices for Parallel Processing, SIAM J. Sci. Comput. 21(6), (2000), 2048-2072.
2. Turhal, B., Solution of Sparse Linear Systems on a Cluster of Workstations Using Graph Partitioning Methods, Msc. Thesis, Ege University, Int. Computer Institute, (2001)
3. P. Krishna et al, A Cluster-based Approach for Routing in Dynamic Networks, ACM SIGCOMM Computer Communication Review, 27 (2), (1997), 49 – 64.
4. Erciyes, K., Marshall, G. : A Cluster based Hierarchical Routing Protocol for Mobile Networks, LNCS, Springer-Verlag, ICCSA(3), (2004), 528-537.
5. Yuanzhu P. C., Liestman, A., L. : A Zonal Algorithm for Clustering Ad Hoc Networks, Int. J. Foundations of Computer Science, 14(2), (2003), 305-322.
6. Schloegel, K, Karypis, G., Kumar, V. : A New algorithm for Multi-objective Graph Partitioning, Tech. Report 99-003, University of Minnesota, Dept. of CS, (1999).
7. Hendrickson, B., Leland, R., Driessche, R., V. : Skewed Graph Partitioning, Proc. of 8th SIAM Conf. Parallel Processing for Scientific Computing, SIAM (1997).
8. Kernighan, B., and Lin, S., An Effective Heuristic Procedure for Partitioning Graphs, The Bell System Technical Journal, (1970), 291-308.
9. Hendrickson, B., Kolda, T., G. : Graph Partitioning Models for Parallel Computing, Parallel Computing. 26, (2000), 1519-1534.
10. Karypis, G., Kumar, V. : A Fast and High Quality Multilevel Scheme for Partitioning Irregular Graphs, Tech. Report 95-035, University of Minnesota, Dept. of CS, (1995).

Two-Layer Planarization:
Improving on Parameterized Algorithmics

Henning Fernau

Universität Tübingen, WSI für Informatik, Sand 13,
72076 Tübingen, Germany
The University of Newcastle, School of Electr. Eng. and Computer Science,
University Drive, Callaghan, NSW 2308, Australia
`fernau@informatik.uni-tuebingen.de`

Abstract. A bipartite graph is *biplanar* if the vertices can be placed on two parallel lines in the plane such that there are no edge crossings when edges are drawn as straight-line segments. We study two problems:

- 2-LAYER PLANARIZATION: can k edges be deleted from a given graph G so that the remaining graph is biplanar?
- 1-LAYER PLANARIZATION: same question, but the order of the vertices on one layer is fixed.

Improving on earlier works of Dujmović *et al.* [4], we solve the 2-LAYER PLANARIZATION problem in $\mathcal{O}(k^2 \cdot 5.1926^k + |G|)$ time and the 1-LAYER PLANARIZATION problem in $\mathcal{O}(k^3 \cdot 2.5616^k + |G|^2)$ time. Moreover, we derive a small problem kernel for 1-LAYER PLANARIZATION.

1 Introduction

In a *2-layer drawing* of a bipartite graph $G = (A, B; E)$, the vertices in A are positioned on a line in the plane, which is parallel to another line containing the vertices in B, and the edges are drawn as straight line-segments. Such drawings have various applications, see [4]. A *biplanar graph* is a bipartite graph that admits a 2-layer drawing with no edge crossings; we call such a drawing a *biplanar drawing*. It has been argued that 2-layer drawings in which all the crossings occur in a few edges are more readable than drawings with fewer total crossings [7]—which gives the CROSSING MINIMIZATION problem(s) [5].

This naturally leads to the definition of the 2-LAYER PLANARIZATION problem (2-LP): given a graph G (not necessarily bipartite), and an integer k called *parameter*, can G be made biplanar by deleting at most k edges? Two-layer drawings are of fundamental importance in the "Sugiyama" approach to multi-layer graph drawing [10]. This method involves (repeatedly) solving the 1-LAYER PLANARIZATION problem (1-LP).

Fixed Parameter Tractability. We develop improved algorithms for 2-LP and for 1-LP that are exponential in the parameter k. This has the following justification: when the maximum number k of allowed edge deletions is small, an algorithm for 1- or 2-LP whose running time is exponential in k but polynomial in the size of the graph may be useful. We expect the parameter k to be small in

M. Bieliková et al. (Eds.): SOFSEM 2005, LNCS 3381, pp. 137–146, 2005.

practice. Instances of the 1- and 2-LP for dense graphs are of little interest from a practical point of view, as the resulting drawing will be unreadable anyway.

This analysis hence fits into the framework of parameterized algorithmics. A parameterized problem with input size n and parameter size k is *fixed parameter tractable*, or in the class $\mathcal{FPT}$, if there is an algorithm to solve the problem in $f(k) \cdot n^\alpha$ time, for some function f and constant α (independent of k).

Our Results. In this paper, we apply so-called kernelization and search tree methods to obtain algorithms for the 1- and 2-LP problems, this way improving earlier results [4] with exponential bases of 3 and 6, respectively. This leads to an $\mathcal{O}(k^2 \cdot 5.1926^k + |G|)$ time algorithm for 2-LP in a graph G. We present a similar second algorithm to solve the 1-LP problem in $\mathcal{O}(k^3 \cdot 2.5616^k + |G|^2)$ time. To this end, we draw connections to HITTING SET problems. The top-down analysis technique presented in [6] is applied to obtain the claimed running times in the analysis of the search tree algorithms. Detailed proofs can be found in http://eccc.uni-trier.de/eccc-reports/2004/TR04-078/index.html.

2   Preliminaries

In this section we introduce notation, recall a characterization of biplanar graphs and formalize the problem statements. All the mentioned results are from [4].

In this paper each graph $G = (V, E)$ is simple and undirected. The subgraph of G induced by a subset E' of edges is denoted by $G[E']$. A vertex with degree one is a *leaf*. If vw is the edge incident to a leaf w, then we say w is a *leaf at v* and vw is a *leaf-edge at v*. The *non-leaf degree* of a vertex v in graph G is the number of non-leaf edges at v in G, and is denoted by $\deg'_G(v)$.

A graph is a *caterpillar* if deleting all the leaves produces a (possibly empty) path. This path is the *spine* of the caterpillar. A *2-claw* is a graph consisting of one degree-3 vertex, the *center*, which is adjacent to three degree-2 vertices, each of which is adjacent to the center and one leaf. A graph consisting of a cycle and possibly some leaf-edges attached to the cycle is a *wreath*. Notice that a connected graph that does not have a vertex v with $\deg'(v) \geq 3$ is either a caterpillar or a wreath.

To prove their kernelization result for 2-LP, Dujmović *et al.* introduced the following potential function. For a graph $G = (V, E)$, define

$$\forall v \in V,\ \Phi_G(v) = \max\{\deg'_G(v) - 2, 0\},\ \text{and}\ \Phi(G) = \sum_{v \in V} \Phi_G(v)\ .$$

Lemma 1. $\Phi(G) = 0$ *if and only if G is a collection of caterpillars and wreaths.*

Biplanar graphs are easily characterized, and there is a simple linear-time algorithm to recognize biplanar graphs, as the next lemma makes clear.

Lemma 2. *Let G be a graph. The following assertions are equivalent: (a) G is biplanar. (b) G is a forest of caterpillars. (c) G is acyclic and contains no 2-claw as a subgraph. (d) G is acyclic and $\Phi(G) = 0$ (with Lemma 1).*

Lemma 2 implies that any biplanarization algorithm must destroy all cycles and 2-claws. The next lemma gives a condition for this situation.

Lemma 3. *If there exists a vertex v in a graph G such that $\deg'_G(v) \geq 3$, then G contains a 2-claw or a 3- or 4-cycle containing v.*

A set T of edges of a graph G is called a *biplanarizing set* if $G \setminus T$ is biplanar. The *bipartite planarization number* of a graph G, denoted by $\mathsf{bpr}(G)$, is the size of a minimum biplanarizing set for G. The 2-LP problem is: given a graph G and an integer k, is $\mathsf{bpr}(G) \leq k$? For a given bipartite graph $G = (A, B; E)$ and permutation π of A, the *1-layer biplanarization number* of G and π, denoted $\mathsf{bpr}(G, \pi)$, is the minimum number of edges in G whose deletion produces a graph that admits a biplanar drawing with π as the ordering of the vertices in A. The 1-LP problem asks if $\mathsf{bpr}(G, \pi) \leq k$.

Lemma 4. *For graphs G with $\Phi(G) = 0$, a minimum biplanarizing set of G consists of one cycle edge from each component wreath.*

Lemma 5. *For every graph G, $\mathsf{bpr}(G) \geq \frac{1}{2}\Phi(G)$.*

3   2-Layer Planarization: Bounded Search Tree

The basic approaches for producing $\mathcal{FPT}$ algorithms are *kernelization* and *bounded search trees* [2]. Based on the preceding lemmas, Dujmović *et al.* showed:

Theorem 1. *Given a graph G and integer k, there is an algorithm that determines if $\mathsf{bpr}(G) \leq k$ in $\mathcal{O}(k \cdot 6^k + |G|)$ time.*

That algorithm consists of two parts: a kernelization algorithm and a subsequent search tree algorithm 2-Layer Bounded Search Tree. The latter algorithm basically looks for a vertex v with $\deg'(v) \geq 3$: if found at most 6 recursive branches are triggered to destroy the forbidden structures described in Lemma 3. After branching, a graph G with $\Phi(G) = 0$ remains, solvable with Lemma 4.

Can we further improve on the running time of the search tree algorithm? Firstly, observe that whenever $\deg'_{G'}(v) \geq \ell$ for any G' obtained from G by edge deletion, then already $\deg'_G(v) \geq \ell$. This means that we can modify the sketched algorithm by collecting *all* vertices of non-leaf degree at least three and, based on this, all *forbidden structures* F, i.e., 2-claws, 3-cycles, or 4-cycles, according to Lemma 3 (which then might interact). For reasons of improved algorithm analysis, we also regard 5-cycles as forbidden structures. By re-interpreting the edges of G as the vertices of a hypergraph $H = (E, F)$, where the hyperedges correspond to the forbidden structures, a 2-LP instance (G, k) is translated into an instance (H, k) of 6-HITTING SET (6-HS).

If we delete all those edges in G that are elements in a hitting set as delivered by a 6-HS algorithm, we arrive at a graph G' which satisfies $\deg'_{G'}(v) < 3$ for all vertices v. Hence, $\Phi(G') = 0$, and Lemma 4 applies.

Unfortunately, we cannot simply take some 6-HS algorithmas described in [8]. Why? The problem is that there may be minimal hitting sets C which are "skipped" due to clever branching, since there exists another minimal solution C' with $|C| \geq |C'|$. However, if we translate back to the original 2-LP instance, we still have to resolve the wreath components, and it might be that we "skipped" the only solution that already upon solving the 6-HS instance was incidentally also destroying enough wreaths. To be more specific, the analysis in [8] is based on the validity of the <u>vertex domination</u> rule: A vertex x is *dominated* by a vertex y if, whenever x belongs to some hyperedge e, then y belongs to e, as well. Then, delete all occurrences of x, since taking y into the hitting set (instead of x) is never worse. As explained, the subsequent wreath analysis destroys the applicability of that rule.

If we insist on enumerating all minimal hitting sets no larger than the given k, this problem can be circumvented, since we can do the wreath component analysis in the leaves of the search tree, but it would gain nothing in terms of time complexity, since examples of hypergraphs having 6^k minimal hitting sets of size at most k can be easily found: just consider k disjoint hyperedges, each of size 6, see [1].

However, our experience with analyzing HITTING SET problems by the help of the so-called top-down approach as detailed in [6] gives the basic ideas for the claimed improvements. The main purpose of the vertex domination rule is that it guarantees the existent of vertices of "sufficiently" high degree in the HITTING SET instance. Our aim is now to provide a more problem-specific analysis which maintains exactly that property. To avoid confusion, in the following, we will stay within the 2-LP formulation and will not translate back and forth between the 2-LP instance and the corresponding 6-HS instance.

In order to maintain the structure of the original graph for the final wreath analysis, we will mark edges that won't be put into a solution during the recursive branching process as *virtual*, but we won't delete them. Hence, along the course of the algorithm we present, there will be built a set M of edges that are marked virtual. A *forbidden structure* f is a set of edges of the graph instance $G = (V, E)$ such that

- f describes a cycle of length up to five or a 2-claw, and
- $f \setminus M \neq \emptyset$.

$c(f) = f \setminus M$ is the *core* of f; $s(f) = |c(f)|$ is the *size* of f.

We will use the following reduction rules:

1. <u>structure domination</u>: A forbidden structure f is *dominated* by another structure f' if $c(f') \subset c(f)$. Then, mark f as dominated.
2. <u>small structures</u>: If $s(f) = 1$, put the only non-virtual edge into the solution that is constructed.

3a <u>isolates</u>: If e is an edge of degree zero, then mark e virtual.

The number of non-dominated forbidden structures to which a specific edge e belongs is also called the *degree* of e. Can we also handle edges of degree one (to a certain extent) by reduction rules? We will discuss this point later on.

Let $C = \{c, w_1, w_2, w_3, x_1, x_2, x_3\}$ be a 2-claw centered at c, such that w_i is neighbored (at least) to c and x_i for $i = 1, 2, 3$. We will call $F_i = \{cw_i, w_ix_i\}$ also a *finger* of C, so that the forbidden structure f_C corresponding to C is partitioned into three disjoint fingers. A 2-claw where one or more edges are virtual is called *injured*. Clearly, in an injured 2-claw with five edges, only one of the fingers actually got injured and two fingers are still *pretty*. In an injured 2-claw with four edges, we still have at least one pretty finger left over.

The second ingredient in the approach to hitting set problems described in [6] are so-called *heuristic priorities*. More specifically, we use the following rules to select forbidden structures and edges to branch at in case of multiple possibilities:

1. Select a forbidden structure f of smallest size that if possible corresponds to a short cycle.
2. If $s(f) \leq 5$ and if there is another forbidden structure f' with $c(f) \cap c(f') \neq \emptyset$ and $s(f') \leq 5$, modify $f := c(f) \cap c(f')$.
3. Select an edge e of maximal degree within f, if possible incident to the center of the 2-claw f, such that e belongs to a pretty finger.

In the following analysis, assume that we have already branched on all cycles up to length five (see the first heuristic priority). Then, we can apply the following reduction rule for (injured) 2-claws:

3b (injured) 2-claws: If e is an edge of degree one in a forbidden structure of size four, five or six corresponding to an (injured) 2-claw, and if e is incident to the center of the corresponding 2-claw, then mark e virtual.

This allows us to state the whole procedure in Alg. 1, where *branch at e* means the following:

if $\mathrm{TLP}(G - e, k - 1, M, D_F)$ **then**
    return YES
else if $G[M \cup \{e\}]$ is acyclic **then**
    return $\mathrm{TLP}(G, k, M \cup \{e\}, D_F)$
end if

To prove the soundness of rule 3b., we have to show that we will never miss out cycles this way. We therefore show the following assertions:

Proposition 1. *At most one edge per finger will turn virtual due to rule 3b.*

Proof. 3b. obviously only affects *one* 2-claw at a time, since only edges of size one are turned virtual. Per 2-claw, the rule triggers at most once per finger. □

Proposition 2. *Cycles of length at least six that only consists of virtual edges can never be created by running Alg. 1.*

To prove Proposition 2, the following observation is crucial.

Property 1. Let $F = \{xy, yz\}$ be one pretty finger of a non-dominated (injured) 2-claw C with center x such that xy occurs only in one forbidden structure, i.e., C. Then, y has degree two.

Algorithm 1 A search tree algorithm for 2-LP, called TLP

Require: a graph $G = (V, E)$, a positive integer k, a set of virtual edges M, a list of dominated forbidden structures D_F

Ensure: YES if there is a biplanarization set $B \subseteq E$, $|B| \le k$ (and it will implicitly produce such a small biplanarization set then) or
NO if no such set exists.

Exhaustively apply the reduction rules 1., 2., and 3a.; the resulting instance is also called (G, k, M, D_F).

if $\Phi(G[E \setminus M]) > 2k$ **then**
    return NO {Lemma 5}
else if $\Phi(G[E \setminus M]) = 0$ **then**
    if $k \ge \#$ component wreaths of $G[E \setminus M]$ **then**
        return YES {Lemma 4}
    else
        return NO
    end if
else
    $\{\exists v \in V$ such that $\deg'_{G[E \setminus M]}(|)v \ge 3\}$
    if possible **then**
        Find a non-dominated cycle C of length at most 5
        Select an edge $e \in C$ and branch at e
    else
        Exhaustively apply all reduction rules
        Select 2-claw C and edge $e \in C$ according to heuristic priorities; branch at e
    end if
end if

Proof. If the conclusion were false, there must be an edge yv in the given 2-LP instance. Hence, there is an (injured) 2-claw C' with center x which is like C, only having z replaced by v. This contradicts that xy has degree one, since xy participates both in C and in C'. $\square$

Now to the time analysis of Alg. 1, following the ideas explained in [6] for 3-HS. $T(k)$ denotes the number of leaves in a worst-case search tree for Alg. 1, which incidentally also is the worst-case for the number of returned solutions. More distinctly, let $T^\ell(k)$ denote the situation of a search tree assuming that at least ℓ forbidden structures in the given instance (with parameter k) have size five. Of course, $T(k) \le T^0(k)$. We analyze the recurrences for T^0, T^1 and T^2.

Lemma 6. $T^0(k) \le T^0(k-1) + T^2(k)$.

Proof. Due to the reduction rule 3b., the 2-LP instance G contains an edge e of degree 2 in a forbidden structure f of size 6, since f represents a 2-claw. Hence, there is another 2-claw corresponding to a forbidden structure f' with $e \in f \cap f'$. One branch is that e is put into the biplanarization set. The size of the corresponding subtree can be estimated by $T^0(k-1)$. If e is not put into the biplanarization set, then e is marked virtual and hence at least two forbidden structures of size five are created: $f \setminus \{e\}$ and $f' \setminus \{e\}$. Therefore, the size of that subtree is upperbounded by $T^2(k)$.   □

Some more involved analysis of the T^1- and T^2-branches as well as some algebra for solving the recursions, shows:

Lemma 7. $T^1(k) \leq 2T^0(k-1) + 2T^1(k-1) + T^2(k-1)$.

Lemma 8. $T^2(k) \leq \max\{2T^1(k-1)+3T^2(k-1), T^0(k-1)+16T^0(k-2), 2T^0(k-1) + 9T^0(k-2), 3T^0(k-1) + 4T^0(k-2), 4T^0(k-1) + T^0(k-2)\}$.

Theorem 2. *Given a graph G and an integer k, Alg. 1 determines if $\mathsf{bpr}(G) \leq k$ in $\mathcal{O}(k^2 \cdot 5.1926^k + |G|)$ time, when applied to the problem kernel derived in [4].*

To prove these results, the following lemma is important, which is also interesting from a structural point of view on its own account; this also explains why we considered 5-cycles as forbidden structures.

Lemma 9. *In a graph without cycles up to length five, each 2-claw is vertex-induced.*

4   1-Layer Planarization: Kernelization Algorithm

The next two results from [4] give important properties for π-biplanar graphs.

Lemma 10. *A bipartite graph $G = (A, B; E)$ with a fixed permutation π of A is π-biplanar if and only if G is acyclic and the following condition holds.*

> *For every path (x, v, y) of G with $x, y \in A$, and for every vertex $u \in A$ between x and y in π, the only edge incident to u (if any) is uv.* $(\star)$

Let $G = (A, B; E)$ be a bipartite graph with a fixed permutation of A that satisfies condition $(\star)$. Let $H = K_{2,p}$ be a complete bipartite subgraph of G with $H \cap A = \{x, y\}$, and $H \cap B = \{v \in B : vx \in E, vy \in E, \deg_G(v) = 2\}$, and $|H \cap B| = p$. Then H is called a *p-diamond*. Every cycle of G is in some p-diamond with $p \geq 2$.

Lemma 11. *If $G = (A, B; E)$ is a bipartite graph and π is a permutation of A satisfying condition $(\star)$ then $\mathsf{bpr}(G, \pi) = \sum_{maximal\ p\text{-}diamonds\ of\ G}(p - 1)$.*

We are now going to derive a kernelization algorithm for 1-LP. Let us say that an edge e of a bipartite graph G *potentially violates condition* $(\star)$ if, using the notation of condition $(\star)$, $e = e_i$ for $i = 1, 2, 3$, where $e_1 = xv$ or $e_2 = vy$ or $e_3 = uz$ for some u strictly between x and y in π such that $z \neq v$. We will also say that e_1, e_2, e_3 (together) *violate condition* $(\star)$.

According to Lemma 10 (as well as the proof of Lemma 11 for the last two rules), the following reduction rules are sound, given an instance $(G = (A, B; E), \pi, k)$ of 1-LP. Analogues to the first three rules are well-known from HITTING SET problems, see [6, 8].

1L-RR-edge: If $e \in E$ does not participate in any cycle and does not potentially violate condition $(\star)$, then remove e from the instance (keeping the same parameter k).

1L-RR-isolate: If $v \in A \cup B$ has degree zero, then remove v from the instance and modify π appropriately (keeping the same parameter k).

1L-RR-large: If $e \in E$ participates in more than k^2 situations that potentially violate condition $(\star)$, then put e into the biplanarization set and modify the instance appropriately (also decreasing the parameter).

Let $E_\star \subseteq E$ be all edges that potentially violate condition $(\star)$. Let $E_\circ \subseteq E$ be all edges that participate in cycles. Let G_{4c} be generated from those edges from $E_\star \setminus E_\star$ that participate in 4-cycles. By construction, G_{4c} satisfies $(\star)$. Lemma 11 shows that the next reduction rule can be applied in polynomial time:

1L-RR-4C: If $\mathsf{bpr}(G_{4c}, \pi) > k$, then NO.

Lemma 12. *Let $G = (A, B; E)$ be a bipartite graph and let π be a permutation of A. Let $v \in B$. Then, there is at most one edge e incident to v that does not potentially violate condition $(\star)$ and participates in cycles of length > 4.*

Theorem 3. *Let $G = (A, B; E)$ be a bipartite graph, π be a permutation of A and $k \geq 0$. Assume that none of the reduction rules applies to the 1-LP instance (G, π, k). Then, $|E| \leq k^3$. The kernel can be found in time $\mathcal{O}(|G|^2)$.**

Proof. Now consider $E_\star$ as vertex set V' of a hypergraph $G' = (V', E')$ and put $\{e_1, e_2, e_3\}$ into E' iff e_1, e_2, e_3 together violate condition $(\star)$. A subset of edges from E whose removal converts $(A, B; E)$ into a bipartite graph which satisfies condition $(\star)$ is in obvious one-to-one correspondence with a hitting set of the hypergraph G'. Niedermeier and Rossmanith have shown [8–Proposition 1] a cubic kernel for 3-HITTING SET, so that at most k^3 edges are in $E_\star$ (else NO). Their reduction rules correspond to our rules 1L-RR-edge and 1L-RR-large.

If $e = xy \in E_\circ \setminus E_\star$ with $y \in B$ does not belong to a 4-cycle, then Lemma 12 shows that there is no other edge $zy \in E_\circ \setminus E_\star$. But since $xy \in E_\circ$, there must be some "continuing edge" zy on the long circle xy belongs to, so that $zy \in E_\star$ follows. We can take zy as a *witness* for xy. By Lemma 12, zy can witness for at most one edge from $E_\circ \setminus E_\star$ incident to y and not participating in a 4-cycle.

This allows us to partition $E_\circ$ into three disjoint subsets: (a) $E_\circ \cap E_\star$, (b) $E_{4c} = \{e \in E_\circ \setminus E_\star \mid e$ participates in a 4-cycle $\}$: there can be at most $4k$ such edges according to 1L-RR-4C and Lemma 11, and (c) $E_\circ \setminus E_{4c}$: according to our preceding reasoning, there are at most $|E_\star|$ many of these edges. $\square$

5   1-Layer Planarization: Bounded Search Tree

Theorem 4. *(Dujmović et al. [4]) Given a bipartite graph $G = (A, B; E)$, a fixed permutation π of A, and integer k, there is an algorithm that determines if $\mathsf{bpr}(G, \pi) \leq k$ in $\mathcal{O}(3^k \cdot |G|)$ time.*

---

* More recently, a quadratic kernel for 3-HITTING SET was derived [9] based on results by Nemhauser and Trotter. Translating the corresponding reduction rules shows that $|E_\star|$ and hence $|E|$ is in fact upperbounded by $\mathcal{O}(k^2)$.

Can we further improve on this algorithm? Firstly, it is clear that we can combine the search tree algorithm with the kernelization algorithm described above. But furthermore, observe that the search tree algorithm basically branches on all members of $E_\star$, trying to destroy the corresponding triples of edges violating condition $(\star)$. This means that we again take ideas stemming from solutions of the naturally corresponding instance of 3-HITTING SET. Unfortunately again, we cannot simply "copy" the currently best search tree algorithm for 3-HITTING SET [6, 8], running in time $\mathcal{O}(k \cdot 2.179^k + |G|)$, since destroying triples of edges violating condition $(\star)$ might incidentally also destroy more or less of the 4-cycles. As explained in the 2-LP case, the problem is again the vertex domination rule. In order to gain anything against the previously sketched algorithm 1-Layer Bounded Search Tree, we must somehow at least *avoid branching on vertices of degree one contained in hyperedges of size three.*

Firstly, we can prove a lemma that shows that, whenever we have branched on all hyperedges of size three in the 3-HITTING SET instance (that correspond to situations violating condition $(\star)$ in the original 1-LP instance) that contain vertices of degree at least two, then we have already destroyed all "large" cycles. Then, we investigate the possible interaction between a cycle of length four and a structure violating $(\star)$, after having "destroyed" all "mutually interacting" structures violating $(\star)$.

Lemma 13. *Let $G = (A, B; E)$ be a bipartite graph and π be a fixed permutation of A. Assume that if $h = \{e_1, e_2, e_3\}$ and $h' = \{e'_1, e'_2, e'_3\}$ are two situations violating $(\star)$, then $h \cap h' = \emptyset$. Let $C = \{ab, bc, cd, da\}$ be a sequence of edges forming a 4-cycle. Then, there is at most one hyperedge h —among the hyperedges modeling situations violating $(\star)$ —such that $C \cap h \neq \emptyset$.*

Hence, after the indicated branching, for each 4-cycle, at most one hyperedge of size three remains such that the corresponding edge sets have non-empty intersection. Since we have to destroy every 4-cycle, the best we then can obviously do is to take out an edge that takes part in the "accompanying" situation violating $(\star)$. This can be done completely deterministically due to the preceding lemma. Finally, the only remaining situations correspond to possibly interacting 4-cycles. These can be solved with Lemma 11.

Theorem 5. *1-LP can be solved in $\mathcal{O}(k^3 \cdot 2.5616^k + |G|^2)$ time.*

6   Conclusion

In this paper we have presented two methods for producing $\mathcal{FPT}$ algorithms in the context of 2-layer and 1-layer planarization. In particular, for fixed k, we have polynomial time algorithms to determine if $\mathsf{bpr}(G) \leq k$ and $\mathsf{bpr}(G, \pi) \leq k$. The smaller exponential bases (in comparison with [4]) are due to the tight relations with HITTING SET, as we exhibited. For small values of k, our algorithms provide a feasible method for the solution of these $\mathcal{NP}$-complete problems.

With the results in [4, 5], we have now good kernelization and search tree algorithms for three types of "layered planarization" problems:

1. For 2-LP, we got an $\mathcal{O}(k^2 \cdot 5.1926^k + |G|)$ algorithm and a kernel size $\mathcal{O}(k)$.[**]
2. For 1-LP, we found an $\mathcal{O}(k^3 \cdot 2.5616^k + |G|^2)$ algorithm and a kernel size $\mathcal{O}(k^3)$.
3. For 1-LAYER CROSSING MINIMIZATION, we obtained an $\mathcal{O}(1.4656^k + k|G|^2)$ algorithm and a kernel size $\mathcal{O}(k^2)$, where k is now the number of crossings.

For 2-LAYER CROSSING MINIMIZATION, the (more general) results of [3] only give an $\mathcal{O}(2^{32(2+2k)^3}|G|)$ algorithm, which should be further improvable.

Acknowledgments. We are grateful for discussion of this topic with V. Dujmović and for the very thoughtful comments of the reviewers.

References

1. P. Damaschke. Parameterized enumeration, transversals, and imperfect phylogeny reconstruction. In R. Downey, M. Fellows, and F. Dehne, editors, *Intern. Workshop on Parameterized and Exact Computation IWPEC 2004*, volume 3162 of *LNCS*, pages 1–12. Springer, 2004.
2. R. G. Downey and M. R. Fellows. *Parameterized complexity.* Springer, 1999.
3. V. Dujmović, M. Fellows, M. Hallett, M. Kitching, G. Liotta, C. McCartin, N. Nishimura, P. Ragde, F. Rosamond, M. Suderman, S. Whitesides, and D. R. Wood. On the parameterized complexity of layered graph drawing. In F. Meyer auf der Heide, editor, *European Symp. on Algorithms ESA*, volume 2161 of *LNCS*, pages 488–499. Springer, 2001.
4. V. Dujmović, M. Fellows, M. Hallett, M. Kitching, G. Liotta, C. McCartin, N. Nishimura, P. Ragde, F. Rosemand, M. Suderman, S. Whitesides, and D. R. Wood. A fixed-parameter approach to two-layer planarization. In P. Mutzel, M. Jünger, and S. Leipert, editors, *Graph Drawing GD 2001*, volume 2265 of *LNCS*, pages 1–15. Springer, 2002.
5. V. Dujmović, H. Fernau, and M. Kaufmann. Fixed parameter algorithms for one-sided crossing minimization revisited. In G. Liotta, editor, *Graph Drawing GD 2003*, volume 2912 of *LNCS*, pages 332–344. Springer, 2004.
6. H. Fernau. A top-down approach to search-trees: Improved algorithmics for 3-hitting set. TR04-073, Electronic Colloquium on Computational Complexity ECCC, 2004.
7. P. Mutzel. An alternative method to crossing minimization on hierarchical graphs. *SIAM J. Optimization*, 11(4):1065–1080, 2001.
8. R. Niedermeier and P. Rossmanith. An efficient fixed-parameter algorithm for 3-hitting set. *Journal of Discrete Algorithms*, 1:89–102, 2003.
9. N. Nishimura, P. Ragde, and D. Thilikos. Smaller kernels for hitting set problems of constant arity. In R. Downey, M. Fellows, and F. Dehne, editors, *Intern. Workshop on Parameterized and Exact Computation IWPEC 2004*, volume 3162 of *LNCS*, pages 121–126. Springer, 2004.
10. K. Sugiyama, S. Tagawa, and M. Toda. Methods for visual understanding of hierarchical system structures. *IEEE Trans. Systems Man Cybernet.*, 11(2):109–125, 1981.

[**] By changing the heuristic priorities in one case and by using a generalization of rule 3b., we can improve the base to 5.1844.

On the Stability of Approximation for Hamiltonian Path Problems*

Luca Forlizzi[2], Juraj Hromkovič[1], Guido Proietti[2,3], and Sebastian Seibert[1]

[1] Department Informatik, ETH Zentrum, CH-8092, Zürich, Switzerland
{jh, seibert}@cs.rwth-aachen.de
[2] Dipartimento di Informatica, Università di L'Aquila, 67010 L'Aquila, Italy
{forlizzi, proietti}@di.univaq.it
[3] Istituto di Analisi dei Sistemi ed Informatica "A. Ruberti", CNR, Roma, Italy

Abstract. We consider the problem of finding a cheapest Hamiltonian path of a complete graph satisfying a relaxed triangle inequality, i.e., such that for some parameter $\beta > 1$, the edge costs satisfy the inequality $c(\{x,y\}) \leq \beta(c(\{x,z\}) + c(\{z,y\}))$ for every triple of vertices x, y, z. There are three variants of this problem, depending on the number of prespecified endpoints: zero, one, or two. For metric graphs, there exist approximation algorithms, with approximation ratio $\frac{3}{2}$ for the first two variants and $\frac{5}{3}$ for the latter one, respectively.

Using results on the approximability of the Travelling Salesman Problem with input graphs satisfying the relaxed triangle inequality, we obtain for our problem approximation algorithms with ratio $\min(\beta^2 + \beta, \frac{3}{2}\beta^2)$ for zero or one prespecified endpoints, and $\frac{5}{3}\beta^2$ for two endpoints.

1 Introduction

It often happens that the hardness of the polynomial-time approximability of a problem varies according to the input instance, and some hard problem becomes relatively easy for certain subclasses of instances. Given an hard optimization problem, and a polynomial-time approximation algorithm for a subclass of input instances, a natural idea is trying to extend the approximation algorithm to a wider class of problem instances. This idea is captured by the notion of *stability of approximation*, which provides a formal framework to study the change of the approximation ratio according to a small change in the specification (some parameter, characteristics) of the set of problem instances considered [1].

One of the most successful application of the concept of stability of approximation concerns the famous Travelling Salesman Problem (TSP). It is well known that TSP is not only NP-hard, but also not approximable in polynomial time with constant approximation ratio. But if one considers Δ-TSP, namely

---

* The work reported in this paper has been partially supported by the Italian MIUR under the project "Web-based management and representation of spatial and geographical data".

M. Bieliková et al. (Eds.): SOFSEM 2005, LNCS 3381, pp. 147–156, 2005.

TSP for complete input graphs satisfying the triangle inequality (i.e., *metric graphs*), one can design a polynomial time $\frac{3}{2}$-approximation algorithm [2]. To extend the class of input graphs for which the TSP is approximable (in polynomial time, with constant approximation ratio), one considers the so called β-*triangle inequality*. For a given $\beta > 1$, a graph (V, E) satisfies the β-triangle inequality if for all vertices u, v, x it is $c(\{u, v\}) \leq \beta(c(\{u, x\}) + c(\{x, v\}))$, where $c : E \mapsto \mathbb{R}^+$ is the cost function of the graph. For every real $\beta > 1$, the Δ_β-TSP is the restriction of the TSP to inputs satisfying the β-triangle inequality.

In the past, several polynomial time approximation algorithms providing constant approximation ratio for Δ_β-TSP, where proposed. Currently, there are three different algorithms which achieve the smallest approximation ratio, each for a distinct range of values of β:

(A) The REFINED T^3 algorithm, providing a $(\beta^2 + \beta)$ approximation ratio [3], which is the best for $2 \leq \beta \leq 3$;
(B) The Bender and Chekuri 4β-approximation algorithm [4], best for $\beta > 3$;
(C) The Path Matching Christofides Algorithm (PMCA) providing a $\frac{3}{2}\beta^2$ approximation ratio [1], best for $1 < \beta \leq 2$.

In this paper, we study how these results can help to design approximation algorithms for the *Hamiltonian Path Problem* (HPP), where one is required to compute a minimum cost Hamiltonian path spanning a complete graph G. There are three natural variants of the HPP, differing in the constraints imposed to the endpoints of the desired path: they can be both arbitrary vertices (HPP$_0$), or one of them can be constrained to be a prespecified vertex s (HPP$_1$), or both of them can be constrained to be prespecified vertices s and t (HPP$_2$). All these variants are easily reducible to TSP, so they are NP-hard too. Concerning their approximability, in [5], Christofides ideas presented in [2] for the Δ-TSP were applied to the Δ-HPP, providing $\frac{3}{2}$-approximation algorithms for Δ-HPP$_0$ and Δ-HPP$_1$, and a $\frac{5}{3}$-approximation algorithm for Δ-HPP$_2$.

In this paper, trying to extend the class of graphs for which HPP is approximable, we consider again β-triangle inequality and investigate whether each of the three approaches for Δ_β-TSP is suitable also for Δ_β-HPP. To this aim, we concentrate on adapting the approaches of (A) and (C), which distinguish themselves by running times $O(n^2)$ and $O(n^3)$, respectively, where n is the number of vertices in G. This is acceptable for practical purposes rather than the $O(n^5)$ running time of (B). We just note why the approach of (B) would need some additional considerations in order to be carried over to HPP. The algorithm of Bender and Chekuri is based on results by Fleischner [6, 7], who proved that the square of a 2-node-connected graph is Hamiltonian, and by Lau [8, 9], who provided an effective procedure for the construction of a Hamiltonian cycle. So, Bender and Chekuri first construct an approximation of the minimum cost 2-node-connected subgraph, and then apply, on the resulting graph, Lau's procedure to obtain a Hamiltonian cycle. The length of a minimum cost 2-node-connected subgraph is a lower bound on the cost of any Hamiltonian cycle, and from this fact the bound on the cost of their solution follows. However, the length of a minimum cost 2-node-connected subgraph is not a lower bound on the cost of a Hamiltonian path. Hence, this approach does not lead immediately to an approximation algorithm for the HPP.

The approaches leading to algorithms (A) and (C) are studied in Sections 2 and 3, respectively. For HPP_0 and HPP_1, they both keep the same ratio bounds as for the TSP. Hence we obtain $\min(\beta^2 + \beta, \frac{3}{2}\beta^2)$-approximation algorithms for Δ_β-HPP_0 and Δ_β-HPP_1. In the case of HPP_2, the $\frac{5}{3}$-approximation ratio for metric graphs is reflected in the $(\frac{5}{3}\beta^2)$-approximation algorithm obtained using the PMCA approach. Instead, with the T^3 approach we have an approximation ratio always worse than that obtained with PMCA. Nevertheless, the T^3 approach is still somehow useful for HPP_2, since it allows to obtain an $O(n^2)$ time 3-approximation algorithm for Δ-HPP_2, faster than the $O(n^3)$ time previously known approximation algorithm.

Following [5], we let P^* denote an optimal Hamiltonian path without prescribed endpoints, P_s^* denote an optimal Hamiltonian path with a single prescribed endpoint s, and P_{st}^* denote an optimal Hamiltonian path with prescribed endpoints s and t. Given a graph G and a collection Π of paths on the nodes of G, we denote by $G \cup \Pi$ the graph obtained by adding to G all the edges of each path in Π. We denote by $\text{EndP}(\Pi)$ the set formed by the endpoints of all the paths contained in Π. Given a graph G and an edge e, we denote by $G - e$ the graph obtained by removing e from G. We call an occurrence of a vertex in a path γ *internal*, if it is not an endpoint of γ. Given a path γ, we say that a subpath γ' of γ is a *terminal subpath* if one of the endpoints of γ' is also an endpoint of γ. A path in a graph is *elementary* if it does not contain the same vertex more than once. An edge e^* is *locally minimal* if there is a vertex $v \in V$ such that e^* is an edge incident to v of minimum cost.

Due to space limitations, we skip theorem proofs and some technical details in the presented algorithms. They can be found in the full version of this paper [10].

2   The REFINED T^3 Algorithm for Hamiltonian Path

In 1960, Sekanina proved that for every tree $T = (V, E)$ the graph

$$T^3 = (V, \{\{x, y\} \mid x, y \in V, \text{ and there is a path from } x \text{ to } y \text{ in } T \text{ of length } \leq 3\})$$

contains a Hamiltonian cycle $H = (u_1, u_2, \ldots, u_n)$. This means that every edge $\{u_i, u_{i+1}\}$ of H has a corresponding unique path P_i in T of length at most 3.

Starting from this result Andreae and Bandelt designed in [11] a $(\frac{3}{2}\beta^2 + \frac{1}{2}\beta)$-approximation algorithm for Δ_β-TSP. In fact, they were able to construct a Hamiltonian cycle H of T^3, such that each edge of T occurs in exactly two of the paths P_i, and that it is the middle edge of at most one path P_i of length 3. The properties of H, imply that expensive edges of T do not occur in $P(H) = P_1, \ldots, P_n$ more often than cheap edges. Then the cost of $P(H)$, and so the cost of H, is bounded by a factor times the cost of T, which, in turn, is a lower bound for the cost of an optimal Hamiltonian cycle.

Andreae and Bandelt result was recently improved by Andreae [3], that presented a $(\beta^2 + \beta)$-approximation algorithm for Δ_β-TSP. The main part of such algorithm is a procedure called $\text{HCT}^3(\text{REFINED})$, which, given a locally minimal edge e^* of T, computes a Hamiltonian cycle H of T^3 containing e^*. The core result obtained by Andreae is the following ([3], Theorem 1): for a tree T with $|T| \geq 3$ and a real number $\beta \geq 1$, suppose T^3 satisfies the β-triangle inequality. Then it is $c(H) \leq (\beta^2 + \beta)c(T)$, and this inequality is strict if $\beta > 1$.

The fact that the cost of the graph constructed is bounded using the cost of T, is particularly interesting for our purposes, since the cost of T is a lower bound for the cost of an optimal Hamiltonian path, too. Indeed, using Andreae's result, we can easily derive $(\beta^2 + \beta)$-approximation algorithms for HPP_0 and HPP_1, by simply removing an appropriate edge from the Hamiltonian cycle computed by $\text{HCT}^3(\text{REFINED})$, see [10] for details.

An interesting feature of $\text{HCT}^3(\text{REFINED})$, is that the input an edge is part of the returned cycle. This feature suggests immediately the following strategy to approximate HPP_2: given the two prespecified endpoints s and t, use $\text{HCT}^3(\text{REFINED})$ to compute a Hamiltonian cycle H_{st} containing $\{s,t\}$, and return the path π_{st} obtained after deleting $\{s,t\}$ from H_{st}. Indeed, this idea leads us to the following approximation algorithm for Δ-HPP_2.

Input: A metric graph $G = (V, E)$ and two vertices $s, t \in V$
1: Find a minimum spanning tree T containing $\{s, t\}$ for G
2: Find a Hamiltonian cycle H_{st} of T^3 by means of HCT^3 with T and $\{s,t\}$ as input
3: Find a Hamiltonian path π_{st} of G by removing edge $\{s,t\}$ from H_{st}
Output: A Hamiltonian path π_{st} of G having s and t as endpoints

---

Algorithm 1. T^3 Metric-HPP_2

Note that in Alg. 1, we use Procedure HCT^3 presented in [11] instead of the improved version $\text{HCT}^3(\text{REFINED})$. The two procedures are similar, and for a metric graph G, given a tree T and an edge e of T, they both compute a Hamiltonian cycle H_e containing e such that $c(H_e) \leq 2c(T) \leq 2c(H^*)$, where H^* is a minimum Hamiltonian cycle of G. There are two advantages in using HCT^3 instead of $\text{HCT}^3(\text{REFINED})$: the former procedure does not require the input edge e to be locally minimal, and it also is more efficient, requiring $O(n)$ time.

Although Alg. 1 has a poor approximation guarantee, it deserves some interest being more efficient than the $O(n^3)$ time algorithm derived in [5] from Christofides one.

Theorem 1. *Let G be a graph satisfying the Δ-inequality. Algorithm 1 is a 3-approximation algorithm for Δ-HPP_2. The algorithm runs in $O(n^2)$ time.*

Unfortunately, Alg. 1 does not provide an approximation guarantee if the input graph does not satisfy the Δ-inequality, because in a general graph the cost of $\{s,t\}$ can not be bounded using $c(P_{st}^*)$.

To extend the Sekanina approach to Δ_β-HPP$_2$, we need another idea. Suppose we have a Hamiltonian path γ spanning G, with cost bounded by a factor times $c(P^*_{st})$. We can transform it into a Hamiltonian path having s and t as endpoints, still having a cost bounded by a factor times $c(P^*_{st})$, as follows. W.l.o.g., let $\gamma = (w, \ldots, s, s_1, \ldots, t_1, t, \ldots, z)$. We first show how to obtain a path γ' having s as endpoint. Consider $\gamma_s = (w, \ldots, s, s_1)$ and let G_s be the subgraph of G induced by the vertices occurring in γ_s. Since γ_s is a tree containing $\{s, s_1\}$, the cost of a minimum spanning tree T_s of G_s containing $\{s, s_1\}$ is a lower bound for $c(\gamma_s)$. Using HCT^3(REFINED), we compute a Hamiltonian cycle H_s containing $\{s, s_1\}$ such that $c(H_s) \leq (\beta^2 + \beta)c(T_s) \leq (\beta^2 + \beta)c(\gamma_s)$. Then, replacing γ_s with H_s in γ, we obtain a graph where s_1 is the only vertex having degree 3. By removing $\{s, s_1\}$, we have the desired path γ'. The same operations can be repeated for the other prescribed endpoint t, leading to the following result:

Theorem 2. *Let G be a graph satisfying the Δ_β-inequality. Algorithm 2 is a $((\beta^2 + \beta)\min(\beta^2 + \beta, \frac{3}{2}\beta^2))$-approximation algorithm for Δ_β-HPP$_2$. The algorithm runs in $O(n^3)$ time.*

Input: A graph $G = (V, E)$ and two vertices $s, t \in V$
1: Compute a Hamiltonian path $\gamma = (w, \ldots, s, s_1, \ldots, t_1, t, \ldots, z)$ for G
2: Let $\gamma_s = (w, \ldots, s, s_1)$, $\gamma_t = (t_1, t, \ldots, z)$, denote by G_s and G_t the subgraphs of G induced by the vertices occurring, respectively in γ_s and γ_t
3: Compute minimum spanning trees T_s of G_s containing $\{s, s_1\}$ and T_t of G_t containing $\{t_1, t\}$
4: Compute Hamiltonian cycles H_s of G_s and H_t of G_t containing respectively $\{s, s_1\}$ and $\{t_1, t\}$
5: Put $\pi_s = H_s - \{s, s_1\}$, $\pi_t = H_t - \{t_1, t\}$
6: Compute π_{st} from γ, by replacing γ_s with π_s and γ_t with π_t
Output: A Hamiltonian path π_{st} of G having s and t as endpoints

---

Algorithm 2. T^3 HPP$_2$

3   The PMCA for Hamiltonian Path

The PMCA is a $(\frac{3}{2}\beta^2)$-approximation algorithm for the Δ_β-TSP problem, inspired by Christofides algorithm for Δ-TSP. The rough idea of both algorithms is the following: first compute a multigraph H with all vertices of even degree, having a cost bounded by $\frac{3}{2}$ times the cost of an optimal Hamiltonian cycle, then compute an Eulerian cycle of H (it has the same cost), and finally transform the Eulerian cycle in a Hamiltonian one by *resolving all conflicts* i.e., by removing repeated occurrences of vertices in the cycle.[1] The final task is trivial in the case of Christofides algorithm, but not for the PMCA. Indeed, given the β-triangle inequality, with $\beta > 1$, the bypassing of some vertices in a path may increase the cost of the path.

To illustrate the conflict resolution performed as last task of the PMCA we need some formal definitions. Let $G = (V, E)$ be a complete graph. A *path matching* for a set of vertices $U \subseteq V$ is a collection Π of edge-disjoint paths having as

---

[1] We recall that a cycle or a path is Eulerian when it uses each edge exactly once.

endpoints vertices of U. The vertices of U which are not endpoints of some path in Π, are said to be *left exposed by* Π. Assume that $p = (u_0, u_1, u_2, \ldots, u_{k-1}, u_k)$ is a path in G, not necessarily simple. A *bypass* for p is an edge $\{u, v\}$ from E, replacing a subpath $(u_i, u_{i+1}, u_{i+2}, \ldots, u_{j-1}, u_j)$ of p from $u = u_i$ to $u_j = v$ $(0 \le i < j \le k)$. Its *size* is the number of replaced edges, i.e. $j - i$. Also, we say that the vertices $u_{i+1}, u_{i+2}, \ldots, u_{j-1}$ *are bypassed*. Given some set of paths Π, a *conflict* according to Π is a vertex which occurs at least twice in Π.

The PMCA succeeds in bounding by a factor β^2 the cost increase due to conflict resolution, by ensuring, with non trivial techniques, that at most 4 consecutive edges of the Eulerian cycle are substituted with a new one. In detail, H is the union of a minimum spanning tree T and a path matching Π for the set of all vertices of odd degree in T. The Eulerian cycle π of H can be seen as a sequence of paths $p_1, q_1, p_2, q_2, \ldots$ such that $p_1, p_2, \ldots$ are paths in T and $q_1, q_2, \ldots \in \Pi$. The conflict resolution process is realized in three steps:

(i). conflicts within Π are resolved obtaining a collection Π' of vertex-disjoint paths;

(ii). some of the conflicts within paths in T are resolved so that the cycle π' obtained by modifying π according to steps (i) and (ii), contains at most 2 occurrences of each vertex;

(iii). all remaining conflicts in π' are resolved, by bypassing at most 2 consecutive vertices.

Combining the ideas of [5] and [1], we obtain an approximation algorithm for the Δ_β-HPP$_x$, $x \in \{0, 1, 2\}$ (see Alg. 3).

Input: A complete graph $G = (V, E)$ with cost function $c : E \mapsto \mathbb{R}^+$ and a set A of k prespecified endpoints $(0 \le k \le 2)$.

1: Construct a minimum spanning tree T of G.

2: Let U be the set composed by vertices in A having even degree in T plus vertices of $V \setminus A$ having odd degree in T; construct a minimal (edge-disjoint) path matching Π for U, leaving $2 - k$ vertex of U exposed. If necessary, remove an edge from T, so that the graph $T \cup \Pi$ has 2 odd degree vertices, which we denote by w and z (observe that any prespecified endpoint is among w and z).

3: Resolve conflicts according to Π (using bypasses of size 2 only), in order to obtain a vertex-disjoint path matching Π' such that z can only occur as an endpoint of a path in Π'.

4: Construct an Eulerian path π of $H = T \cup \Pi'$ having w and z as endpoints (π can be considered as a sequence of alternating paths from T and Π', where $p_1, p_2, \ldots$ are the paths in T and $q_1, q_2, \ldots \in \Pi'$).

5: Resolve conflicts inside the paths $p_1, p_2, \ldots$ obtaining the modified paths $p'_1, p'_2, \ldots$ and the modified path matching Π', so that T is divided into a forest T_f of trees of degree at most 3, w and z are the endpoints of π', and z is not a conflict (conflict resolution in this step is done using bypasses of size 2 only).

6: Resolve every remaining conflicts in π' using bypasses of overall size 4 (where overall means that a bypass constructed in any previous step counts for 2 edges), obtaining a Hamiltonian path π'' having w and z as endpoints.

Output: A Hamiltonian path π'' of G having w and z as endpoints.

Algorithm 3. PMCA-HPP$_k$

Similarly to the PMCA, Alg. 3 starts by computing a multigraph H with all vertices but 2 of even degree. The 2 odd degree vertices include any prespecified endpoint. Since H is the union of 2 graphs, between a pair of vertices there can be at most 2 edges, one from T and one from Π'. In the following descrption, it will be clear from the context whether edges we refer to are contained in T or in Π'. Successively, Alg. 3 constructs an Eulerian path π of H, having the odd degree vertices as endpoints. Finally, conflicts are resolved obtaining a Hamiltonian path.

Here, the conflict resolution process can not be realized as in the PMCA. In particular, in step (iii) of the conflict resolution process in PMCA, for each conflict there is complete freedom in choosing which of the 2 vertex occurrences to bypass. To avoid that more than 2 consecutive vertices of π' are bypassed, PMCA relies exactly on this freedom. In our problem, we loose part of such freedom, since it may happen that the endpoints of π' are conflicts: in this case, we are not allowed to bypass the occurrences which are endpoints of π', hence we are forced to bypass the 2 internal ones. Although the problem regards only two vertices, it may render impossible to resolve all conflicts bypassing at most 2 consecutive vertices, as the following example shows.

Fig. 1. Impossibility of conflict resolution bypassing at most 2 consecutive vertices

In Fig. 1, w_1, w_2, (as well as z_1, z_2 and v_1, v_2) denote distinct occurrences in π of the same vertex. Since we are forced to bypass both w_2 and z_1, no matter which one of v_1, v_2 we bypass, there would be 3 consecutive bypassed vertices in the Hamiltonian path, causing the cost to increase more than a factor β^2. To avoid such situations, and resolve all conflicts in π' by bypassing at most 2 consecutive vertices, we have to change the whole conflict resolution process, as described in the following.

Step 1 of Alg. 3 is trivial, while Step 2 is described in [1]. After Step 2, the multigraph $T \cup \Pi$ is connected and has two or zero odd-degree vertices. The latter case occurs only if: there is a single prespecified endpoint s, s has even degree in T (so it belongs to U), and s is left exposed by Π. In this case we remove an arbitrary edge of T incident to s. Let w and z be the two odd-degree vertices in the obtained multigraph. It can be easily seen that any prespecified endpoint is contained in $\{z, w\}$. Given a vertex $v \in V$ we define the *distance in T of v from z*, as the number of edges in the unique elementary path existing in T from v to z, prior to the possible removal, discussed above, of an edge incident to s from T. We denote by y the unique neighbor of w in T whose distance in T from z is less than the distance in T of w from z. The remaining steps of Alg. 3 deserve a detailed description.

Step 3. To perform Step 3 of the algorithm, i.e., to modify path matching Π into a vertex-disjoint one, we use a strategy different from the one employed in the PMCA. The reason is that we have the additional requirement that at least one of the two odd-degree vertices that exist in $T \cup \Pi$ after Step 2, say z, does not have internal occurrences on paths in Π'. As in Procedure 1 of [1], we process each connected component of the forest formed by Π separately. To this aim, here we use Algorithm Decompose-Tree (see Alg. 4) which, given a set of edge-disjoint paths computes a new set of paths with the same set of endpoints, such that on each new path there is at most one bypass of size 2, and on one of the new paths there is no bypass. More precisely, we prove the following lemma.

Lemma 1. *Let S be a set of edge-disjoint paths with distinct endpoints, forming a tree T_S, and let x be a vertex occurring in some of the paths in S. Algorithm Decompose-Tree computes a set S' of vertex-disjoint paths such that:*

(i).$\mathrm{EndP}(S) = \mathrm{EndP}(S')$;
(ii).each path in S' is obtained peeking a path from the tree T_S and applying to it at most one bypass of size 2;
(iii).vertex x occurs on a path in S' obtained peeking a path from T_S (with no bypass applied).

Input: A vertex x and a set of edge-disjoint paths $S = \{q_1, \ldots, q_l\}$ with distinct endpoints such that $q_1, \ldots, q_l$ form a subtree of G and q_1 contains x
Let $S' = S$ and $q'_1 = q_1$
While there is at least one conflict in q'_1 **do**
    Let v be a conflict in q'_1 having maximum distance from x. W.l.o.g. assume $q'_1 = (u_a, \ldots, x, \ldots, v, u_b, \ldots, u_c)$ where the nodes $u_b, \ldots, u_c$ are not conflicts
    Let $\{q_{i_1}, \ldots, q_{i_h}\}$ be the paths forming the connected component of $S \setminus \{q'_1\}$ such that q_{i_1} contains v
    Call recursively Decompose-Tree with vertex v and set $\{q_{i_1}, \ldots, q_{i_h}\}$ as input, obtaining as result the set of paths $\{q'_{i_1}, \ldots, q'_{i_h}\}$
    Exchange in S' paths $q_{i_1}, \ldots, q_{i_h}$ with $q'_{i_1}, \ldots, q'_{i_h}$
    If v is internal to q'_{i_1} **then** bypass v from q'_{i_1}
    else assuming w.l.o.g. $q'_{i_1} = (y, \ldots, y', v)$ set $q'_{i_1} = (y, \ldots, y', u_b, \ldots, u_c)$ and $q'_1 = (u_a, \ldots, x, \ldots, v)$
Output: A set of vertex-disjoint paths $S' = \{q'_1, \ldots, q'_l\}$ such that paths in S' have the same set of endpoints as those in S and q'_1 contains x

---

Algorithm 4. Decompose-Tree

Step 3 is realized by applying Algorithm Decompose-Tree to each connected component of the forest formed by Π. Property (iii) shown in Lemma 1 is used to ensure that no internal occurrence of z exists on any path in Π' (see [10]).

Step 4. In $H = T \cup \Pi'$, w and z are the only vertices of odd degree, hence it is possible to build an Eulerian path of H having such vertices as endpoints. How to construct such an Eulerian path is a well-studied task. However, to allow the conflict resolution performed in Steps 5 and 6 we need an Eulerian path π with a specific structure. In general, there are several occurrences of z and w in an Eulerian path, but we need that the ones which are endpoints of π satisfy proper conditions. More precisely, for any of z and w, we need that if it occurs as endpoint of a path in Π', then such an occurrence is one of the endpoints of π. Note that when z and w are endpoints of the *same* path in Π', only one of such two occurrences can be endpoint of π, so we choose to let the occurrence of z be endpoint of π. In such a case, as well as if w does not occur at all as endpoint of a path in Π', we are forced to have the occurrence of w as endpoint of π, be endpoint of a path p in T. Then we need that any occurrence of w internal to π which is contained in a path p_i in T, is the vertex of p_i having minimum distance in T from z.

It is not difficult to build a path π with the desired properties, see [10] for details. Path π can be considered as an alternating sequence of the form $p_1, q_1, p_2, q_2, \ldots$ or $q_1, p_1, q_2, p_2, \ldots$, where $p_1, p_2, \ldots$ are paths in T and $q_1, q_2, \ldots \in \Pi'$. Note that since T is a tree and π is an Eulerian path, paths $p_1, p_2, \ldots$ are elementary. The following lemma proves some properties of π.

Lemma 2. *Let Π' be the vertex-disjoint path matching obtained at the end of Step 3 and π be the Eulerian path constructed in Step 4. Then:*

- *every vertex $v \in V$ different from w, occurs at most once as endpoint of a path in T;*
- *z occurs as endpoint of either a path in T or a path in Π';*
- *if the occurrence of w which is endpoint of π, is endpoint of a path p_l in T, then each occurrence of w internal to π which is contained in a path p in T, is the vertex of p with the minimum distance in T from z.*

Step 5. The main part of Step 5, namely the conflict resolution inside the paths $p_1, p_2, \ldots$ in T, is realized by a procedure derived from Procedure 2 of PMCA [1], with modifications in order to ensure that there is exactly one occurrence of z in π', and that such an occurrence is indeed an endpoint of π'. In this way, situations like the one illustrated in Figure 1 are not possible, allowing to complete in Step 6 the conflict resolution process by bypassing at most 4 consecutive edges.

The rough idea is the following. First, z is picked as root of T. Then, we consider a path p_i in T which, under the orientation with respect to z, will go up and down. The two edges immediately before and after the turning point are bypassed. One possible view of this procedure is that the minimal spanning tree is divided into several trees, since each bypass building divides a tree into two.

Lemma 3. *Consider the Eulerian path π' obtained at the end of Step 5. The endpoints of π' are w and z. In π', each vertex $v \in V$ occurs either once or twice, and z occurs exactly once.*

Step 6. We first state a crucial property used to prove that bypasses, at the end of the whole algorithm, have size at most 4. The proof goes along the way of [1], but there are many technical differences (see [10]).

Lemma 4. *In the path π', between each two bypasses there is at least one vertex that is not a conflict.*

Step 6 derives from Procedure 3 of PMCA [1], with the only change that if w is a conflict, it is the first one to be resolved. To avoid that more than two consecutive vertices of π' are bypassed, the procedure realizing Step 6 iterates through conflicts bypassing one of the two vertex instances according to the following rule: immediately after bypassing a vertex instance v, resolve, as *not* bypassed, an unresolved conflict adjacent to v, if any.

Lemma 5. *Step 6 terminates after resolving all conflicts, and it generates bypasses of size at most 4 overall, i.e., taking into account that some edges of the input path π' may be bypasses of size 2 themselves. The endpoints of the returned Hamiltonian path π'' are w and z.*

Next theorem analyzes approximation ratio and time complexity of Alg. 3.

Theorem 3. *For every β, there is a $(\frac{3}{2}\beta^2)$-approximation algorithm for Δ_β-HPP_0 and Δ_β-HPP_1, and a $(\frac{5}{3}\beta^2)$-approximation algorithm for Δ_β-HPP_2. The algorithms run in $O(n^3)$ time.*

References

1. Böckenhauer, H.J., Hromkovič, J., Klasing, R., Seibert, S., Unger, W.: Towards the notion of stability of approximation for hard optimization tasks and the traveling salesman problem. Theoretical Computer Science **285** (2002) 3–24
2. Christofides, N.: Worst-case analysis of a new heuristic for the traveling salesman problem. Technical report, Graduate School of Industrial Administration, Carnegy–Mellon University (1976)
3. Andreae: On the traveling salesman problem restricted to inputs satisfying a relaxed triangle inequality. Networks: An International Journal **38** (2001) 59–67
4. Bender, M., Chekuri, C.: Performance guarantees for the TSP with a parameterized triangle inequality. In Dehne, F.K.H.A., Gupta, A., Sack, J.R., Tamassia, R., eds.: Algorithms and Data Structures, 6th International Workshop, WADS '99, Vancouver, British Columbia, Canada, August 11-14, 1999, Proceedings. Volume 1663 of Lecture Notes in Computer Science., Springer (1999) 80–85
5. Hoogeveen, J.A.: Analysis of christofides' heuristic: Some paths are more difficult than cycles. Operational Research Letters **10** (1991) 291–295
6. Fleischner, H.: The square of every two-connected graph is hamiltonian. Journal of Combinatorial Theory **16** (1974) 29–34
7. Fleischner, H.: On spanning subgraphs of a connected bridgeless graph and their application to dt graphs. Journal of Combinatorial Theory **16** (1974) 17–28
8. Lau, H.: Finding a Hamiltonian cycle in the square of a block. PhD thesis, McGill University (1980)
9. Lau, H.: Finding eps-graphs. Monatshefte für Math. **92** (1981) 37–40
10. Forlizzi, L., Hromkovič, J., Proietti, G., Seibert, S.: On the stability of approximation for hamiltonian path problems. Technical Report TR 025/2004, Computer Science Department, University of L'Aquila (2004)
11. Andreae, T., Bandelt, H.J.: Performance guarantees for approximation algorithms depending on parametrized triangle inequalities. SIAM Journal on Discrete Mathematics **8** (1995) 1–16

Robustness of Composed Timed Systems

Hacène Fouchal[1], Antoine Rollet[2], and Abbas Tarhini[1,*]

[1] GRIMAAG,
Université des Antilles et de Guyane,
F-97157 Pointe-à-Pitre, Guadeloupe, France
Hacene.Fouchal@univ-ag.fr
[2] CReSTIC/LICA,
Université de Reims Champagne-Ardenne,
BP 1039 F-51687 Reims Cedex, France
{Antoine.Rollet, Abbas.Tarhini}@univ-reims.fr

Abstract. In this study we present a technique for testing robustness of Real-time systems described as Component-Based System having timing constraints. Each component is modeled as a Timed Input-Output Automaton (TIOA). For robustness issue, we handle two specifications : a nominal one (the more detailed specification) and a degraded one (considering only vitale functionnalities). We derive test sequences from the nominal specification of each component. We proceed to a mutation technique on these sequences in order to simulate hostile environments. Then we present a detailled algorithm for the application of test sequences on the Implementation of the system. This is done by means of an adequate test architecture consisting of the Implementation Under Test (IUT) of components, and a distributed tester that consists of a set of coordinating testers. Each tester is dedicated to test a single component.

Keywords: Real-Time System, Timed Automata, Component based System, Testing, Robustness.

1   Introduction

The complexity of systems becomes higher and higher. One reason of this complexity increase is the fact that systems consist of many independent distributed components running concurrently on heterogeneous networks. On other hand, traditional testing methods are not able to take care of all features of such systems like time constraints, distribution, integration of a component in a complex system. Such integration may lead to architectural mismatches when assembling components with incorrect behavior [1], leaving the system in a hostile environment. The criticality of such systems requires the creation of software components that can function correctly even when faced with improper usage or stressful environmental conditions. The degree of tolerance to such situations is referred to as a component's robustness.

---

* Also at Lebanese American University, Beirut, Lebanon.

M. Bieliková et al. (Eds.): SOFSEM 2005, LNCS 3381, pp. 157–166, 2005.

In this paper, we will show how to model a Real-Time Component-Based Systems (RTCBS). Then how to test robustness on such systems. The mutation method inserts hazards to test sequences generated from the component's nominal (original) specification. The mutated test sequences are used to test the system's robustness according to a corresponding degraded specification which describes the minimal authorized behavior in case of unexpected situations.

The execution of timed-mutated test sequences on the IUT is divided into two phases. In the first phase, the tester executes a set of test sequences to test the robustness of each component in isolation. In the second phase, the tester also executes and monitor timed test sequences, but to test the robustness of interaction between components integrated in a RTCBS and thus testing the system's robustness.

In section 2 we give a review on related work done on testing distributed and component based systems. Section 3 provides a background of different modeling techniques used to represent CBS. In Section 4 we focus on our method for testing Real-Time Component Based Systems. In Section 5, we detail the testing architecture as well as the test execution process. Finally, we conclude in Section 6.

2   Related Work

As far as we know, robustness testing of RTCBS is not widely developed, but there some interesting works done on distributed testing of real-time systems [2]; however, only few of them, [3], [4] dealt with testing component based systems, and others [5], [6], [7] studied the construction of component based systems.

Distributed testing was studied with different test architectures. [2] proposed a distributed tester, with the distribution of global test sequence (GTS) into local test sequences executed over local testers, where the synchronization and the fault detectability problems are discussed.

Robustness testing has been studied in many areas. It is based on insertion of hazards into test sequences. In the following, we give an overview of selected works on robustness testing and fault injection methods.

A fault injection approach called Software Implemented Fault Injection (SWIFI) has been developped in many studies. Among, theses studies the FIAT system ([8]) modifies the binary image of a process in memory (fault injection at compile time). FTAPE ([9]) combines system-wide fault injection with a controllable workload in order to create high stress conditions and fault propagation for the machine.

[10] proposed an object oriented approach to test software robustness-based on parameter data types rather than component functionality. It is a black box approach, ie the tester uses only the description of the component interface in terms of parameters and data types. They consider that most robustness failures of software are caused by code that simply neglects to test for invalid inputs.

Castanet and all present in [11], a study on formal robustness testing. It mainly consider all possible faults that a system can execute. The designer has to insert them into the specification and finally derives test sequences from the modified specification. In the following section we discus how a real-time component-based system is best modelled.

3   Formal Description of Systems

Software architecture (SA) dynamics that represent the individual components and their interactions are used in building component-based systems. [12] distinguished between the following perspectives of CBS:

- the individual component *implementation*: the executable realization of a component,
- the component *interface (interaction)*: summarizes the properties of the component that are externally visible to the other parts of the system,
- real-time component: component technology for real-time systems should support specification and prediction of timing.

In fact, the time factor problem in reactive real-time component based systems (RTCBS) is not enough investigated [12]. The time issue should be included in the model of such systems, where an action from the environment is modeled as an input event, while the reaction from the system is an output event. A Timed Input Output Automata (TIOA) is able to specify a sequence of inputs and their corresponding outputs; moreover, it shows timing constraints on events occurring in the system. Consequently, a TIOA best represents the (SA) dynamics of a RTCBS. Time is considererd as continous. But this model does not consider data exchanges.

Even this model is well known for state explosion in industrial systems, we have chosen it to describe our systems. It has been defined by Alur-Dill [13] as a finite set of states and a finite set of clocks which are real-valued variables. All clocks proceed at the same rate and measure the amount of time that has elapsed since they were started or reset. Each transition of the system might reset some of the clocks, and has an associated enabling condition (EC) which is a constraint on the values of the clocks. A transition is enabled only if its corresponding state is reached and the current clock values satisfy its enabling condition.

4   Test Sequence Mutation

The purpose of our testing method is to check how the system reacts to hazards, and consequently to some stressful situations. A hazard could be defined as any unexpected event from the environment. In our study we handle external hazards. External hazards are modeled as actions received by the system or erroneous timing constraints. The automatic generation and integration of hazards in the test sequences can be respectively found in [14] and [15].

Each component of the RTCBS is described by a nominal and a degraded specification. Each of these specifications is represented by input-complete, reduced, strongly connected timed input-output automata TIOA.

For each component, we derive a test sequence for each controllable state. The generation method uses the test purpose technique [16]. The generated test sequences start with an input event and end with output event. For each controllable state we derive a sequence able to characterize this state among all others.

A state s of a TIOA is said controllable, if all reached states by s are connected to it by transitions labelled by only input actions.

Fig.1 shows an example of a nominal specification with initial state s_1 and a degraded specification with initial state s'_1. A transition is represented by an arrow between two states and labeled by (action; EC; Cs). The set of actions A={?temperatureReq,!temperature,?posReq,!pos,?moveMode,?endMoveMode} and the set of states S={s_1, s_2, s_3, s_4} and S'={s'_1, s'_2, s'_3}

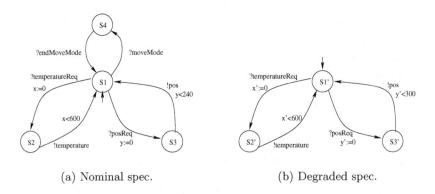

(a) Nominal spec.   (b) Degraded spec.

Fig. 1. A nominal and a degraded specification

4.1   Hazard Injection

Hazards are inserted in the generated test sequences to simulate a hostile environment.

In fact, we will modify only some input actions in test sequences. Since we check the ability of the system to react correctly in presence of relevant actions, we do not allow to modify any input action of the degraded specification.

Suppose that $M = (S, A, C, T, s_0)$ is the nominal specification, and $M' = (S', A', C', T', s'_0)$ is the degraded one.

The designer decides which scenario he wants to integrate in a sequence based on some previous experiences. He may have already a collection of known scenari of non expected events from the environment. The insertion may be done automatically based on a probabilistic choice or manually. Let $T = \{Tr_1.Tr_2...Tr_n\}$ this sequence, where each Tr_i from M. Let $T' = \{Tr'_1.Tr'_2...Tr'_{n'}\}$ the mutated sequence. In the following we detail the most well known situations:

1. Replacing an input action

   We simulate the fact that another component sends an unexpected action to the tested component. The designer chooses a transition $Tr_i = (a_i; EC_i; C_{si})$

of T such that $a_i \in A_I - A'_I$. Then we change the action a_i with an action $a' \in A_I - A'_I$. Thus, the result T' is $T' = \{Tr_1.Tr_2... Tr_{i-1}.Tr'_i.Tr_{i+1}...Tr_n\}$ with $Tr'_i = (a'; EC_i; C_{si})$.

2. Changing the instant of an input action occurrence

   Here, we simulate the fact that another component sends the expected action, but not at the right moment. This could happen for example in case of heavy CPU processing. The designer chooses a transition $Tr_i = (a_i; EC_i; C_{si})$ of T such that $a_i \in A_I - A'_I$ and such that EC_i is bounded. Then we change the enabling condition EC_i with EC' such that EC' is the complementary of EC_i. In fact, we just delay the occurrence of the transition with an amount of time δ, such that the occurrence arrives later than expected. Thus, the result $T'= \{Tr_1.Tr_2... Tr_{i-1}.Tr'_i.Tr_{i+1}...Tr_n\}$ where $Tr'_i = (a_i; EC'; C_{si})$.

3. Exchanging two input actions

   We simulate the situation where other components have scheduling troubles. The designer chooses two transitions $Tr_i = (a_i; EC_i; C_{si})$ and $Tr_j = (a_j; EC_j; C_{sj})$ of T such that $a_i \in A_I - A'_I$ and $a_j \in A_I - A'_I$. Then we exchange the actions a_i and a_j. Thus, the result T' is $T' = \{Tr_1.Tr_2...Tr'_i...Tr'_j ...Tr_n\}$ with $Tr'_i = (a_j; EC_i; C_{si})$ and $Tr'_j = (a_i; EC_j; C_{sj})$.

4. Adding an unexpected transition

   Here, we simulate the fact that a component of the whole system sends an unexpected additional action. This may be caused for example by some troubles with any sensor. The designer chooses a transition $Tr_i = (a_i; EC_i; C_{si})$ of T such that $a_i \in A_I$. He also chooses an action $b \in A_I - A'_I$. We will add a transition in the sequence with the input action b without changing the timing conditions of the whole sequence. In fact, we insert the new action respecting the timing conditions allowed for Tr_i. Finally the resulting sequence becomes:
   $T' = \{Tr_1.Tr_2...Tr_{i-1}.Tr'.Tr'_i.Tr_{i+1}...Tr_n\}$ with $Tr' =(b; EC'; -)$ such that $EC' \subset EC_i$ and $Tr'_i =(a_i; EC_i; C_{si})$

5. Removing a transition

   In this part, we simulate the fact that an information is lost in the system. This could happen for many reasons, such that a problem in communication channels. The designer chooses a transition $Tr_i = (a_i; EC_i; C_{si})$ of T such that $a_i \in A_I - A'_I$ and we just remove it from the sequence. Thus the the resulting sequence becomes: $T' =\{Tr_1.Tr_2...Tr_{i-1}.Tr_{i+1}...Tr_n\}$.

5   Test Execution

Fig. 2 illustrates the test architecture we use. It consists of a set of distributed local testers. For each component, P_m, of the system, a dedicated Tester, T_m,is assigned. Each tester is also detailled with its modules : TEU (Test Executer Unit) and TMU (Test Monitor Unit).

This part is inspired by the work done in [17] dedicated to an adapted test architecture for distributed timed systems.

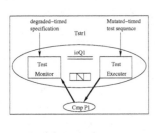

(a) unit tester

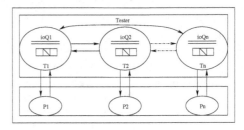

(b) main architecture

Fig. 2. Test Architecture

5.1   Tester Coordination

Components communicate with each other through their corresponding testers. An output from component P_i to component P_j is achieved by sending the output to tester T_i which in turns sends to tester T_j, next, T_j forwards the output to P_j. In this model we ignore the time taken by the communication between components through testers (we only cnsider the amount of time taken by direct component communication).

Testers communicate with each other through the input/output queues (ioQ). The output transition from tester T_i, sent as an input, to tester T_j will wait on queue ioQ_j until this transition is enabled in T_j. On the other hand, the execution of a communication test sequence CTs_j by tester T_j will pause if this test sequence requires P_j to wait on an input from component P_i. The execution in tester T_j will resume after receiving the needed input from tester T_i; and thus, testers give a higher priority to handle inputs received from the components over inputs received from other testers that are stored in local testers' queues.

In order to cover time space, for any input action,"?A", the corresponding tester sends "A" as soon as the enabling condition EC of A satisfies the instant "t" reached by the local tester's clock, and another experiment is performed at the latest instant satisfying EC.

5.2   Test Execution Algorithm

The testing execution process is done in two phases. The first phase tests the robustness of each component separately, and the second phase tests the robustness of communication among components.

In both phases, each tester T_m executes, using TEU, the corresponding mutated test sequences and records, using TMU, the corresponding feedbacks from each component into execution traces without instantaneous evaluation. We check the robustness of each recorded execution trace from each component with its corresponding degraded specification. The theoretical framework of this issue is described in [18]

In the first phase, we ignore all communication requests from other components; and thus, all input actions from testers are sent at the instant where those inputs are needed in the nominal specification S^{P_m} without considering the communication from any other tester T_i , and therefore, here we are checking the robustness of the component in isolation, without taking any communication input from other components.

An execution of a mutated test sequence Ts_m by T_m gives a verdict success, iff, in its corresponding recorded execution trace, the reception of outputs from P_m are accepted in the degraded specification $S_d^{P_m}$.

In the second phase, all testers T_m execute, using TEU, the same communication test sequence ITs generated from the communication nominal specification. Tester T_i executes only its corresponding events (transitions) in ITs ignoring all transitions for other testers. An execution of tester T_m to a communication test sequence ITs_m gives a verdict success, iff , in its corresponding recorded execution trace, the sending of inputs and reception of outputs $O_m^{T_i}$ from P_m or T_i are accepted in the communication degraded specification $S_d^{C_m}$.

Algorithm 1 : Test execution
This code is executing on component / tester indexed "m".
Input: S^{P_m} Test sequences: Ts_m, ITs_m ; ITs_m initially is empty
Output: verdict pass, fail: set of failed components.

Phase 1:

1 For all components P_m
2   For all test sequences Ts_m
3   Select a test sequence Ts_j^m
4   Respecting timing constraints, apply each event of Ts_j^m
        via tester T_m to P_m.
5   Record the inputs and the corresponding feedbacks into traces
        for each component.
6   Check the $accept_{\mathcal{R}}$ of each recorded trace with its degraded specification
7   Assign the appropriate verdict (Robust — NonRobust)

Phase 2:

9 For all components:
This code is executing on component / tester indexed "m".
10   For all test sequences ITs_m
11   Select a test sequence ITs_j^m from ITs_m
12   For all transitions Tr_k of ITs_j^m
13   If (current event of ITs_j^m is a local transition:Tr_k^m)
14   Respecting timing constraints, T_m applies current event
        of ITs_j^m on component P_m.
15   Record the inputs and the corresponding feedbacks into traces
        for each P_m,

taking into consideration the update of enabling conditions with time delays.

16   T_m forwards the *appropriate* output of $O_m^{T_i}$ to io-queue ioQ_i of tester T_i.

17   If (tester T_i is waiting on this input)

18   Tester T_m Signals Tester T_i to resume execution:**signal**(T_i).

19   else if current event is not local transition: Tr_k^i

20   if transition Tr_k^i is found in (already sent to) current local queue ioQ_m

21   Respecting timing constraints, of Tr_k^i, T_m applies Tr_k^i to P_m.

22   Record the corresponding traces for each component.
          taking into consideration the update of enabling conditions with time delays.

23   else if the other tester T_i is not waiting on an input from this tester

24   pause execution of T_m until it gets input from T_i: **wait** (T_m).

25   else if (the other tester T_i is waiting on an input from this tester T_m)

26   Supply T_m with the appropriate input from S^{P_m} to be applied on P_m.

27   Record the corresponding traces for each component taking into consideration the update of enabling conditions with time delays. Identify a deadlocks state on test sequence ITs_j^m.

28 Signal all testers waiting on T_m.

29 Check the $accept_R$ of each recorded trace with the degraded . specification of the corresponding component.

30 Assign the appropriate verdict (Robust — NonRobust)

Comments on the algorithm First phase is locally testing robustness of each component and also identifies interoperable sequences. In the second phase, each tester T_m selects one of its sequences in ITs_m, and applies it, while the other testers parallely do the same with other sequences. At the end of the algorithm, all the possible combinations of sequences has been tested. Supposing that each tester has an average of k sequences, and N components, this gives about k^N test cases.

For each output transition of sequence ITs_m, tester T_m has to send this action to the corresponding tester, and eventually to signal it resuming of execution. If it is an input, the tester identifies if it comes from the environment or from a tester. In the last case, T_m applies the corresponding action only an other componant has already sent this action to P_m.

If a tester T_i is waiting on an input from T_j, it pauses the execution until the corresponding action arrives. However, if tester T_i is waiting on an input from T_j and reciproqually, then a deadlock case is identified.

6   Conclusion and Future Work

Robustness testing for real-Time component-based systems is discussed in this paper. To the best of our knowledge, quite a few works has been done in this field. In this paper, we present a methodology for testing robustness of real-time component-based systems using fault injection and adequate distributed test architecture. Each component is described by its nominal specification and its degraded one. Three main contributions are noticeable in this method:

The first is the automatic generation of test sequence set for each component from its nominal specification. Then relevant faults are inserted in these sequences in order to simulate hostile environment.

The second contribution is that by the end of the first phase of the test execution, we are able to tell about all robust-implemented stand-alone components. In this phase, we experiment mutated test sequences on each component of the IUT and we record the results traces. These later are checked on the degraded specification. Each component is said to be robust if the verdict of experimentation of recorded traces on the degraded specification gives the verdict success.

A third contribution is the test-execution algorithm that executes and synchronizes test sequence execution on local testers. The synchronization is done via two atomic statements, Signal() and Wait(), and a set of input-output queues. A queue is attached to each tester. The robustness of the whole IUT is deduced if the communication events between components are accepted by the degraded specifications of all components.

As a future work, we intend to investigate more realistic hazard insertion by using metrics produced by real case studies. We need also to experiment this methodology in a real case study such as industrial control software or complex embedded systems.

We intend to use other real-time modeling languages as UML-RT or statecharts to describe RTCBS. We will reduce the state explosion problem inherent to the timed automata model.

References

1. Garlan, D., Allen, R., Ockerbloom, J.: Architectural mismatch: Why reuse is so hard. IEEE software (1995)
2. Khoumsi, A.: Timing issues in testing distributed systems. In: 4th IASTED International Conference on Software Engineering and Applications, SEA2000, November 2000. (2000)
3. Bertolino, A., Corradini, F., Inveradi, P., Muccini, H.: Deriving test plans from architectural descriptions. In: ACM Proceedings, International Conference on Software Engineering ICSE2000, June 2000. (2000)
4. Zimmerer, P.: Test architectures for testing distributed systems. In: 12th International software quality week (QW'99), May 1999. (1999)
5. Schmidt, H.: Trustworthy components-compositionality and prediction. The Journal of Systems and Software 65 (2003) 215–225
6. Zalewski, J.: Developing component-based software for real-time systems. In: 27th Euromicro Conference 2001: A Net Odyssey (euromicro'01), September 2001. (2001)

7. Tesanovic, A., Nystrom, D., Hansson, J., Norstrom, C.:   Towards aspectual component-based development of real-time systems. In: Proceeding of the 9th International Conference on Real-Time and Embedded Computing Systems and Applications (RTCSA 2003), February 2003. (2003)
8. Barton, J.H., Czeck, E.W., Segall, Z.Z., Siewiorek, D.P.: Fault injection experiments using fiat. IEEE Trans. Comput. **39** (1990) 575–582
9. Tsai, T.K., Iyer, R.K., Jewitt, D.: An approach towards benchmarking of fault-tolerant commercial systems. In: Symposium on Fault-Tolerant Computing. (1996) 314–323
10. Kropp, N.P., Jr., P.J.K., Siewiorek, D.P.: Automated robustness testing of off-the-shelf software components. In: Symposium on Fault-Tolerant Computing. (1998) 230–239
11. Castanet, R.: Les enjeux du test de robustesse. In: Journées du Réseau Thématique Prioritaire SECC. (2002)
12. Brinksma, E., Coulson, G., Crnkovic, I.: (Project ist-2001-34820 - artist- advanced real-time systems. roadmap: Component-based design and integration platforms) http://www.systemes-critiques.org/ARTIST/.
13. Alur, R., Dill, D.: A theory of timed automata. Theoretical Computer Science **126** (1994) 183–235
14. Fouchal, H., Rollet, A.: Embedded system testing. In: Proceedings of 7th International Conference on Principles of Distributed Systems (OPODIS 2003), December 10-13 2003, La Martinique, France. Lecture Notes in Computer Science, Springer-Verlag (2003)
15. Rollet, A.: Testing robustness of real-time embedded systems. In: Proceedings of Workshop On Testing Real-Time and Embedded Systems (WTRTES), Satellite Workshop of FM 2003 Symposium, Pisa, Italy - September 13, 2003. (2003)
16. Fouchal, H., Petitjean, E., Salva, S.: An User-Oriented Testing of Real Time Systems. In: Proceedings of the International Workshop on Real-Time Embeded Systems RTES'01 (London), IEEE Computer Society. (2001)
17. Khoumsi, A.: Testing distributed real-time systems in the presence of inaccurate clock synchronization. Journal of Information Soft. Technology (IST) **45** (2003)
18. Tarhini, A., Rollet, A., Fouchal, H.: A pragmatic approach for robustness testing on real time component based systems. In: The 3rd ACS/IEEE International Conference on Computer Systems and Application (AICSSA05), January 2-5 2005, Cairo, Egypt. (2004) accepted.

Topology Generation for Web Communities Modeling⋆

György Frivolt and Mária Bieliková

Institute of Informatics and Software Engineering,
Faculty of Informatics and Information Technologies,
Slovak University of Technology in Bratislava,
Ilkovičova 3, 842 16 Bratislava, Slovakia
{frivolt, bielik}@fiit.stuba.sk

Abstract. In this paper we present a model of Web communities which constitute a part of the Web structure. The proposed model is aimed at characterization of the topology behind the Web communities. It is inspired by small world graphs that show behaviors similar to many natural networks. We model Web communities as clusters of Web pages using graph grammars. Graph grammars allow us to simulate the structural properties of Web communities including their growth and evolution. An example of a grammar is presented. We discuss possibilities for utilization of the proposed model for research into Web communities, their properties and identification.

1   Introduction

As the Web grows, effective searching for information becomes more and more important. Present Web search engines typically crawl the Web pages in order to build indexes and/or local copies for further analysis. The search is based mainly on analysis of the content gathered. Several search engines use the hyperlink structure to provide additional information regarding the quality of the results (using for example the PageRank algorithm [13]). Knowledge of the structure of the Web graph dramatically improves the search results, in particular ranking of the search results. However, most current search engines consider the Web as a network at a rather low level of abstraction in which the vertices represent Web pages and the edges are associated with hyperlinks that connect the information content of the pages. To capture the features of the Web at a higher level of abstraction, considering a collection of Web pages created by individuals, or any kind of associations that have a common interest on a specific topic (web communities), instead of the Web pages per se, is a challenging task. However, it would enable reasoning at a higher level of abstraction, with the potential for improving the efficiency and accuracy of the information search, and also for improving the search results.

⋆ This work has been partially supported by the Grant Agency of Slovak Republic grant No. VG1/ 0162/03.

M. Bieliková et al. (Eds.): SOFSEM 2005, LNCS 3381, pp. 167–177, 2005.

Groupings can be observed in various natural networks. People, companies, etc., which form collections (or clusters) represented by vertices, are often denoted as communities. Edges represent interactions such as social relations between people in a social network, or trade relationships in a business network. Communities exist not only in the physical world. Research into the Web showed that they emerge in the virtual world as well [7].

Since it is our intention to model communities on the Web, we concentrate on clusters formed by Web pages. The primary understanding of the Web communities comes from sociology [10]. Similar to the connections found in human society, there exist connections between Web pages created by those who share a common interest (so that the content of the pages is oriented towards a specific topic).

We distinguish two main sources of knowledge that can be extracted from the Web: (i) the Web page content, and (ii) the topology of the network. There is a growing amount of work directed at the identification of Web communities according to the topology of the network based on hyperlink structure, i.e., it is supposed that Web pages which share similar themes, or similar interests of the authors, are interconnected, or that they belong to the same cluster [5, 4, 12]. Due to the large size of the Web, the topology of the Web network is largely unknown or unexplored. A significant step towards using the topology of the Web for reasoning about its content is the PageRank algorithm [13]. The idea behind PageRank is that it is possible to extract the quality of a Web page based on the references (or hyperlinks) leading to it, i.e., from its position on the network. A different method, but one still relying on topology, is introduced by Jon Kleinberg [5, 6]. In [7, 10] the authors show structures observable on the Web and explain the motivations for searching among communities on the Web.

Our aim is to define a model which captures the concept of Web communities. The proposed model is tested by constructing an example of a grammar and analyzing selected properties of the graphs generated according to the defined grammar. In a way similar to the approaches mentioned above we rely on the topology of the Web network and assume that the quality of the Web page content is correlated with the incoming/outgoing links of the page.

The rest of the paper is structured as follows. In section 2 we describe small-world graphs, which form a viable alternative for Web modeling. Section 3 discusses the proposed model based on the graph grammar system. In Section 4 we give an example of a grammar together with generated graphs. The properties of the generated graphs are described. The paper concludes with a discussion, a summary, and a description of future directions of our research.

2   Small-World Graphs and Web Networks

Different types of natural networks share some specific features. Despite their random character the topology of the graphs representing these networks has a number of universal scale-free characteristics and displays a high degree of clustering. The graphs show the so called *small-world effect*, possessing average vertex-to-vertex distances which increase only logarithmically with the total number of vertices, together with a clustering effect (which is missing in a random graph) [11]. Small-world networks can be observed in many spheres of nature. The networks of neurons in the brain, genetic networks, social networks of people, networks of words in natural languages, the Internet at the router or domain levels, and networks of Web pages, all share the features mentioned [14, 9, 2].

Ordered and random networks differ in two seemingly opposed ways. Ordered networks exhibit high clustering, i.e., neighboring vertices share several common neighbors. On the other hand, the average distance between any two vertices in an ordered network is high. Random networks show significant differences from ordered networks in these two properties. The growth of a random network with a given coordination number (average number of neighbors of each vertex) results in a decrease of the number of common neighbors. Furthermore, any two vertices can be connected by a relatively short path.

The difference in scale between ordered and random networks is large. Models for scaling the transition from ordered to random networks are studied in [11, 15]. Networks called small-world networks share the interesting properties of both random and ordered networks: high levels of clustering and low relative distances between the vertices. These properties for small-world networks are as follows.

Average Vertex Distance. The average distance ℓ between any two vertices in a small-world network logarithmically depends on the size N of the network:

$$\ell \approx \log{(N)}$$

Logarithmic dependence allows the average distance between the vertices to be quite small even in very large networks. The precise definition of the average distance between vertices in a small-world network is still a matter of debate, but it is accepted that ℓ should be comparable with the value it would have on a random graph [11].

Clustering. Vertices in the same area are connected to each other. The clustering coefficient C_v for a vertex v with k_v neighbors is

$$C_v = \frac{2E_v}{k_v(k_v - 1)}$$

where E_v is the number of edges between the k_v neighbors of v.

Empirical results indicate that C_v averaged over all nodes is significantly higher for most real networks than for a random network, and the clustering coefficient of real networks is to a high degree independent of the number of nodes in the network [14].

Several authors have studied big portions of the Web network (with vertices representing the Web pages and connections representing hyperlinks pointing from one page to another) and demonstrated its small-world properties. In [11, 2] the average diameter for a Web network with $N = 8 * 10^8$ vertices is shown to be $\ell_{web} = 18.59$, i.e., two randomly chosen pages on the Web are on average

19 clicks away from each other. The logarithmic dependence of average distance between the Web pages on the number of the pages is important to the future potential for growth of the Web. For example, the expected 1 000% increase in the size of the Web over the next few years will change ℓ_{web} to only 21 [2].

3   Web Topology Generation Using Graph Grammars

As already mentioned, the Web graph shows the characteristics of a scale-free network. However, empirical measurements have also shown its hierarchical topology [14]. The modular organization of the Web is related to the high clustering coefficient. The Web model should reflect these characteristics.

We have proposed to model this kind of the pattern using graph generating L-systems. L-systems are a class of string rewriting mechanisms, originally developed by Lindenmayer [8] as a mathematical theory of plant development. With an L-system, a sequence of symbols (string) can be rewritten into another sequence, by replacing all symbols in the string in parallel by other symbols, using so-called rewriting rules (also called production rules).

L-systems are capable of generating fractal-like structures. Self-similarity was observed also in the Web [3]. General properties of the Web topology discussed in Section 2 can also be found in its parts. We expect that the proposed approach is also capable of generating networks that capture the growth of the Web, together with its Web communities large scale topology with the properties of scale-free networks with a high clustering.

Definition 1. *We call a tuple $Gr = (R, \sigma)$ a graph generating L-system, where σ is the initial graph and R is the finite set of production rules written in the form $LHS \rightarrow RHS$.*

The production rules of a graph grammar are mappings of the vertices. The application of a production rule to a vertex of the graph means replacing the vertex with the vertices defined on the right hand side of the rule. We do not distinguish between terminal and non-terminal states.

The LHS of a graph generation rule represents a vertex. The RHS of the rule consists of (a) a list of vertices together with related mappings of edges incident to these vertices and the LHS vertex, and (b) a list of edges joining the mapped vertices defined in the RHS.

Definition 2. *We denote a production rule as:*

$$v \rightarrow \left\{ \begin{array}{l} (v_1, \mu_1, p_1), \\ (v_2, \mu_2, p_2), \\ \dots \end{array} \right\}, \eta$$

where
$p_i \in [0, 1]$ is probability of mapping the vertex v to v_i;
$\mu_i \in [0, 1]$ is probability of overtaking an incident edge to the vertex v and v_i;
η is a subset of edges joining the vertices $v_i \in \{v_1, v_2, \dots\}$ such that

$$\eta \subset \{(_ov, _iv, p)|_ov, _iv \in \{v_1, v_2, \ldots\}, p \in [0, 1]\}.$$

where p is probability of generating an incident edge to the vertex $_ov$ and $_iv$.

An L-system grows the graph starting with the initial graph by applying production rules. The rule application means a replacement of a vertex with the vertices mapped by the right hand side of the rule. The rule application is called an expansion.

Definition 3. *Let $r \in R$ be a production rule, G a graph, $v \in G(V)$ a vertex, and $e_1, e_2, \ldots \in G(E)$ edges incident to the vertex v. We call an expansion a mapping:*

$$ApplyRule : G \times G(V) \times R \mapsto G'$$

The result of the application of $ApplyRule(G, v, r) = G'$ is:

$$G'(V) = (G(V) \setminus \{v\}) \cup \{p_1(v_1), p_2(v_2), \ldots\}$$
$$G'(E) = \{\mu_1(e_1), \mu_1(e_2), \ldots, \mu_2(e_1, \ldots), \ldots\} \cup \eta(v_1, v_2, \ldots)$$

where
$p_i : \{v_i\} \mapsto \{v_i, \perp\}$, $\mu_i : \{e_i\} \mapsto \{e_i, \perp\}$ is a mapping giving items from the set $\{v_1, \ldots\}$, $\{e_1, \ldots\}$ with probability p_i, resp. μ_i;
η is deduced from p_i as $\eta : 2^{\{p_i(v_1), \ldots\}} \mapsto \{_{out}v, _{in}v|_{out}v, _{in}v \in \{p_1(v_1), \ldots\}\}$.

The graph grows by repeated expansion. The inference step in a grammar is executed by the application of randomly chosen rule on every vertex.

Definition 4. *Let R be a set of rules, and G_1 and G_2 graphs. We say that G_2 is inferred from G_1 if a sequence $s = (v_1, r_1), (v_2, r_1), \ldots (v_n, r_n)$ exists, where*

- *$\forall v \in G_1(V) \exists i \leq n \exists r \in R : s_i = (v, r)$ and $\forall i, j \leq n, i \neq j : v(s_i) \neq v(s_j)$ [1]*
- *$\coprod_{i=1}^n ApplyRule(G_1, s_i) = G_2$ [2]*

$G_1 \overset{\cdot}{\leadsto}_R G_2$ *denotes that G_2 was inferred from G_1 in one inference step using R. If there exists a sequence of inference steps $G_1 \overset{\cdot}{\leadsto}_R G_2, G_2 \overset{\cdot}{\leadsto}_R G_2, \ldots G_{n-1} \overset{\cdot}{\leadsto}_R G_n$, we say that G_n can be inferred from G_1 and denote it as $G_1 \leadsto_R G_n$.*

We note that every vertex is mapped during one inference step once and only once. Finally we define the language generated by a grammar.

Definition 5. *Let $Gr = (R, \sigma)$ be a graph generating L-system. We call set of graphs L a language generated by the grammar Gr if every graph contained in L can be inferred from the initial graph σ using the rules from the finite set R:*
$L = \{G|\sigma \leadsto_R G\}$.

---

[1] $v(s_i)$ is the first (vertex) item of the tuple.
[2] $\coprod_{i=1}^3 ApplyRule(G_1, s_i) = ApplyRule(ApplyRule(ApplyRule(G_1, s_1), s_2), s_3)$

4   Graph Grammar Application

We have used the proposed language in our experiments to generate a topology with properties similar to the Web network. Our approach is demonstrated by a simple grammar containing three rules. We measure two properties of the generated graphs: the clustering coefficient and the graph diameter. We show that the formalism presented in the previous section is strong enough to generate graphs with properties that resemble small-world networks. Although the generated graphs are directed, in our measurements of the clustering coefficient and network diameter we consider them as undirected, which suffices for the purposes of evaluating the characteristics of the generated graphs. We developed a software prototype for graph generation using the specified grammar, in the Python programming language. The visualization of the generated graphs was performed by the BioLayout software[3].

4.1   Definition of the Example Grammar

The example grammar contains three rules generating three kinds of structures:

- *hierarchies:* a vertex is mapped to one central and several child vertices;
- *bipartite graphs:* generated vertices are divided into two sets such that no edge connects vertices in the same set;
- *cliques:* a vertex is mapped to the graph where a majority of vertices is connected.

Fig. 1 shows examples of the first expansion of each rule: a hierarchy with three child vertices, a bipartite structure with two sets by three vertices and a five clique cluster. Fig. 2 presents graphs generated by several expansions using again each rule. The grammar of every example produces graphs from an initial graph of a single vertex: $Gr = (R, \sigma = G(\{v\}, \emptyset))$. The rules R are defined thereinafter.

Fig. 1. Illustration of one expansion for (a) hierarchy, (b) bipartite structures and (c) cliques

Hierarchies. Hierarchical organization can be observed in several real complex networks including the Web. A graph theoretical discussion related to the fact that the hierarchy is a fundamental characteristic of many complex systems can be found in [14].

---

[3] http://www.ebi.ac.uk/research/cgg/services/layout/

Fig. 2. Structures generated by the hierarchy production rule after 4 inference steps (a), bipartite (b) and clique (c) generation rule after 2 inference steps

An example of a hierarchy generation rule is defined as follows:

$$v \rightarrow \left\{ \begin{array}{l} (v_{central},\ 1.0,\ 1.0), \\ (v_{child_1},\ 0.2,\ 0.8), \\ (v_{child_2},\ 0.2,\ 0.8), \\ (v_{child_3},\ 0.2,\ 0.8) \end{array} \right\} , \left\{ \begin{array}{l} (v_{central}, v_{child_i}, 0.8), \\ (v_{child_i}, v_{central}, 0.2), \\ (v_{child_i}, v_{child_j}, 0.2)|i, j \leq 3, i \neq j \end{array} \right\}$$

The hierarchy generation rule of our grammar produces a structure containing one central and a maximum of three child vertices. The central vertex is with high probability connected with the child vertices. We set a lower probability for generating connections between the child vertices.

The graph generated by four inference steps has a clustering coefficient of 0.475. The diameter of the largest component is 8, and the total number of vertices and edges is 93 and 169, respectively (see Fig. 2a).

Bipartite Graphs. Bipartite structure models service-provider relationships, which occur on the current Web quite often. Web communities in this case are formed implicitly, i.e., the community is formed by unconnected vertices (an actual example of this is where providers' pages on similar topics do not provide links to each other).

The rule defined below generates a bipartite graph $K_{3,3}$. The clustering coefficient of the structure after the first expansion is 0. The clustering remains low after two inference steps. The graph in Fig. 2b has clustering coefficient 0.111.

$$v \rightarrow \left\{ \begin{array}{l} (v_{service_1},\ 0.3,\ 0.8), \\ (v_{service_2},\ 0.3,\ 0.8), \\ (v_{service_3},\ 0.3,\ 0.8), \\ (v_{customer_1}, 0.3,\ 0.8), \\ \dots \\ (v_{customer_5},\ 0.3,\ 0.8) \end{array} \right\} , \left\{ \begin{array}{l} (v_{service_i}, v_{customer_j}, 0.8), \\ (v_{customer_j}, v_{service_i}, 0.8)|i \leq 3, j \leq 5 \end{array} \right\}$$

Cliques. The clique structure models mutually interconnected Web pages. This kind of structure can be found, for example, in Web portals such as corporate Web sites or home pages. An example of a clique generation rule is defined as follows:

$$v \rightarrow \left\{ \begin{array}{l} (v_1, 0.6, 0.8), \\ (v_2, 0.6, 0.8), \\ \ldots \\ (v_5, 0.6, 0.8) \end{array} \right\}, \{(v_i, v_j, 0.8)|i, j \le 5, i \ne j\}$$

The clustering coefficient of the graph in Fig. 2c generated by two inference steps is 0.619.

4.2   Mixed Structures

We have generated various mixed structures using a grammar consisting of the three rules defined above. The rules are applied randomly, each vertex is mapped by one of the three rules in every step of inference. Two exampes of graph evolution are illustrated in Fig. 3. The measured values are listed in Tab. 1.

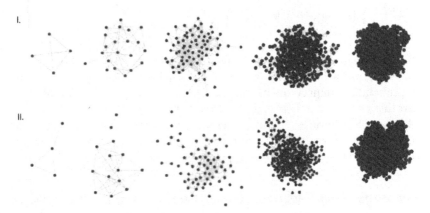

Fig. 3. Illustration of graph evolutions. Evolution of the graph *I* is started by a clique structure, whereas graph evolution *II* starts as a hierarchical structure. The initial shape of the graph persists over the growth, however after several iterations the two graphs become similar in shape and properties (see Tab. 1)

5   Discussion and Conclusions

The main contribution of this paper is to propose a formalism capable of modeling the topology of Web communities. The results in Tab. 1 support our aim to generate graphs with small-world effects. The clustering coefficient is much higher than in random graphs. However, more experiments are needed in order to tune the parameters defined within the production rules, or to define new useful production rules that would improve the small-world characteristics of

Table 1. Properties of generated graphs illustrated in Fig. 3/I in the first half and fig. 3/II in the second half of the table

inf. steps	$|G(V)|$	$|G(E)|$	clustering	diameter	avg. out deg.	avg. in deg.
I-1	4	11	1.0	1	2.75	2.75
I-2	21	100	0.3657	3	5.0	4.76
I-3	81	423	0.3542	5	5.42	5.29
I-4	370	2 352	0.3254	7	6.60	6.44
I-5	1 719	11 997	0.2788	10	7.18	7.01
I-6	7 856	60 206	0.2744	14	7.88	7.73
II-1	4	4	0.5833	2	1.33	1.33
II-2	14	56	0.6418	4	4.31	4.31
II-3	71	371	0.2788	7	5.46	5.3
II-4	330	2 063	0.2844	9	6.47	6.29
II-5	1 519	10 852	0.2841	11	7.37	7.2
II-6	7 094	55 272	0.2841	14	8.01	8.85

the generated graphs. One such extension is to introduce edges between distant vertices.

Naturally several directions for future work emerge. We give a list of possible usages of the formalism presented in this paper.

Analysis of Graph Properties Based on the Rules. The results presented in Tab.1 show that although the initial properties of the graphs differ, after several iterations the resulting generated graphs have similar clustering coefficients and diameters. These properties depend on the rules of the grammar. So we assume that the properties can be computed without the inference of the graphs, which can save considerable resources when experimenting with appropriate rules for Web topology generation.

Modeling Interactions Between Web Pages. Currently we map only one vertex to a set of the vertices. By mapping more vertices we could model also interactions between Web sites. Such a model requires also modeling of attributes of the Web pages and a definition of strategies for identification of those vertices, which repose in the LHS of the rules. Although our current model produces expanding graphs, a set of rules extended by the possibility of a definition of more vertices on the RHS could also decrease the number of vertices or edges.

Definition of Scalable Models. Models introduced in [1, 11] are scalable. Similar scale parameters can be introduced into the formalism proposed here. Tweaking of these parameters would result in grammars with different properties.

Graph Pattern Recognition. The proposed formalism can be used for testing or modeling some aspects of natural networks. A tool for generating networks similar to natural ones can be useful for testing algorithms for identification of the structure of the network, which was our main intention. However, another aspect that we also found interesting was the recognition of patterns defined on the RHS of grammar rules of a natural network. We expect to be able to identify a network's structure by working backwards through the inference sequence using recognition of RHS patterns. This process is far from simple. We should at least ensure the continuous backward chaining and effective recognition of isomorphic graphs.

The proposed model extends classical L-systems by defining probabilities of the mapping of vertices and edges. By not using exact patterns we hope to decrease the complexity of the problem of computation. We believe that the work presented in this paper can be of great help in the analysis of Web communities. The characteristics of generated graphs are promising in the sense that they possess similar properties to those expected of actual Web graphs. Generated graphs could serve as a basis for identification of Web communities and their use in searching for information and recommending of high quality.

References

1. Albert-László Barabási and Réka Albert. Emergence of scaling in random networks. *Science*, 286:509–512, October 1999. www.sciencemag.org.
2. Albert-László Barabási, Réka Albert, and Hawoong Jeong. Scale-free characteristics of random networks: the topology of the World-Wide Web. *Physica A*, 281:69–77, 2000.
3. Stephen Dill, Ravi Kumar, Kevin S. McCurley, Sridhar Rajagopalan, D. Sivakumar, and Andrew Tomkins. Self-similarity in the Web. 2:205–223, August 2002.
4. Gary Flake, Steve Lawrence, and C. Lee Giles. Efficient identification of Web communities. In *Sixth ACM SIGKDD International Conference on Knowledge Discovery and Data Mining*, pages 150–160, Boston, MA, August 2000.
5. David Gibson, Jon M. Kleinberg, and Prabhakar Raghavan. Inferring Web communities from link topology. In *Proc. of the 9th ACM Conf. on Hypertext and Hypermedia*, pages 225–234, 1998.
6. Jon M. Kleinberg. Authoritative sources in a hyperlinked environment. *Journal of the ACM*, 46:604–632, September 1999.
7. Ravi Kumar, Prabhakar Raghavan, Sridhar Rajagpalan, and Andrew Tomkins. Trawling the Web for emerging cyber-communities. In *Proc. of the 8th World Wide Web Conference*, pages 1481–1493, 1999.
8. Aristid Lindenmayer. Mathematical models for cellular interaction in development. 18:280–315, 1968.
9. Mária Markošová. Language as a small world network. In J. Kelemen and V. Kvasnička, editors, *Proc. of Conf. on Cognition and Artificial Life*.
10. Pınar Yolum and Munindar P. Singh. Dynamic communities in referral networks. *Web intelligence and agent systems*, 1:105–116, December 2003.
11. Mark E. J. Newman. Models of small world (a review). *Physical Review Letters*, May 2000. cond-mat/0001118.
12. Mark E. J. Newman and Michelle Girvan. Finding and evaluating community structure in networks. *Physical Review Letters*, 2004. cond-mat/026113.
13. Lawrence Page, Sergey Brin, Rajeev Motwani, and Terry Winograd. The Pagerank citation ranking: Bringing order to the Web. Standford Digital Libraries, Technologies Project, 1998.

14. Erzsébet Ravasz and Albert-László Barabási. Hierarchical organization in complex networks. *Physical Review Letters*, September 2002. cond-mat/0206130.
15. Duncan J. Watts. *Small Worlds: The Dynamics of Networks between Order and Randomness*. Princeton University Press, 2003.

Recursion Versus Replication in Simple Cryptographic Protocols

Hans Hüttel* and Jiří Srba**

BRICS***, Department of Computer Science, University of Aalborg
Fredrik Bajersvej 7B, 9220 Aalborg East, Denmark

Abstract. We use some recent techniques from process algebra to draw several conclusions about the well studied class of ping-pong protocols introduced by Dolev and Yao. In particular we show that all nontrivial properties, including reachability and equivalence checking wrt. the whole van Glabbeek's spectrum, become undecidable for a very simple recursive extension of the protocol. The result holds even if no nondeterministic choice operator is allowed. We also show that the extended calculus is capable of an implicit description of the active intruder, including full analysis and synthesis of messages in the sense of Amadio, Lugiez and Vanackère. We conclude by showing that reachability analysis for a replicative variant of the protocol becomes decidable.

Note: full proofs are available in [11].

1 Introduction

Process calculi have been suggested as a natural vehicle for reasoning about cryptographic protocols. In [1], Abadi and Gordon introduced the spi-calculus and described how properties such as secrecy and authenticity can be expressed via notions of observational equivalence (like may-testing). Alternatively, security questions have been studied using reachability analysis [3, 5, 9].

We provide a basic study of expressiveness and feasibility of cryptographic protocols. We are interested in two verification approaches: *reachability analysis* and *equivalence (preorder) checking*. In reachability analysis the question is whether a certain (bad or good) configuration of the protocol is reachable from a given initial one. In equivalence checking the question is whether a protocol implementation is equivalent (e.g. bisimilar) to a given specification (optimal behaviour). These verification strategies can be used even in the presence of an *active intruder* (in the Dolev-Yao style), i.e., an agent with capabilities to listen to any communication, to perform analysis and synthesis of communicated messages according to the actual knowledge of compromised keys, and to actively

* hans@cs.auc.dk
** srba@cs.auc.dk, supported in part by the GACR, grant No. 201/03/1161.
*** Basic Research in Computer Science,
    Centre of the Danish National Research Foundation.

M. Bieliková et al. (Eds.): SOFSEM 2005, LNCS 3381, pp. 178–187, 2005.

participate in the protocol behaviour by transmitting new messages. This can be naturally implemented not only into the reachability analysis (see e.g. [4]) but also into the equivalence checking approach (see e.g. [10]).

A number of security properties are decidable for finite protocols [3, 14]. In the case of an unbounded number of protocol configurations, the picture is more complex. Durgin et al. showed in [8] that security properties are undecidable in a restricted class of so-called bounded protocols (that still allows for infinitely many reachable configurations). In [2] Amadio and Charatonik consider a language of tail-recursive protocols with bounded encryption depth and name generation; they show that, whenever certain restrictions on decryption are violated, one can encode two-counter machines in the process language. On the other hand, Amadio, Lugiez and Vanackère show in [4] that the reachability problem is in PTIME for a class of protocols with iteration.

In this paper we focus solely on ping-pong based behaviours of recursive and replicative protocols (perhaps the simplest behaviour of all studied calculi) in order to draw general conclusions about expressiveness and tractability of formal verification of cryptographic protocols. The class of *ping-pong protocols* was introduced in 1983 by Dolev and Yao [7]. The formalism deals with memory-less protocols which may be subjected to arbitrarily long attacks. Here, the secrecy of a finite ping-pong protocol can be decided in polynomial time. Later, Dolev, Even and Karp found a cubic-time algorithm [6]. The class of protocols studied in [4] contains iterative ping-pong protocols and, as a consequence, secrecy properties remain polynomially decidable even in this case.

In the present paper we continue our study of recursive and replicative extensions of ping-pong protocols. In [12] we showed that the recursive extension of the calculus is Turing powerful, however, the nondeterministic choice operator appeared to be essential in the construction. The question whether the calculus is Turing powerful even without any explicit way to define nondeterministic processes was left open. Here we present a radically new reduction from multi-stack automata and strengthen the undecidability results to hold even for protocols without nondeterministic choice. We prove, in particular, that both reachability and equivalence checking for all equivalences and preorders between trace equivalence/preorder and isomorphism of labelled transition systems (which includes all equivalences and preorders from van Glabbeek's spectrum [15]) become undecidable. These results are of general importance because they prove the impossibility of automated verification for essentially all recursive cryptographic protocols capable of at least the ping-pong behaviour.

In the initial study from [12], the question of active attacks on the protocol was not dealt with. We shall demonstrate that a complete notion of the active intruder (including analysis and synthesis of messages in the sense of Amadio, Lugiez and Vanackère [4]) can be explicitly encoded into our formalism.

Finally, we study a replicative variant of the calculus. Surprisingly, such a calculus becomes decidable, at least with regard to reachability analysis. We use a very recent result from process algebra (decidability of reachability for weak process rewrite systems by Křetínský, Řehák and Strejček [13]) in order to derive the result.

2   Basic Definitions

2.1   Labelled Transition Systems with Label Abstraction

In order to provide a uniform framework for our study of ping-pong protocols, we define their semantics by means of labelled transition systems. A *labelled transition system* (LTS) is a triple $\mathcal{T} = (S, \mathcal{A}ct, \longrightarrow)$ where S is a set of *states* (or *processes*), $\mathcal{A}ct$ is a set of *labels* (or *actions*), and $\longrightarrow \subseteq S \times \mathcal{A}ct \times S$ is a *transition relation*, written $\alpha \xrightarrow{a} \beta$, for $(\alpha, a, \beta) \in \longrightarrow$. As usual we extend the transition relation to the elements of $\mathcal{A}ct^*$. We also write $\alpha \longrightarrow^* \beta$, whenever $\alpha \xrightarrow{w} \beta$ for some $w \in \mathcal{A}ct^*$.

The idea is that the states represent *global configurations* of a given protocol and the transitions describe the *information flow*. Labels on the transitions moreover represent the messages (both plain-text and cipher-text) which are being communicated during the state changes.

The explicit possibility to observe the full content of messages is sometimes not very realistic; it means that an external observer of such a system can e.g. distinguish between two different messages encrypted by the same encryption key, without the actual knowledge of the key.

In order to restrict capabilities of the observer we introduce a so called *label abstraction function* $\phi : \mathcal{A}ct \mapsto \mathcal{A}ct$. Given a LTS $\mathcal{T} = (S, \mathcal{A}ct, \longrightarrow_{\mathcal{T}})$ and a label abstraction function ϕ we define a new LTS $\mathcal{T}_\phi \stackrel{\text{def}}{=} (S, \mathcal{A}ct, \longrightarrow_{\mathcal{T}_\phi})$ where $\alpha \xrightarrow{\phi(a)}_{\mathcal{T}_\phi} \beta$ iff $\alpha \xrightarrow{a}_{\mathcal{T}} \beta$ for all $\alpha, \beta \in S$ and $a \in \mathcal{A}ct$. We call $\mathcal{T}_\phi$ a *labelled transition system with label abstraction*.

Let us now focus on the messages (actions). Assume a given set of encryption keys $\mathcal{K}$. The set of all messages over $\mathcal{K}$ is given by the following abstract syntax

$$m ::= k \mid k \cdot m$$

where k ranges over $\mathcal{K}$. Hence every element of the set $\mathcal{K}$ is a *(plain-text) message* and if m is a message then $k \cdot m$ is a *(cipher-text) message* (meaning that the message m is encrypted by the key k). Given a message $k_1 \cdot k_2 \cdots k_n$ over $\mathcal{K}$ we usually[1] write it only as a word $k_1 k_2 \cdots k_n$ from $\mathcal{K}^*$. Note that k_n is the plain-text part of the message and the outermost encryption key is always on the left (k_1 in our case). In what follows we shall identify the set of messages and $\mathcal{K}^*$, and we denote the extra element of $\mathcal{K}^*$ consisting of the empty sequence of keys by ϵ.

The level of abstraction we may select depends on the particular studied property we are interested in. Nevertheless, it seems reasonable to require at least the possibility to distinguish between plain-text and cipher-text messages.

---

[1] In our previous work on ping-pong protocols [12] we denoted a message m encrypted by a key k as $\{m\}_k$. We changed the notation in order to improve the clarity of the proofs. In particular, when messages like $k_1 k_2 \cdots k_n$ are used, the previous syntax described the keys in a reversed order, which was technically inconvenient.

We say that a label abstraction function ϕ is *reasonable* iff $\phi(k) \neq \phi(k'w)$ for all $k, k' \in \mathcal{K}$ and $w \in \mathcal{K}^+$.

2.2   A Calculus of Recursive Ping-Pong Protocols

We shall now define a calculus which captures exactly the class of ping-pong protocols by Dolev and Yao [7] extended (in a straightforward manner) with recursive definitions.

Let $\mathcal{K}$ be a set of encryption keys. A *specification* of a recursive ping-pong is a finite set of process definitions Δ such that for every *process constant* P (from a given set *Const*) the set Δ contains exactly one process definition of the form

$$P \stackrel{\text{def}}{=} \sum_{i_1 \in I_1} v_{i_1} \triangleright . \overline{w_{i_1} \triangleright}.P_{i_1} + \sum_{i_2 \in I_2} v_{i_2}.P_{i_2} + \sum_{i_3 \in I_3} \overline{w_{i_3}}.P_{i_3}$$

where I_1, I_2 and I_3 are finite sets of indices such that $I_1 \cup I_2 \cup I_3 \neq \emptyset$, and v_{i_1}, v_{i_2}, w_{i_1} and w_{i_3} are messages (belong to $\mathcal{K}^*$) for all $i_1 \in I_1$, $i_2 \in I_2$ and $i_3 \in I_3$, and $P_i \in Const \cup \{\mathbf{0}\}$ for all $i \in I_1 \cup I_2 \cup I_3$ such that $\mathbf{0}$ is a special constant called the *empty process*. We moreover require that v_{i_2} and w_{i_3} for all $i_2 \in I_2$ and $i_3 \in I_3$ are different from the empty message ϵ. (Observe that any specification Δ contains only finitely many keys.)

Summands continuing in the empty process constant $\mathbf{0}$ will be written without the $\mathbf{0}$ symbol and process definitions will often be written in their unfolded form using the *nondeterministic choice operator* '+'. An example of a process definition is e.g. $P \stackrel{\text{def}}{=} k_1 \triangleright . \overline{k_2 \triangleright}.P_1 + k_1 \triangleright . \overline{k_3 \triangleright} + k_1 k_2.P_1 + \overline{k_1 k_1} + \overline{k_1 k_2}.P_2$.

The intuition is that each summand of the form $v_{i_1} \triangleright . \overline{w_{i_1} \triangleright}.P_{i_1}$ can receive a message encrypted by a sequence v_{i_1} of outermost keys, decrypt the message using these keys, send it out encrypted by the sequence of keys w_{i_1}, and finally behave as the process constant P_{i_1}. The symbol $\triangleright$ stands for the rest of the message after decrypting it with the key sequence v_{i_1}. This describes a standard ping-pong behaviour of the process. (Note that the symbol $\triangleright$ is equivalent to our $\{x\}$ notation from [12]).

In addition to this we may have summands of the forms $v_{i_2}.P_{i_2}$ and $\overline{w_{i_3}}.P_{i_3}$, meaning simply that a message is received and forgotten or unconditionally transmitted, respectively. This is a small addition to the calculus we presented in [12] in order to allow for discarding of old messages and generation of new messages. These two features were not available in the earlier version of the calculus but they appear to be technically convenient when modeling an explicit intruder and for strengthening the positive decidability results in Section 5. Nevertheless, the undecidability results presented in Section 3 are valid even without this extension since only the standard ping-pong behaviour is used in the constructions. A feature very similar to the forgetful input operation can be also found in [4].

A *configuration* of a ping-pong protocol specification Δ is a parallel composition of process constants, possibly preceded by output messages. Formally the set *Conf* of configurations is given by the following abstract syntax

$$C ::= \mathbf{0} \mid P \mid \overline{w}.P \mid C \parallel C$$

where $\mathbf{0}$ is the empty configuration, $P \in Const \cup \{\mathbf{0}\}$ ranges over process constants including the empty process, $w \in \mathcal{K}^*$ ranges over the set of messages, and '$\|$' is the operator of parallel composition.

We introduce a structural congruence relation $\equiv$ which identifies configurations that represent the same state of the protocol. The relation $\equiv$ is defined as the least congruence over configurations ($\equiv \subseteq Conf \times Conf$) such that $(Conf, \|, \mathbf{0})$ is a commutative monoid and $\overline{\epsilon}.P \equiv P$ for all $P \in Const$. In what follows we shall identify configurations up to structural congruence.

Remark 1. We let $\overline{\epsilon}.P \equiv P$ because the empty message should never be communicated. This means that when a prefix like $k \triangleright .\overline{\triangleright}.P$ receives a plain-text message k and tries to output $\overline{\epsilon}.P$, it simply continues as the process P.

We shall now define the semantics of ping-pong protocols in terms of labelled transition systems. We define a set $Conf_S \subseteq Conf$ consisting of all configurations that do not contain the operator of parallel composition and call these *simple configurations*. We also define two sets $In(C, m), Out(C, m) \subseteq Conf_S$ for all $C \in Conf_S$ and $m \in \mathcal{K}^+$. The intuition is that $In(C, m)$ ($Out(C, m)$) contains all configurations which can be reached from the simple configuration C after receiving (resp. outputting) the message m from (to) the environment. Formally, $In(C, m)$ and $Out(C, m)$ are the smallest sets which satisfy:

- $Q \in In(P, m)$ whenever $P \in Const$ and $m.Q$ is a summand of P
- $\overline{w\alpha}.Q \in In(P, m)$ whenever $P \in Const$ and $v \triangleright .\overline{w\triangleright}.Q$ is a summand of P such that $m = v\alpha$
- $P \in Out(\overline{m}.P, m)$ whenever $P \in Const \cup \{\mathbf{0}\}$
- $Q \in Out(P, m)$ whenever $P \in Const$ and $\overline{m}.Q$ is a summand of P.

A given protocol specification Δ determines a labelled transition system $T(\Delta) \stackrel{\text{def}}{=} (S, Act, \longrightarrow)$ where the states are configurations of the protocol modulo the structural congruence ($S \stackrel{\text{def}}{=} Conf/\equiv$), the set of labels (actions) is the set of messages that can be communicated between the agents of the protocol ($Act \stackrel{\text{def}}{=} \mathcal{K}^+$), and the transition relation $\longrightarrow$ is given by the following SOS rule (recall that '$\|$' is commutative).

$$\frac{m \in \mathcal{K}^+ \quad C_1, C_2 \in Conf_S \quad C_1' \in Out(C_1, m) \quad C_2' \in In(C_2, m)}{C_1 \| C_2 \| C \stackrel{m}{\longrightarrow} C_1' \| C_2' \| C}$$

This means that (in the context C) two simple configurations (agents) C_1 and C_2 can communicate a message m in such a way that C_1 outputs m and becomes C_1' while C_2 receives the message m and becomes C_2'.

For further discussion and examples of recursive ping-pong protocols we refer the reader to [12].

2.3   Reachability and Behavioural Equivalences

One of the problems that is usually studied is that of *reachability analysis*: given two configurations $C_1, C_2 \in Conf$ we ask whether C_2 is reachable from C_1, i.e., if $C_1 \longrightarrow^* C_2$. In this case the set of labels is irrelevant.

As the semantics of our calculus is given in terms of labelled transition systems (together with an appropriate label abstraction function), we can also study the *equivalence checking* problems. Given some behavioural equivalence or preorder $\leftrightarrow$ from van Glabbeek's spectrum [15] (e.g. strong bisimilarity or trace, failure and simulation equivalences/preorders just to mention a few) and two configurations $C_1, C_2 \in Conf$ of a protocol specification Δ, the question is to decide whether C_1 and C_2 are $\leftrightarrow$-equivalent (or $\leftrightarrow$-preorder related) in $T(\Delta)$, i.e., whether $C_1 \leftrightarrow C_2$.

3   Recursive Ping-Pong Protocols Without Explicit Choice

In this section we strengthen the undecidability result from [12] and show that the reachability and equivalence checking problems are undecidable for ping-pong protocols without an explicit operator of nondeterminism and using classical ping-pong behaviour only, i.e., for protocols without any occurrence of the choice operator '+' and where every defining equation is of the form $P \stackrel{\text{def}}{=} v \triangleright . \overline{w} \triangleright . P'$ such that $P' \in Const$.

We moreover show that the negative results apply to all behavioural equivalences and preorders between trace equivalence/preorder and isomorphism of LTS (which preserves labelling) with regard to all reasonable label abstraction functions as defined in Section 2.

These results are achieved by showing that recursive ping-pong protocols can step-by-step simulate a Turing powerful computational device, in our case a computational model called multi-stack machines.

A *multi-stack machine* R with ℓ stacks ($\ell \geq 1$) is a triple $R = (Q, \Gamma, \longrightarrow)$ where Q is a finite set of *control-states*, Γ is a finite *stack alphabet* such that $Q \cap \Gamma = \emptyset$, and $\longrightarrow \subseteq Q \times \Gamma \times Q \times \Gamma^*$ is a finite set of *transition rules*, written $pX \longrightarrow q\alpha$ for $(p, X, q, \alpha) \in \longrightarrow$.

A *configuration* of a multi-stack machine R is an element from $Q \times (\Gamma^*)^\ell$. We assume a given initial configuration $(q_0, w_1, \ldots, w_\ell)$ where $q_0 \in Q$ and $w_i \in \Gamma^*$ for all i, $1 \leq i \leq \ell$. If some of the stacks w_i are empty, we denote them by ϵ.

A *computational step* is defined such that whenever there is a transition rule $pX \longrightarrow q\alpha$ then a configuration which is in the control-state p and has X on top of the i'th stack (the tops of the stacks are on the left) can perform the following transition: $(p, w_1, \ldots, Xw_i, \ldots, w_\ell) \longrightarrow (q, w_1, \ldots, \alpha w_i, \ldots, w_\ell)$ for all $w_1, \ldots, w_\ell \in \Gamma^*$ and for all i, $1 \leq i \leq \ell$.

It is a folklore result that multi-stack machines are Turing powerful. Hence (in particular) the following problem is easily seen to be undecidable: given an initial configuration $(q_0, w_1, \ldots, w_\ell)$ of a multi-stack machine R, can we reach the configuration $(h, \epsilon, \ldots, \epsilon)$ for a distinguished halting control-state $h \in Q$ such that all stacks are empty? Without loss of generality we can even assume that a configuration in the control-state h is reachable iff all stacks are empty.

Let $R = (Q, \Gamma, \longrightarrow)$ be a multi-stack machine. We define the following set of keys of a ping-pong specification Δ: $\mathcal{K} \stackrel{\text{def}}{=} Q \cup \Gamma \cup \{k_p \mid p \in Q\} \cup \{t, k_*\}$. Here t is a special key such that every communicated message is an encryption of the plain-text key t. The reason for this is that it ensures that the protocol never communicates any plain-text message. The key k_* is a special purpose locking key and it is explained later on in the construction.

We shall construct a ping-pong protocol specification Δ as follows.

- For every transition rule $pX \longrightarrow q\alpha$ we have a process constant $P_{pX \longrightarrow q\alpha}$ with the following defining equation: $P_{pX \longrightarrow q\alpha} \stackrel{\text{def}}{=} pX \triangleright . \overline{k_q \alpha \triangleright} . P_{pX \longrightarrow q\alpha}$.
- For every state $p \in Q$ we have two process constants T_p and T'_p.

$$T_p \stackrel{\text{def}}{=} k_p \triangleright . \overline{k_* \triangleright} . T'_p$$

$$T'_p \stackrel{\text{def}}{=} k_* \triangleright . \overline{p \triangleright} . T_p \quad \text{if } p \in Q \smallsetminus \{h\}, \text{ and} \quad T'_h \stackrel{\text{def}}{=} h \triangleright . \overline{h \triangleright} . T'_h$$

Recall that $h \in Q$ is the halting control-state.

- Finally, we define a process constant B (standing for a buffer over a fixed key k_*): $B \stackrel{\text{def}}{=} k_* \triangleright . \overline{k_* \triangleright} . B$.

  In this defining equation the key k_* locks the content of the buffer such that it is accessible only by some T'_p.

Note that Δ does not contain any choice operator '+' as required.

Let $(q_0, w_1, \ldots, w_\ell)$ be an initial configuration of the multi-stack machine R. The corresponding initial configuration of the protocol Δ is defined as follows (the meta-symbol Π stands for a parallel composition of the appropriate components).

$$\left(\prod_{(r,A,s,\beta) \in \longrightarrow} P_{rA \longrightarrow s\beta} \right) \parallel \left(\prod_{p \in Q \smallsetminus \{q_0\}} T_p \right) \parallel T'_{q_0} \parallel \left(\prod_{j \in \{1, \ldots, \ell\}} \overline{k_* w_j t} . B \right) \quad (1)$$

The following invariants are preserved during any computational sequence starting from this initial configuration:

- at most one T'_p for some $p \in Q$ is present as a parallel component (the intuition is that this represents the fact that the machine R is in the control-state p), and
- plain-text messages are never communicated.

Theorem 1. *The reachability problem for recursive ping-pong protocols without an explicit choice operator is undecidable.*

Theorem 2. *The equivalence checking problem for recursive ping-pong protocols without an explicit choice operator is undecidable for any behavioral equivalence/preorder between trace equivalence/preorder and isomorphism (including all equivalences and preorders from van Glabbeek's spectrum [15]) and for any reasonable label abstraction function.*

4   The Active Intruder

In the literature on applying process calculi to the study of cryptographic protocols, there have been several proposals for explicit modelling the active intruder (environment). Foccardi, Gorrieri and Martinelli in [10] express the environment within the process calculus, namely as a process running in parallel with the protocol. In [4] Amadio, Lugiez and Vanackère describe a tiny process calculus similar to ours, except that they use replication instead of recursion. Moreover, the environment is described in the semantics of the calculus. Transitions are of the form $(C, T) \to (C', T')$ where C and C' are protocol configurations and T and T' denote the sets of messages known to the environment (all communication occurs only by passing messages through these sets).

The environment is assumed to be hostile; it may compute new messages by means of the operations of analysis and synthesis and pass these on to the process. Let $\mathcal{K}$ be a set of encryption keys as before. The *analysis* of a set of messages $T \subseteq \mathcal{K}^*$ is the least set $\mathcal{A}(T)$ satisfying

$$\mathcal{A}(T) = T \cup \{w \mid kw \in \mathcal{A}(T),\ k \in \mathcal{K} \cap \mathcal{A}(T)\}. \tag{2}$$

The *synthesis* of a set of messages $T \subseteq \mathcal{K}^*$ is the least set $\mathcal{S}(T)$ satisfying

$$\mathcal{S}(T) = \mathcal{A}(T) \cup \{kw \mid w \in \mathcal{S}(T),\ k \in \mathcal{K} \cap \mathcal{S}(T)\}. \tag{3}$$

We can now design an environment sensitive semantics for our calculus close in style to that of [4]. We define the reduction relation $\to$ by the following set of axioms (here $x \in P$ means that x is a summand in the defining equation of the process constant P).

$$
\begin{array}{lll}
(P \parallel C, T) \to (\overline{wa}.P' \parallel C, T) & \text{if } (v \triangleright . \overline{w}\triangleright.P') \in P \text{ and } v\alpha \in \mathcal{S}(T) & \text{(A1)} \\
(P \parallel C, T) \to (P' \parallel C, T) & \text{if } (v.P') \in P \text{ and } v \in \mathcal{S}(T) & \text{(A2)} \\
(\overline{w}.P \parallel C, T) \to (P \parallel C, T \cup \{w\}) & & \text{(A3)} \\
(P \parallel C, T) \to (P' \parallel C, T \cup \{w\}) & \text{if } (\overline{w}.P') \in P & \text{(A4)}
\end{array}
$$

We show that this semantics can be internalized in our calculus within our existing semantics. The construction is nontrivial as (on the contrary with stronger calculi like spi-calculus) we can use only a very limited set of operations. The details are in the full paper.

Theorem 3. *For any recursive ping-pong protocol, we can define its new parallel component which enables all the attacks described by axioms (A1) – (A4).*

5   Replicative Ping-Pong Protocols

In this section we shall define a replicative variant of our calculus for ping-pong protocols. We will then show that this formalism is not Turing powerful because the reachability problem becomes decidable.

Let us now define *replicative ping-pong protocols*. Let $\mathcal{K}$ be the set of encryption keys as before. The set *Conf* of protocol configurations is given by the following abstract syntax

$$C ::= \mathbf{0} \mid v \triangleright . \overline{w \triangleright} \mid v \mid \overline{w} \mid !(v \triangleright . \overline{w \triangleright}) \mid !(v) \mid !(\overline{w}) \mid C \parallel C$$

where $\mathbf{0}$ is the symbol for the empty configuration, v and w range over $\mathcal{K}^*$, and ! is the bang operator (replication). As before, we shall introduce structural congruence $\equiv$, which is the smallest congruence over *Conf* such that $(Conf, \parallel, \mathbf{0})$ is a commutative monoid; $\epsilon \parallel C \equiv C \equiv \overline{\epsilon} \parallel C$; $!(\epsilon) \equiv \mathbf{0} \equiv !(\overline{\epsilon})$; and $!(C) \equiv C \parallel !(C)$. A labelled transition system determined by a configuration (where states are configurations modulo $\equiv$ and labels are non-empty messages as before) is defined by the following SOS rules (recall the replicative axiom $!(C) \equiv C \parallel !(C)$ and the fact that '$\parallel$' is commutative).

$$\frac{m \in \mathcal{K}^+}{\overline{m} \parallel m \parallel C \xrightarrow{m} C} \qquad \frac{m \in \mathcal{K}^+ \qquad m = v\alpha}{\overline{m} \parallel v \triangleright . \overline{w \triangleright} \parallel C \xrightarrow{m} \overline{w\alpha} \parallel C}$$

We can now show that the reachability problem for general replicative ping-pong protocols is decidable. We reduce our problem to reachability of weak process rewrite systems (wPRS) which was very recently proven to be decidable [13].

Theorem 4. *The reachability problem for replicative ping-pong protocols is decidable.*

6   Conclusion

We have seen that ping-pong protocols extended with recursive definitions have full Turing power. This is the case even in the absence of nondeterministic choice operator '+'. A result like this implies that any reasonable property for all richer calculi cannot be automatically verified.

We also presented an explicit description of the active intruder in the syntax of recursive ping-pong protocols.

Finally, we showed that reachability analysis for a replicative variant of the protocol becomes feasible. Our proof uses very recent results from process algebra [13] and can be compared to the work of Amadio, Lugiez and Vanackère [4] which establishes the decidability of reachability for a similar replicative protocol capable of ping-pong behaviour. Their approach uses a notion of a pool of messages explicitly modelled in the semantics and reduces the question to a decidable problem of reachability for prefix rewriting. In our approach we allow spontaneous generation of new messages which is not possible in their calculus. Moreover, we can distinguish between replicated and once-only behaviours (unlike in [4] where all processes have to be replicated).

Last but not least we hope that our approach can be possibly extended to include other operations as the decidability result for replicative protocols uses only a limited power of wPRS (only a parallel composition of stacks). Hence there is a place for further extensions of the protocol syntax while preserving a decidable calculus (e.g. messages of the form $\overline{k_1(k_2\,op\,k_3)k_4}$ for some extra composition operation op on keys can be easily stored in wPRS as $k_1.(k_2\,\|\,k_3).k_4$). Such a study is left for future research.

References

1. M. Abadi and A.D. Gordon. A bisimulation method for cryptographic protocols. *Nordic Journal of Computing*, 5(4):267–303, 1998.
2. R.M. Amadio and W. Charatonik. On name generation and set-based analysis in the Dolev-Yao model. In *Proc. of CONCUR'02*, vol. 2421 of *LNCS*, 499–514. Springer-Verlag, 2002.
3. R.M. Amadio and D. Lugiez. On the reachability problem in cryptographic protocols. In *Proc. of CONCUR'00*, vol. 1877 of *LNCS*, 380–394. Springer-Verlag, 2000.
4. R.M. Amadio, D. Lugiez, and V. Vanackère. On the symbolic reduction of processes with cryptographic functions. *TCS*, 290(1):695–740, October 2002.
5. M. Boreale. Symbolic trace analysis of cryptographic protocols. In *Proc. of ICALP'01*, vol. 2076 of *LNCS*, 667–681. Springer, 2001.
6. D. Dolev, S. Even, and R.M. Karp. On the security of ping-pong protocols. *Information and Control*, 55(1–3):57–68, 1982.
7. D. Dolev and A.C. Yao. On the security of public key protocols. *Transactions on Information Theory*, IT-29(2):198–208, 1983.
8. N. Durgin, P. Lincoln, J. Mitchell, and A. Scedrov. Undecidability of bounded security protocols. In N. Heintze and E. Clarke, editors, *Proc. of FMSP'99*, 1999.
9. M. Fiore and M. Abadi. Computing symbolic models for verifying cryptographic protocols. In *Proc. of CSFW'01*, 160–173. IEEE, 2001.
10. R. Focardi, R. Gorrieri, and F. Martinelli. Non interference for the analysis of cryptographic protocols. In *Proc. of ICALP'00*, vol. 1853 of *LNCS*, 354–372. Springer-Verlag, 2000.
11. H. Hüttel and J. Srba. Recursion vs. replication in simple cryptographic protocols. Technical Report RS-04-23, BRICS Research Series, 2004.
12. H. Hüttel and J. Srba. Recursive ping-pong protocols. In *Proc. of WITS'04*, 129–140, 2004.
13. M. Křetínský, V. Řehák, and J. Strejček. Extended process rewrite systems: Expressiveness and reachability. In *Proc. of CONCUR'04*, vol. 3170 of *LNCS*, 355–370. Springer-Verlag, 2004.
14. M. Rusinowitch and M. Turuani. Protocol insecurity with a finite number of sessions and composed keys is NP-complete. *TCS*, 299, 2003.
15. R.J. van Glabbeek. The linear time - branching time spectrum I: The semantics of concrete, sequential processes. In *Handbook of Process Algebra*, chapter 1, 3–99. Elsevier Science, 2001.

Modeling Data Integration with Updateable Object Views

Piotr Habela[1], Krzysztof Kaczmarski[2], Hanna Kozankiewicz[3],
and Kazimierz Subieta[1,3]

[1] Polish-Japanese Institute of Information Technology, Warsaw, Poland
habela@pjwstk.edu.pl
[2] Warsaw University of Technology, Warsaw, Poland
kaczmars@mini.pw.edu.pl
[3] Institute of Computer Science PAS, Warsaw, Poland
{hanka, subieta}@ipipan.waw.pl

Abstract. Recently, a range of applications of views increases. Views are not anymore tightly related to classical databases – there are proposals to use them as means of data transformation and integration in distributed environment. Despite many aspects of view applications, there is still lack of suitable graphical notation that would help designers by providing clear notions from the very early stage of system development. Therefore, our objective is to propose a suitable extension of UML supporting the design process. We focus on modeling object-oriented updateable views in the context of data integration. We believe it is one of the most prominent appliances of views. The paper describes assumed general features of updateable object views and fits them into an object-oriented metamodel. Based on this, we suggest the necessary view-specific notation elements, and present some examples of view modeling.

1 Introduction

Views constitute one of the fundamental database mechanisms, which provide virtual images of data stored in a database. Views are an important component of many applications as they provide abstraction and generalization over data, transformation of data, access and merging data from multiple sources, etc. Views can be used in Web applications as means of integration of heterogeneous resources stored at remote sites into a unified ontology. Traditionally such applications were implemented in lower level languages like C or Java. An advantage of using query language to describe integration is a higher level of abstraction what in turns reduces the time required for development and the cost of maintenance. Although the idea to use views as a mean of data integration is not new (e.g. [2]) and views are an important component of many applications, there are still no satisfactory means to model them. The problem of view's modeling has already appeared in literature [1, 10]. Several useful notions were discussed in the context of relational databases, views, and mapping between object diagrams and database tables. The mentioned guides propose constructs for virtual objects (certain stereotype), view dependencies (dependency) and read-only attributes. However, they are not advanced due to limited capabilities of views in relational databases. They allow to represent flat, read-only views to stored data whereas modern databases (relational, object-relational, object-oriented, XML-oriented) require much more advanced modeling features.

M. Bieliková et al. (Eds.): SOFSEM 2005, LNCS 3381, pp. 188–198, 2004.

Earlier approaches to view modeling have presented structural view definition in contrast to operational definition used in classical systems, where a view is defined by a single query [5]. The idea was to provide a representation for resulting objects and their properties instead of defining how they are created. In this sense, our solution is rather structural. However, it is necessary to note that our object and view models are much more sophisticated than the ones assumed by the ODMG standard [3, 11].

We propose a new extension of UML that allows to model integration of data through updateable views [6, 8]. The view features covered by this notation are inspired by the implementation we have developed, based on the Stack-Based Approach. However, we believe that the notation can be also used for modeling views defined within other approaches e.g. in SQL using *instead-of triggers* like in Oracle and MS SQL Server. Our objective was to analyze the ways of extending existing modeling standards to support more advanced capabilities of database systems. We believe that a better view modeling would:

- create a reliable communication tool between business and DBMS designers;
- simplify the process of view design and implementation;
- clarify and standardize views documentation;
- allow to better comprehend the dependencies between objects and views;
- provide uniform expressiveness for all system components.

The proposed notions are dedicated for a design phase of a software lifecycle. We are skeptical about any automatic translation between view models and view implementation because graphical notations are so far less descriptive than programming/query languages. Therefore, we do not assume the elimination of the implementation phase.

The rest of the paper is structured as follows. Section 2 describes our approach to updateable views that is a basis for this paper. In Section 3, we summarize the existing UML notions related to view modeling and indicate their limitations. Section 4 presents the main contribution of this paper i.e., the extensions to the UML metamodel and the respective notation elements proposed to effectively support the updateable object views modeling. In Section 5, we briefly outline the position of view modeling within the software development process. Section 6 concludes.

2   The Approach to Updateable Views

In this section we shortly present our approach to updateable views [7], which is a motivation for the graphical notation presented in this paper. The view mechanism is based on the Stack-Based Approach (SBA) [13], which assumes that query languages are a special kind of programming languages and can be formally described in a similar manner. SBA defines its own query language – Stack-Based Query Language [14].

A database view definition is not a single query (as it is in SQL), but it is a more complex structure. It consists of two parts: the first one determines the so-called *seeds* (the values or references to stored objects that are the basis for building up virtual objects), and the second one redefines the generic operations on virtual objects. The first part of the view definition is an arbitrarily complex functional procedure. The *seeds* it returns are passed as parameters for the operations on virtual objects. The operations have the form of procedures that override default updating operations. We identified four generic operations that can be performed on virtual objects:

1. **Updating**, which assigns a new value to the virtual object. A parameter the procedure accepts is the new value to be assigned.
2. **Deletion**, which deletes the virtual object.
3. **Insertion**, which inserts a new object into the given virtual object. The object to be inserted is provided as a parameter.
4. **Dereference**, which returns the value of the given virtual object.

For a given view an arbitrary subset of these operations can be defined. If any operation is not defined, it means it is forbidden (we assume no updating through side effects, e.g. by references returned by a view invocation).

Moreover, a view definition may contain nested views, defined within the containing view's environment. Thus, arbitrarily nested complex objects can be constructed.

When a view is invoked in a query, it returns a set of virtual identifiers (that are counterparts of the identifiers of stored objects). Next, when a system tries to perform update operation with a virtual identifier as an l-value, it recognizes that it deals with the virtual object and calls a proper update operation from the view definition. To enable that, a virtual identifier must contain both a seed and the identifier of the view definition. The whole process of view updating is internal to the proposed mechanism and is invisible to view users, who deal with virtual objects in the same manner as with real objects (this feature is known as a view *transparency*).

3   Available Modeling Notions

3.1   UML Versus Object-Oriented Databases

The UML object model is strongly inspired by programming languages like C++ and Java. Thus, the most straightforward to model are applications written with one of the mainstream general purpose programming languages. For other applications like e.g. data modeling for relational databases, specialized profiles are required [1]. One would expect that for modeling object databases the core UML constructs could suffice. In fact, this is not true: we need to go further and reconsider the issue of object's features accessibility and visibility.

The first problem of modeling such database is object relativism, which allows arbitrarily nested object compositions. This feature differs from traditional programming languages, where the structure of objects is physically "flat", since it does not allow object members to be themselves complex objects. Fortunately, UML is not equally restrictive. Nested objects can be represented as class-typed attributes. The detailed structure can be shown using the composition symbol. However, in this case the visual notation becomes rather inconvenient. One needs to choose between the "nested" composition notation, where the class of the subobject is shown inside the class of its owner, and the regular composition (see Fig. 1). CASE tools seldom support the former construct. UML does not allow to draw associations from the nested class symbol, neither. In this sense, it makes the inner object "encapsulated". The latter approach provides the necessary flexibility, at the cost of expressiveness, as the nested object's features are no longer shown inside the super-object's class symbol, and may be thus mistreated as an association.

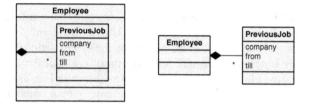

Fig. 1. UML notations available for object composition

Another problem is describing the manipulations allowed for particular object's feature. The visibilities (public, protected, private) definitely do not provide the complete description. Abstracting from the declarations available in current popular programming languages, we find the need for specifying the applicability for the following generic operations (as supported by the previously presented view mechanism): updating, dereferencing[1], inserting and deleting (if a feature's multiplicity allows). The UML metamodel already defines a meta-attribute of similar purpose. This is the *changeability* attribute defined in the *StructuralFeature* metaclass. However its allowed values are enumerated as: *changeable*, *frozen* and *addOnly*, which does not cover all the cases (2^4=16) resulting from the presence or absence of a given generic operation. Simply extending this enumeration to 16 values would be rather impractical. Instead, four independent boolean attributes seem to be more suitable.

Although four generic operations appear in the context of virtual objects, it seems reasonable to apply analogous constraints (marking a given feature e.g. read-only, removable etc.) also to regular (concrete) objects, to preserve view transparency.

3.2   Describing Derived Data

The well-known UML "derived" symbol (represented by "/" character) allows to mark any model element as derived from other element or elements and thus serving redundant data. In practice, as suggested e.g. by the *UML Notation Guide* (part of the official specification [11]), the symbol is applicable to attributes and association ends, to indicate that their contents can be computed from other data. This feature is suitable to mark the database features served by object views. On the other hand, due to genericity of the notion, the way of specifying the source features of a given derived attribute or association is not precise, and requires additional comments to be associated. Taking into account the importance of data source traceability in virtual views modeling, we suggest introducing a kind of dependency relationship dedicated for indicating the source data in a more detailed form of class diagrams.

---

[1] This could be named "reading" as it is intended to return a value representing a given object. However, we chose "dereferencing" to note that even without this operation provided, it is still possible to navigate into a given object (if it is a complex object). For details see [9, 12].

4   Proposed Extensions to the Modeling Notions

4.1   Modeling Virtual Objects

Database Global Objects. It is necessary to decide how the top-level database features[2] should be shown in class diagrams. In UML, *structural features* designate, where class instances may occur. Thus, relying on the *extent* notion can be avoided.

Following this style would require introduction of e.g. *Database* pseudo-class, in order to "anchor" the global object declarations (as shown in Fig. 2a). In typical cases (but not all) it can be perceived as overly formalistic: global object declarations are the only place where the instances of their classes occur, names of those global objects (*instance names*) are usually fixed by classes. Fig. 2b shows a less formal notation for those situations. The presence of *Employee* concrete objects and *Clerk* virtual objects in database's global scope is assumed implicitly.[3] Also the "/" sign marking the derived feature has been moved from the feature name to the class name compartment. The same can be done with changeability symbols discussed later.

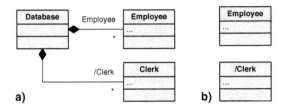

Fig. 2. Traditional (a) and simplified (b) notation for top-level database features

Objects Interfaces. Externally visible features (interfaces) of virtual objects require constructs for proper:

- Distinction of the composition of nested objects and references among objects.
- Marking the derived (virtual) features.
- Showing the changeability allowed for particular (derived or concrete) objects.

The first two problems can be solved with the standard UML notation, provided that there is an agreement on the semantics of the composition relationship. However, changeability flags would need the following symbols (see Fig. 3):

- *isUpdateable* – represented by the exclamation mark ("!");
- *isDereferencable* – represented by the question mark ("?");
- *isRemovable* – represented by the caret mark ("^");
- *isInsertable* – represented by the "greater than" mark (">");

The symbols can appear before a feature name (or before a class name in the simplified syntax suggested in Fig. 3). The changeability symbols are shown within the

---

[2] Usually objects from which we start navigation in queries.
[3] Similarly for the lower levels of object hierarchy, that is, for nested objects, we tend to suppress their composition role name, showing only their class name.

curly brackets in order to allow suppressing changeabilities (by showing no brackets, to distinguish from declaring a feature with none of the changeabilities allowed).

Dependency Illustration. For the most detailed diagrams, the notation presented above can be accompanied with the view dependency symbols, based on the generic UML dependency relationship and using the same graphical notation (labeled «view dependency» if necessary). Notice that for pragmatic reasons we simplify the nota-tion. Although the view dependencies span between structural features (as shown in Fig. 3), the dependency arrows are drawn rather between their classes. In contrast to the regular dependency arrow, view dependency can additionally indicate (using keywords within curly brackets, as shown in Fig. 3), the selectivity and aggregation property (*selection* and *aggregation* keywords respectively). To indicate that particu-lar complex view (that is, a view containing other views) preserves the structure of its source object (mapping the features of the latter), we use the *stem* keyword in the properties representing sub-views. Section 4.2 shows and explains notions introduced in the metamodel.

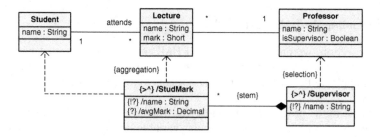

Fig. 3. Exemplary complex view with its data dependency specifications. Assume the data is restructured according to the needs of some external system (e.g. a statistical analysis subsys-tem), which should not have any access to the identities of the students. *StudMark* and *Supervi-sor* show also the changeability notation. *Selection* means, that to provide *Supervisor* virtual objects only certain *Professor* source objects are selected. *Aggregation* indicates that a number of *Lecture* objects is used to create a single *StudMark* object. Stem label indicates preserving the structure (and dependency) of source objects. Here *StudMark* depends on *Supervisor* as *Lecture* is connected to *Professor*

Data Integration. Recently one of the most important tasks of object views is data in-tegration. Fig. 4 presents how example integration can be modeled within our nota-tion. Let us assume the following case. Data about students is distributed among three locations: Warsaw, Gdansk, Cracow, and Radom. All students are identified by their IDs. In Warsaw some personal data: students' names and addresses are kept; in Ra-dom – information about students' scholarships; whereas in Gdansk the information about their supervisors is stored. Additionally, we need to incorporate complete data of other students, provided from Cracow. We would like to gather all these informa-tion and present them as if they were located in one place.

Merge and *join* labels show relationship between dependencies rather than rela-tionships between virtual and concrete objects. This is another field in which UML must be extended. Clearly, while presenting integration of data from multiple sites one can also use the discussed earlier properties of view dependencies like *aggregation*, or *selection*.

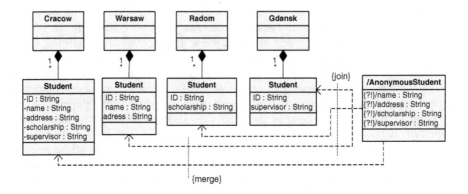

Fig. 4. Integration of distributed data

4.2   Extending the UML Metamodel

In this subsection, we present an extended UML metamodel provided with the features necessary to describe view definitions. The nature of the proposed extension (universal applicability of derived features) seems to justify a modification of the core metamodel.

In contrast to the programming languages, where the class declarations remain independent on their instances (e.g. particular variable declarations), database class (or interface) are often related to particular extent and may determine its name used when referring to the object of that class. In other words, there could be practically one-to-one relation between a class and the place where its instances occur. However, to better align with the UML style and not to limit the object model flexibility, we decided to locate the view-related notions within the feature definition rather than within a class. All the relevant features are shown in Fig. 5.

Modifications of the UML Metamodel. The only change into existing UML notions is the replacement of the *changeability* attribute from the *StructuralFeature* class. As explained in Section 4.1, it would be also possible to keep this attribute and only extend the enumeration of values allowed for it. However, with total number of 16 possible changeabilities, we suggest introducing four boolean attributes as a more intuitive solution. As already explained, we use the following names: *isUpdateable*, *isRemovable*, *isInsertable* and *isDereferencable*.

Additions to the UML Metamodel. We assume that those structural features, which posses (standard-defined) tagged value "derived" represent virtual objects[4] and may therefore be the subject of data dependency specifications.

---

[4]   In our approach we currently deal only with virtual (not materialized) object views. Thus, a feature marked as derived is assumed it to provide virtual objects. A more general approach would require an additional flag to distinguish virtual views from materialized ones.

The dependencies point other features to indicate that they are used as sources for virtual object represented by particular feature. Although it is not possible to precisely describe visually how a given virtual object is computed, some information can be easily provided concerning the characteristics of a view dependencies and relations between them, which are in fact data integration patterns.

- **View Dependency Properties** (mutually orthogonal) modeled by flags in *ViewDependency*: *Selection* (Source data is used to select only the objects meeting a given criteria); *Aggregation* (This property indicates that a given virtual object realizes a many-to-one mapping of the source data).
- **Integration Patterns** modeled by binding two or more dependencies (of complex features) with the *formedWith* association link (Fig. 5). We have suggested: *Merge* (integration dealing with horizontally fragmented data) and, *Join* (used with vertically fragmented data (see example in Section 4.1 and Fig. 4).

We also choose to add a *stem* mark to the composition, which makes the dependency graph simpler (as explained in the previous section and Fig. 3). The necessary meta-attribute *isStem* was located within the *Feature*. This is consistent since each nested view belongs to exactly one composition.

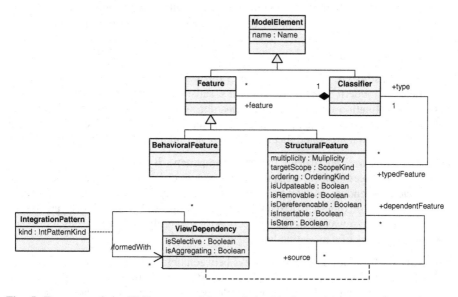

Fig. 5. Fragment of the UML metamodel extended with the notions supporting updateable object views

The proposed dependency properties are not exhaustive, as it is not possible to cover with such notions the whole expressiveness of even the most typical queries that may be used as view definitions. However, it provides some hint concerning the intent of a given view, with the level of detail that is feasible to show on a diagram.

Integration description usually requires choosing an integration key to join the fragmented data. Although it would be possible to specify such key without further extending the metamodel (by introducing yet another *IntegrationPattern* kind), we do not describe it due to the inherent detail limitation of such diagram. Complete specification of view definitions the query language statements are inexplicable.

Note that the information stored using the abovementioned metamodel extensions is very detailed as it may indicate the origin of every elementary data item of a virtual object. However, it is hardly feasible and rather impractical to show such a detailed dependency network on a diagram. Thus, we assume that in most cases, the lowest level of view definitions (that is, the level providing primitive objects) would be "collapsed" using the UML's attribute notation and in consequence, the dependencies on this level would not be shown.

5   View Modeling in a Development Process

View modeling in our approach can be a subject of different modeling perspectives commonly used by practitioners during software development [4]:

- Basic level (analysis) perspective shows only general idea of data reorganization by description of resulting virtual object properties via interface feature. In this stage of system development, too many details could obscure important ideas.
- In design perspective the number of details may be adjusted by a modeler accordingly to specific needs. Data dependencies can be shown, including indication of some typical transformation kinds applied and changeability flags.

As already stressed, the main concern of this research is modeling data integration. This field of modern system design is not satisfactorily covered by existing techniques. This situation may lead to problems during development and could cause delays or unpredicted complications and thus additional costs. Extended modeling constructs based on the presented metamodel, give advantages for data integration efforts.

Data Modeling and Software Change Management. By explicitly documenting dependencies between objects analysts outline required data transformation. Specification of view dependencies may help not only in early estimations of a view complexity, but also in predictions of database change impact. Virtual objects are treated exactly in the same way as normal objects – following the object relativism principle clarifies semantics of a modeling language.

Integration Modeling. Dependency links between objects help recognizing necessary data transfer if objects are distributed. Design diagrams uniformly describe data and data integration paths (plus relationships between sources), thus more completely document the system. Extended information about virtual objects supports reflection and may be dynamically used in more advanced applications. For example, design diagrams may generate the templates for view's implementation.

Integration Verification. Extended view dependency links also help in verification of view implementation completeness. Using uniform modeling notions described in the metamodel helps tracking user requirements through all the development stages.

6   Conclusions and Future Work

In our opinion, the current lack of adequate notation for view modeling in UML is a serious drawback. Therefore, in this paper we proposed an extension of UML that supports view modeling. The presented notation seems to be consistent and well fitted into the UML metamodel. It may be a useful notation that supports view modeling at analysis and design stages of application development.

Presented notation allows to model integration of object-oriented or XML data. It supports modeling of (possibly nested) views and allows to describe relationships between stored and virtual data. Some limitations come from the style UML uses for representing classes and their features, especially in case of nested objects. We tried to propose an optimum solution, providing a valuable description without introducing significant changes into the UML. The notation supports description of access rights to virtual objects at any level of view's hierarchy. The presented notation can be also useful for automatic generation of skeletons of views.

Taking into account that Grid applications are recently in focus of researchers all over the world, we claim that UML should support modeling in this area. Therefore, our future works include development of methodology for data-intensive Grid development that is based on our UML view notation.

References

1. Scott W. Ambler: Agile Database Techniques. Effective Strategies for the Agile Software Developer. John Wiley & Sons 2003
2. Z. Bellahsene: Extending a View Mechanism to Support Schema Evolution in Federated Database Systems. Proc. of DEXA 1997, 573-582
3. R. Cattel, D. Barry. (eds.) The Object Data Standard: ODMG 3.0. Morgan Kaufmann, 2000
4. M. Fowler. UML Distilled, Addison-Wesley Pub Co, ISBN: 0321193687
5. W. Heijenga. View definition in OODBS without queries: a concept to support schema-like views. In Doct. Cons. 2nd Intl. Baltic Wg on Databases and Information Systems, Tallinn (Estonia), 1996.
6. K. Kaczmarski, P. Habela, K. Subieta. Metadata in a Data Grid Construction. Proc. of the 13th IEEE International Workshops on Enabling Technologies: Infrastructures for Collaborative Enterprises (WETICE-2004), Modena, Italy, 2004
7. H. Kozankiewicz, J. Leszczyłowski, K. Subieta. Updateable XML Views. Proc. of ADBIS'03, Springer LNCS 2798, 2003, 385-399
8. H. Kozankiewicz, K. Stencel, K. Subieta. Integration of Heterogeneous Resources through Updatable Views. Proc. of the 13th IEEE International Workshops on Enabling Technologies: Infrastructures for Collaborative Enterprises (WETICE-2004), Italy, 2004
9. H. Kozankiewicz and K. Subieta. SBQL Views – Prototype of Updateable Views. Local Proc. of ADBIS'04, Budapest, Hungary, 2004
10. E. Naiburg, R. A. Maksimchuk. UML for Database Design. Addison-Wesley, 2001
11. Object Management Group: Unified Modeling Language (UML) Specification. Version 1.5, March 2003 [http://www.omg.org].

12. K.Subieta. Theory and Construction of Object-Oriented Query Languages. Editors of the Polish-Japanese Institute of Information Technology, 2004, ISBN: 83-89244-28-4, pp. 522.
13. K. Subieta, C. Beeri, F. Matthes, and J. W. Schmidt. A Stack Based Approach to Query Languages. Proc. of 2nd Intl. East-West Database Workshop, Klagenfurt, Austria, September 1994, Springer Workshops in Computing, 1995.
14. K. Subieta, Y. Kambayashi, and J. Leszczyłowski. Procedures in Object-Oriented Query Languages. Proc. of 21-st VLDB Conf., 182-193, 1995

Fixed-Parameter Tractable Algorithms for Testing Upward Planarity

Patrick Healy and Karol Lynch

CSIS Department, University of Limerick, Limerick, Ireland
{patrick.healy, karol.lynch}@ul.ie

Abstract. We consider the problem of testing a digraph $G = (V, E)$ for upward planarity. In particular we present two fixed-parameter tractable algorithms for testing the upward planarity of G. Let $n = |V|$, let t be the number of triconnected components of G, and let c be the number of cut-vertices of G. The first upward planarity testing algorithm we present runs in $O(2^t \cdot t! \cdot n^2)$-time. The previously known best result is an $O(t! \cdot 8^t \cdot n^3 + 2^{3 \cdot 2^c} \cdot t^{3 \cdot 2^c} \cdot t! \cdot 8^t \cdot n)$-time algorithm by Chan. We use the *kernelisation technique* to develop a second upward planarity testing algorithm which runs in $O(n^2 + k^4(2k + 1)!)$ time, where $k = |E| - |V|$. We also define a class of non upward planar digraphs.

1 Introduction

A drawing of a digraph is *planar* if no edges cross and *upward* if all edges are monotonically increasing in the vertical direction. A digraph is *upward planar*(UP) if it admits a drawing that is both *upward* and *planar*. Fig. 1(a) shows a planar (but not upward) drawing of a digraph, Fig. 1(b) shows an upward (but not planar) drawing of the same digraph, while Fig. 1(c) shows an UP drawing of a different digraph. Testing if a digraph is UP is a classical problem in the field of graph drawing. Graph drawing is concerned with constructing geometric representations of graphs and is surveyed by Di Battista et al. [1]. The upward planarity of digraphs has been much studied and many interesting results have been published including a proof that upward planarity testing (UPT) is an *NP-complete* problem [2], a linear time algorithm to test whether a given drawing is UP [3], a quadratic time testing algorithm for embedded digraphs [4], and a linear time testing algorithm for single-source digraphs [5].

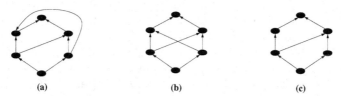

Fig. 1. A planar drawing (a); An upward drawing (b); An upward planar drawing (c)

M. Bieliková et al. (Eds.): SOFSEM 2005, LNCS 3381, pp. 199–208, 2005.
© Springer-Verlag Berlin Heidelberg 2005

Parameterised complexity theory tries to tell apart problems that are well-behaved with respect to one or more parameters from those that are not and is surveyed by Downey and Fellows [6]. Just like polynomial time algorithms are a central idea in the classical formulation of computational complexity, *fixed-parameter tractable* algorithms are a central idea in parameterised complexity. A fixed-parameter algorithm with input size s and parameter size p is said to be *fixed-parameter tractable*, or in the class *FPT*, if it has running time $f(p) \cdot s^{\alpha}$ where f is an arbitrary function and α is a constant. Chan has developed a fixed-parameter tractable algorithm for testing the upward planarity of an arbitrary digraph G that runs in $O(t! \cdot 8^t \cdot n^3 + 2^{3 \cdot 2^c} \cdot t^{3 \cdot 2^c} \cdot t! \cdot 8^t \cdot n)$-time, where n is the number of nodes of G, t is the number of triconnected components of G, and c is the number of cut-vertices of G [7]. As the running time of Chan's algorithm is exponential in terms of the parameters, but polynomial in terms of the size of the graph, it is efficient for a small, but hopefully useful, range of the parameters. Perhaps the first applications of FPT algorithms in graph drawing were to an assortment of *NP*-complete layered graph drawing problems [8, 9, 10].

The three main contributions of this paper are numbered **1)** to **3)** as follows: **1)** We develop an FPT algorithm (henceforth referred to as Algorithm 1) for the parameterised problem of testing a digraph G with t triconnected components for upward planarity which is an improvement on the previous best algorithm for this parameterised problem; **2)** We use the *kernelisation* technique to develop the first FPT algorithm (henceforth referred to as Algorithm 2) for the parameterised problem of testing a digraph G with $k = |E| - |V|$ for upward planarity; **3)** We give an upper bound on the number of edges in an arbitrary UP digraph (Theorem 5). Although Algorithms 1 and 2 are both designed to test digraphs for upward planarity the fact that their parameters are different means that they will each run efficiently on a different set of digraphs. The remainder of this paper is organised as follows. After definitions and preliminaries in Section 2 we present Algorithm 1 in Section 3, Algorithm 2 in Section 4, Theorem 5 in Section 5 and conclusions in Section 6.

2   Preliminary Definitions

We assume all digraphs are connected, acyclic and planar. This does not affect the generality of our algorithms because a disconnected digraph is UP if and only if all its components are UP. Also planarity and acyclicity are necessary conditions for upward planarity and can be detected in linear time. Let G be a digraph. We denote the node set of G by $V(G)$ and the edge set of G by $E(G)$. For any $v \in V(G)$ we denote the number of incoming (resp., outgoing) edges incident on v by $d^-(v)$ (resp., $d^+(v)$) and we denote $d^-(v) + d^+(v)$ by $d(v)$. For any digraph G we let $V_2(G) = \{v \in V(G) : d(v) = 2\}$ and $V_h(G) = \{v \in V(G) : d(v) \geq 3\}$. We refer to nodes with degree greater than 2 as *heavy nodes*. An *upward planar straight-line* (UPSL) drawing of a digraph G is an upward planar drawing of G in which every edge is represented by a straight line segment (see Fig. 1(c)). An *st-digraph* is an acyclic digraph with exactly one source, exactly one sink, and with an edge from the source to the sink. We will need Theorem 1 by Di Battista and Tamassia [11] and Kelly [12], later.

Theorem 1 (Di Battista and Tamassia; Kelly). *For any digraph G the following statements are equivalent.*

1. *G is UP;*
2. *G is the spanning subgraph of a planar st-digraph;*
3. *G admits an UPSL drawing.*

A cut-set of size two (resp., one) is referred to as a *separation pair* (resp., *cut-vertex*). A *split pair* of G is either a separation pair or a pair of adjacent nodes. A *block* (biconnected component) of a digraph G is a maximal connected subgraph B of G such that no node of B is a cut-vertex of B. Thus B is either a maximal biconnected subgraph of G or else the underlying graph of B is K_2. We use Hopcroft and Tarjan's [13] definition of the triconnected components of a biconnected graph which was rephrased by Di Battista and Tamassia [14] as follows. If B is triconnected, then B itself is the unique triconnected component of B. Otherwise, let (u, v) be a separation pair of B. We partition the edges of B into two disjoint subsets E_1 and E_2 ($|E_1|, |E_2| \geq 2$), such that the subgraphs B_1 and B_2 induced by them have only nodes u and v in common. We continue the decomposition process recursively on $B_1' = B_1 + (u, v)$ and $B_2' = B_2 + (u, v)$ until no decomposition is possible. The resulting graphs are each either a triconnected simple graph, or a set of three multiple edges (triple bond), or a cycle of length three (triangle). The triconnected components of B are obtained from such graphs by merging the triple bonds into maximal sets of multiple edges (bonds), and the triangles into maximal simple cycles (polygons). The triconnected components of B are unique. An *embedded digraph* G_ϕ is an equivalence class of planar drawings of a digraph G with the same clockwise orderings, ϕ, of the edges incident upon each node. Such a choice ϕ for a clockwise ordering of the edges incident on each node is called an *embedding* of G. We use $\phi(v)$ to denote the clockwise ordering of the edges incident on v in G_ϕ. An embedded digraph is *upward planar* if it contains an UP drawing.

Theorem 2 (Bertolazzi et al. [4]). *An embedded digraph G_ϕ, with n nodes can be tested for upward planarity in $O(n^2)$ time.*

The *angles* of an embedded digraph G_ϕ are ordered triples $\langle a, v, b \rangle$, where a and b are edges and v is a node incident on both a and b, such that either a directly precedes b in $\phi(v)$ or v is a node of degree 1. An angle $\langle a, v, b \rangle$ of G_ϕ is said to be *incident on* the node v. An angle $\langle a, v, b \rangle$ is said to be an *S-angle* (resp., *T-angle*) if both a and b *leave* (resp., *enter*) v; and an *I-angle* if one of the edges a, b leaves v and the other enters v. The angles of G_ϕ are mapped to geometric angles in an UPSL drawing Γ of G_ϕ. Let $\langle a, v, b \rangle$ be an angle of Γ. If $a \neq b$ the size of the corresponding geometric angle of $\langle a, v, b \rangle$ in Γ equals the number of radians one has to rotate a in the clockwise direction around v in order to reach b. If $a = b$ the size of the corresponding geometric angle of $\langle a, v, b \rangle$ is 2π. An angle of Γ is said to be *large* (resp., *small*) if its corresponding geometric angle is greater (resp., smaller) than π. An *SPQR-tree* is a data structure that represents the decomposition of a biconnected graph with respect to its triconnected components. Due to space constraints we don't include a definition of SPQR-trees here, and instead cite two of the many papers that contains a definition of SPQR-trees [14, 15]. A familiarity with SPQR-trees is necessary to fully understand Section 3.

3   An FPT Algorithm Based on an SPQR-Tree Bound

Let G be a digraph with n nodes, c cut-vertices, and t triconnected components. In this section we outline an FPT algorithm for testing the upward planarity of G where the parameter is t. Chan has developed a parameterized algorithm for UPT that runs in $O(t! \cdot 8^t \cdot n^3 + 2^{3 \cdot 2^c} \cdot t^{3 \cdot 2^c} \cdot t! \cdot 8^t \cdot n)$-time [7]. In this section we present an algorithm (which we refer to as Algorithm 1) that runs in $O(2^t \cdot t! \cdot n^2)$-time and thus improves on Chan's algorithm [7]. Algorithm 1 works by dividing G into its blocks and testing each block separately for upward planarity subject to certain conditions. A decomposition strategy is described [16, 17] that allows the blocks of G to be tested separately with certain extra conditions applied to them. G can be decomposed into its blocks and the conditions deduced for each block in $O(n^2)$-time. If B is a block of G then the conditions applied to B using this decomposition strategy take the following forms:

1. B must be UP.
2. B must have an UP drawing whose external face contains a specified node.
3. B must have an UP drawing whose external face contains an angle of a certain type (i.e., S-angle, T-angle, or I-angle) and size (i.e., *large* or *small*) incident on a specified node.

If the underlying graph of B is K_2 then it is trivially true that B has an UP drawing satisfying all of these conditions. Otherwise B is a biconnected simple graph. Bertolazzi et al.'s algorithm [4] for testing the upward planarity of an embedded digraph can be tailored to test an embedded biconnected digraph for the aforementioned conditions in quadratic time [17]. Thus one can test if B has an UP drawing satisfying any of the above conditions by enumerating all its embeddings and testing each in $O(|V(B)|^2)$-time.

3.1   Bounding the Number of Embeddings of a Block

In this subsection we show that the number of embeddings of B is bounded by a function of the number of triconnected components it contains. Let $\mathcal{T}$ be the SPQR-tree of B. $\mathcal{T}$ is a rooted ordered tree whose nodes are of four types, S-nodes, P-nodes, Q-nodes and R-nodes. Lemmas 1 and 2 are used in our proof of Lemma 3 which gives an upper bound on the number of embeddings of a biconnected graph in terms of its triconnected components.

Lemma 1 (Di Battista and Tamassia [14]). *Two P-nodes cannot be adjacent.*

Lemma 2. *A P-node cannot be adjacent to more than one Q-node.*

Proof. As B is a simple graph at most one split component of any split pair $\{u, v\}$ is an edge. Thus if the split pair $\{u, v\}$ corresponds to a P-node μ in $\mathcal{T}$ then at most one child of μ is a Q-node. But all P-nodes in $\mathcal{T}$ correspond to a split pair in B. It follows that a P-node is adjacent to at most one Q-node.   □

Lemma 3. *A simple biconnected graph B with t triconnected components has at most $2^t \cdot t!$ embeddings.*

Proof. Let r, p and s represent the number of R-nodes, P-nodes, and S-nodes in $\mathcal{T}$ respectively. The triconnected components of B are in one-to-one correspondence with the S-nodes, P-nodes and R-nodes of the SPQR-tree $\mathcal{T}$ of B [14]. Thus B has $t = r + p + s$ triconnected components. Label the P-nodes of $\mathcal{T}$ $P_1, \ldots, P_p$. Let c_i represent the number of children of P_i has in $\mathcal{T}$, for $i = 1, \ldots, p$. It follows from Lemmas 1 and 2 that all but one child of a P-node is either an S-node or an R-node. Thus $\sum_{i=1}^{p}(c_i - 1) \leq r + s$. As $t = r + s + p$ it follows that $r + s = t - p$. It is also true that $\sum_{i=1}^{p}(c_i - 1) = \sum_{i=1}^{p}(c_i) - p$. Therefore $\sum_{i=1}^{p}(c_i) - p \leq t - p$ which means $\sum_{i=1}^{p} c_i \leq t$. Therefore $\prod_{i=1}^{p} c_i! \leq t!$. Di Battista and Tamassia [15] describe how the number of embeddings of B equals $2^r \prod_{i=1}^{p} c_i!$. Clearly $2^r \leq 2^t$. Therefore the number of embeddings of B is less than or equal to $2^t \cdot t!$ □

Thus there are $O(2^t \cdot t!)$ embeddings of B each of which can be tested in $(|V(B)|^2)$-time. Thus it can be tested if B has an upward planar drawing satisfying the conditions specified during the decomposition process in $O(2^t \cdot t! \cdot |V(B)|^2)$-time. It follows that G can be tested for upward planarity in $O(2^t \cdot t! \cdot n^2)$-time.

4   An FPT Algorithm for Sparse Acyclic Digraphs

In this section we develop an FPT algorithm for testing the upward planarity of a digraph G where the parameter is $k = |E(G)| - |V(G)|$. We use a standard technique for developing FPT algorithms called *kernelisation*, which involves reducing a parameterised problem instance I to an "equivalent" instance I_{kr} where the size of I_{kr} is bounded by some function of the parameter. Then the instance I_{kr} is solved and from this a solution to the original instance I follows. Following preliminary definitions in Subsection 4.1 we present reductions for obtaining the *kernel graph* of G in Subsection 4.2 and in Subsection 4.3 we show how to test the kernel graph of G for upward planarity.

4.1   Preliminaries

A *path* P in a digraph G is a sequence $P = \langle v_0, e_1, v_1, e_2, \ldots, v_{n-1}, e_n, v_n \rangle$ of distinct nodes (except possibly v_0 and v_n) and edges of G, such that $e_i = \langle v_i, v_{i-1} \rangle$ or $e_i = \langle v_{i-1}, v_i \rangle$ for $i = 1, \ldots, n$. We refer to a node v in P as an *internal node* of P if $v \notin \{v_1, v_n\}$. A *chain* of G is a path of G, with length ≥ 1, whose internal nodes (if any) have degree 2 and whose end-nodes have degree greater than 2. If c is a chain we denote the set of nodes in c by $V(c)$. We partition the set of chains into four classes. A $\mathcal{V}$-chain (resp. Λ-chain) is a chain whose first edge

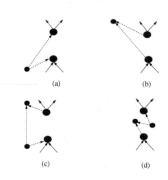

enters (resp. leaves) its first node and whose last edge enters (resp. leaves) its last node. A $\mathcal{C}$-chain (resp. $\mathcal{N}$-chain) is a chain whose first edge enters (resp. leaves) its first node and whose last edge leaves (resp. enters) its last node. These chains are illustrated in Fig. 2. Two or more chains of a digraph G are said to be *parallel* if they have the same class and the same first node and the same last node. A *closed chain* is a chain whose first node and last node are identical. Let C be a chain which contains no node v such that $d^-(v) = d^+(v) = 1$. We denote the length of C by $len(C)$. By *contracting* C we mean the operation of deleting any pair of adjacent nodes of degree 2 in C (one of which

Fig. 2. $\mathcal{V}$-chain (a); Λ-chain (b); $\mathcal{C}$-chain (c); $\mathcal{N}$-chain (d)

is a source and one of which is a sink) and adding a directed edge from the node that was previously adjacent to the deleted sink to the node that was previously adjacent to the deleted source. No matter which pair of adjacent nodes are deleted during the contraction the result is the same (up to isomorphism).

Theorem 3 (Healy-Lynch [17]). *Let G be a digraph with fewer than three heavy nodes. Then G is UP if and only if G is acyclic.*

4.2   Obtaining the Kernel Graph

Given a digraph G to test for upward planarity reductions are performed on G in order to obtain a digraph G_{kr}, called the *kernel graph* of G which is UP if and only if G is UP. The number of nodes and the number of embeddings of G_{kr} are each bounded by a (different) function of k. The following four reductions are upward planarity invariant.

R1. Remove all nodes of degree 1;
R2. Replace any node v whose in-degree and out-degree equals 1 and its two incident edges, $\langle u, v \rangle$ and $\langle v, w \rangle$ with the edge $\langle u, w \rangle$;
R3. Remove all parallel chains (by leaving only the shortest chain whenever two or more chains are parallel);
R4. Remove all closed chains.

Perform reductions **R1** - **R4** until they are no longer possible. Call the resulting digraph G'. Then perform **R5** until it is no longer possible. The resulting digraph is called the kernel graph of G and denoted by G_{kr}.

R5. Contract all chains with length $l \geq 10k + 2$, where $k = |E(G')| - |V(G')|$.

It is not hard to see that reductions **R1 - R2** are upward planarity invariant. We have shown previously that reductions **R3 - R5** are upward planarity invariant for acyclic digraphs [17]. It can be shown that the kernel graph may be found in $O(|V(G)|^2)$-time. We use $\mathcal{K}_i$ to denote the family of digraphs with the property that $F \in \mathcal{K}_i$ implies that F is the kernel graph of some acyclic digraph F' and that $|E(F)| - |V(F)| = i$. Observe that $|E(G_{kr})| - |V(G_{kr})| \leq |E(G)| - |V(G)|$ holds because reductions **R1**, **R2**, and **R5** remove as many edges as nodes, and reductions **R3** and **R4** remove one more edge than node.

4.3   Testing the Kernel Graph

Bertolazzi et al. have shown that if G is a digraph with n nodes, then one can test if a given embedding of G has an UP drawing in $O(n^2)$ time (Theorem 2). This suggests an upward planarity testing algorithm which works by enumerating all embeddings of G and testing each individually. The drawback with this approach is that the number of embeddings of G, even for fixed k, can be exponential in terms of n. We show that the number of embeddings of the kernel graph of G, G_{kr} is bounded by a function of k however, where $k = |E(G_{kr})| - |V(G_{kr})|$ (Lemma 7). We also bound the size of G_{kr} in terms of the parameter k (Theorem 4). We first develop some properties of the kernel graph involving the parameter k though. Lemma 4 bounds the number of heavy nodes G_{kr}, Lemma 5 bounds the degree of a node in G_{kr}, and Lemma 6 gives a formula for the number of internally disjoint chains in G_{kr}.

Lemma 4. $1 \leq |V_h(G_{kr})| \leq 2k$.

Proof. It follows from **R1** that $V(G_{kr}) = V_h(G_{kr}) \cup V_2(G_{kr})$. Thus,

$$\sum_{v \in V(G_{kr})} (d(v)-2) = \sum_{v \in V_h(G_{kr})} (d(v)-2) + \sum_{v \in V_2(G_{kr})} (d(v)-2) = \sum_{v \in V_h(G_{kr})} (d(v)-2)$$

Also observe that,

$$\sum_{v \in V(G_{kr})} (d(v) - 2) = 2|E(G_{kr})| - 2|V(G_{kr})| = 2(|V(G_{kr})| + k) - 2|V(G_{kr})| = 2k$$

Thus, $\sum_{v \in V_h(G_{kr})} (d(v) - 2) = 2k$ (1).

Since $2k \geq 2$ it follows that $|V_h(G_{kr})| \geq 1$. And since $d(v) - 2 \geq 1, \forall v \in V_h(G_{kr})$ it follows that $|V_h(G_{kr})| \leq 2k$. Therefore, $1 \leq |V_h(G_{kr})| \leq 2k$.   □

Lemma 5. If $v \in V(G_{kr})$ then $d(v) \leq 2k - |V_h(G_{kr})| + 3$.

Proof. Equation 1 can be rewritten as $\sum_{v \in V_h(G_{kr})} d(v) = 2k + 2|V_h(G_{kr})|$. But since $d(v) \geq 3, \forall v \in V_h(G_{kr})$ it follows that

$$d(v) \leq 2k + 2|V_h(G_{kr})| - 3(|V_h(G_{kr})| - 1) = 2k + 3 - |V_h(G_{kr})|, \quad \forall v \in V_h(G_{kr})$$

Since $1 \leq |V_h(G_{kr})|$ (by Lemma 4), it follows that $2k - |V_h(G_{kr})| + 3 \leq 2k + 2$. Therefore $d(v) \leq 2k + 3 - |V_h(G_{kr})| \leq 2k + 2, \quad \forall v \in V_h(G_{kr})$.   □

Lemma 6. G_{kr} *contains* $|V_h(G_{kr})| + k$ *edge-disjoint chains.*

Proof. The underlying undirected graph of G_{kr} is homeomorphic to a unique multigraph that contains no nodes of degree 2 and which we call the *reduced multigraph* of G_{kr}, and denote as $\mathcal{M}(G_{kr})$. There is a one-to-one correspondence between the edges of $\mathcal{M}(G_{kr})$ and the chains of G_{kr} and between the nodes of $\mathcal{M}(G_{kr})$ and the heavy nodes of G_{kr}. Equation 1 can be rewritten as $\sum_{v \in V_h(G_{kr})} d(v) = 2k + 2|V_h(G_{kr})|$ which clearly holds for undirected graphs also. Thus, $\sum_{v \in V(\mathcal{M}(G_{kr}))} d(v) = 2k + 2|V_h(G_{kr})|$. Since $|E(\mathcal{M}(G_{kr})| = k + |V_h(G_{kr})|$, G_{kr} contains $k + |V_h(G_{kr})|$ edge-disjoint chains. $\square$

Theorem 4. $|V(G_{kr})| \leq 30k^2 + 2k.$

Proof. G_{kr} contains exactly $|V_h(G_{kr})| + k$ edge-disjoint chains (from Lemma 6) none of which chain has length greater than $10k + 1$ (from **R5**). Therefore no chain contains more than $10k$ nodes of degree two. Therefore, $|V_2(G_{kr})| \leq 10k(|V_h(G_{kr})| + k) = 10k|V_h(G_{kr})| + 10k^2$. Now $|V(G_{kr})| = |V_2(G_{kr})| + |V_h(G_{kr})| \leq 10k|V_h(G_{kr})| + 10k^2 + |V_h(G_{kr})|$. But since $|V_h(G_{kr})| \leq 2k$ it follows that $|V(G_{kr})| \leq 20k^2 + 10k^2 + 2k = 30k^2 + 2k.$

We use $\#(G_{kr})$ to represent the number of embeddings of G_{kr}.

Lemma 7. $\#(G_{kr}) \leq (2k+1)!$, *where* $k = |E(G_{kr})| - |V(G_{kr})|.$

Proof. It follows from the definition of an embedding and the fact that a objects have $(a-1)!$ distinct clockwise orderings that $\#(G_{kr}) \leq \prod_{v \in V(G_{kr})} (d(v) - 1)!$. Observe that 2 objects have only one distinct clockwise ordering. So, $\prod_{v \in V_2(G_{kr})} (d(v) - 1)! = 1$ and therefore $\#(G_{kr}) \leq \prod_{v \in V_h(G_{kr})} (d(v) - 1)!$. In an effort to improve readability henceforth we use $\mathcal{P}$ to denote the expression $\prod_{v \in V_h(G_{kr})} (d(v) - 1)!$ and let $n_h = |V_h(G_{kr})|$. Equation 1 can be rewritten as $\sum_{v \in V_h(G_{kr})} d(v) = 2k + 2n_h$. It follows that $\sum_{v \in V_h(G_{kr})} d(v) - 1 = 2k + n_h$.

Therefore $\mathcal{P}$ is the product of exactly $2k + n_h$ integers, n_h of which equal 1 and n_h of which equal 2. So there are $2k - n_h$ integers greater than 2 in the product $\mathcal{P}$. Consider the maximum value of $\mathcal{P}$. Let y be the largest integer in $\mathcal{P}$. Lemma 5 states that $d(v) \leq 2k - n_h + 3$, $\forall v \in V_h(G_{kr})$, and so $y \leq 2k - n_h + 2$.

If $y = 2k - n_h + 2$, then the $2k - n_h$ integers greater than 2 in the $\mathcal{P}$ are $3, ..., 2k - n_h + 2$ (because if x is an integer in the product $\mathcal{P}$ then $x - 1, x - 2, ..., 2$ must also in it). Therefore if $y = 2k - n_h + 2$ then, $\mathcal{P} = (2k - n_h + 2)! \cdot 2^{n_h - 1}$.

If $y < 2k - n_h + 2$ then the $2k - n_h$ integers greater than 2 in $\mathcal{P}$ are $3, ..., y$ and $(2k - n_h - y + 2)$ other integers all of which are less than or equal to y. But the integers $3, ..., 2k - n_h + 2$ are the integers $3, ..., y$ and $(2k - n_h - y + 2)$ other integers all of which are greater than y. Therefore if $y < 2k - n_h + 3$, then $\mathcal{P} < (2k - n_h + 2)! \cdot 2^{n_h - 1}$. Therefore $\mathcal{P} \leq (2k - n_h + 2)! \cdot 2^{n_h - 1}$.

Lemma 4 states that $1 \leq n_h \leq 2k$. It is easy to show that $(2k - n_h + 2)! \cdot 2^{n_h - 1}$ decreases as n_h increases from 1 to $2k$. Thus its maximum value in the range $n_h = 1, \ldots, 2k$ is at $n_h = 1$. Therefore, $\mathcal{P} \leq (2k - n_h + 2)! \cdot 2^{n_h - 1} \leq (2k + 1)!$. $\square$

It follows that G_{kr} can be tested for upward planarity in $O\left((2k+1)! \cdot |V(G_{kr})|\right)$-time. But $|V(G_{kr})| \leq 30k^2 + 2k$ (Theorem 4) so G_{kr} can be tested in $O\left((2k+1)! \cdot k^2\right)$-time. As G_{kr} can be obtained from G in $O(n^2)$ time it follows that G can be tested for upward planarity in $O\left(n^2 + (2k+1)! \cdot k^2\right)$-time.

5   Polynomial Time Preprocessing Steps

In this section we use Euler's upper bound on the number of edges in a planar graph to give an upper bound on the number of edges in an UP digraph. We also give an alternative proof of a theorem stating that for an acyclic digraph $k \in \{-1, 0, 1\}$ guarantees upward planarity, that is shorter than the original [18].

Theorem 5. *Let $G(V,E)$ be a simple connected digraph with s sources and t sinks. Then G is UP only if $|E| \leq 3|V| - 5 - max\{s,t\}$.*

Proof. G is UP if and only if it is a spanning subgraph of a planar st-digraph (Theorem 1). As an st-digraph has exactly one source an edge must be added to G to cancel out each "surplus" source in G. Therefore at least $s - 1$ edges must added to G. If $s - 1$ edges are added to G then the resulting digraph G' will have $|V|$ nodes and $|E| + s - 1$ edges. But Euler has shown that a graph with n nodes and m edges is planar only if $m \leq 3n - 6$. It follows that G' is planar, and G is UP, only if, $|E| + s - 1 \leq 3|V| - 6$, which can be rewritten as $|E| \leq 3|V| - s - 5$. An analogous argument can be used to show that G is UP only if $|E| \leq 3|V| - t - 5$.   □

Theorem 6 (Healy-English [18]). *Let G be a connected acyclic digraph. If $|E(G)| - |V(G)| \in \{-1, 0, 1\}$ then G is upward planar.*

Proof. Let G be a DAG with $|E(G)| \leq |V(G)| + 1$ and let G_{kr} be the kernel of G. Thus $|E(G_{kr})| \leq |V(G_{kr})| + 1$. Suppose that $|E(G_{kr})| - |V(G_{kr})| = 1$. Thus $1 \leq |V_h(G_{kr})| \leq 2$ (Lemma 4). Therefore G_{kr} is upward planar if $|E(G_{kr})| - |V(G_{kr})| = 1$ (Theorem 3). Thus all digraphs in $\mathcal{K}_1$ are UP. Each digraph in $\mathcal{K}_i$, where $i < 1$, is a subgraph of some digraph in $\mathcal{K}_1$. Therefore, all digraphs in $\mathcal{K}_j$, where $j \leq 1$, are UP. It follows from the fact that reductions **R1 - R5** are upward planarity invariant that every DAG G with $|E(G)| \leq |V(G)| + 1$ is upward planar.   □

6   Conclusions

In this paper we have presented two FPT algorithms for testing the upward planarity of DAGs. Moreover as their parameters are different they are each efficient for a different subset of digraphs than the other. We have also identified a class of non UP digraphs.

References

1. Battista, G.D., Eades, P., Tamassia, R., Tollis, I.G.: Graph Drawing: Algorithms for the Visualization of Graphs. Prentice-Hall (1999)
2. Garg, A., Tamassia, R.: On the computational complexity of upward and rectilinear planarity testing. SIAM Journal Comput. **31** (2001) 601 – 625
3. Battista, G.D., Liotta, G.: Upward planarity checking: "Faces are more than polygons" (Extended Abstract). In Whitesides, S.H., ed.: Proceedings of the 6th International Symposium on Graph Drawing. Volume 1547 of Lecture notes in computer science., Springer-Verlag (1998) 72–86
4. Bertolazzi, P., Battista, G.D., Liotta, G., Mannino, C.: Upward drawings of tri-connected digraphs. Algorithmica **6** (1994) 476–497
5. Bertolazzi, P., Battista, G.D., Mannino, C., Tamassia, R.: Optimal upward planarity testing of single-source digraphs. SIAM Journal on Computing **27** (1998) 132–169
6. Downey, R.G., Fellows, M.R.: Parameterized Complexity. Monographs in Computer Science. Springer (1997)
7. Chan, H.: A parameterized algorithm for upward planarity testing. In: ESA(to appear). (2004)
8. Dujmović, V., Fellows, M., Hallett, M., Kitching, M., Liotta, G., McCartin, C., Nishimura, N., Ragde, P., Rosamond, F., Suderman, M., Whitesides, S., Wood, D.R.: A fixed-parameter approach to two-layer planarization. In: Proceedings of the 9th International Symposium on Graph Drawing. Volume 2265 of Lecture Notes in Computer Science. (2001) 1–15
9. Dujmović, V., Fellows, M., Hallett, M., Kitching, M., Liotta, G., McCartin, C., Nishimura, N., Ragde, P., Rosamond, F., Suderman, M., Whitesides, S., Wood, D.R.: On the parameterized complexity of layered graph drawing. In: Proceedings of the 9th Annual European Symposium on Algorithms. (2001) 488–499
10. Dujmović, V., Fernau, H., Kaufmann, M.: Fixed parameter algorithms for one-sided crossing minimization revisited. In Liotta, G., ed.: Proceedings of the 11th International Symposium on Graph Drawing. Volume 2912 of Lecture notes in computer science., Springer-Verlag (2004) 332–344
11. Di Battista, G., Tamassia, R.: Algorithms for plane representations of acyclic digraphs. Theoretical Computer Science **61** (1988) 175 – 198
12. Kelly, D.: Fundamentals of planar ordered sets. Discrete Math. **63** (1987) 197–216
13. Hopcroft, J., Tarjan, R.E.: Dividing a graph into triconnected components. SIAM Journal on Computing **2** (1973) 135–158
14. Battista, G.D., Tamassia, R.: On-line maintenance of triconnected components with spqr-trees. Algorithmica **15** (1996) 302–318
15. Battista, G.D., Tamassia, R.: On-line planarity testing. SIAM Journal on Computing **25** (1996) 956–997
16. Healy, P., Lynch, K.: Building blocks of upward planar digraphs. In: Proceedings of the 12th International Symposium on Graph Drawing(to appear). Lecture notes in computer science, (Springer-Verlag)
17. Healy, P., Lynch, W.K.: Investigations into upward planar digraphs. Technical Report TR-04-02, Dept. of CSIS, University of Limerick, http://www.csis.ul.ie/Research/TechRpts.htm (2004)
18. Healy, P., English, M.: Upward planarity of sparse graphs. In Brankovic, L., Ryan, J., eds.: Proceedings of the Eleventh Australasian Workshop on Combinatorial Algorithms. (2000) 191 – 203

Read/Write Based Fast-Path Transformation for FCFS Mutual Exclusion

Prasad Jayanti[1], Srdjan Petrovic[1], and Neha Narula[2]

[1] Department of Computer Science,Dartmouth College, Hanover, NH 03755, USA
{prasad, spetrovic}@cs.dartmouth.edu
[2] Google Inc., 1600 Amphitheatre Parkway, Mountain View, CA 94043, USA
neha@google.com

Abstract. Lamport observed that in practical systems processes rarely compete for the entry into the Critical Section [1]. This led to research on *fast* mutual exclusion algorithms that, in the absence of contention, allow a process to enter and exit the Critical Section in $O(1)$ steps. Anderson and Kim designed a general transformation that can turn any mutual exclusion algorithm $\mathcal{A}$ into a new algorithm $\mathcal{A}'$ that is fast [2]. Their transformation, however, does not preserve the fairness property FCFS. The main result of this paper is the design of a new transformation which works similarly as Anderson and Kim's, but additionally preserves FCFS. Our transformation, like theirs, requires only read/write registers.

1 Introduction

In the *N-process mutual exclusion problem* [3], each asynchronous process repeatedly cycles through four sections of code—the Remainder Section, Entry Section, Critical Section (CS) and Exit Section. In its basic version, the problem is to design code for the Entry and Exit Sections so that the following safety and liveness properties hold:

(M1). Mutual Exclusion: At any time, at most one process is in the CS.

(M2). Lockout-Freedom: A process in the Entry Section eventually enters the CS, and a process in the Exit Section eventually enters the Remainder Section.

Lockout freedom only states that all requests for entry to the CS are *eventually* satisfied. To ensure greater fairness, Lamport proposed an additional property: requests for entry to the CS are satisfied in the order in which they are made [4]. This *first-come-first-served* (FCFS) property is formalized by requiring that the Entry Section consists of two fragments of code: a *doorway* followed by a *waiting room*. The doorway should be such that any process can execute the doorway to completion within a bounded number of its own steps. Then, the FCFS property is stated as follows [4]:

(M3). FCFS: If a process q completes the doorway before a process p enters the doorway, then p does not enter the CS before q.

Lamport [1] observed that in practical systems two or more processes rarely compete for the entry into the CS. Thus, the common case is one of no contention:

M. Bielikova et al. (Eds.): SOFSEM 2005, LNCS 3381, pp. 209–218, 2005.

a process executes the mutual exclusion algorithm while all other processes are in the Remainder Section. Lamport therefore proposed the *fastness property* which, intuitively, requires the algorithm to be optimized for the common case. More precisely:

(M4). Fastness: In the absence of contention, a process executes the Entry and the Exit Sections in a constant number of steps, where the constant is independent of the maximum number N of processes for which the algorithm is designed [1].

When designing mutual exclusion algorithms for multiprocessor machines, one must take into account an important trend in hardware technology—the steadily growing gap between high processor speeds and the low speed/bandwidth of the processor-memory interconnect [5]. Consequently, minimizing the number of *remote memory references* has become the central goal of recent research on mutual exclusion algorithms (in NUMA machines, a reference to a shared variable X is remote if X is at a memory module of a different processor; in Cache-Coherent machines, a reference is remote if it is not satisfied by the cache or if it must invalidate another processor's cache entry). Specifically, the following two properties have become important:

(M5). Local-spin: Remote variables are not accessed in busywait loops.

(M6). Adaptivity: The *RMR-complexity* is the number of remote memory references that a process generates when executing the Entry or the Exit Section. The *contention* at time t is the number of processes executing the algorithm at time t (*i.e.*, the number of processes that are outside the Remainder Section at time t). The *adaptivity property* requires that the RMR-complexity of an execution of the Entry or Exit Section depends only on the maximum contention during that execution, and not on the maximum number N of processes for which the algorithm is designed.

In recent years, a number of algorithms have been designed to satisfy Properties M5 and M6 [2, 6, 7, 8, 9, 10, 11, 12, 13, 14]. A few of these algorithms [8, 9, 11, 13], in fact, satisfy an even stronger property than M6, stated as follows:

(M7). Constant-RMR: The RMR-complexity is $O(1)$ (*i.e.*, the complexity does not depend on the contention nor does it depend on N).

Notice that constant-RMR (M7) implies adaptivity (M6) which, in turn, implies fastness (M4).

1.1   The Result and Its Applications

The main result of this paper is an algorithm that transforms any mutual exclusion algorithm $\mathcal{A}$ into a *fast* mutual exclusion algorithm $\mathcal{A}'$ with the following properties: (1) $\mathcal{A}'$ preserves properties M1–M3 and M5–M7 of $\mathcal{A}$, and (2) in the absence of contention, a process executes only reads and writes in $\mathcal{A}'$. The transformation uses $O(N)$ bounded read/write registers. This result is significant in two ways:

(1). The transformation gives a general method to derive register-based fair and fast mutual exclusion algorithms: by instantiating $\mathcal{A}$ with each of the known FCFS algorithms, we can obtain different fast FCFS algorithms. For example, if we instantiate $\mathcal{A}$ with Lamport's Bakery algorithm [4], we get a fast FCFS algorithm that has $O(N)$ space complexity. In comparison, the only known register-based fast FCFS algorithm, due to Afek, Stupp and Touitou [6], has a worst-case space complexity of $O(N^4)$. (However, their algorithm is adaptive on cache-coherent machines, and ours is not.) Instantiating $\mathcal{A}$ with Lycklama and Hadzilacos' FCFS algorithm [15], which uses only bounded registers, we get a fast FCFS algorithm that uses only bounded registers. To the best of our knowledge, this is the first such algorithm.

(2). All known constant-RMR algorithms (Mellor-Crummey and Scott [13], Craig [11], Anderson [9], and Anderson and Kim [8]) are based on strong synchronization primitives, which are known to take significantly longer to execute than simple reads and writes [16]. Therefore, it is desirable to eliminate the use of such instructions, especially in the common case of no contention. Our transformation helps attain this goal. Specifically, if we apply our transformation to any existing constant-RMR $\mathcal{A}$, we get an algorithm $\mathcal{A}'$ where, in the absence of contention, a process executes only $O(1)$ steps all of which are simple reads and writes. Furthermore, $\mathcal{A}'$ retains all of the good properties of $\mathcal{A}$, including FCFS and constant-RMR.

There is an earlier algorithm, due to Anderson and Kim, that transforms any mutual exclusion algorithm into a fast mutual exclusion algorithm [2]. Their transformation, however, does not preserve the FCFS property. Our transformation is patterned after theirs: it has a similar structure and relies on the resettable splitter implementation that they invented. The difference is that our transformation is equipped with an additional mechanism that helps preserve FCFS.

The remainder of the paper is organized as follows. In Section 2, we define the primitives used in our transformation. Anderson and Kim's transformation, which is the starting point of our transformation, is described in Section 3, where we also describe why their transformation fails to preserve the FCFS property. Our transformation is described in Section 4.

2   Primitives Used in Our Transformation

Presence-Detectable FCFS Mutual Exclusion: A *registered waiter* is a process that has completed the doorway, but has not yet entered the Critical Section. A *presence-detectable FCFS mutual exclusion algorithm* allows a process in the Critical Section to determine whether any registered waiters are currently present. More specifically, it is an FCFS algorithm that supports the function *RWpresent*(), which can be called by a process in the Critical Section. A process should be able to execute this function in a bounded number of its own steps. There are two requirements on this function, which are stated below.

Let $\mathcal{I}$ denote a time interval during which the function *RWpresent*() is executed (by a process in the Critical Section). The two requirements are:

(RW1). If some process q is a registered waiter during the entire interval $\mathcal{I}$, then the function returns *true*.

(RW2). If no process is in the Entry Section (*i.e.*, doorway or waiting room) during the entire interval $\mathcal{I}$, then the function returns *false*.

These requirements are weak enough that FCFS algorithms can be easily enhanced to support the function $RWpresent()$. In the full version of the paper [17], we describe how this enhancement can be done for several known algorithms [4, 8, 9, 11, 12, 13, 15] without affecting their RMR-complexity.

Resettable Splitter: A *resettable splitter* object, which extends Lamport's splitter [1], is due to Anderson and Kim [2] and may be described as follows. The object supports two operations—*Capture*, which returns a boolean, and *Release*, which has no return value. When a process wants to gain the ownership of the (resettable) splitter, it invokes the Capture operation. If this operation returns true, the process becomes the owner and remains so until it subsequently invokes the Release operation. Regardless of whether Capture returns *true* or *false*, a process is required to invoke the Release operation before invoking Capture again. Below we specify the behavior of a splitter after a few simple definitions.

A process is *active* from the time it begins a Capture operation until the time it completes the subsequent Release operation. A Capture operation is *successful* if it returns *true*. A process *owns* a splitter from the time it completes a successful Capture operation until the time it begins the subsequent Release operation. The two properties of a splitter are now stated as follows:

(RS1). At most one process owns the splitter at any time.

(RS2). Let $\mathcal{I}$ denote a time interval during which a process executes the Capture operation. If no process is active at any time in the interval $\mathcal{I}$, then the Capture operation succeeds.

Anderson and Kim [2] give an efficient register based implementation of a splitter, stated as follows:

Theorem 1 ([2]). *It is possible to implement a resettable splitter, shared by N processes, using only read/write operations, under the assumption that at most one process executes the Release operation at any time. A process completes a Capture or a Release operation in $O(1)$ steps, regardless of the speeds of other processes. The space complexity is $O(N)$.*

Dynamic 2-Process FCFS Mutual Exclusion: In a system of N processes, a *dynamic 2-process mutual exclusion algorithm* allows at most two processes to execute the algorithm concurrently, but the identities of the two processes are not fixed (*i.e.*, the processes that execute the algorithm at time t need not be the same as the processes that execute the algorithm at a different time). The details are as follows.

The algorithm consists of four procedures, namely, Entry_2(i) and Exit_2(i), where i is 0 or 1. A process p executes the algorithm by first assuming an identity i from $\{0, 1\}$, and then executing Entry_2(i), the CS, and Exit_2(i). At any time, at most one process may have the assumed identity of i, $i \in \{0, 1\}$.

Anderson and Young [18] give an efficient register-based dynamic 2-process mutual exclusion algorithm, stated as follows:

Theorem 2 ([18]). *There exists a dynamic 2-process mutual exclusion algorithm that uses only registers, and satisfies properties M1–M7. The space complexity of the algorithm is $O(N)$.*

Shared variables
  infast: boolean;
Underlying Algorithms
  X : Resettable Splitter implemented by Anderson and Kim's algorithm
      by Theorem 1 (supports *Capture* and *Release* operations)
  2-mutex : Dynamic 2-process mutual exclusion algorithm that satisfies Properties
      M1 and M2 (consists of *Entry_2* and *Exit_2* procedures)
  N-mutex : N-process mutual exclusion algorithm that satisfies Properties M1
      and M2 (consists of *Entry_N* and *Exit_N* procedures)
Initialization
  infast = *false*;

loop		FastPath()		SlowPath()	
0:	Remainder Section	6:	infast = *true*	12:	Entry_N(p)
1:	if ¬Capture(p, X)	7:	Entry_2(0)	13:	Entry_2(1)
2:	SlowPath()	8:	CS	14:	CS
3:	else if infast	9:	Release(p, X)	15:	Release(p, X)
4:	SlowPath()	10:	Exit_2(0)	16:	Exit_2(1)
5:	else FastPath()	11:	infast = *false*	17:	Exit_N(p)
	forever				

Fig. 1. Anderson and Kim's algorithm that transforms any mutual exclusion algorithm into a fast mutual exclusion algorithm

3 Anderson and Kim's Algorithm

Anderson and Kim's algorithm [2], which transforms any mutual exclusion algorithm into a fast mutual exclusion algorithm, is the foundation for our main algorithm. In this section, we therefore briefly review their algorithm and explain why it does not preserve the FCFS property.

The basic idea of Anderson and Kim's algorithm [2], presented in Figure 1, is as follows.[1] A process first determines if it is executing alone. If it is, then it executes the fast path, otherwise it executes the slow path. Multiple processes that concurrently execute the slow path compete via the N-mutex algorithm. The winner q of this competition proceeds to compete with the fast-path process r through the 2-mutex algorithm. The winner between q and r enters the CS. Below we informally describe how this idea is implemented with the help of a resettable splitter X and a shared variable infast.

A process p attempts to become the owner of the splitter X (Line 1). If it is unsuccessful, p is certain that some other process is also active and therefore takes the slow path (Line 2). Even if p acquires the ownership of the splitter, it does not immediately enter the fast path because another process q may be already in the fast path (this is possible because q releases the splitter, on Line 9, while it is still in the fast path). So p inspects the infast variable to determine

---

[1] Anderson and Kim presented their algorithm with the splitter implementation integrated into the rest of the code. In Figure 1, we presented their algorithm modularly, abstracting the splitter out as an object.

if another process is in the fast path (Line 3). If there is such a process, there is clearly contention and so p feels justified in taking the slow path (Line 4). Otherwise, the fast path is clear and p enters it (Line 5).

In the fast path, p first sets the **infast** variable to announce that the fast path is occupied (Line 6). As previously explained, p competes with the winner of the slow-path processes by participating in the 2-**mutex** algorithm (Line 7). When successful, it enters the CS (Line 8). Then, on Line 9, p releases the splitter when it still has exclusive access to the CS. This action ensures that p's Release operation is not concurrent with any other Release, as required by the splitter implementation (see Theorem 1). Finally, p exits the 2-**mutex** algorithm (Line 10) and unsets the **infast** variable to indicate that the fast path is no longer occupied (Line 11).

If p takes the slow path, it competes with other slow-path processes via the N-**mutex** algorithm (Line 12). When successful, p enters another competition, where it competes with the fast-path process via the 2-**mutex** algorithm (Line 13). Whenever it succeeds, it enters the CS (Line 14). Then, on Line 15, p releases the splitter when it still has exclusive access to the CS. As mentioned earlier, this action ensures that p's Release operation is not concurrent with any other Release, as required by the splitter implementation. Finally, it exits the 2-**mutex** algorithm and then the N-**mutex** algorithm (Lines 16 and 17).

Why Anderson and Kim's Algorithm Does Not Preserve the FCFS Property: Below we give a scenario in which Anderson and Kim's algorithm violates FCFS, even if the underlying N-**mutex** and 2-**mutex** algorithms satisfy FCFS. Our argument rests on the observation that the doorway, regardless of how it is chosen, cannot extend past Line 12 because Entry_N is an unbounded section of code.

Suppose that a process p enters the CS through the fast path. While p is in the CS, suppose that a process q enters and proceeds through the slow path up to the completion of Entry_N. Thus, q has completed executing the doorway. Now suppose that p exits the CS, goes back to the Remainder Section, and then reenters the algorithm. An important observation is that, even though q is active, the definition of the splitter (and its implementation in Anderson and Kim's algorithm) allows p to successfully capture the splitter. Accordingly, p enters the fast path and executes Entry_2(0). Recall that q has not yet begun executing Entry_2(1). Consequently, supposing that p runs alone, it completes Entry_2(0) and enters the CS. This, however, violates FCFS: even though q had completed the doorway before p reentered the algorithm, p entered the CS before q.

Intuitively, FCFS is violated because p failed to observe that there is already a process q in the slow path beyond the doorway. To prevent such a scenario, our algorithm in the next section will include a mechanism by which entering processes such as p can detect the presence of processes in the slow path that have advanced past the doorway.

Shared variables
   infast: boolean; rw: **array** $[0 .. 1]$ **of** boolean;
Underlying Algorithms
   X : Resettable Splitter implemented by Anderson and Kim's algorithm
      by Theorem 1 (supports *Capture* and *Release* operations)
   2-mutex : Dynamic 2-process FCFS mutual exclusion algorithm that satisfies
      Properties M1–M3 (consists of *Entry_2* and *Exit_2* procedures)
   N-mutex : Presence-detectable N-process FCFS mutual exclusion algorithm that
      satisfies Properties M1–M3 (consists of *Doorway_N*, *WaitingRoom_N*,
      Exit_N and *RWpresent_N* procedures)
Initialization
   infast $=$ *false*; rw$[0]$ $=$ *false*; rw$[1]$ $=$ *false*

loop		FastPath()		SlowPath()	
0:	Remainder Section	10:	infast $=$ *true*	16:	Doorway_N(p)
1:	if ¬Capture(p, X)	11:	Entry_2(0)	17:	rw$[0]$ $=$ *true*
2:	SlowPath()	12:	CS	18:	rw$[1]$ $=$ *true*
3:	else if infast	13:	Release(p, X)	19:	WaitingRoom_N(p)
4:	SlowPath()	14:	Exit_2(0)	20:	Entry_2(1)
5:	else if rw$[0]$	15:	infast $=$ *false*	21:	CS
6:	SlowPath()			22:	if ¬RWpresent_N(p)
7:	else if rw$[1]$			23:	rw$[0]$ $=$ *false*
8:	SlowPath()			24:	if ¬RWpresent_N(p)
9:	else FastPath()			25:	rw$[1]$ $=$ *false*
	forever			26:	Release(p, X)
				27:	Exit_2(1)
				28:	Exit_N(p)

Fig. 2. The new algorithm that transforms any FCFS mutual exclusion algorithm into a fast FCFS mutual exclusion algorithm

4   Our Algorithm

Our algorithm is presented in Figure 2. It transforms any presence-detectable FCFS mutual exclusion algorithm into a fast FCFS mutual exclusion algorithm. This algorithm has a similar structure as Anderson and Kim's, but employs two additional variables, rw$[0]$ and rw$[1]$. These variables are used to keep track of whether there are any registered waiters in the slow path (recall that a registered waiter is a process that has completed the doorway, but has not yet entered the CS). In the following, we first state two crucial properties that these variables satisfy, and then show how these properties help achieve FCFS.

We begin by specifying the doorway with respect to which our algorithm satisfies the FCFS property.

Definition of the doorway: For a process entering the CS by the fast path, the doorway consists of Lines 1–10 and the doorway of the Entry_2(0) section. For a process entering the CS by the slow path, the doorway consists of Lines 1–8 and Lines 16–18.

Properties of rw Variables: The boolean variables rw[0] and rw[1] keep track of whether there are any registered waiters in the slow path. Specifically, they satisfy the following two properties:

(P1). If $\mathcal{I}$ is a time interval during which a process is a registered waiter in the slow path, either rw[0] is true throughout $\mathcal{I}$ or rw[1] is true throughout $\mathcal{I}$.

(P2). If there is no process in the slow path (*i.e.*, there is no process between Lines 16 and 28), then both rw[0] and rw[1] have the value *false*.

For an intuitive understanding of why the above two properties hold, we turn to the implementation of the slow path and examine how the variables rw[0] and rw[1] are manipulated.

Property P1 requires that at least one of the rw variables remains true during the entire interval in which a process is a registered waiter in the slow path. To help satisfy this property, a process p sets both rw variables to *true* in the doorway (Lines 17 and 18). However, Property P2 requires that both rw variables be false whenever the slow path is unoccupied. To help satisfy this property, after a process exits the CS, it sets the rw variables to false. But, performing this action unconditionally can cause both rw variables to become false even when there is already a registered waiter, thus violating P1. For this reason, before setting either rw variable to false (on Line 23 or 25), p calls *RWpresent_N()* to check that there are no registered waiters in the slow path. (Process p can perform this check by calling *RWpresent_N()* because, by the definition of the doorway, a registered waiter in the slow path is also a registered waiter of N-mutex.) It is a subtle feature of the algorithm that the same check is performed twice, once before setting rw[0] and once more before setting rw[1] (Lines 22–25). This feature helps ensure Property P1, as the next paragraph explains.

We now informally describe why P1 holds. Suppose that a process p writes *true* in rw[0] and rw[1] (Lines 17 and 18) and then enters the waiting room to become a registered waiter. Before p leaves the N-mutex waiting room (Line 19), suppose that one of the rw variables is overwritten with a value of *false*. Assume that rw[0] is the first variable to be so overwritten and q is the process that overwrites it by executing Line 23 (the argument would be analogous if rw[1] were the first variable to be overwritten). The important feature of the algorithm is that no process will be able to overwrite the other rw variable, namely, rw[1], with *false*, before p leaves the N-mutex waiting room. This is because any overwriting of rw[1] (Line 25) is preceded by a call to *RWpresent_N()* that occurs after q has overwritten rw[0]. Thus, this call to *RWpresent_N()* occurs after p has completed the N-mutex doorway (Line 16) and before p leaves the N-mutex waiting room. As a result, *RWpresent_N()* returns *true*, preventing Line 25 from being executed. In conclusion, at most one rw variable is overwritten with *false* before p leaves the N-mutex waiting room (Line 19). While p executes Line 20, neither rw variable is overwritten with *false* because, by the mutual exclusion property of the N-mutex algorithm, no other process is between Lines 20–27. Hence, we have Property P1.

We now turn to Property P2. This property states that if there is no process in the slow path, then both $rw[0]$ and $rw[1]$ have the value *false*. Intuitively, this property holds because the last slow-path process p to leave the CS finds that there are no processes in the slow-path waiting room (more precisely, p's calls to *RWpresent_N()* on Lines 22 and 24 return *false*) and, hence, p sets $rw[0]$ and $rw[1]$ to *false* (on Lines 23 and 25).

How FCFS is Ensured: We now explain informally why the algorithm satisfies FCFS. Suppose that FCFS does not hold, i.e., that the following statement is true:

Statement **S**: There exist some processes p and q such that q is a registered waiter when p enters the doorway, and yet p enters the CS before q.

We consider two cases, corresponding to whether q is a fast-path process or a slow-path process. Consider the case that q is a fast-path processes. Since q owns the splitter when p enters the doorway, p enters the CS via slow path. Further, since q is a registered waiter in the fast path when p enters the doorway, it follows that q has completed the 2-mutex doorway of Line 11 before p executes Line 1 and, therefore, before p enters the 2-mutex doorway of Line 20. Then, by the FCFS property of the 2-mutex algorithm, p does not enter the CS on Line 21 before q enters the CS on Line 12, contradicting **S**.

Consider the case that q is a slow-path process. By Property P1, one of $rw[0]$ and $rw[1]$ holds true until q enters the CS. Consequently, when p reads $rw[0]$ on Line 5 and $rw[1]$ on Line 7, it finds one of them to be true, and, as a result, takes the slow path. Since q is a registered waiter in the slow path when p enters the doorway, it follows that q has completed the N-mutex doorway on Line 16 before p executes Line 1 and, therefore, before p enters the N-mutex doorway on Line 16. Then, by the FCFS property of the N-mutex algorithm, p does not enter the CS on Line 21 before q enters the CS on Line 21, contradicting **S**.

The Main Theorem: Below, we state the theorem that summarizes the main properties of our algorithm. The proof of the theorem is presented in the full version of the paper [17].

Theorem 3. *Let $\mathcal{A}$ be a presence-detectable mutual exclusion algorithm satisfying properties M1–M3. Let $\mathcal{B}$ be Anderson and Young's 2-process mutual exclusion algorithm from Theorem 2. Let $\mathcal{A}'$ be the algorithm obtained by replacing N-mutex and 2-mutex in Figure 2 with $\mathcal{A}$ and $\mathcal{B}$, respectively. Then, the following statements are true:*

1. *$\mathcal{A}'$ satisfies properties M1–M3.*
2. *$\mathcal{A}'$ is fast (i.e., $\mathcal{A}'$ satisfies property M4).*
3. *If $\mathcal{A}$ satisfies any of properties M5–M7, $\mathcal{A}'$ also satisfies that property.*
4. *In the absence of contention, a process executes only reads and writes in $\mathcal{A}'$.*

Acknowledgments. We thank the anonymous SOFSEM referees for their valuable comments on an earlier version of this paper.

References

1. Lamport, L.: A fast mutual exclusion algorithm. ACM Transactions on Computer Systems **5** (1987) 1–11
2. Anderson, J., Kim, Y.J.: A new fast-path mechanism for mutual exclusion. Distributed Computing **14** (2001) 17–29
3. Dijkstra, E.: Solution of a problem in concurrent programming control. Communications of the ACM **8** (1965) 569
4. Lamport, L.: A new solution of Dijkstra's concurrent programming problem. Communications of the ACM **17** (1974) 453–455
5. Culler, D., Singh, J., Gupta, A.: Parallel Computer Architecture: A Hardware/Software Approach. Morgan-Kaufmann (1998)
6. Afek, Y., Stupp, G., Touitou, D.: Long-lived and adaptive collect with applications. In: Proceedings of the 40th Annual IEEE Symposium on Foundations of Computer Science. (1999) 262–272
7. Anderson, J., Kim, Y.J.: Adaptive mutual exclusion with local spinning. In: Proceedings of the 14th International Symposium on Distributed Computing. (2000) 29–43
8. Anderson, J., Kim, Y.J.: Local-spin mutual exclusion using fetch-and-ϕ primitives. Unpublished manuscript (2002)
9. Anderson, T.: The performance of spin lock alternatives for shared memory multiprocessors. IEEE Transactions on Parallel and Distributed Systems **1** (1990) 6–16
10. Attiya, H., Bortnikov, V.: Adaptive and efficient mutual exclusion. In: Proceedings of the 19th Annual Symposium on Principles of Distributed Computing. (2000)
11. Craig, T.S.: Queuing spin-lock algorithms to support timing predictability. In: Proceedings of the Real Time Systems Symposium. (1993) 148–157
12. Jayanti, P.: f-arrays: implementation and applications. In: Proceedings of the 21st Annual Symposium on Principles of Distributed Computing. (2002) 270–279
13. Mellor-Crummey, J.M., Scott, M.L.: Algorithms for scalable synchronization on shared-memory multiprocessors. ACM Transactions on Computer Systems **9** (1991) 21–65
14. Scott, M.L.: Non-blocking timeout in scalable queue-based spin locks. In: Proceedings of the 21st Annual Symposium on Principles of Distributed Computing. (2002)
15. Lycklama, E., Hadzilacos, V.: A first-come-first-served mutual-exclusion algorithm with small communication variables. ACM Transactions on Programming Languages and Systems **13** (1991) 558–576
16. Bershad, B.N.: Practical considerations for non-blocking concurrent objects. In: Proceedings of the 13th IEEE International Conference on Distributed Computing Systems. (1993) 264–273
17. Jayanti, P., Petrovic, S., Narula, N.: Read/write based fast-path transformation for FCFS mutual exclusion. Technical Report TR 2004 522, Dartmouth College Computer Science Department (2004)
18. Yang, J.H., Anderson, J.: A fast, scalable mutual exclusion algorithm. Distributed Computing **9** (1995) 51–60

Adjustment of Indirect Association Rules for the Web

Przemysław Kazienko and Mariusz Matrejek

Wrocław University of Technology, Department of Information Systems,
Wybrzeże S. Wyspiańskiego 27, 50-370 Wrocław, Poland
kazienko@pwr.wroc.pl, matrejekm@adapt.pl

Abstract. Indirect association rules are the extension of classic association rules that enables to discover indirect relationships existing between objects. To estimate the importance of individual parameters of the indirect association rules mining, experiments were carried out on historical web user sessions coming from an e-commerce portal. The influence of parameters of standard direct rules: direct support and direct confidence thresholds, was studied and it was proved that greater values of these two thresholds could significantly decrease the final quantity of indirect rules. This reduction may be additionally strengthened by the introduction of additional threshold to complete or partial indirect confidence. The choice of calculation method for partial indirect confidence was also examined and the multiplication method was selected as the most discriminative.

1   Introduction

Indirect association rules are the extension of standard, direct association rules. Association rules in classic meaning are one of the most popular data mining methods, well described in many papers and there are several algorithms for rule discovering like apriori [2], Eclat [16], FP Growth [4]. The first algorithms were not able to solve some specific problems such as maintaining rules for continuously changing data sets. In consequence incremental algorithms were proposed e.g. FUP [3] or DLG [15].

The main application domain of association rules is the market basket analysis but they are also useful in the web environment for discovering regularities in user behaviors hidden in web logs [1, 10, 14]. Original association rule method, implemented to the web, was expanded to indirect association rules concept in [12, 13] but another approach to indirect associations was presented in [5, 6, 7]. This enabled to discover relationships between web pages, which are not taken into consideration by direct rules. Indirect rules are useful especially in case of short ranking lists in e-commerce recommender systems, when direct rules deliver to few suggestions [7].

2   Direct and Indirect Association Rules

Definition 1. Let d_i be an independent web page (document) and D be web site content (web page domain) that consists of independent web pages $d_i \in D$.

Definition 2. A set X of pages $d_i \in D$ is called *a pageset X*. Pageset does not contain repetitions: $\forall (d_i, d_j \in D)\ (d_i, d_j \in X \Rightarrow d_i \neq d_j)$. The number of pages in a pageset is called *the length of the pageset*. A pageset with the length k is denoted by *k-pageset*.

M. Bieliková et al. (Eds.): SOFSEM 2005, LNCS 3381, pp. 219–228, 2004.

Definition 3. The i-th user session S_i is the pageset containing all pages viewed by the user during one visit in the web site; $S_i \subseteq D$. S^s is the set of all user sessions gathered by the system, $S_i \in S^s$. Each session must consist of at least two pages $card(S_i) \geq 2$. A session S_i contains the pageset X if and only if $X \subseteq S_i$.

Sessions correspond to transactions in typical data mining approach [3, 11]. Note that pagesets and user sessions are unordered and without repetitions – we turn navigational sequences (paths) into sets. Additionally, user sessions may also be filtered to omit too short ones, which are not representative enough [8, 9].

Definition 4. A direct association rule is the implication $X \rightarrow Y$, where $X \subseteq D$, $Y \subseteq D$ and $X \cap Y = \varnothing$. A direct association rule is described by two measures: support and confidence. The direct association rule $X \rightarrow Y$ has the support $sup(X \rightarrow Y) = sup(X \cup Y)/card(S^s)$; where $sup(X \cup Y)$ is the number of sessions S_i containing both X and Y; $X \cup Y \in S_i$. The confidence con for direct association rule $X \rightarrow Y$ is the probability that the session S_i containing X also contains Y: $con(X \rightarrow Y) = sup(X \cup Y)/sup(X)$; $sup(X)$ – the number of sessions that contain the pageset X. The pageset X is the body and Y is the head of the rule.

Direct association rules represent regularities discovered from a large data set [2]. The problem of mining association rules is to extract rules that are strong enough and have the support and confidence value greater than given thresholds: minimal direct support *supmin* and minimal direct confidence *conmin*. In this paper, we consider dependencies only between 1-pagesets – single web pages, so the 1-pageset X including d_i ($X = \{d_i\}$) will be denoted by d_i and a direct association rule from d_i to d_j is $d_i \rightarrow d_j$.

Definition 5. Partial indirect association rule $d_i \rightarrow^{P\#} d_j, d_k$ is the indirect implication from d_i to d_j with respect to d_k, for which exist two direct association rules: $d_i \rightarrow d_k$ and $d_k \rightarrow d_j$ with $sup(d_i \rightarrow d_k) \geq supmin$, $con(d_i \rightarrow d_k) \geq conmin$ and $sup(d_k \rightarrow d_j) \geq supmin$, $con(d_k \rightarrow d_j) \geq conmin$, where $d_i, d_j, d_k \in D$; $d_i \neq d_j \neq d_k$. The page d_k, in the partial indirect association rule $d_i \rightarrow^{P\#} d_j, d_k$, is called the transitive page (Fig. 1).

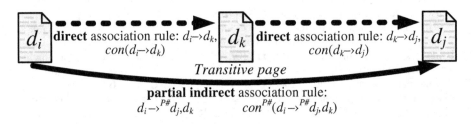

partial indirect association rule:
$$d_i \rightarrow^{P\#} d_j, d_k \qquad con^{P\#}(d_i \rightarrow^{P\#} d_j, d_k)$$

Fig. 1. Indirect association between two web pages

Please note that for the chosen pair of pages d_i, d_j there may be many transitive pages d_k and as a result many partial indirect association rules $d_i \rightarrow^{P\#} d_j, d_k$.

Each indirect association rule is described by the partial indirect confidence $con^{P\#}(d_i \rightarrow^{P\#} d_j, d_k)$, as follows:

$$con^{P\#}(d_i \rightarrow^{P\#} d_j, d_k) = con(d_i \rightarrow d_k) * con(d_k \rightarrow d_j) \tag{1}$$

Pages d_i, d_j in $d_i \rightarrow^{P\#} d_j, d_k$ do not need to have any common sessions, but in (1) we respect only "good" direct associations to ensure that indirect associations are based on sensible grounds. From questionable or uncertain direct knowledge we should not derive reasonable indirect knowledge. In consequence, it was assumed rules $d_i \rightarrow d_k$ and $d_k \rightarrow d_j$ must be "strong" enough, so that $con(d_i \rightarrow d_k)$ and $con(d_k \rightarrow d_j)$ exceed *conmin*.

The partial indirect confidence is calculated using direct confidence without access to source user sessions, so the computational complexity of partial indirect rule mining is much less than for direct ones.

Definition 6. The set of all possible transitive pages d_k for which partial indirect association rules from d_i to d_j exists, is called T_{ij} (Fig. 2).

Note that T_{ij} is not the same set as T_{ji}.

Definition 7. Complete indirect association rule $d_i \rightarrow^{\#} d_j$ aggregates all partial indirect association rules from d_i to d_j with respect to all existing transitive pages $d_k \in T_{ij}$ and it is characterized by complete indirect confidence - $con^{\#}(d_i \rightarrow^{\#} d_j)$:

$$con^{\#}(d_i \rightarrow^{\#} d_j) = \frac{\sum_{k=1}^{card(T_{ij})} con^{P\#}\left(d_i \rightarrow^{P\#} d_j, d_k\right)}{max_T} \tag{2}$$

where $max_T = \max_{d_i, d_j \in D}\left(card\left(T_{ij}\right)\right)$. This is the normalized sum of all existing partial rules.

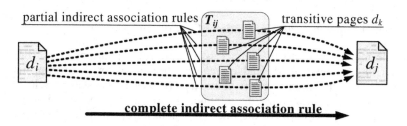

partial indirect association rules T_{ij} transitive pages d_k

d_i d_j

complete indirect association rule

Fig. 2. Complete indirect association rule

Only indirect rules with complete indirect confidence greater than the given confidence threshold - *iconmin* are accepted. A complete indirect association rule from d_i to d_j exists if and only if it exists at least one partial indirect association rule from d_i to d_j. Note that complete indirect association rules are not symmetric: the rule $d_i \rightarrow^{\#} d_j$ may exists but the reverse one $d_j \rightarrow^{\#} d_i$ not necessarily. It results from features of partial indirect associations and direct associations, which also are not symmetric.

To enable the usage of both direct and indirect association rules e.g. for recommendation of web pages, the joined, complex association rules are introduced [6, 7]. They combine main parameters of direct and indirect rules – confidences.

Definition 8. Complex association rule $d_i \rightarrow^{*} d_j$ from d_i to d_j exists, if direct $d_i \rightarrow d_j$ or complete indirect $d_i \rightarrow^{\#} d_j$ association rule from d_i to d_j exists. A complex association rule is characterized by *complex confidence* - $con^{*}(d_i \rightarrow^{*} d_j)$, as follows:

$$con^{*}(d_i \rightarrow^{*} d_j) = \frac{1}{2} * \left(\frac{con\left(d_i \rightarrow d_j\right)}{avg(con)} + \frac{con^{\#}\left(d_i \rightarrow^{\#} d_j\right)}{avg(con^{\#})} \right) \tag{3}$$

where $avg(con)$ and $avg(con^\#)$ are average values of all $con(d_i{\rightarrow}d_j)$ and $con^\#(d_i{\rightarrow}^\#d_j)$, respectively. Normalization based on average values was introduced to make domains of both direct and indirect confidence more comparable. It results from significant differences between values of direct and indirect confidence. Values of indirect confidence are smaller than direct ones according to (1) and (2), see Tab. 1.

Values of $con^*(d_i{\rightarrow}d_j)$ may exceed 1 and to transfer the domain of complex confidence into the range [0,1], we would need to normalize (3) with the maximum value of complex confidence. However, the normalization is not necessary for the research presented in following sections and for that reason it was not included in (3).

3   Adjustment of Mining of Indirect Rules

The main research has been conducted on real web session logs coming from one of the largest polish online computer store. The session data extracted from text files were cleaned to exclude 1-page sessions and enormous big sessions so that HTTP requests from search engine spiders would be removed. Finally, the set of 4,200 pages and about 100,000 user sessions was obtained. However, only 22,000 sessions containing first 500 pages were used for further experiments to shorten the time of experiments. An average session consisted of 9 pages. Indirect rules were mined using IDARM algorithm described in [5].

3.1   Minimal Support – *supmin*, Minimal Confidence – *conmin*

Direct rule mining is a part of indirect association rule mining process. Threshold values for direct rules: *supmin* and *conmin* can be crucial for discovering indirect associations. The minimum support value *supmin* is the first step in selection of candidates for direct rules, while the minimum confidence value *conmin* determines the "strength" and usefulness of the rule. They both have a significant influence on the quantity of discovered complete indirect rules (Fig. 3).

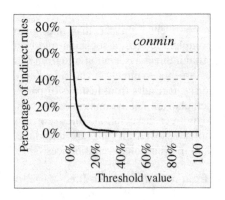 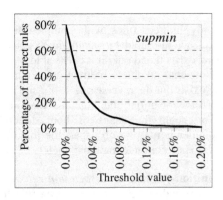

Fig. 3. Percentage of all possible (249,500) complete indirect rules discovered in relation to *supmin* (left) and *conmin* values (right)

3.2 Methods of Calculation of Partial Indirect Confidence

In accordance with definition 5, the confidence of a partial indirect association rule respects both direct rules, which are involved in an indirect association. Their contribution is fixed by (1) to the simple multiplication, but we suggest considering other formulas for estimation of partial indirect confidence.

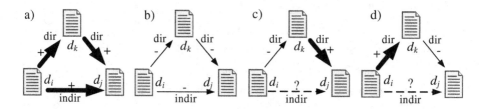

Fig. 4. Possible cases of relationships between opening and closing rule in partial indirect association rule. The width of the line and a "+" or "-" sign correspond to the confidence value

We can distinguish several cases of relations between confidence values of both direct rules engaged in indirect association (Fig. 4). Two kinds of rules were considered: "strong" rules, which confidence values are greater than the average, and "weak" rules with confidence values smaller than the average. Strong and weak rules can provide following cases in the estimation of partial indirect confidence:

- both component direct rules are strong (the "+" sign and a thick line on the Fig. 4): the partial indirect rule is also strong (Fig. 4a);
- both direct rules are weak ("-", a thin line on the Fig. 4): the partial indirect rule is very weak and it is about to be cut by the threshold *iconmin* (Fig. 4b);
- the first (opening) direct rule is strong and the second (closing) one is weak or inversely: the partial indirect rule can be either weak or strong depending on the concrete values of direct confidences (Fig. 4c,d).

Table 1. Values of partial indirect confidence for different methods of calculation

Association	Direct con. first, second	Method of estimation of partial indirect confidence con''				
		Multipl.	Arith.mean	Max	Min	Weighted
	0.20, 0.20	0.04	0.20	0.20	0.20	0.20
	0.75, 0.20	0.15	0.48	0.75	0.20	0.57
	0.20, 0.75	0.15	0.48	0.75	0.20	0.39
	0.75, 0.90	0.68	0.83	0.90	0.75	0.80
	1.00, 0.50	0.50	0.75	1.00	0.50	0.83
	0.50, 1.00	0.50	0.75	1.00	0.50	0.66

Following methods of estimation of partial indirect confidence can be considered (see examples in Tab.1):

- multiplication – both direct associations have the same contribution (1),
- maximal value – the weaker rule has no matter:

$$con^{P\#}(d_i \to^{P\#} d_j, d_k) = \max(con(d_i \to d_k); con(d_k \to d_j)), \tag{4}$$

- minimal value – the stronger rule has no matter:

$$con^{P\#}(d_i \to^{P\#} d_j, d_k) = \min(con(d_i \to d_k); con(d_k \to d_j)), \tag{5}$$

- arithmetical mean – both direct associations have the same contribution:

$$con^{P\#}(d_i \to^{P\#} d_j, d_k) = 0.5*(con(d_i \to d_k) + con(d_k \to d_j)), \tag{6}$$

- weighted mean – direct associations have different contributions depended on their positions in the indirect association:

$$con^{P\#}(d_i \to^{P\#} d_j, d_k) = \frac{2}{3}(con(d_i \to d_k) + \frac{1}{3} con(d_k \to d_j)) \tag{7}$$

Please note that multiplication provides the smallest values of partial indirect confidences (Tab.1) and it was confirmed for real test data (Fig. 5): the smallest average confidence value (0.035) but the best discrimination abilities – the standard deviation (0.081) doubles the average. This last feature of multiplication determined that (1) appeared to be the most appropriate for further processing.

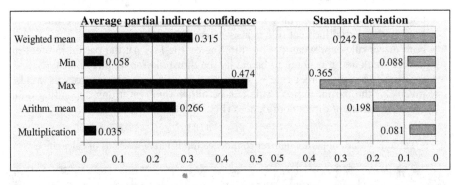

Fig. 5. The average partial indirect confidence and its standard deviation calculated using methods from Tab. 1. Minimal support *supmin*=0.0002 and minimal confidence *conmin*=0.01

Arithmetical mean and weighted mean deliver information about same characteristics of test data set. The average value for weighted mean (0.315) is greater than for arithmetical mean (0.266). This reveals that the direct rule opening indirect association has statistically greater direct confidence value than the closing one.

3.3 Minimal Complete Indirect Confidence – *iconmin*

The minimal complete indirect confidence *iconmin* is a threshold corresponding to minimal direct confidence *conmin* in direct rules. The significant problem at indirect association rules mining is similar like at direct associations: to find appropriate values for thresholds or otherwise results could be useless. The value of complete confidence threshold *iconmin* should not be less than the square of direct confidence threshold *conmin* – see (1). Additionally, according to (2), there is no point in setting *iconmin* with the value less than $conmin^2/max_r$, because no rule would be rejected. The introduction of minimal indirect confidence *iconmin* significantly reduces the amount of complete indirect rules discovered (Fig. 6). The simultaneous usage of all 3 thresholds: *supmin*, *conmin* and *iconmin* allow reducing the result set even more. This should shorten the time needed to perform the whole process of indirect association rules mining. The threshold *iconmin* significantly influences on the number of direct rules, which are strengthened in complex rules (3). Its influence on the quantity of weakened rules is minimal (Fig. 7).

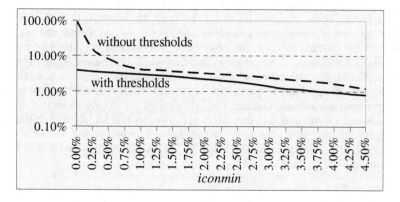

Fig. 6. Percentage of all possible (249,500) complete rules discovered in relation to *iconmin*; without thresholds (dotted line) or with *supmin*=0.008% and *conmin*=2% (constant line)

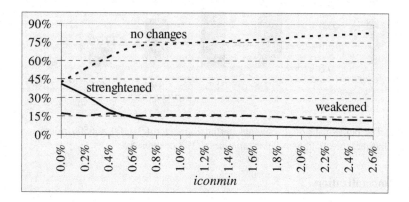

Fig. 7. Percentage of all discovered direct rules (4,500), which were strengthened or weakened in relation to minimal indirect confidence *iconmin*; *supmin*=0.02% and *conmin*=0.2%

3.4 Cut Point

During indirect association rules mining two kinds of indirect association rules are processed. The former are partial and the latter are complete indirect association rules. The number of partial rules can be very large. Performed experiments revealed that there could be even 370,000 partial indirect associations, mostly with very small partial confidence, derived from only 6,000 direct rules. The question about exchanging complete indirect confidence threshold with partial indirect confidence threshold seems to be legitimate. If we look closer at the process of discovering complete indirect rules, we can see that all partial confidences are estimated before the *iconmin* threshold is applied. This approach provides the information about all partial indirect rules, which can support complete association, even if the partial ones' confidences are very small. Yet another approach is to introduce partial indirect confidence threshold *piconmin* and to cut most of partial rules as soon as possible. Obviously, the value of *piconmin* should be from the range [0;1] Such threshold can reduce the time of discovering complete rules, but there is a danger that we loose too many "small" partial rules. On the other hand, why should we eliminate partial rules with small confidences, if we already cut many weak association using direct support and direct confidence thresholds? Such filtering could have been even more selective than the use of threshold to partial indirect rules. Both indirect thresholds (*iconmin* and *piconmin*) are actually similar. They operate on previously generated and filtered direct association rules set and they do not require any recalculation of this set.

The introduction of partial indirect threshold *piconmin* to the test data decreased significantly the number of partial rules: by 12.3% for *piconmin*=1% and 92.3% for *piconmin*=20% (Fig. 8).

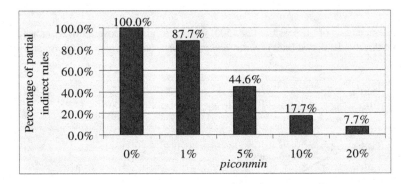

Fig. 8. The percentage of discovered partial indirect rules (100%=6500 rules) in relation to partial indirect threshold *piconmin*; *supmin*=0,02%; *conmin*=5%

3.5 Normalization

Complete indirect confidence (2) is the sum of all partial indirect confidences normalized with the given factor u. We strongly recommend to normalize the values of complete indirect confidence to avoid situations it exceeds out of the range [0;1]. In original indirect association rules mining u is the size of the bigger transitive pages set $u=max(card(T_{ij}))$ [6, 7].

In different environments, the mining of indirect associations may result in smaller, or larger number of transitive pages. It has to be decided, whether associations with many transitive pages and smaller partial confidence are better or worse than associations with only several transitive pages but with a greater confidence. We can use different methods of normalization in (2) to "promote" or "demote" various types of discovered associations:

1. Global normalization using constant value of u, common for the whole base set of elements and sessions – used in (1). Sum of all partial values is divided by the size of the largest transitive page set. It is essential for this approach that the factor u is different for every environment (every set of indirect rules). The change of any threshold – *supmin*, *conmin* or *iconmin* changes the value of u.
2. Global normalization using constant value among the same base set of pages and sessions. In this normalization factor u is independent from any parameter of the method, so *supmin*, *conmin*, *iconmin* can be adjusted without influence on u value. An example of such factor is the number of all possible transitive pages. Naturally, the factor u changes for sets with different cardinality.
3. Global normalization using constant value among different base sets u=const. This kind of normalization uses constant factor u, independent from cardinality of base set and any other parameter, e.g. u=5000. Such normalization is questionable since factor u could not be great enough to ensure results from the range [0;1] in some specific environments.
4. Local normalization using variable. Different values of factor u for each single complete indirect association rule $d_i \rightarrow^{\#} d_j$ are used in this approach. An example of such factor is the cardinality of transitive page set for current complete association $u=card(T_{ij})$. Using this way of normalization we loose information about the number of transitive pages for individual indirect association rule.

The normalization by the local variable or the global constant (cases 4 and 1) delivers us greater values of the average indirect complete confidence. It comes from the usually greater value of the factor u for the two other methods (cases 2 and 3).

4   Conclusions and Future Work

Indirect association rule mining is a new and promising data mining technique. Research results presented in the paper provide some useful information about crucial features of indirect associations. Firstly, the number, and in consequence also the quality of discovered indirect rules can be adjusted by either minimal direct support and confidence thresholds (Fig. 3) or minimal indirect confidence (Fig. 6) or minimal partial indirect confidence (Fig. 8) separately or by using all of these four thresholds simultaneously. Secondly, the multiplication as the method of estimation of partial indirect confidence (1) is the most discriminative among all tested formulas (Fig. 5).

Some research on the utility of indirect rules in recommendation systems has been conducted in [7]. This revealed that indirect rules significantly extend short ranking lists. Additionally, the presented above results proofed that indirect rules are capable of strengthening or weakening classic direct rules (Fig. 7). For that reason, indirect rules can be helpful in selection of the best (most amplified) direct associations.

Direct rules applied to web logs reflect typical navigation patterns and usually only confirm "hard hyperlinks" existing on web pages. In opposite, indirect rules "go outside" of these typical user paths and provide some non-trivial knowledge. Such knowledge can be utilized in any case when patterns delivered by direct rules are too obvious for the end user.

The potential usefulness of combination of direct and indirect rules, e.g. in the form of complex rules (3), can be further studied using the recommendation system incorporated into an existing web site and running with real web users who could outright estimate the relevance of suggestions.

References

1. Adomavicius G., Tuzhilin A., Using Data Mining Methods to Build Customer Profiles. IEEE Computer, Vol. 34, No. 2, 2001, 74-82.
2. Agrawal R., Imieliński T., Swami A., Mining association rules between sets of items in large databases. ACM SIGMOD Int. Conference on Management of Data, ACM Press, 1993, 207-216.
3. Cheung D.W.L., Han J., Ng V., Wong C.Y., Maintenance of Discovered Association Rules in Large Databases: An Incremental Updating Technique. Twelfth Int. Conference on Data Engineering, IEEE Computer Society, 1996, 106–114.
4. Han J., Pei J., Yin Y., Mining Frequent Patterns without Candidate Generation. ACM SIGMOD Int. Conference on Management of Data, ACM, 2000, 1-12.
5. Kazienko P., IDARM - Mining of Indirect Association Rules. 2005, to appear.
6. Kazienko P., Multi-agent Web Recommendation Method Based on Indirect Association Rules. KES'2004, 8th Int. Conference on Knowledge-Based Intelligent Information & Engineering Systems, Wellington, New Zealand, September 20-25, 2004, Part II, LNAI 3214, Springer Verlag, 1157-1164.
7. Kazienko P., Product Recommendation in E-Commerce Using Direct and Indirect Confidence for Historical User Sessions. DS'04. 7th Int. Conf. on Discovery Science. Padova, Italy, October 2-5, 2004, Springer Verlag, LNAI 3245, pp. 255-269.
8. Kazienko P., Kiewra M., Link Recommendation Method Based on Web Content and Usage Mining. Proc. of the International IIS: IIPWM´03 Conference, Zakopane, Advances in Soft Computing, Springer Verlag, 2003, 529-534.
9. Kazienko P., Kiewra M., Personalized Recommendation of Web Pages. Chapter 10 in: Nguyen T. (ed.) Intelligent Technologies for Inconsistent Knowledge Processing. Advanced Knowledge International, Adelaide, South Australia, 2004, 163-183.
10. Mobasher B., Cooley R., Srivastava J., Automatic Personalization Based on Web Usage Mining. Communications of the ACM, Volume 43, Issue 8, August, 2000, 142-151.
11. Morzy T., Zakrzewicz M., Data mining. Chapter 11 in Błażewicz J., Kubiak W., Morzy T., Rubinkiewicz M (eds): Handbook on Data Management in Information Systems. Springer Verlag, Berlin Heidelberg New York, 2003, 487-565.
12. Tan P.-N., Kumar V., Mining Indirect Associations in Web Data. WEBKDD 2001. LNCS 2356 Springer Verlag, 2002, 145-166.
13. Tan P.-N., Kumar V., Srivastava J., Indirect Association: Mining Higher Order Dependencies in Data. PKDD 2000, LNCS 1910, Springer Verlag (2000) 632-637.
14. Yang H., Parthasarathy S., On the Use of Constrained Associations for Web Log Mining. WEBKDD 2002, LNCS 2703, Springer Verlag, 2003, 100 – 118.
15. Yen S.J., Chen A.L.P., An Efficient Approach to Discovering Knowledge from Large Databases. 4th Int.Conf. on Parallel and Distributed Information Systems, IEEE Computer Society, 1996, 8–18.
16. Zaki M.J., Parathasarathy S., Li W., A Localized Algorithm for Parallel Association Mining. SPAA'97, 9th ACM Symposium on Parallel Algorithms and Architectures, 1997, 321-330.

Anonymous Communication with On-line and Off-line Onion Encoding*

Marek Klonowski, Mirosław Kutyłowski, and Filip Zagórski

Institute of Mathematics, Wrocław University of Technology
Miroslaw.Kutylowski@pwr.wroc.pl,
{Marek.Klonowski, Filip.Zagorski}@im.pwr.wroc.pl

Abstract. Anonymous communication with onions requires that a user application determines the whole routing path of an onion. This scenario has certain disadvantages, it might be dangerous in some situations, and it does not fit well to the current layered architecture of dynamic communication networks.

We show that applying encoding based on universal re-encryption can solve many of these problems by providing much flexibility – the onions can be created on-the-fly or in advance by different parties.

Keywords: anonymous communication, onion, universal re–encryption.

1 Introduction

Anonymous Communication. Providing anonymity of communication in public networks is one of the most serious problems in computer security. Many interesting ideas have been presented so far, but still we are far away from solving the problem completely. Perhaps the most prominent proposals are Chaum's DC-networks and MIX networks [3, 2]. Later Rackoff and Simon [14] proposed a protocol in which each user chooses a random route for his message and encrypts the route and the message within a structure that resembles an onion. Due to a cryptographic encoding, the messages that meet in the same node are indistinguishable when leaving this node. This effect is called "mixing" [2] or a "conflict" [1]. So, the idea of anonymity with onions is that many messages travel around the network, meet each other and are recoded so that an adversary gradually looses control over origin points of the messages. The idea of onions became the basic component of Babel [12] and of Onion Routing [7]. (In fact, the name onion was introduced in [7].) Recently, it has been used in the TOR protocol [4].

It may happen that some ingredients of the network might be controlled or monitored by an adversary that tries to break anonymity. Many adversary models have been considered - some of the models allow an adversary to perform only passive traffic analysis based on information obtained from nodes and links under his control [1, 11]. Other models [12, 13] allow active attacks based on adding or delaying messages by an adversary.

---

* Partially supported by the EU within the 6th Framework Programme under contract 001907 (DELIS).

M. Bieliková et al. (Eds.): SOFSEM 2005, LNCS 3381, pp. 229–238, 2005.

Connection based protocols such as Onion Routing face a fundamental problem that breaking one connection at a time effects traffic along exactly one path and therefore once a path disappears it may betray the path used. No fully satisfactory solution addressing this problem has been found yet. In this paper we are concerned with protocol aimed for sending short messages for which security proofs with respect to traffic analysis do exist.

As it was pointed in [1], anonymity level in an onion based protocol is strongly correlated with the number of messages processed by the network and the probability of a conflict/mixing of two or more messages in one node. It turns out that if an adversary may control only a fraction of links, possibilities of traffic analysis are quite limited [1, 11]. On the other hand, if an adversary controls the whole traffic, then the onions provide a low level of anonymity in the case light traffic. Simply, the onions do not meet frequently in this case, so the paths can be easily recovered by the adversary.

The regular onion encoding, as proposed in [14] and used in the later papers, has another disadvantage: a user has to know the whole network in order to be able to choose a truly random path. This assumption is unrealistic in dynamic networks. Moreover, anonymity is in a serious danger if different parties participating in the protocol use different sets of servers for intermediate nodes on the onion paths. We are aware of certain attacks possible in this situation.

Papers [5, 10] introduce new encoding techniques to onions based on universal re-encryption schemes [9]. The idea is that the components processed by the servers can be re-encrypted without any knowledge of the contents and the recipient. However, even then a protocol should be checked carefully - the scheme from [5] has been broken very fast.

New Results. In our paper we explore new possibilities for design of anonymous communication protocols based on onions encoded with universal re-encryption schemes [10]. First we propose an off-line protocol that allows to prepare a route of a message in advance – the onion routes (or their parts) are created by third parties as a kind of general service. Then, if an application process has to send a message, and it does not know the topology of the network, it can ask for the service mentioned. This solution is aimed for the layered communication architectures. It is also useful in a LAN if there are specialized servers responsible for anonymization messages sent to external locations.

In the second proposal (online merge onions), we show how to move responsibility of determining onion paths to specialized servers that make decisions dynamically (for instance based on the traffic load). It enables to adjust quickly to network conditions. Another important feature of this construction is that it decreases overhead of a message volume due to onion encapsulation.

2   Onions

2.1   Classical Onions

In this section we briefly recall construction of onions. We assume that a network consists of n servers (called nodes); each of them has its own widely accessible public key and the corresponding secret key. Moreover, we assume that each server can communicate directly with any other server.

A basic onion protocol looks as follows: in order to send a message m to node R, node S chooses at random intermediate nodes $J_1, \ldots, J_\lambda$, and encodes m as an *onion*:

$$\mathrm{Enc}_{J_1}(\mathrm{Enc}_{J_2}(\ldots(\mathrm{Enc}_{J_\lambda}(\mathrm{Enc}_R(m), R), J_\lambda)\ldots), J_3), J_2)$$

(Enc_X stands for encryption with the public key of X). This onion is sent by S to J_1. Node J_1 decrypts the message - the plaintext obtained consists of two parts: the second part is J_2, the first one is an onion with one layer peeled off:

$$\mathrm{Enc}_{J_2}(\ldots(\mathrm{Enc}_{J_\lambda}(\mathrm{Enc}_R(M), R), J_\lambda)\ldots), J_3) \, .$$

Then J_1 sends this onion to J_2. Nodes $J_2, \ldots, J_\lambda$ work similarly, the onion is gradually "peeled off" until it is finally received by node R.

In fact, additional countermeasures are necessary in order to avoid some simple attacks on the onion protocol (for details see for instance [1]):

- We have to use a probabilistic encryption scheme. Otherwise an adversary could establish a permutation between the input and the output of a node by a simple encryption of the whole output batch.
- The size of the onions should be fixed. A kind of padding can be used.

2.2   Universal Re-encryption

We recall universal re-encryption scheme from [9] based on ElGamal encryption. Let G be a cyclic group of order p such that the discrete logarithm problem is hard for G. Let g be a generator of G. Then a private key is a random $x < p$; the corresponding public key is $y = g^x$.

Encryption: In order to encrypt a message m for Alice, Bob generates uniformly at random values k_0 and k_1. Then, the following quadruple is a ciphertext of m:

$$(\alpha_0, \beta_0; \alpha_1, \beta_1) := (m \cdot y^{k_0}, g^{k_0}; y^{k_1}, g^{k_1})$$

In fact, (α_0, β_0), and (α_1, β_1) are ElGamal ciphertexts of, respectively, m and 1.

Decryption: Alice computes $m_0 = \frac{\alpha_0}{\beta_0^x}$ and $m_1 = \frac{\alpha_1}{\beta_1^x}$, and accepts a message $m = m_0$, if and only if $m_1 = 1$.

As for the ElGamal scheme, this is a probabilistic cryptosystem – if we encrypt the same message twice, we get two different ciphertexts. Moreover, given two ciphertexts, it is impossible to say whether they were encrypted under the same key, provided that the private key is unknown. This property is called *key-privacy* (see [9]).

ElGamal cryptosystem has another important feature. We can re-encrypt a ciphertext (α, β) so that any relation between the old and the new ciphertext (α', β') is hidden for the observer that has no access to the private key. For the scheme presented above even a **public key** is not necessary – for this reason, it is called *universal re-encryption*, or *URE* for short. The re-encryption procedure looks as follows: First, random values k_0' and k_1' are chosen. Then a re-encrypted version of a ciphertext $(\alpha_0, \beta_0, \alpha_1, \beta_1)$ is obtained as:

$$\left(\alpha_0 \cdot \alpha_1^{k_0'}, \beta_0 \cdot \beta_1^{k_0'}; \alpha_1^{k_1'}, \beta_1^{k_1'}\right) \, .$$

Let $URE_x(m)$ stand for a ciphertext of m obtained with universal re-encryption scheme, where x is the private decryption key.

2.3   Onions Based on Universal Re-encryption

In [10] the following method of encoding a message m going from A to $B = J_{\lambda+1}$ through path $J_1, \ldots, J_\lambda$ is presented. Let (y_i, x_i) be the pair of public and private key of J_i for $i \leq \lambda+1$. An URE-onion consists of ciphertexts $URE_{x_1}(J_1)$, $URE_{x_1+x_2}(J_2), \ldots,$ $URE_{x_1+\ldots+x_\lambda}(J_\lambda)$ and $URE_{x_1+\ldots+x_{\lambda+1}}(m)$. These ciphertexts are obtained with the public keys, respectively, $y_1, y_1 \cdot y_2, \ldots, y_1 \cdot \ldots \cdot y_{\lambda+1}$. After creating, these ciphertexts are permuted at random.

Processing an URE-onion consists of two phases: a partial decryption and a re-encryption phase. For instance, J_1 performs the following steps: each URE-ciphertext $(\alpha_0, \beta_0, \alpha_1, \beta_1)$ is replaced during partial decryption by

$$(\alpha_0/\beta_0^{x_1}, \beta_0, \alpha_1/\beta_1^{x_0}, \beta_1) \ .$$

It is easy to see that if $(\alpha_0, \beta_0, \alpha_1, \beta_1)$ is $URE_{x_1+\ldots+x_i}(w)$, then after the partial description we get $URE_{x_2+\ldots+x_i}(w)$. One of the ciphertexts obtained is in fact of the form $(J_2, \beta_0, 1, \beta_1)$ and indicates the next destination. Then all ciphertexts (including this of J_2) are re-encrypted and permuted at random before being sent to J_2.

It is easy to see that the encoding described above guarantees that only processing along the path chosen by A guarantees delivery of the URE-onion. Any malicious processing (re-direction, detours, changing the contents) can be detected with high probability and the malicious server can be identified.

3   Features and Protocols Based on URE-Onions

In this section we present several important features of URE-onions that can be used in design of anonymity protocols.

3.1   Basic Features

Plaintext Insertion After Encryption. Universal re-encryption inherits the remarkable property of ElGamal encryption scheme: the plaintext may be determined after essential part of encryption computation. Indeed, first we prepare a ciphertext of 1. It has the form $\left(1 \cdot y^{k_0}, g^{k_0}; y^{k_1}, g^{k_1}\right)$. Then we can convert it to a ciphertext of m simply by multiplying the first component by m.

Navigators. An onion encoding a special void message $-$, with a starting point A and destination B can be used to encode only a "path". If a node obtains after decoding an onion a message "$-$", it knows that the onion has reached the end of its path. Such an onion will be called a *navigator from A to B* and denoted $Nav[A, B]$.

Navigators are particularly handy for URE-onions: a so-called *URE-navigator* consists of two parts: the first one is a navigator, say $Nav[A, B]$, the second part is an URE-ciphertext obtained with some public key (not necessarily the key of the destination node B) encoding some additional information. Immediately after creation of an

URE-navigator the ciphertext encodes 1. Afterwards, when the URE-navigator is used and re-coded, we can replace 1 with an arbitrary message m, as described above. We use notation $\text{Nav}[A, B] \, \text{URE}_x(m)$ for such a URE-navigator, where x is the decryption key of the ciphertext of m.

Let us remark that for traditional onions we can add external layers to a navigator $\text{Nav}[A, B]$: afterwards the path of the onion would lead from a chosen C to A, and then follow the route defined by $\text{Nav}[A, B]$. For standard onion constructions such a modification is possible even, if we get $\text{Nav}[A, B]$ from a third party and we cannot disassemble it. For URE-onions such a manipulation is impossible.

3.2   Plain Off-line Scheme

In order to send anonymously a message m from S to R we can simply send an URE-navigator $\text{Nav}[S, R] \, URE_{x_R}(m)$. Subsequent servers from the path "peel off" the navigator and re-encrypt message m. Node R can decrypt the ciphertext and retrieve m.

Such a URE-navigator can be called an *off-line onion*, since an empty navigator can be created in advance and as soon as a message m to be sent is ready at application level, an URE-navigator encoding m is created by inserting m into the URE-ciphertext, as described above, and by re-encrypting all ciphertexts of the URE-navigator immediately afterwards (in order to hide m and the navigator used from the party that constructed the navigator).

Replacement Attack. Assume that an active adversary controls (actively) the beginning and the end of a path encoded in the navigator. At the beginning of the path, he replaces the URE-ciphertext of the off-line onion by an URE-ciphertext of a random string r encrypted with his own key. Of course, he can trace such a modified onion while it moves through the network. Simply, he decrypts all URE-ciphertexts of the onions with his decryption key – re-encryption does not prevent retrieving r. Once the message arrives at the end of the path, the adversary replaces the URE-ciphertext of r back by the original one and re-encrypts it. The destination node obtains a proper ciphertext and has no idea that the connection was under attack.

One can prevent this attack: instead of $\text{URE}_{x_R}(m)$ the sender transmits ciphertext $\text{URE}_{x_1+\ldots+x_\lambda}(m)$, where $x_1, \ldots, x_\lambda$ are private keys of the subsequent nodes on the path from S to R. Now, each intermediate node has to decrypt partially (and re-encrypt) the URE-ciphertext obtained. It is easy to see that after this modification the attack described above fails - the URE-ciphertext must be processed by all intermediate servers indicated in the navigator. So the destination node would retrieve a different message. Also due to the partial decryption, the adversary would not detect its own message inserted at the beginning of the path. Indeed, it is difficult to detect any connection between the ciphertexts of the form $(m \cdot (yz)^k, g^k)$ and $(m \cdot z^{k'}, g^{k'})$ knowing the public keys y, z only.

Advantages of the Scheme. The main point is that the scheme separates encoding the message from encoding the route. It may be useful in many ways:

1. The onions can be prepared in advance.
2. If a sender does not know topology of the network or its knowledge is not up to date, it is better to use navigators offered by trusted servers. In this way we can delegate the chores of creating the routes to a special well protected and administered server. This is quite advantageous since if some users choose intermediate servers in a different way than the rest of the world, then traffic analysis might become easy. Note that all results on traffic analysis [14, 1, 11] require that the intermediate nodes are chosen by all users with the same probability distribution.
3. An empty off-line onion (i. e. one encoding the message 1) can be delivered as a regular message to any node. Then this node can use it as anonymous return-address and send it back without knowing the address of the request source. Of course, such an anonymous reply scheme is possible also with the traditional onions [1], however the present solution does not require the intermediate servers to memorize any values.

The main disadvantage of the scheme is that the server preparing a navigator has to know all pairs (source, destination) used (of course, the user can fetch much more navigators that it uses and in this way hide a particular connection). Hence the solution might be suited for a company, but it is not aimed for a general use.

3.3   Merging Navigators

Using plain off-line onions becomes dangerous, when navigators were created by a server cooperating with an adversary. Even if a direct identification of a navigator in the traffic transmitted is impossible, traffic analysis might provide valuable information. This would be facilitated by the fact that the adversary might know all random paths encoded in the navigators generated by a certain server.

In order to avoid such a situation we propose *merge onions* (MO for short); our protocol shows how to combine navigators from different sources into an onion with a longer path. If the navigators come from different and non-cooperating sources, the resulting onion cannot be traced by an adversary collaborating with only some of these sources.

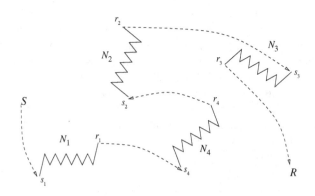

Fig. 1. Composing a merge-onion path from navigators N_1, N_2, N_3, N_4

Creating MO. For the sake of simplicity we describe how to compose a MO from two parts (Fig. 1 presents the case in which 4 navigators are used). A sender S wishing to transmit a message m to destination R executes the following steps:

- it chooses two navigators at hand, say $\mathrm{Nav}[J_1, J_\delta], \mathrm{Nav}[L_1, L_\delta]$.
- it composes a merge-onion containing the following components:

$$\mathrm{Nav}[J_1, J_\delta], \mathrm{URE}_x(L_1), \mathrm{URE}_x(\mathrm{Nav}[L_1, L_\delta]), \mathrm{URE}_{x+y}(R), \mathrm{URE}_{x+y+x_R}(m)$$

where, respectively, x and y are the sums of the description keys related to the navigators $\mathrm{Nav}[J_1, J_\delta], \mathrm{Nav}[L_1, L_\delta]$, and x_R is the decryption key of R.

The way of processing such an onion is clear: first it is sent to J_1. Then it is processed according to the navigator $\mathrm{Nav}[J_1, J_\delta]$; at each step all remaining components are partially decrypted and re-encrypted. This lasts until we reach the end of the first navigator. Then the second component reveals L_1 and the last three components are sent to L_1. Then the message follows the route encoded by the navigator $\mathrm{Nav}[L_1, L_\delta]$ until it reaches L_δ. Then R is retrieved and the last component (which is a ciphertext of m with the decryption key x_R) is sent to R.

For the protocol described, the adversary can see what is the number of remaining navigators to be used until the end of the path. In order to hide this information we may introduce a simple modification of the protocol. The server, which retrieves the next navigator to be used, does not remove its ciphertext, but re-encrypts it and moves behind the last ciphertext of a navigator.

Advantages of Merge Onions. The size of MO grows moderately with the number of navigators used. Each navigator (except one) is represented by a single URE-ciphertext of the navigator and a URE-ciphertext of the starting node of the next navigator.

3.4   Online Merge Onions

Online Merge Onion scheme (OMO), in contrast to Merge Onion scheme, demands from the sender knowledge of a few stable servers in the network that remain working all the time. Navigators are chosen online by the servers selected by the sender. We can think about OMO as a scheme in which sender "asks" some servers to provide anonymity of his message by sending it along routes with many conflicts.

Creating OMO. A sender S wishing to transmit a message m to R executes the following steps:

- it chooses k servers $A_1, A_2, ..., A_k$ at random (from a common public list), and creates a navigator $N = \mathrm{Nav}[A_1, ..., A_k]$ encoding the path $A_1, ..., A_k$;
- it inserts a message "to R" into N,
- it creates $\mathrm{URE}_{x_R}(m)$, where x_R is the decryption key of R,
- it chooses an URE-navigator U from a set of available navigators and inserts a message: "to A_1" into it,
- finally, it sends a message:

$$U(\text{to } A_1), \mathrm{URE}_{x_{A_1}}(\mathrm{Nav}[A_1, ..., A_k](\text{to } R)), \mathrm{URE}_{x_R}(m)$$

to the starting node of the navigator U.

Processing OMO. There are two cases. If a server D receiving the onion is not on the list $(A_1, ..., A_k)$, then it processes it according to the navigator standing in front of the message and re-encrypts the remaining parts. If $D = A_i$, then

- it decrypts $\text{URE}_{x_{A_i}}(\text{Nav}[A_i, A_{i+1}, ..., A_k])$, so it gets $\text{Nav}[A_{i+1}, ..., A_k])$ and the message: "to A_{i+1}",
- it encrypts the navigator obtained with the public key of A_{i+1}, that is, it gets $\text{URE}_{x_{A_{i+1}}}(\text{Nav}[A_{i+1}, ..., A_k])$,
- it chooses an URE-navigators M and inserts the message "to A_{i+1}" into it.
- it re-encrypts the last part of the message, which is $\text{URE}_{x_R}(m)$,
- it sends concatenation of these parts:

$$M(\text{"to } A_{i+1}\text{"}), \text{URE}_{A_{i+1}}(Nav[A_{i+1}, ..., A_k]), \text{URE}_{x_B}(m)$$

to the starting node of the navigator M.

Protocol Features.

Adapting Onion Length to Network Load. Servers can dynamically adopt their behavior to message traffic independently of the senders. It is often believed that using dummy messages to increase the traffic to the maximal amount and keeping lengths of the onion paths fixed is a proper answer to network dynamics. However, proving resilience to traffic analysis when a fraction of servers and lines is malicious depends on how often the onions are processed through honest servers and links[11]. So dummies do not help that much, as one may hope at the first look.

The protocol OMO gives the freedom to adopt the lengths of the paths on-the-fly. So, as shown on picture below, server A can assign to packet m arriving at time t not only a different path in a navigator, but also a different path length. For instance, on the figure below a packet corresponding to the same message m, but arriving at time, say t', will reach server B in 5 steps instead of 11. Moreover, due to re-encryption, those two packets look completely different.

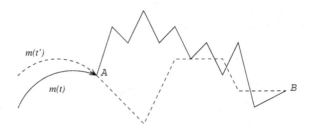

Fig. 2. Adjusting the path length by OMO

Traffic Reduction. If a message m to be transmitted is small, then the volume of routing information contained in an onion containing m might be high compared to the volume of m. This disadvantage can be relaxed somewhat through online merge onions: while the total length of the path along which a message is processed is long (preventing a traffic analysis), all the time the message transmitted contains only two (much shorter) navigators.

Enforcing Conflicts and Constructing Navigators. The way of choosing the mix servers is the following. The procedure requires a global list $\mathcal{A}$ of mix servers, known to every protocol participant. When a server A_i has to choose a navigator leading to a server A_{i+1}, it uses an approximation of the number of onions in the network to determine the length of the navigator, say t. The number t can be found through observation of the traffic passing through in the preceding moments. Assume that a mix can process at most z onions at once. Then A_i takes s such that $z = \Theta(t/s + \log s / \log \log s)$. The mix servers for the navigator constructed by A_i are chosen uniformly at random from the prefix of list $\mathcal{A}$ of length s. Standard "bin and balls" arguments may be applied here to show that a large number of conflicts at mix servers would be generated in this way. We skip a detailed analysis here.

Conclusions

We have shown that universal re-encryption provides many new interesting features:

- possibility to prepare onions in advance,
- adaptiveness to network traffic,
- size reduction of the auxiliary parts of onion messages,
- possibility to process the onions through arbitrary chosen mixes,
- implementing onions in a layered architecture of a distributed, dynamic system.

Let us compare the parameters used by the schemes. Necessary path length λ for each of the schemes depends on assumptions about adversary model. If an adversary can corrupt only a constant fraction of navigator sources, essentially the same analysis applies as in the case of [11]. So we consider the same (global) path length λ for 3 schemes considered below.

	Classical Onions	Merge Onions	Online Merge Onions
message size	$O(\lambda+|m|)$	$O(\lambda+|m|)$	$O(k+\lambda/k+|m|)$
end-user encoding cost	$O(enc(\lambda|m|))$	$O(k{\cdot}enc(\lambda/k)+enc(|m|))$	$O(enc(\lambda/k+|m|))$
preprocessing possible	no	yes	partially
processing cost at a server	$O(enc(\lambda+|m|))$	$O(enc(\lambda+|m|))$	$O(enc(\lambda/k+|m|))$
messages tracing*	easy	easy	hard
repetitive attack**	easy	easy	harder
traffic change	–	moderate increase	decrease
required knowledge of network topology	full	none	limited
traffic adaptiveness	no	no	yes

* at low traffic, by a passive adversary who controls all links

** at any traffic, by an active adversary

References

1. Berman R., Fiat A., Ta-Shma A.: Provable Unlinkability Against Traffic Analysis, Financial Cryptography 2004, LNCS , Springer-Verlag
2. Chaum, D.: Untraceable Electronic Mail, Return Addresses, and Digital Pseudonyms, CACM 24(2) (1981), 84-88
3. Chaum, D.: The Dining Cryptographers Problem: Unconditional Sender and Recipient Untraceability, Journal of Cryptology 1.1 (1988), 65-75
4. Dingledine R., Mathewson N., Syverson P., *Tor: the Second Generation Onion Router*, USENIX Security, 2004
5. Fairbrother, P.: An Improved Construction for Universal Re-encryption, Privacy Enhancing Technologies'2004, LNCS , Springer-Verlag.
6. Frankling, M., Haber, S.: Joint Encryption and Message-Efficient Secure Computation, Journal of Cryptology 9.4 (1996), 217-232.
7. Goldschlag, D.M., Reed, M.G., Syverson, P.F.: Hiding Routing Information. Information Hiding '1996, LNCS 1174, Springer-Verlag, 137-150.
8. Golle, P.: Reputable Mix Networks, Privacy Enhancing Technologies '2004, LNCS , Springer-Verlag.
9. Golle, P., Jakobsson, M., Juels, A., Syverson, P.: Universal Re-encryption for Mixnets, RSA-CT'2004, 163-178.
10. Gomułkiewicz, M., Klonowski, M., Kutyłowski, M.: Anonymous Communication Immune against Repetitive Attack, Workshop on Information Security Applications (WISA)'2004, LNCS , Springer-Verlag, to appear.
11. Gomułkiewicz, M., Klonowski, M., Kutyłowski, M.: Provable Unlinkability Against Traffic Analysis already after $\mathcal{O}(\log(n))$ Steps!, Information Security Conference (ISC)'2004, LNCS 3225, Springer-Verlag, 354-366.
12. Gülcü, C., Tsudik, G.: Mixing E-mail with BABEL, ISOC Symposium on Network and Distributed System Security, IEEE 1996, 2-16.
13. Kesdogan D., Egner J., Büschkes R.: Stop-and-Go-MIXes Providing Probabilistic Anonymity in an Open System, Information Hiding '98, LNCS 1525, Springer-Verlag, 83-98.
14. Rackoff, C., Simon, D.R.: Cryptographic Defense Against Traffic Analysis, ACM STOC25 (1993), 672-681.

Characteristic Patterns for LTL[*]

Antonín Kučera and Jan Strejček

Faculty of Informatics, Masaryk University,
Botanická 68a, 602 00 Brno, Czech Republic
{tony, strejcek}@fi.muni.cz

Abstract. We give a new characterization of those languages that are definable in fragments of LTL where the nesting depths of X and U modalities are bounded by given constants. This brings further results about various LTL fragments. We also propose a generic method for decomposing LTL formulae into an equivalent disjunction of "semantically refined" LTL formulae, and indicate how this result can be used to improve the functionality of existing LTL model-checkers.

1 Introduction

Linear temporal logic (LTL) [1] is a popular formalism for specifying properties of (concurrent) programs. The syntax of LTL is given by the following abstract syntax equation:

$$\varphi ::= \mathtt{tt} \mid a \mid \neg\varphi \mid \varphi_1 \wedge \varphi_2 \mid X\varphi \mid \varphi_1 \, U \, \varphi_2$$

Here a ranges over a countable set $\Lambda = \{a, b, c, \ldots\}$ of *letters*. We also use $F\varphi$ to abbreviate $\mathtt{tt} \, U \, \varphi$, and $G\varphi$ to abbreviate $\neg F\neg\varphi$. The set of all letters which appear in a given formula φ is denoted $\Lambda(\varphi)$.

The semantics of LTL is defined in terms of languages over infinite words. An *alphabet* is a finite set $\Sigma \subseteq \Lambda$. An ω-*word* over Σ is an infinite sequence $\alpha = \alpha(0)\alpha(1)\alpha(2)\ldots$ of letters from Σ. The set of all finite words over Σ is denoted by Σ^*, and the set of all ω-words by Σ^ω. The length of a given $u \in \Sigma^*$ is denoted $|u|$. In the rest of this paper we use $a, b, c, \ldots$ to range over Σ, $u, v, \ldots$ to range over Σ^*, and $\alpha, \beta, \ldots$ to range over Σ^ω. For every $i \in \mathbb{N}_0$ we denote by α_i the ith suffix of α, i.e., the word $\alpha(i)\alpha(i{+}1)\ldots$.

Let Σ be an alphabet. The *validity* of a formula φ for $\alpha \in \Sigma^\omega$ is defined as follows:

$$
\begin{aligned}
&\alpha \models \mathtt{tt} \\
&\alpha \models a && \text{iff} && a = \alpha(0) \\
&\alpha \models \neg\varphi && \text{iff} && \alpha \not\models \varphi \\
&\alpha \models \varphi_1 \wedge \varphi_2 && \text{iff} && \alpha \models \varphi_1 \wedge \alpha \models \varphi_2 \\
&\alpha \models X\varphi && \text{iff} && \alpha_1 \models \varphi \\
&\alpha \models \varphi_1 \, U \, \varphi_2 && \text{iff} && \exists i \in \mathbb{N}_0 : \alpha_i \models \varphi_2 \wedge \forall 0 \leq j < i : \alpha_j \models \varphi_1
\end{aligned}
$$

For each alphabet Σ, a formula φ defines the ω-language $L_\varphi^\Sigma = \{\alpha \in \Sigma^\omega \mid \alpha \models \varphi\}$.

---

[*] This work has been supported by GAČR, grant No. 201/03/1161.

M. Bieliková et al. (Eds.): SOFSEM 2005, LNCS 3381, pp. 239–249, 2005.

For every LTL formula φ and every modality $M \in \{X, U\}$ we define the *nesting depth* of M in φ, denoted $M\text{-}depth(\varphi)$, inductively as follows.

$$M\text{-}depth(\text{tt}) = M\text{-}depth(a) = 0$$

$$M\text{-}depth(\neg\varphi) = M\text{-}depth(\varphi)$$

$$M\text{-}depth(\varphi_1 \wedge \varphi_2) = \max\{M\text{-}depth(\varphi_1), M\text{-}depth(\varphi_2)\}$$

$$M\text{-}depth(X\varphi) = \begin{cases} M\text{-}depth(\varphi) + 1 & \text{if } M = X, \\ M\text{-}depth(\varphi) & \text{otherwise.} \end{cases}$$

$$M\text{-}depth(\varphi_1 \cup \varphi_2) = \begin{cases} \max\{M\text{-}depth(\varphi_1), M\text{-}depth(\varphi_2)\} + 1 & \text{if } M = U, \\ \max\{M\text{-}depth(\varphi_1), M\text{-}depth(\varphi_2)\} & \text{otherwise.} \end{cases}$$

For all $m, n \in \mathbb{N}_0 \cup \{\infty\}$, the symbol $\text{LTL}(\mathsf{U}^m, \mathsf{X}^n)$ denotes the set of all LTL formulae φ such that $\mathsf{U}\text{-}depth(\varphi) \leq m$ and $\mathsf{X}\text{-}depth(\varphi) \leq n$. To simplify our notation, we omit the "∞" superscript. Hence, e.g., $\text{LTL}(\mathsf{U}^3, \mathsf{X})$ is a shorthand for $\text{LTL}(\mathsf{U}^3, \mathsf{X}^\infty)$.

A lot of research effort has been invested into characterizing the expressive power of LTL and its fragments. A concise survey covering basic results about LTL expressiveness can be found in [2]. A more recent survey [3] contains also results concerning some of the LTL fragments. In this paper, we give a new characterization of ω-languages that are definable in $\text{LTL}(\mathsf{U}^m, \mathsf{X}^n)$ for given $m, n \in \mathbb{N}_0$.[1] Roughly speaking, for each alphabet Σ and all $m, n \in \mathbb{N}_0$ we design a finite set of (m, n)-*patterns*[2], where each (m, n)-pattern is a finite object representing an ω-language over Σ so that the following conditions are satisfied:

- Each $\alpha \in \Sigma^\omega$ is represented by exactly one (m, n)-pattern (consequently, the sets of ω-words represented by different patterns are disjoint).
- ω-words which are represented by the same (m, n)-pattern cannot be distinguished by any formula of $\text{LTL}(\mathsf{U}^m, \mathsf{X}^n)$.
- For each (m, n)-pattern p we can effectively construct a formula $\psi \in \text{LTL}(\mathsf{U}^m, \mathsf{X}^n)$ so that for each $\alpha \in \Sigma^\omega$ we have that $\alpha \models \psi$ if and only if α is represented by the pattern p.

Thus, the semantics of each formula $\varphi \in \text{LTL}(\mathsf{U}^m, \mathsf{X}^n)$ is fully characterized by a finite subset of (m, n)-patterns, and vice versa. Intuitively, (m, n)-patterns represent *exactly* the information about ω-words which determines the (in)validity of $\text{LTL}(\mathsf{U}^m, \mathsf{X}^n)$ formulae. The patterns are defined inductively on m, and the inductive step brings some insight into what is actually gained (i.e., what new properties can be expressed) by increasing the nesting depth of U by one.

Characteristic patterns can be used as a tool for proving further results about the logic LTL and its fragments. In particular, they can be used to construct a short proof of a (somewhat simplified) form of stutter invariance of $\text{LTL}(\mathsf{U}^m, \mathsf{X}^n)$ languages introduced in [4]. This, in turn, allows to construct simpler proofs for some of the results presented in [4] (like, e.g., the strictness of the $\text{LTL}(\mathsf{U}^m, \mathsf{X})$, $\text{LTL}(\mathsf{U}, \mathsf{X}^n)$, and

---

[1] The expressiveness of these fragments has already been studied in [4]. In particular, it has been proven that the classes of languages definable by two syntactically incomparable fragments of this form are also incomparable.

[2] Let us note that (m, n)-patterns have nothing to do with the forbidden patterns of [3].

LTL(U^m, X^n) hierarchies). An interesting question (which is left open) is whether one could use characteristic patterns to demonstrate the decidability of the problem if a given ω-regular language L is definable in LTL(U^m, X) for a given m.

Another application area for characteristic patterns is LTL model-checking. We believe that this is actually one of the most interesting parts of our work, and therefore we explain the idea in greater detail.

An instance of the LTL model-checking problem is a system and an LTL formula (called "specification formula") which defines desired properties of the system. The question is whether all runs of the system satisfy the formula. This problem can dually be reformulated as follows: for a given system and a given formula φ (representing the negation of the desired property), decide whether the system has at least one run satisfying φ. Characteristic patterns can be used to decompose a given LTL formula φ into an equivalent disjunction $\varphi \equiv \psi_1 \vee \ldots \vee \psi_n$ of mutually exclusive formulae (i.e., we have $\psi_i \Rightarrow \bigwedge_{j \neq i} \neg \psi_j$ for each i). Roughly speaking, each ψ_i corresponds to one of the patterns which define the semantics of φ. Hence, the ψ_i formulae are not necessarily smaller or simpler than φ from the syntactical point of view. The simplification is on semantical level, because each ψ_i "cuts off" a dedicated subset of runs that satisfy φ. Another advantage of this method is its scalability—the patterns can be constructed also for those n and m that are larger than the nesting depths of X and U in φ. Thus, the patterns can be repeatedly "refined", which corresponds to decomposing the constructed ψ_i formulae. Another way of refining the patterns is enlarging the alphabet Σ.

The decomposition technique enables the following model-checking strategy: First try to model-check φ. If this does not work (because of, e.g., memory overflow), then decompose φ into $\psi_1 \vee \ldots \vee \psi_n$ and try to model-check the $\psi_1, \ldots, \psi_n$ formulae. This can be done sequentially or even in parallel. If at least one subtask produces a positive answer, we are done (there is a "bad" run). Similarly, if all subtasks produce a negative answer, we are also done (there is no "bad" run). Otherwise, we go on and decompose those ψ_i for which our model-checker did not manage to answer.

Obviously, the introduced strategy can only lead to better results than checking just φ, and it is completely independent of the underlying model-checker. Moreover, some new and relevant information is obtained even in those cases when this strategy does not lead to a definite answer—we know that if there is a bad run, it must satisfy some of the subformulae we did not manage to model-check. The level of practical usability of the above discussed approach can only be measured by outcomes of practical experiments which are beyond the scope of this (mainly theoretical) paper.[3] Here we concentrate on providing basic results and identifying promising directions for applied research.

Let us note that similar decomposition techniques have been proposed in [5] and [6]. In [5], a specification formula of the form $G\varphi$ is decomposed into a set of formulae $\{G(x{=}v_i \Rightarrow \varphi) \mid v_i$ is in the range of the variable $x\}$. This decomposition technique has been implemented in the SMV system together with methods aimed at reducing the range of x. This approach has then been used for verification of specific types of infinite-state systems (see [5] for more details). In [6], a given specification formula φ is model-checked as follows: First, a finite set of formulae $\psi_1, \ldots, \psi_n$ of the form

---

[3] A practical implementation of the method is under preparation.

$\psi_i = G(x{\neq}v_0 \Rightarrow x{=}v_i)$ is constructed such that the verified system satisfies $\psi_1 \vee \ldots \vee \psi_n$. The formulae $\psi_1, \ldots, \psi_n$ are either given directly by the user, or constructed automatically using methods of static analysis. The verification problem for φ is then decomposed into the problems of verifying the formulae $\psi_i \Rightarrow \varphi$. Using this approach, the peak memory in model checking has been reduced by 13–25% in the three case studies included in the paper.

It is worth mentioning that characteristic patterns could potentially be used also in a different way: we could first extract *all* patterns that can be exhibited by the system, and then check whether there is one for which φ holds. Unfortunately, the set of all patterns exhibited by a given system seems to be computable only in restricted cases, e.g., when the system has just a single path (see [7] for more information about model checking of these systems and [8] for a pattern-based algorithm).

The paper is organized as follows. Section 2 provides a formal definition of (m, n)-patterns together with basic theorems. Section 3 is devoted to detailed discussion of the indicated decomposition technique. Conclusions and directions for future research are given in Section 4. Other applications of characteristic patterns in the area of LTL model checking as well as all proofs (which were omitted due to space constraints) can be found in [8].

2   Characteristic Patterns

To get some intuition about characteristic patterns, let us first consider the set of patterns constructed for the alphabet $\Sigma = \{a, b, c\}$, $m = 1$, and $n = 0$ (as we shall see, the m and n correspond to the nesting depths of U and X, respectively). Let $\alpha \in \Sigma^\omega$ be an ω-word. A letter $\alpha(i)$ is *repeated* if there is $j < i$ such that $\alpha(j) = \alpha(i)$. The $(1,0)$-pattern of α, denoted $pat(1, 0, \alpha)$, is the finite word obtained from α by deleting all repeated letters (for reasons of consistent notation, this word is written in parenthesis). For example, if $\alpha \underline{a}\underline{a}bb\underline{b}baabababab\underline{c}abccacab\ldots$, then $pat(1, 0, \alpha) = (abc)$. So, the set of all $(1,0)$-patterns over the alphabet $\{a, b, c\}$, denoted $Pats(1, 0, \{a, b, c\})$, has exactly 15 elements which are the following:

$$(abc), (acb), (bac), (bca), (cab), (cba), (ab), (ba), (ac), (ca), (bc), (cb), (a), (b), (c)$$

Thus, the set $\{a, b, c\}^\omega$ is divided into 15 disjoint subsets, where each set consists of all ω-words that have a given pattern. It remains to explain why these patterns are interesting. The point is that $\text{LTL}(\mathsf{U}^1, \mathsf{X}^0)$ formulae can actually express just the order of non-repeated letters. For example, the formula $a\,\mathsf{U}\,b$ says that either the first non-repeated letter is b, or the first non-repeated letter is a and the second one is b. So, this formula holds for a given $\alpha \in \{a, b, c\}^\omega$ iff $pat(1, 0, \alpha)$ is equal to (b), (ba), (bc), (bac), (bca), (ab), or (abc). We claim (and later also prove) that ω-words of $\{a, b, c\}^\omega$ which have the same $(1, 0)$-pattern cannot be distinguished by *any* $\text{LTL}(\mathsf{U}^1, \mathsf{X}^0)$ formula. So, each $\varphi \in \text{LTL}(\mathsf{U}^1, \mathsf{X}^0)$, where $\Lambda(\varphi) \subseteq \{a, b, c\}$, is fully characterized by a subset of $Pats(1, 0, \{a, b, c\})$. Moreover, for each $p \in Pats(1, 0, \{a, b, c\})$ we can construct an $\text{LTL}(\mathsf{U}^1, \mathsf{X}^0)$ formula φ_p such that for every $\alpha \in \{a, b, c\}^\omega$ we have that $\alpha \models \varphi_p$ iff $pat(1, 0, \alpha) = p$. For example, $\varphi_{(abc)} = a \wedge (a\,\mathsf{U}\,b) \wedge ((a \vee b)\,\mathsf{U}\,c)$.

To indicate how this can be generalized to larger m and n, we show how to extract a $(2,0)$-pattern from a given $\alpha \in \{a, b, c\}^\omega$. We start by considering an infinite word over the alphabet $Pats(1, 0, \{a, b, c\})$ constructed as follows:

$$pat(1, 0, \alpha_0) \; pat(1, 0, \alpha_1) \; pat(1, 0, \alpha_2) \; pat(1, 0, \alpha_3) \dots$$

For example, for $\alpha = aabaca^\omega$ we obtain the sequence $(abc)(abc)(bac)(ac)(ca)(a)^\omega$. The pattern $pat(2, 0, \alpha)$ is obtained from the above sequence by deleting repeated letters (realize that now we consider the alphabet $Pats(1, 0, \{a, b, c\})$). Hence, $pat(2, 0, \alpha) = ((abc)(bac)(ac)(ca)(a))$. Similarly as above, it holds that those ω-words of $\{a, b, c\}^\omega$ which have the same $(2, 0)$-pattern cannot be distinguished by any $LTL(U^2, X^0)$ formula. Moreover, for each $p \in Pats(2, 0, \{a, b, c\})$ we can construct an $LTL(U^2, X^0)$ formula φ_p such that for every $\alpha \in \{a, b, c\}^\omega$ we have that $\alpha \models \varphi_p$ iff $pat(2, 0, \alpha) = p$.

Formally, we consider *every* finite sequence of $(1, 0)$-patterns, where no $(1, 0)$-pattern is repeated, as a $(2, 0)$-pattern. This makes the inductive definition simpler, but in this way we also introduce patterns that are not "satisfiable". For example, there is obviously no $\alpha \in \{a, b, c\}^\omega$ such that $pat(2, 0, \alpha) = ((a)(ab))$.

The last problem we have yet not addressed is how to deal with the X operator. First note that the X operator can be pushed inside using the following rules (see, e.g., [2]):

$$\mathsf{X}tt \equiv tt \quad \mathsf{X}\neg\varphi \equiv \neg\mathsf{X}\varphi \quad \mathsf{X}(\varphi_1 \wedge \varphi_2) \equiv \mathsf{X}\varphi_1 \wedge \mathsf{X}\varphi_2 \quad \mathsf{X}(\varphi_1 \, \mathsf{U} \, \varphi_2) \equiv \mathsf{X}\varphi_1 \, \mathsf{U} \, \mathsf{X}\varphi_2$$

Note that this transformation does not change the nesting depth of X. Hence, we can safely assume that the X operator occurs in LTL formulae only within subformulae of the form $\mathsf{X}\mathsf{X} \dots \mathsf{X}a$. This is the reason why we can handle the X operator in the following way: the set $Pats(m, n, \Sigma)$ is defined in the same way as $Pats(m, 0, \Sigma)$. The only difference is that we start with the alphabet Σ^{n+1} instead of Σ.

Definition 1. *Let Σ be an alphabet. For all $m, n \in \mathbb{N}_0$ we define the set $Pats(m, n, \Sigma)$ inductively as follows:*

- $Pats(0, n, \Sigma) = \{w \in \Sigma^* \mid |w| n+1\}$
- $Pats(m+1, n, \Sigma) = \{(p_1 \dots p_k) \mid k \in \mathbb{N}, \; p_1, \dots, p_k \in Pats(m, n, \Sigma),$
  $p_i \neq p_j \text{ for } i \neq j\}$

The size of $Pats(m, n, \Sigma)$ and the size of its elements are estimated in our next lemma (the proof follows directly from definitions).

Lemma 1. *For every $i \in \mathbb{N}_0$, let us define the function $fac_i : \mathbb{N}_0 \to \mathbb{N}_0$ inductively as follows: $fac_0(x) = x$, $fac_{i+1}(x) = (fac_i(x) + 1)!$. The number of elements of $Pats(m, n, \Sigma)$ is bounded by $fac_m(|\Sigma|^{n+1})$, and the size of each $p \in Pats(m, n, \Sigma)$ is bounded by $(n+1) \cdot \Pi_{i=0}^{m-1} fac_i(|\Sigma|^{n+1})$.*

The bounds given in Lemma 1 are non-elementary in m. This indicates that all of our algorithms are computationally unfeasible from the asymptotic complexity point of view. However, LTL formulae that are used in practice typically have a small nesting depth of U (usually not larger than 3 or 4), and do not contain any X operators. In this light, the bounds of Lemma 1 can actually be interpreted as "good news", because even a relatively small formula φ can be decomposed into a disjunction of many formulae which refine the meaning of φ.

To all $m, n \in \mathbb{N}_0$ and $\alpha \in \Sigma^\omega$ we associate a unique pattern of $Pats(m, n, \Sigma)$. This definition is again inductive.

Definition 2. *Let* $\alpha \in \Sigma^\omega$. *For all* $m, n \in \mathbb{N}_0$ *we define the* characteristic (m, n)- *pattern of* α, *denoted* $pat(m, n, \alpha)$, *and* (m, n)-*pattern* word *of* α, *denoted* $patword(m, n, \alpha)$, *inductively as follows:*

- $pat(0, n, \alpha) = \alpha(0) \ldots \alpha(n)$
- $patword(m, n, \alpha) \in Pats(m, n, \Sigma)^\omega$ *is defined by* $patword(m, n, \alpha)$ $(i) = pat(m, n, \alpha_i)$
- $pat(m{+}1, n, \alpha)$ *is the finite word (written in parenthesis) obtained from* $patword(m, n, \alpha)$ *by deleting all repeated letters*

Words $\alpha, \beta \in \Sigma^\omega$ *are* (m, n)-equivalent, *written* $\alpha \sim_{m,n} \beta$, *iff* $pat(m, n, \alpha) = pat(m, n, \beta)$.

Example 1. Let us consider a word $\alpha = abbbacbac(ba)^\omega$. Then

$$pat(0, 0, \alpha) = a$$
$$patword(0, 0, \alpha) = abbbacbac(ba)^\omega = \alpha$$
$$pat(1, 0, \alpha) = (abc)$$
$$patword(1, 0, \alpha) = (abc)(bac)(bac)(bac)(acb)(cba)(bac)(acb)(cba)((ba)(ab))^\omega$$
$$pat(2, 0, \alpha) = ((abc)(bac)(acb)(cba)(ba)(ab))$$

$$pat(0, 1, \alpha) = \underline{ab}$$
$$patword(0, 1, \alpha) = \underline{ab}\,\underline{bb}\,\underline{bb}\,\underline{ba}\,\underline{ac}\,\underline{cb}\,\underline{ba}\,\underline{ac}\,\underline{cb}(\underline{ba}\,\underline{ab})^\omega$$
$$pat(1, 1, \alpha) = (\underline{ab}\,\underline{bb}\,\underline{ba}\,\underline{ac}\,\underline{cb}) \qquad \square$$

Theorem 1. *Let* Σ *be an alphabet. For all* $m, n \in \mathbb{N}_0$ *and every* $p \in Pats(m, n, \Sigma)$ *there effectively exists a formula* $\varphi_p \in \mathrm{LTL}(\mathsf{U}^m, \mathsf{X}^n)$ *such that for every* $\alpha \in \Sigma^\omega$ *we have that* $\alpha \models \varphi_p$ *iff* $pat(m, n, \alpha) = p$.

Example 2. Let $\alpha = abbabaaabb(ac)^\omega$. Then the formula φ_p, where $p = pat(2, 0, \alpha) = ((abc)(bac)(ac)(ca))$ is constructed (according to the proof of the previous theorem) as follows:

$$\varphi_{(abc)} = \mathsf{G}(a \vee b \vee c) \wedge a \wedge (a\,\mathsf{U}\,b) \wedge ((a \vee b)\,\mathsf{U}\,c)$$
$$\varphi_{(bac)} = \mathsf{G}(b \vee a \vee c) \wedge b \wedge (b\,\mathsf{U}\,a) \wedge ((b \vee a)\,\mathsf{U}\,c)$$
$$\varphi_{(ac)} = \mathsf{G}(a \vee c) \wedge a \wedge (a\,\mathsf{U}\,c)$$
$$\varphi_{(ca)} = \mathsf{G}(c \vee a) \wedge c \wedge (c\,\mathsf{U}\,a)$$
$$\varphi_p = \mathsf{G}(\varphi_{(abc)} \vee \varphi_{(bac)} \vee \varphi_{(ac)} \vee \varphi_{(ca)}) \wedge \varphi_{(abc)} \wedge (\varphi_{(abc)}\,\mathsf{U}\,\varphi_{(bac)}) \wedge$$
$$\wedge ((\varphi_{(abc)} \vee \varphi_{(bac)})\,\mathsf{U}\,\varphi_{(ac)}) \wedge ((\varphi_{(abc)} \vee \varphi_{(bac)} \vee \varphi_{(ac)})\,\mathsf{U}\,\varphi_{(ca)}) \qquad \square$$

Let us note that the size of φ_p for a given $p \in Pats(m, n, \Sigma)$ is exponential in the size of p. However, if φ_p is represented by a circuit (DAG), then the size of the circuit is only linear in the size of p.

Theorem 2. *Let Σ be an alphabet and let $m, n \in \mathbb{N}_0$. For all $\alpha, \beta \in \Sigma^\omega$ we have that α and β cannot be distinguished by any $\mathrm{LTL}(\mathsf{U}^m, \mathsf{X}^n)$ formula if and only if $\alpha \sim_{m,n} \beta$.*

In other words, Theorem 2 says that the information about α which is relevant with respect to (in)validity of all $\mathrm{LTL}(\mathsf{U}^m, \mathsf{X}^n)$ formulae is exactly represented by $pat(m, n, \alpha)$. Thus, characteristic patterns provide a new characterization of $\mathrm{LTL}(\mathsf{U}^m, \mathsf{X}^n)$ languages which can be used to prove further results about LTL. In particular, a simplified form of (m, n)-stutter invariance of $\mathrm{LTL}(\mathsf{U}^m, \mathsf{X}^n)$ languages (see [4]) follows easily from the presented results on characteristic patterns:

Theorem 3. *Let $m, n \in \mathbb{N}_0$, $u, v \in \Sigma^*$ and $\alpha \in \Sigma^\omega$. If v is (m, n)-redundant in $uv\alpha$, then $uv\alpha \sim_{m,n} u\alpha$.*

Theorem 3 provides the crucial tool which was used in [4] to prove that, e.g., the $\mathrm{LTL}(\mathsf{U}^m, \mathsf{X})$, $\mathrm{LTL}(\mathsf{U}, \mathsf{X}^n)$, and $\mathrm{LTL}(\mathsf{U}^m, \mathsf{X}^n)$ hierarchies are strict, that the class of ω-languages which are definable both in $\mathrm{LTL}(\mathsf{U}^{m+1}, \mathsf{X}^n)$ and $\mathrm{LTL}(\mathsf{U}^m, \mathsf{X}^{n+1})$ is strictly larger than the class of languages definable in $\mathrm{LTL}(\mathsf{U}^m, \mathsf{X}^n)$, and so on. The proof of Theorem 3 is shorter than the one given in [4].

3   Applications in Model Checking

In this section, we expand the remarks about formula decomposition and pattern refinement that were sketched in the introduction. We also discuss potential benefits and drawbacks of these techniques, and provide examples illustrating the presented ideas.

Definition 3. *Let $p \in Pats(m, n, \Sigma)$ be a pattern and $\varphi \in \mathrm{LTL}(\mathsf{U}^m, \mathsf{X}^n)$ be a formula. We say that p satisfies φ, written $p \models \varphi$, iff for every ω-word $\alpha \in \Sigma^\omega$ we have that if $pat(m, n, \alpha) = p$, then $\alpha \models \varphi$.*

Note that Theorem 2 implies the following: if $p \not\models \varphi$, then for every ω-word α such that $pat(m, n, \alpha) = p$ we have $\alpha \not\models \varphi$.

Theorem 4. *Given an (m, n)-pattern p and an $\mathrm{LTL}(\mathsf{U}^m, \mathsf{X}^n)$ formula φ, the problem whether $p \models \varphi$ can be decided in time $\mathcal{O}(|\varphi| \cdot |p|)$.*

In the rest of this section we consider the variant of LTL where formulae are built over *atomic propositions (At)* rather than over letters. The only change in the syntax is that a ranges over At. The logic is interpreted over ω-words over an alphabet $\Sigma \subseteq 2^{At}$, where $\alpha \models a$ iff $a \in \alpha(0)$. The formula $\mathsf{F}\varphi$ is to be understood just as an abbreviation for $\mathsf{tt}\,\mathsf{U}\,\varphi$, and $\mathsf{G}\varphi$ as an abbreviation for $\neg\mathsf{F}\neg\varphi$.

Let $\varphi \in \mathrm{LTL}(\mathsf{U}^m, \mathsf{X}^n)$ be the negation of a property we want to verify for a given system. If our model-checker fails to verify whether the system has a run satisfying φ or not (one typical reason is memory overflow), we can proceed by decomposing the formula φ in the following way. First, we compute the set $P = \{p \in Pats(m, n, 2^{At(\varphi)}) \mid p \models \varphi\}$. Then, each $p \in P$ is translated into an equivalent LTL formula.

Example 3. We illustrate the decomposition technique on a formula $\varphi = \mathsf{FG}\neg a$ which is the negation of a typical liveness property $\mathsf{GF}a$. The alphabet is $\Sigma = 2^{\{a\}} =$

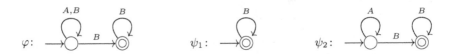

Fig. 1. Büchi automata corresponding to formulae φ, ψ_1, and ψ_2 of Example 3

$\{\{a\}, \emptyset\}$. To simplify our notation, we use A and B to abbreviate $\{a\}$ and $\emptyset$, respectively. The elements of $Pats(2, 0, (\{A, B\})$ are listed below (unsatisfiable patterns have been eliminated). All patterns which satisfy φ are listed in the second line.

$$((A)), ((BA)(A)), ((AB)(BA)), ((BA)(AB)), ((AB)(BA)(A)), ((BA)(AB)(A))$$
$$((B)), ((AB)(B)), ((BA)(AB)(B)), ((AB)(BA)(B))$$

So, the formula φ is decomposed into a disjunction $\psi_1 \vee \psi_2 \vee \psi_3 \vee \psi_4$ of formulae corresponding to the patterns listed in the second line, respectively[4]:

$$\psi_1 = \mathsf{G}\neg a \qquad\qquad \psi_3 = \neg a \wedge \mathsf{F}(a \wedge \mathsf{F}\neg a) \wedge \mathsf{FG}\neg a$$
$$\psi_2 = a \wedge a\,\mathsf{U}\,\mathsf{G}\neg a \qquad \psi_4 = a \wedge \mathsf{F}(\neg a \wedge \mathsf{F}a) \wedge \mathsf{FG}\neg a \qquad\qquad \square$$

Thus, the original question whether the system has a run satisfying φ is decomposed into k questions of the same type. These can be solved using standard model-checkers.

We illustrate potential benefits of this method in the context of automata-theoretic approach to model checking [9]. Here the formula φ is translated into a corresponding Büchi automaton A_φ. Then, the model-checking algorithm computes another Büchi automaton called *product automaton*, which accepts exactly those runs of the verified system which are accepted by A_φ as well. The model-checking problem is thus reduced to the problem whether the language accepted by the product automaton is empty or not. The bottleneck of this approach is the size of the product automaton.

Example 4. Let us suppose that a given model-checking algorithm does not manage to check the formula φ of Example 3. The subtasks given by the ψ_i formulae constructed in Example 3 can be more tractable. Some of the reasons are given below.

- The size of the Büchi automaton for ψ_i can be smaller than the size of A_φ. In Example 3, this is illustrated by formula ψ_1 (see Fig. 1). The corresponding product automaton is then smaller as well.
- The size of the product automaton for ψ_i can be smaller than the one for φ, even if the size of A_{ψ_i} is larger than the size of A_φ. This can be illustrated by the formula ψ_2 of Example 3; the automata for φ and ψ_2 are almost the same (see Fig. 1), but the product automaton for ψ_2 can be much smaller as indicated in Fig. 2. $\square$

It is of course possible that some of the ψ_i formulae in the constructed decomposition remain intractable. Such a formula ψ_i can further be decomposed by a technique

---

[4] For notation convenience, we simplified the formulae obtained by running the algorithm of Theorem 1 into a more readable (but equivalent) form.

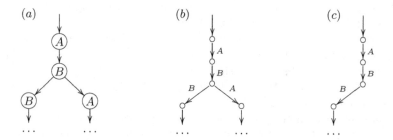

Fig. 2. An example of a verified system (a) and product automata (b) and (c) corresponding to φ and ψ_2 of Example 3, respectively

called *refinement* (since ψ_i corresponds to a unique pattern $p_i \in Pats(m, n, \Sigma)$, we also talk about pattern refinement). We propose two basic ways how to refine the pattern p_i. The first possibility is to compute the set of (m', n')-patterns, where $m' \geq m$ and $n' \geq n$, and identify all patterns satisfying the formula ψ_i.

Example 5. The formula ψ_3 of Example 3 corresponding to the $(2, 0)$-pattern $((BA)(AB)(B))$ can be refined into two $LTL(U^3, X^0)$ formulae given by the $(3, 0)$-patterns $(((BA)(AB)(B))((AB)(BA)(B))((AB)(B))((B)))$ and $(((BA)(AB)(B))$ $((AB)(B))((B)))$. $\qquad\square$

The other refinement method is based on enlarging the alphabet before computing the patterns. We simply add a new atomic proposition to the set of atomic propositions that occur in φ. The choice of the new atomic proposition is of course important. By a "suitable" choice we mean a choice which leads to a convenient split of system's runs into more manageable units. An interesting problem (which is beyond the scope of this paper) is whether suitable new propositions can be identified effectively.

Example 6. Let us consider the formula ψ_2 of Example 3 corresponding to the $(2, 0)$-pattern $((AB)(B))$. The original set of atomic propositions $At(\varphi) = \{a\}$ generates the alphabet $\Sigma = \{A, B\}$, where $A = \{a\}, B = \emptyset$. If we enrich the set of atomic propositions with b, we get a new alphabet $\Sigma' = \{C, D, E, F\}$, where $C = \{a, b\}, D = \{a\}, E = \{b\}, F = \emptyset$. Hence, the original letters A, B correspond to the pairs of letters C, D and E, F, respectively. Thus, the formula ψ_2 is refined into $LTL(U^2, X^0)$ formulae given by 64 $(2, 0)$-patterns $((CE)(E)), ((CDE)(DE)(E)), ((CDE)(DCE)(CE)$ $(E)), \dots$. $\qquad\square$

Some of the subtasks obtained by refining intractable subtasks can be tractable. Others can be refined again and again. Observe that even if we solve only some of the subtasks, we still obtain a new piece of relevant knowledge about the system—we know that if the system has a "bad" run satisfying φ, then the run satisfies one of the formulae corresponding to the subtasks we did not manage to solve. Hence, we can (at least) classify and repeatedly refine the set of "suspicious" runs.

We finish this section by listing the benefits and drawbacks of the presented method.

+ The subtasks are formulated as standard model-checking problems. Therefore, the method can be combined with all existing algorithms and heuristics.

+ With the help of the method, we can potentially verify some systems which are beyond the reach of existing model-checkers.
+ Even if it is not possible complete the verification task, we get partial information about the structure of potential (undiscovered) bad runs. We also know which runs of the system have been successfully verified.
+ The subtasks can be solved simultaneously in a distributed environment with a very low communication overhead.
+ When we verify more formulae on the same system, the subtasks occurring in decompositions of both formulae are solved just once.
− Calculating the decomposition of a given formula can be expensive. On the other hand, this is not critical for formulae with small number of atomic propositions and small nesting depths of U and X.
− Runtime costs of the proposed algorithm are high. It can happen that all subtasks remain intractable even after several refinement rounds and we get no new information at all.

4   Conclusions and Future Work

The aim of this paper was to introduce the idea of characteristic patterns, develop basic results about these patterns, and indicate how they can be used in LTL model-checking. An obvious question is how the presented algorithms work in practice. This can only be answered by performing a set of experiments. We plan to implement the presented algorithms and report about their functionality in our future work.

Acknowledgement. We thank Michal Kunc for providing crucial hints which eventually led to the definition of characteristic patterns.

References

1. Pnueli, A.: The temporal logic of programs. In: Proceedings of the 18th IEEE Symposium on the Foundations of Computer Science (FOCS'77), IEEE Computer Society Press (1977) 46–57
2. Emerson, E.A.: Temporal and modal logic. In: Handbook of Theoretical Computer Science. Volume B: Formal Models and Semantics. Elsevier (1990) 995–1072
3. Wilke, T.: Classifying discrete temporal properties. In: Annual Symposium on Theoretical Aspects of Computer Science (STACS'99). Volume 1563 of LNCS., Springer (1999) 32–46
4. Kučera, A., Strejček, J.: The stuttering principle revisited: On the expressiveness of nested X and U operators in the logic LTL. In: 11th Annual Conference of the European Association for Computer Science Logic (CSL'02). Volume 2471 of LNCS., Springer (2002) 276–291
5. McMillan, K.L.: Verification of infinite state systems by compositional model checking. In: Correct Hardware Design and Verification Methods (CHARME'99). Volume 1703 of LNCS., Springer (1999) 219–237
6. Zhang, W.: Combining static analysis and case-based search space partitioning for reducing peak memory in model checking. Journal of Computer Science and Technology **18** (2003) 762–770

7. Markey, N., Schnoebelen, P.: Model checking a path (preliminary report). In: Proc. 14th Int. Conf. Concurrency Theory (CONCUR'03). Volume 2761 of LNCS., Springer (2003) 251–265
8. Kučera, A., Strejček, J.: Characteristic patterns for LTL. Technical Report FIMU-RS-2004-10, Faculty of Informatics, Masaryk University Brno (2004)
9. Vardi, M.Y., Wolper, P.: An automata-theoretic approach to automatic program verification. In: Proceedings of the First Annual IEEE Symposium on Logic in Computer Science (LICS'86), IEEE Computer Society Press (1986) 332–344

Planar Straight-Line Drawing in an $\mathcal{O}(n) \times \mathcal{O}(n)$ Grid with Angular Resolution $\Omega(1/n)$

Maciej Kurowski[*]

Institute of Informatics, Warsaw University,
Banacha 2, 02-097 Warsaw, Poland
kuros@mimuw.edu.pl

Abstract. We study a problem of straight-line drawings for plane graphs. We show that each plane graph can be drawn in an $\mathcal{O}(n) \times \mathcal{O}(n)$ grid with angular resolution $\Omega(1/n)$.

1   Introduction

The most intensively studied area in graph drawing is devoted to algorithms for computing aesthetic drawings of graphs. For an extensive survey on the subject see [1].

When dealing with planar graphs we typically require the edges to be mapped into internally non-intersecting curves in the plane. Additionally, edges' drawings are required to have a simple structure.

The most important measures of quality of the drawing are its area (assuming all the vertices to have integer coordinates) and angular resolution (the minimum angle between drawings of incident edges). The angular resolution of the drawing computed by the algorithm can be measured in two ways: as a function of the maximum degree Δ of the input graph or the number n of its vertices.

The asymptotically optimal angular resolution $\Theta(1/\Delta)$ and the quadratic size of the drawing can be achieved simultaneously if one uses polylines to draw the edges. Many papers study the algorithms computing such drawings: [11] (three bends per edge, size $(2n-5) \times (\frac{3}{2}n - \frac{7}{2})$, resolution $2/d$), [10] (two bends per edge, size $(20n-48) \times (10n-24)$, the angles around vertex v are at least $\Omega(1/\deg(v))$), [2, 3] (one bend per edge, size $\mathcal{O}(n) \times \mathcal{O}(n)$, resolution $\Theta(1/\Delta)$), [5, 6] (one bend per edge, size $(5n) \times (5n/2)$ resolution $1/(2d)$).

In this paper we focus on the particularly interesting case when all the edges are drawn as internally non-intersecting straight-line segments. This kind of drawings are called *straight-line drawings*.

There are several known straight-line drawing algorithms, based on *the shift method* [7, 8, 4] or on *the baricentric method* [14, 15]. They draw n-vertex plane graphs on the $(n-2) \times (n-2)$ grid with the angular resolution of order $\Omega(1/n^2)$. The quadratic area of the drawing is proved to be asymptotically optimal [7, 8].

---

[*] Research supported by KBN grant 4T11C04425.

M. Bieliková et al. (Eds.): SOFSEM 2005, LNCS 3381, pp. 250–258, 2005.

It is shown that every plane graph has a straight-line drawing with angular resolution $\Omega(1/c^{\Delta})$, where $c > 1$ is a constant [12, 13]. However in these kind of drawings not only the angular resolution decreases exponentially with Δ but also the sizes of the drawings are exponentially large (even for small values of Δ).

The second type of measure of the angular resolution is less restrictive. Garg and Tamassia [9] study a dependence between the angular resolution and the area of the drawing. They show, that for every $n \geq \Delta > 6$, there exists an n-vertex planar graph G of degree Δ, such that every straight-line drawing of G with the angular resolution p has the area at least c^{pn}, where $c > 1$ is a constant. Subsequently, when we require the drawing to be of a polynomial size, it is not possible to achieve angular resolution $\Omega(1/n^{\varepsilon})$ for any constant $\varepsilon < 1$, even for degree-7 graphs.

In this paper we show a complementary positive result to the latter bound. Namely, we present a variant of shift algorithm which runs in $\mathcal{O}(n)$ time and computes a straight-line drawing on the $(3n-7) \times (\frac{3}{2}n - \frac{7}{2})$ grid with the angular resolution at least $\frac{\sqrt{2}}{3\sqrt{5}n}$.

2 Preliminaries

Let $G = (V, E)$ be a graph with the vertex set V and the edge set E. We consider only simple graphs without loops and multiple edges. The degree of vertex v is defined to be the number of the neighbors of v in G and denoted by $\deg(v)$. A connected graph is said to be *biconnected* when removal of any vertex doesn't disconnect it.

Every *planar drawing* of G is a function ϕ mapping the vertices of G into distinct points in the plane and the edges of G into internally non-intersecting, continuous curves, such that for each edge (v, w), the corresponding curve $\phi(v, w)$ has ends $\phi(v)$ and $\phi(w)$. If additionally ϕ maps all the edges into straight-line segments we say that it is a *straight-line drawing*. For brevity we use the same denotations for graph elements (vertices, edges) and geometrical objects assigned to them (points, straight-line segments). The meaning will be always clear from the context. Planar drawing divides the plane into a number of arcwise connected components called *faces*. The only unbounded region is referred to as the *external face* and the other ones are referred to as *internal faces*. A *plane graph* is a planar graph with a fixed *combinatorial embedding* – a collection of cyclic orders of edges around each vertex achievable in a planar drawing.

Let $\mathrm{Dist}(A, s)$ denote the distance between point A and straight line s and $\mathrm{Dist}(A, CD)$ be the distance from point A to a straight line through points C and D. The area of triangle T is denoted by $\mathrm{Area}(T)$, coordinates of a point A are denoted by $x(A)$ and $y(A)$, respectively. The length of a straight-line segment with endpoints A and B is denoted by $|AB|$.

3   Algorithm

Fraysseix, Pach and Pollack [8] introduced the concept of *the canonical order* of vertices in a graph. They also proposed a straight-line drawing algorithm based on so called *shift method*. To make the paper self-contained we will sketch these concepts.

3.1   Canonical Order

Describing an algorithm that computes drawings of plane graphs we can often restrict ourselves to the case when the input graph is a triangulation. This is because one can triangulate the input graph adding some dummy edges, compute the drawing of the resulting triangulation and eventually remove the dummy edges from the picture. So from now on we assume that the input graph is a triangulation.

Let G be a fixed plane triangulation and p, q, r three consecutive vertices, in clockwise order, incident with the external face. The labeling $v_1 = p$, $v_2 = q$, $v_3, \ldots, v_n = r$ is called a canonical order when the following conditions are satisfied (G_k denotes the subgraph of G induced by vertices $v_1, \ldots v_k$, C_k denotes the cycle bounding the external face in G_k).

(i) $G_3, \ldots, G_n$ are biconnected.
(ii) $C_2, \ldots, C_n$ contain edge (p, q).
(iii) In graph G_k, vertex v_k lies on cycle C_k and has at least two neighbors which are consecutive vertices of the cycle C_{k-1}, for $k = 4, \ldots, n$.

3.2   The Shift Algorithm

Let G be an input triangulation. We compute its straight-line drawing sequentially by adding vertices to the picture according to their canonical order. In the first three steps we draw graph G_3 (a triangle). Vertices v_1, v_2, v_3 are assigned the points: $(1, 1)$ – the bottom-left corner, $(3, 1)$ – the bottom right corner, $(2, 2)$, respectively. In the k-th step of the algorithm, for $k = 4, \ldots, n$, we add to the picture vertex v_k and draw the edges between v_k and its neighbors in C_{k-1}.

The algorithm preserves the following invariants ((i), (ii), (iii) are the same as in [8]). After the k-th step of the algorithm:

(i) A planar straight-line drawing of graph G_k is given, all vertices have integer coordinates.
(ii) The polyline bounding the picture from above is called *a contour*. The successive vertices in the contour, from the left to the right, are the vertices of C_k in order: $v_1 = w_1, w_2, \ldots, w_{m-1}, w_m = v_2$. Additionally $x(w_1) = 1 < x(w_2) < \ldots < x(w_m)$.
(iii) Each edge (w_i, w_{i+1}) has slope equal $+1$ or -1.
(iv) All the angles between the incident edges are at least θ/n, where θ is a constant that will be defined later (see Theorem 1).

Each time we add a new vertex we have to increase the width of the drawing and move some vertices in the contour to make space for the new one. Throughout the process of drawing each vertex w_i in the contour maintains its *shifting set* $M_k(w_i)$. After the k-th step of the algorithm the shifting sets satisfy the following conditions:

(i) $M_k(w_{i+1}) \subset M_k(w_i)$.
(ii) $w_i \in M_k(w_i)$ and $w_i \notin M_k(w_{i-1})$.
(iii) If the vertices from set $M_k(w_i)$ are shifted right by any positive distance δ the drawing remains planar.

It is shown in [8] that shifting sets can be described by a recurrent equation. Let w_l and w_r denote respectively the leftmost and the rightmost neighbor of v_k in the contour. After addition, vertex v_k is assigned a shifting set $M_k(v_k) = \bigcup_{i=l+1}^{r} M_{k-1}(v_i)$. Furthermore, we add v_k to the shifting sets of all vertices preceding v_k in the contour. Formally $M_k(v_i) = M_{k-1}(v_i) \cup \{v_k\}$ for $i \leq l$ and $M_k(v_i) = M_{k-1}(v_i)$ for $i \geq r$.

3.3   Providing Good Angular Resolution

Let us define an auxiliary transformation of the drawing called *edge relaxation*. When an edge (w_{i-1}, w_i) is relaxed in the k-th step of the algorithm, the vertices from $M_k(w_i)$ are moved by one unit to the right.

Assume that we are in the k-th phase of algorithm and we are about to add vertex $v = v_k$ to the drawing. Let $w_l, w_{l+1}, w_{l+2}, \ldots, w_{r-1}, w_r$ be the neighbors of v_k in C_{k-1} listed in order of their appearance in the contour, from the left to the right. In the original algorithm [8], before v_k is added to the picture, only edges (w_l, w_{l+1}) and (w_{r-1}, w_r) are relaxed. Now we relax all the edges: $(w_l, w_{l+1}), (w_{l+1}, w_{l+2}), \ldots, (w_{r-1}, w_r)$. If the number of the relaxed edges (i.e. $r - l$) is odd the last edge (w_{r-1}, w_r) is relaxed once again.

Finally v is added to the drawing in such a way that edges (w_l, v_k) and (v_k, w_r) have slopes $1, -1$ respectively. Observe that the total number of the relaxations during each step is even and therefore v has integer coordinates.

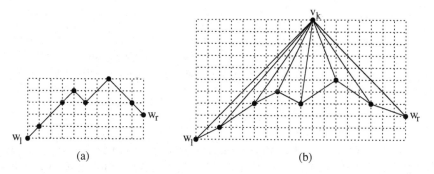

(a)   (b)

Fig. 1. (a) Before adding a new vertex; (b) After that

Note that the number of relaxations during the k-th step of the algorithm is at most $2 + c_k$, where c_k denotes the number of the vertices that disappear from the contour. Hence the total number of relaxations during steps $4, \ldots n$ is at most $\sum_{k=4}^{n} c_k + 2 * (n - 3) = 3(n - 3)$ and the dimensions of the grid at the end of the algorithm are at most $W \times H$ where $W = 2 + 3(n - 3)$ and $H = \lfloor \frac{1}{2} W \rfloor$.

An example of computations performed by our algorithm is presented in Figure 2. The pictures show the drawings after adding each successive vertex.

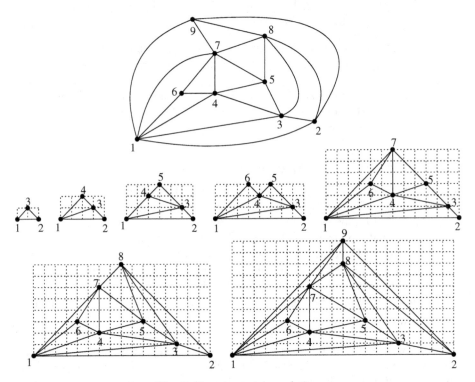

Fig. 2. Drawing a triangulation

3.4   Complexity

A naive implementation of the algorithm from the previous section runs in $\mathcal{O}(n^2)$ time. To reduce the complexity to $\mathcal{O}(n)$ we can use the technique presented in [4]. Since the necessary modifications in the algorithm from [4] are straightforward we omit the details.

3.5   Correctness

In this section we show that all the angles that appear in the picture throughout the process of drawing are of order $\Omega(1/n)$. We start with several geometric lemmas.

Lemma 1. *Let $c \geq b \geq a$ be the lengths of sides of a triangle T with $\text{Area}(T) \geq \delta b$, where δ is a constant. Then all the angles in T are at least $\frac{2\delta}{c}$.*

Proof. Let α be the angle between the sides of lengths b and c. Obviously α is the smallest angle in T.

$$\alpha \geq \sin \alpha = \frac{2}{bc} \text{Area}(T) \geq \frac{2\delta}{c}$$

Observe that if T is embedded in the $W \times H$ grid, where $W, H = \mathcal{O}(n)$, then Lemma 1 implies that all the angles in T are of order $\frac{2\delta}{\sqrt{W^2+H^2}} = \mathcal{O}(1/n)$.

Lemma 2. *Let T be an acute or a right triangle embedded in the $W \times H$ grid. Then all the angles in T are at least $\frac{\sqrt{3}/2}{\sqrt{W^2+H^2}}$.*

Proof. Let γ be the greatest angle in T and a, b ($a \leq b$) the lengths of the sides incident with γ. As T is acute or right we have $\frac{\pi}{2} \geq \gamma \geq \frac{\pi}{3}$ and thus $\sin \gamma \geq \frac{\sqrt{3}}{2}$. $\text{Area}(T) \geq \frac{\sqrt{3}}{4} ab \geq \frac{\sqrt{3}}{4} b$. Now we can apply Lemma 1.

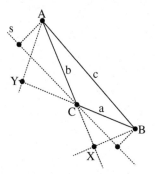

Fig. 3. Proof of Lemma 3

Lemma 3. *Let T be a triangle with vertices A, B, C embedded in the $W \times H$ grid. Let s be a straight line incident with vertex C but disjoint with the other points of T (see Figure 3). Let δ be a positive constant. If angle $\angle ACB$ is obtuse, $\text{Dist}(A, s) > \delta$ and $\text{Dist}(B, s) > \delta$ then all the angles in T are at least $\frac{\delta}{\sqrt{W^2+H^2}}$.*

Proof. As T is embedded in the $W \times H$ grid, the lengths of its sides are bounded by $\sqrt{W^2 + H^2}$. Let BX and AY be perpendicular to AC and BC respectively. Observe that $|BX| \geq \delta$ and $|AY| \geq \delta$. Let us denote $a = |BC|$ and $b = |AC|$. Since $\text{Area}(T) \geq \frac{\delta}{2} a$ and $\text{Area}(T) \geq \frac{\delta}{2} b$ we can apply Lemma 1.

Lemma 4. *Consider the situation directly after vertex v_k is added, where $k = 4 \ldots n$. Let T be an arbitrary triangle incident with v_k. Then all the angles in T are at least $\frac{\sqrt{2}/2}{\sqrt{W^2+H^2}}$.*

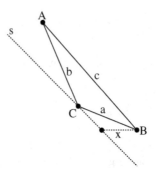

Fig. 4. Proof of Lemma 4

Proof. Points assigned to v_k and its neighbors in C_{k-1} are denoted by A, B, C respectively. We can assume that s has slope -1. The other case is symmetric. The situation is presented in Figure 4. If T is an acute triangle or a right triangle we are done due to Lemma 2. Assume that T is obtuse. Notice that $\angle BAC \leq \frac{\pi}{2}$. There are two cases to be considered. If $\angle CBA$ is obtuse then AC is the longest side in T. Since $y(B) + 1 \leq y(C)$ we have $\mathrm{Dist}(B, AC) \geq 1$. Finally $\mathrm{Area}(T) \geq |AC|/2$ and we can apply Lemma 1.

Now assume that $\angle ACB$ is obtuse. Since $x \geq 1$ then $\mathrm{Dist}(B, s) \geq \sqrt{2}/2$. As AB has slope at most -1, then also $\mathrm{Dist}(A, s) \geq \mathrm{Dist}(B, s) \geq \sqrt{2}/2$ and we can apply Lemma 3.

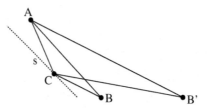

Fig. 5. Shifting does not spoil the angles – T remains obtuse

To complete the correctness proof we shall show that shifting of vertices during edge relaxations cannot decrease the angles too much. Consider the drawing after the k-th step of the algorithm. Let $w_l, w_{l+1}, \ldots, w_{r-1}, w_r$ be consecutive neighbors of v_k in the contour. Notice that vertices $w_{l+1}, \ldots, w_{r-1}, w_r$ are in shifting set $M_k(v_k)$. Subsequently all three vertices of triangles: $\triangle v_k w_{l+1} w_{l+2}$, $\triangle v_k w_{l+2} w_{l+3}, \ldots, \triangle v_k w_{r-2} w_{r-1}$ belong to the same shifting sets. Thus the only triangles which can be deformed in the further process of drawing are $\triangle v_k w_l w_{l+1}$, called *a left-side triangle*, and $\triangle v_k w_{r-1} w_r$, called a *right-side triangle*.

Let T be an arbitrary right-side triangle (see Figure 5), the proof for left-side triangles is symmetric. The only deformation that T can be subjected to is move of vertex B to the right. Le us denote triangle T after such a transformation by T'.

Assume first that T is an obtuse triangle (at the moment of its appearance in the picture). Notice that $Dist(B', s) \geq Dist(B, s)$. Since Lemma 3 can be applied to triangle T we can apply it also to T'.

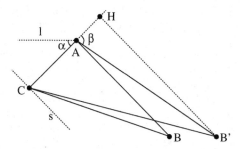

Fig. 6. Shifting does not spoil the angles – T is acute or right

Now assume that T is an acute or a right triangle (see Figure 6). We can also assume that T' is obtuse because otherwise we can apply Lemma 2. Observe that $x(B) > x(A) > x(C)$ and $y(B) < y(A)$, $y(C) < y(A)$. Let us denote the straight line crossing A and parallel to the x axis by l, the height of T' with endpoint B' by $B'H$ and its length by h. Let β denote $\angle B'AH$, and α be the angle between AC and l.

Observe that $h = |B'A| \sin \beta \geq |B'A| \sin \alpha$. Thus we can bound the area of T' as follows:

$$\text{Area}(T') = \frac{1}{2}|AC|h \geq \frac{1}{2}|B'A||AC| \sin \alpha \geq \frac{1}{2}|B'A|.$$

The last inequality follows from $|AC| \sin \alpha = Dist(C, l) \geq 1$. Since $y(A) > y(B')$ we have also $h \geq Dist(B', l) \geq 1$ and consequently: $\text{Area}(T') = \frac{1}{2}|AC|$. Lemma 1 implies that all the angles in T' are at least $\frac{1}{\sqrt{W^2+H^2}}$.

We have proved the following theorem which bounds the angular resolution provided by the algorithm presented in the previous section.

Theorem 1. *For each n-vertex plane graph there exists a straight-line drawing in the $W \times H$ grid with angular resolution $\frac{\sqrt{2}/2}{\sqrt{W^2+H^2}}$, where $W = 2 + 3(n-3)$ and $H = \lfloor \frac{1}{2}W \rfloor$.*

References

1. G. Di Battista, P. Eades, R. Tamassia and I.G. Tollis, *Graph Drawing: Algorithms for the Visualization of Graphs*, Prentice-Hall, 1999.
2. C. C. Cheng, C A. Duncan, M. T. Goodrich and S. G. Kobourov, *Drawing Planar Graphs with Circular Arcs*, In Proc. of the 7th Int. Symp. on Graph Drawing, pp. 117-126, 1999.

3. C. C. Cheng, C A. Duncan, M. T. Goodrich and S. G. Kobourov, *Drawing Planar Graphs with Circular Arcs*, Discrete & Computational Geometry 25(3), pp. 405-418, 2001.

4. M. Chrobak and T. Payne, *A linear time algorithm for drawing a planar graph on a grid*, Information Processing Letters 54, pp. 241-246, 1995.

5. C. A. Duncan and S. G. Kobourov, *Polar Coordinate Drawing of Planar Graphs with Good Angular Resolution*, In Proc. of the 9th Int. Symp. on Graph Drawing, pp. 407-421, 2001.

6. C. A. Duncan and S. G. Kobourov, *Polar Coordinate Drawing of Planar Graphs with Good Angular Resolution*, Journal of Graph Algorithms and Applications (JGAA) 7(4), pp. 311-333, 2003.

7. H.Fraysseix, J.Pach and R.Pollack *Small sets supporting Fary embeddings of planar graphs*, In Proc. 20th ACM Sympos. Theory Comput., pp. 426-433, 1988.

8. H.Fraysseix, J.Pach and R.Pollack, *How to Draw a Planar Graph on a Grid*, Combinatorica 10, pp. 41-51, 1990.

9. A.Garg and R.Tamassia, *Planar Drawings and Angular Resolution: Algorithms and Bounds*, European Symposium on Algorithms, pp. 12-23, 1994.

10. M. T. Goodrich and C. G. Wagner, *A Framework for Drawing Planar Graphs with Curves and Polylines*, In Proc. of the 6th Int. Symp. on Graph Drawing, pp. 153-166, 1998.

11. C. Gutwenger and P. Mutzel, *Planar Polyline Drawings with Good Angular Resolution*, In Proc. of the 6th Int. Symp. on Graph Drawing, pp. 167-182, 1998.

12. S. Malitz, *On the angular resolution of planar graphs*, In Proc. 24th Annual Symposium on Theory of Computing, pp. 527-538, 1992.

13. S. Malitz and A.Papakostas, *On the angular resolution of planar graphs*, SIAM J.Discrete Math. 7, pp. 172-183, 1994.

14. W. Schnyder, *Embedding planar graphs in the grid*, In Proc. 1st Annual ACM-SIAM Symp. on Discrete Algorithms, pp. 138-147, 1990.

15. W. Schnyder and W. Trotter, *Convex drawings of planar graphs*, Abstracts of the AMS 13, 5, 1992.

Modeling Nested Relationships in XML Documents Using Relational Databases

Olli Luoma

Department of Information Technology, University of Turku, Finland
olli.luoma@it.utu.fi

Abstract. Structural joins, i.e., operations that determine all occurrences of parent/child or ancestor/descendant relationships between node sets, are at the heart of XML management systems. To perform these joins, the systems exploit the information about the nested relationships between elements, attributes, and pieces of text in XML documents. Since performing the structural joins is often the most time-consuming phase in query evaluation, the method chosen to model the nested relationships has a considerable impact on the overall effectiveness of any XML management system. In this paper, we discuss four different methods for modeling the nested relationships using relational databases. We also propose a novel modeling method and present the results of our comprehensive performance experiments.

1 Introduction

Since its advent, XML [1], a self-describing markup language recommended by the World Wide Web Consortium, has rapidly been adopted as the standard for data representation and interchange in computer networks. In recent years, XML has increasingly been employed to perform more and more duties; in modern Web service environments, for example, XML can be used as a means to model software components which automatically construct themselves around the information expressed in XML [2]. The importance of XML has also passed over to the area of databases, where XML serves as a format to store heterogeneous information which cannot easily be organized into tables, columns, and rows [3]. Storing, querying and updating XML documents presents an interesting research problem and a plethora of work has been done on the subject.

From a technical viewpoint, an XML management system can be built in several ways. The first option is to build a specialized XML data manager. By building the data manager from scratch one is able to tailor indexing, storage management, and query optimization specifically to suit XML data. Obviously, this is a tempting option and many *native XML databases*, such as Lore [4] and NATIX [5], have been developed. However, native XML databases often suffer from scalability and stability problems and it will still take years before they reach the maturity that can be expected from a practicable XML management system.

M. Bieliková et al. (Eds.): SOFSEM 2005, LNCS 3381, pp. 259–268, 2005.

Another option is to build an *XML management system on top of an object-oriented database* [6] [7]. At first glance, object-oriented databases with their rich data modeling capabilities may seem to be a perfect solution to the problem of XML data management, but this is not quite the case. Navigating large object hierarchies is a rigorous task, and hence the scalability of XML management systems built on top of an object-oriented database often leaves a lot to be desired. Relational databases, on the contrary, provide maturity, stability, portability, and scalability, so the third alternative, an *XML management system on top of a relational database*, is a very viable option. This alternative also allows the XML data and the relational data coexist in the same database, which is advantageous, since it is unlikely that XML databases can completely replace the existing relational database technology in any application area [3] [8].

Since relational databases were originally designed to support non-hierarchical data, a method for mapping XML data into relational schemas is needed. The existing mapping methods can roughly be divided into two categories [9]. In the *structure-mapping approach*, the database schemas are designed to represent the logical structure or the DTDs of the documents. The basic method is to create one relation for each element type [8], but more sophisticated methods in which the database schema is designed based on detailed analysis of the document DTDs have also been proposed [10]. Nonetheless, retaining the optimality of such a schema can be a rigorous task, since inserting new documents with different DTDs may result in redesigning the schema and rebuilding the relations.

The other method is the *model-mapping approach* in which the database schemas represent the generic constructs of the XML data model, i.e., element, attribute, and text nodes. The schemas designed according to the model-mapping approach are fixed, so documents can be stored without any information of their DTDs. Furthermore, having a fixed schema simplifies the transformation of XPath queries into SQL queries [11]. For the aforementioned reasons, we believe that the model-mapping approach will yield better results than its less generic counterpart.

When the model-mapping approach is pursued, an incoming document is first represented as an *XML tree*, a partially ordered, labeled tree in which each element, attribute, and text node corresponds to an element, attribute, or piece of text in the document, respectively; the ancestor/descendant relationships between the nodes correspond to the nested relationships between elements, attributes, and pieces of text [12]. This tree is then stored into the database and queried using the query facilities provided by the database management system.

In this paper, we focus on modeling the ancestor/descendant relationships. We discuss four different methods: parent/child index, pre-/postorder encoding, ancestor/descendant index, and our own proposal, the *ancestor/leaf index*, which maintains the ancestor information for the leaf nodes only. Obviously, this information can be used to perform structural joins between leaf nodes and inner nodes; structural joins between inner nodes are performed by checking whether common leaf node descendants for these inner nodes can be found.

The remainder of this paper is organized as follows. In section 2, we briefly review the related work, and in section 3, we present the different methods for modeling the nested relationships. The results of our experiments are presented in section 4; section 5 concludes this article and discusses our future work.

2   Related Work

From the vast amount XML-to-SQL query translation literature [11], the papers concentrating on mapping XML data into relations pursuing the model-mapping approach are relevant to our work. Previously, many such methods have been proposed, two of which are most essential here, namely XRel [9] and XParent [13]. In XRel, the nested relationships are modeled using region coordinates; a similar method was also presented in [3]. According to the authors, XRel provides better overall performance than Edge [8], a structure-mapping method proposed by Florescu and Kossmann. However, the performance evaluation was carried out using only one set of XML documents, so the scalability of this approach is still somewhat in doubt.

XParent, on the contrary, models the structural relationships between nodes using a parent/child index. The authors compared the effectiveness of their method against Edge and XRel, but again, the performance evaluation was carried out using one set of documents and a very limited set of queries. The authors also discussed using an ancestor/descendant index, but did not present any results from this alternative.

3   Modeling Nested Relationships

In this section, we present the relational schemas used in our tests. Our relational schemas are designed according to the XPath data model, and thus the element, attribute, and text nodes[1] are stored in three relations **Element**, **Attribute**, and **Text**, respectively. The nodes are identified using their preorder numbers which also preserve the information of the order among the nodes.

Since path expressions regularly appear in XPath queries, we preserve the information about the label paths of the nodes using a **Path** table which allows fast retrieval of the nodes according to their label paths. Similar decomposition of the nodes is used also in both XRel and XParent as well as in many native XML databases [4].

3.1   Parent/Child Index

The most obvious method for modeling the structural relationships between nodes is to build a *parent/child index* which, for a given set of nodes, allows fast

---

[1] According to the original XPath recommendation, there are seven different node types, but for simplicity, we have omitted root, comment, namespace, and processing instruction nodes.

retrieval of their parent or child nodes. Since each node has at most one parent, the identifier of the parent node can be inlined into the same relation with the node itself. Hence, this approach results in the following relational schema:

```
Path(PathId, PathExp)
Element(DocId, PreId, ParId, PathId)
Attribute(DocId, PreId, ParId, PathId, Value)
Text(DocId, PreId, ParId, PathId, Value)
```

In the above schema, the database attributes `PathId` and `PathExp` represent the path identifier and path expression, respectively. For technical reasons [9], the labels in a path are separated using "#/" instead of "/". In relations `Element`, `Attribute`, and `Text`, the database attributes `DocId`, `PreId`, `ParId`, and `PathId` represent the document identifier, node identifier, the identifier of the parent node, and path identifier, respectively. The database attribute `Value` represents the value of an attribute or text node. In each relation, the underlined set of attributes serves as the primary key.

However, if we are to retrieve all ancestors of a given node, we need recursive joins. This can be done using the linear recursion queries of the SQL3 standard, but if the database management system does not provide such feature, we can determine the number of needed joins by constructing the SQL queries in accordance with the document DTDs. Obviously, this can lead to very complicated queries and query generation, since each document may have a different DTD. Furthermore, XML documents often have to be managed without any DTDs at all, and thus this method lacks the genericity provided by the other three alternatives discussed in this paper.

3.2   Pre-/Postorder Encoding

The information about ancestor/descendant relationships can also be implicitly encoded by using *region coordinates* [9] or *pre-/postorder encoding* [14]. The structural joins can be performed by taking advantage of the following simple property of preorder and postorder numbes: for any two nodes n_1 and n_2, n_1 is an ancestor of n_2, iff the preorder number of n_1 is smaller than the preorder number of n_2 and the postorder number of n_1 is greater than the postorder number of n_2. Since both preorder and postorder numbers are needed to perform the structural joins, we also include the postorder number `PostId` in the primary key although attributes `DocId` and `PreId` would be enough to identify a node.

```
Path(PathId, PathExp)
Element(DocId, PreId, PostId, PathId)
Attribute(DocId, PreId, PostId, PathId, Value)
Text(DocId, PreId, PostId, PathId, Value)
```

For brevity, we have omitted the actual algorithms for translating XPath queries into SQL queries; for a detailed description we refer the reader to [9]. The idea is to translate the XPath queries first into query trees which are then translated into SQL queries. For example, the query tree corresponding to the XPath query //a/b[//c='ecm']/d translates into the following SQL query:

```
SELECT DISTINCT e3.DocId, e3.PreId
FROM Element e1, Element e3, Text t2, Path p1, Path p2, Path p3
-- Match path expressions:
WHERE p1.PathExp LIKE '#%/a#/b'
AND p2.PathExp LIKE '#%/a#/b#%/c'
AND p3.PathExp LIKE '#%/a#/b#/d'
-- Retrieve nodes:
AND e1.PathId = p1.PathId
AND t2.PathId = p2.PathId
AND e3.PathId = p3.PathId
-- Match values:
AND t2.Value = 'ecm'
-- Perform structural joins:
AND t2.DocId = e1.DocId
AND t2.PreId > e1.PreId
AND t2.PostId < e1.PostId
AND e3.DocId = e1.DocId
AND e3.PreId > e1.PreId
AND e3.PostId < e1.PostId
```

Notice that the structural joins are performed using nonequijoins on preorder and postorder numbers, which can lead to scalability problems when querying large XML documents. However, splitting the data into many small documents can be expected to help, since part of the structural joins can then be performed using equijoins on document identifiers.

3.3   Ancestor/Descendant Index

By building an *ancestor/descendant index*, i.e., by calculating the transitive closure over the parent/child relation, we can perform the structural joins completely without the expensive nonequijoins. To maintain the ancestor/descendant information, we employ a new relation AncDesc.

```
Path(PathId, PathExp)
Element(DocId, PreId, PathId)
Attribute(DocId, PreId, PathId, Value)
Text(DocId, PreId, PathId, Value)
AncDesc(DocId, AncId, DescId)
```

Obviously, this rather extreme approach can result in a very large AncDesc table. However, since no nonequijoins are needed, this approach should usually perform better than the pre-/postorder encoding, and thus in applications where the query performance is of paramount importance, it might just pull its weight. If this approach is pursued, the XPath query //a/b[//c='ecm']/d translates into the following SQL query:

```
SELECT DISTINCT e3.DocId, e3.PreId
FROM Element e1, Element e3, Text t2, Path p1, Path p2, Path p3,
AncDesc d2, AncDesc d3
-- Match path expressions as in previous query.
-- Retrieve nodes as in previous query.
-- Match values as in previous query.
-- Perform structural joins:
AND t2.DocId = e1.DocId
AND d2.DocId = t2.DocId
AND d2.AncId = e1.PreId
AND d2.DescId = t2.PreId
AND e3.DocId = e1.DocId
AND d3.DocId = e3.DocId
AND d3.AncId = e1.PreId
AND d3.DescId = e3.PreId
```

3.4   Ancestor/Leaf Index

We can easily represent the ancestor/descendant information in a more compact manner by using an *ancestor/leaf index* which, essentially, is an ancestor/descendant built on the leaf nodes only. More formally, an ancestor/leaf index for an XML tree is a set of pairs (n_1, n_2), where n_2 is a leaf node and n_1 is an element node located on the path from n_2 to the root of the tree. According to this definition, nodes n_1 and n_2 do not have to be distinct, so the leaf nodes of type element serve a dual purpose. We maintain the ancestor/descendant information for the leaf nodes in relation AncLeaf.

```
Path(PathId, PathExp)
Element(DocId, PreId, PathId)
Attribute(DocId, PreId, PathId, Value)
Text(DocId, PreId, PathId, Value)
AncLeaf(DocId, AncId, LeafId)
```

The leaf nodes are joined with element nodes as they are joined using the ancestor/descendant index. The structural joins between element nodes can be performed by checking whether the nodes have common leaf node descendants. We must also be able to determine which set of the element nodes contains the ancestors. In many cases, this information can be deduced based on the label paths of the nodes, but there are some situations where this is not the case[2]. However, to simplify the query translation, we always use the lengths of the path expressions to determine which one of two sets of element nodes contains the ancestor or descendant nodes. Hence, the XPath query //a/b[//c='ecm']/d translates into the following SQL query:

---

[2] One example of such an instance would be evaluating query //a[a] using document <a><a><a/></a></a>. Without checking the heights, the ancestor/leaf index would also, incorrectly, return the leaf element <a/>.

```
SELECT DISTINCT e3.DocId, e3.PreId
FROM Element e1, Element e3, Text t2, Path p1, Path p2, Path p3,
AncLeaf f2, AncLeaf f3, AncLeaf f4
-- Match path expressions as in previous query.
-- Retrieve nodes as in previous query.
-- Match values as in previous query.
-- Perform structural joins:
-- Structural join between leaf nodes and element nodes:
AND t2.DocId = e1.DocId
AND f2.DocId = t2.DocId
AND f2.AncId = e1.PreId
AND f2.LeafId = t2.PreId
-- Structural join between element nodes:
AND LENGTH(p1.PathExp) < LENGTH(p3.PathExp)
AND e3.DocId = e1.DocId
AND f3.DocId = e3.DocId
AND f3.AncId = e1.PreId
AND f4.DocId = f3.DocId
AND f4.AncId = e3.PreId
AND f4.DescId = f3.DescId
```

Notice that the nonequijoin is now performed using the Path table which usually contains only a small number of rows, so this join is not as expensive as the nonequijoins performed using the pre-/postorder encoding.

4   Experimental Results

Because of the lack of genericity in the parent/child approach, we conducted the performance evaluation only for the pre-/postorder encoding (PP), the ancestor/descendant index (AD), and the ancestor/leaf index (AL). We used three different sets of XML documents: the 7.65 MB collection of Shakespeare's plays [15], a synthetic 111 MB XMark document generated using XMLgen [16], and a 127 MB XML document generated from the DBLP database [17]. The Shakespeare collection consisted of 37 documents and the other collections consisted of only one document.

Table 1. Database sizes for PP

Relation	Shakespeare Tuples	Size(MB)	XMark Tuples	Size(MB)	DBLP Tuples	Size(MB)
Path	57	0	548	0	145	0
Element	179618	8	1666315	73	3332130	135
Attribute	0	0	381878	27	404276	28
Text	147383	12	1188922	139	3005857	201
Total	327058	20	3237663	239	6742408	364

Table 2. Database sizes for AD

Relation	Shakespeare Tuples	Size(MB)	XMark Tuples	Size(MB)	DBLP Tuples	Size(MB)
Path	57	0	548	0	145	0
Element	179618	6	1666315	58	3332130	107
Attribute	0	0	381878	22	404276	23
Text	147385	11	1188922	129	3005857	175
AncDesc	1406939	44	16807315	520	16243275	513
Total	1733999	61	20044978	729	22985683	818

Table 3. Database sizes for AL

Relation	Shakespeare Tuples	Size(MB)	XMark Tuples	Size(MB)	DBLP Tuples	Size(MB)
Path	57	0	548	0	145	0
Element	179618	6	1666315	58	3332130	107
Attribute	0	0	381878	22	404276	23
Text	147383	11	1188922	129	3005857	175
AncLeaf	729554	22	9278809	289	9904635	319
Total	1056612	39	12516472	498	16647043	624

Table 4. Query evaluation times (in seconds)

#	Query	PP	AD	AL	Tuples
1	//ACT/TITLE	0.00	0.00	0.00	185
2	//ACT[//SPEAKER='EDMUND']	0.48	0.03	0.02	5
3	//ACT[//STAGEDIR='Aside']	5.94	0.14	0.08	89
4	//ACT[//SPEAKER='EDMUND']/TITLE	4.16	4.76	21.48	5
5	//people//profile	0.17	0.16	0.16	12832
6	//item[//location='Finland']	2.91	1.54	1.39	16
7	//item[@featured='yes']	>300	1.13	0.45	2210
8	//item[@featured='yes']//location	>300	11.06	16.03	2210
9	//article/author	3.65	3.30	3.30	221465
10	//article[@rating='SUPERB']	8.39	2.00	2.00	11
11	//article[author='Jukka Teuhola']	15.34	0.22	0.09	27
12	//article[author='Donald E. Knuth']/year	61.91	2.05	1.94	55

We stored these collections into MySQL databases pursuing approaches PP, AD, and AL, and built indexes on Element(PathId), Attribute(PathId), Text(PathId), AncDesc(DocId, DescId), and AncLeaf(DocId, LeafId). We also built indexes on first three characters of the attribute Value in relations Attribute and Text. The database sizes for PP, AD, and AL are presented in Tables 1-3. According to these experiments, AD results in three times and AL in two times larger database than PP; the size of AncLeaf table is roughly half of the size of the AncDesc table.

We evaluated the query performance of the three approaches by using four queries for each collection; the queries and the query evaluation times are presented in Table 4. Queries 1-4 were evaluated using the Shakespeare collection, queries 5-8 using the XMark document, and queries 9-12 using the DBLP document. Queries 1, 5, and 9 are simple path expression queries which do not involve structural joins, and thus all three approaches perform well.

Queries 2, 3, 6, 7, 10, and 11 involve a structural join between leaf node and inner node, so these queries provide information especially about the performance of PP against AD and AL. On large documents (queries 6, 7, 10, and 11), AD and AL quite clearly outperform PP, since in PP, the structural joins completely have to be performed using expensive nonequijoins on pre- and postorder numbers. However, when the collection is splitted into many documents (queries 2 and 3 on the Shakespeare collection), PP performs relatively well. Queries 7 and 8 involve structural joins between massive node sets of thousands of nodes, so PP performs very poorly. Thus, it can be argued that the scalability of PP leaves a lot to be desired.

Queries 4, 8, and 12 involve structural joins also between inner nodes, so these queries can be used to compare AD against AL. Overall, AL seems to perform almost as well as AD, but as an interesting detail, we found that both AD and PP outperform AL on query 4. Thus, although AL does not suffer from the severe scalability problems of PP demonstrated by queries 7 and 8, joining node sets with large number of leaf node descendants pursuing AL is still rather expensive. This finding also suggests that splitting large documents into smaller entities before inserting them into the database would lead to considerable performance gains in PP, since part of the structural joins can then be carried out using equijoins on document identifiers.

5   Concluding Remarks

In this paper, we discussed four different methods for modeling the nested relationships between elements, attributes, and pieces of text in XML documents. We also proposed a new approach, namely the ancestor/leaf index. We presented the relational schemas designed according to these models and presented our experimental results which clearly demonstrated the trade-off between storage consumption and query performance. When building an XML management system on a relational database, one should consider both this trade-off and the requirements of the application area. For example, if the management system will be used to store only web pages, using ancestor/descendant index or ancestor/leaf index would be a waste of space, since web pages written in XML are usually only kilobytes in size.

One interesting detail in our experimental results was the relatively good performance of pre-/postorder encoding when many small documents were queried. Considering that this approach only consumes half of the disk space consumed by the ancestor/leaf approach, it might be worthwhile to develop methods for splitting large XML trees before inserting them into the database. In this case, we obviously need ways to perform structural joins between node sets that reside in separate trees, which presents an interesting research problem.

References

1. World Wide Web Consortium. Extensible Markup Language (XML) 1.0. http://www.w3c.org/TR/REC-xml, 2000.
2. D. Florescu, A. Grünhagen, and D. Kossmann. An XML Programming Language for Web Service Specification and Composition. *IEEE Data Engineering Bulletin*, 24(2): 48-56, 2001.
3. A. B. Chaudri, A. Rashid, and R. Zicari. *XML Data Management: Native XML and XML-Enabled Database Systems.* Addison-Wesley, 2003.
4. J. McHugh, S. Abiteboul, R. Goldman, R. Quass, and J. Widom. Lore: A database management system for semistructured data. *SIGMOD Record*, 26(3): 54-66, 1997.
5. C.C. Kanne and G. Moerkotte. Efficient storage of XML data. Poster abstract in *Proc. of the 16th Intl Conf. on Data Engineering*, page 198, 2000.
6. V. Christophides, S. Abiteboul, S. Cluet, and M. Scholl. From structured documents to novel query facilities. In *Proc. of 1994 ACM SIGMOD Intl Conf. on Management of Data*, pages 313-324, 1994.
7. R. v. Zvol, P. Apers, and A. Wilschut. Modelling and querying semistructured data with MOA. In *Proc. of the Workshop on Query Processing for Semistructured Data and Non-Standard Data Formats.*, 1999.
8. D. Florescu and D. Kossmann. A performance evaluation of alternative mapping schemes for storing XML data in a relational database. Technical Report 3684, INRIA, 1999.
9. M. Yoshikawa, T. Amagasa, T. Shimura, and S. Uemura. XRel: A path-based approach to storage and retrieval of XML documents using relational databases. *ACM Transactions on Internet Technologies*, 1(1): 110-141, 2001.
10. J. Shanmugasundaram, K. Tufte, G. He, C. Zhang, D. J. DeWitt, and J. F. Naughton. Relational databases for querying XML documents: limitations and opportunities. In *Proc. of the 25th Intl Conf. on Very Large Databases*, pages 302-314, 1999.
11. R. Krishnamurthy, R. Kaushik, J.F. Naughton. XML-to-SQL query translation literature: the state of the art and open problems. In *Proc. of the 1st Intl XML Database Symposium*, pages 1-18, 2003.
12. World Wide Web Consortium. XML Path Language (XPath) Version 1.0. http://www.w3c.org/TR/xpath, 2000.
13. H. Jiang, H. Lu, W. Wang, and J. Xu Yu. Path materialization revisited: an efficient storage model for XML data. In *Proc. of the 13th Australasian Database Conf.*, pages 85-94, 2002.
14. P.F. Dietz. Maintaining order in a linked list. In *Proc. of the 14th ACM Symp. on Theory of Computing*, pages 122-127, 1982.
15. J. Bosak. The complete plays of Shakespeare marked up in XML. http://www.ibiblio.org/xml/examples/shakespeare
16. R. Busse, M. Carey, D. Florescu, M. Kersten, I. Manolescu, A. Schmidt, and F. Waas. XMark - an XML benchmark project. http://monetdb.cwi.nl/xml/index.html
17. M. Ley. Digital bibliography library project. http://dblp.uni-trier.de/

RAQ: A Range-Queriable Distributed Data Structure

Hamid Nazerzadeh and Mohammad Ghodsi

Computer Engineering Department,
Sharif University of Technology, Tehran, Iran

Abstract. Different structures are used in peer-to-peer networks to represent their inherently distributed, self-organized, and decentralized memory structure. In this paper, a simple range-queriable distributed data structure, called RAQ, is proposed to efficiently support exact match and range queries over multi-dimensional data. In RAQ, the key space is partitioned among the network with n nodes, in which each element has links to $O(\log n)$ other elements. We will show that the look-up query for a specified key can be done via $O(\log n)$ message passing. Also, RAQ handles range-queries in at most $O(\log n)$ communication steps.

1   Introduction

Distributed and peer-to-peer networks are significant components of recent research on networking. There is a simple idea behind the peer-to-peer networks: each node maintains its own index and searching mechanism compared to the traditional client-server architecture with global information. The significant growth in the scale of such networks, (e.g. Gnutella [4]), reveals the critical emerging need to develop decentralized searching methods. A peer-to-peer storage system can be considered as a large scale distributed decentralized data structure. We use the term *Queriable Distributed Data Structure* (QDS) to denote such a self-organized, decentralized, distributed, internet-scale structure which provides searching and data transferring services. New file sharing systems such as Scour, FreeNet, Ohaha, Kazaa and Jungle Monkey are examples QDS from current internet systems.

In QDS, every node of the network is an element of the whole structure, which provides decentralized searching services over the data scattered among the nodes of the network.

Distributed Hash Table (DHT) [15, 11] can be viewed as a QDS. In DHT systems, keys and data are stored in the nodes of the network using a *hash function*, in which data can be the addressing information (e.g. IP address of the server containing the data), rather than its actual data. Searching mechanism in these systems consists of two main phases: (1) hashing the key, and (2) querying the network to find the node that contains the key. This node handles the query by providing the actual data or its addressing information.

M. Bieliková et al. (Eds.): SOFSEM 2005, LNCS 3381, pp. 269–277, 2005.
© Springer-Verlag Berlin Heidelberg 2005

Similarly, some other systems like *SkipNet* [6] are designed based on more theoretical data structures like *skip graphs* [2], allows more flexibility on the location of the data on the nodes of the network.

In this paper, we propose RAQ, a range-queriable distributed data structure to handle exact match and range queries over multi-dimensional data efficiently. In RAQ, the key space is partitioned among the n nodes of the network, in which each element has links to $O(\log n)$ other elements of the network. We will show that the look-up query for a specified key can be done, in our structure, via $O(\log n)$ message passing. The bound on the out-degree of the nodes and the exact-match query cost are both comparable to those in DHT systems like Chord [15], CAN [11], Pastry [13] and Viceroy [9].

The main contribution of RAQ is that it is simple and can handle range-queries in multi-dimensional space. Our data structure supports such queries in at most $O(\log n)$ communication steps. *Split the Space, Duplicate the Query* is a novel approach used by the RAQ to resolve range-queries. This method anticipates the answer space of the query at the source and spreads the query only through the appropriate nodes by duplicating the query meanwhile each of the new queries addresses a reduced subspace.

Most other QDS systems do not support multi-dimensional range-queries, because they mostly use one-dimensional key space. CAN [11] supports multi-dimensional key space, but despite of its similarity to RAQ's basic structure, the out-degree of node and its routing cost depend on the dimension of the key space. For a d dimensional space, the average routing path length in CAN is $(d/4)(n^{1/d})$ hops and individual nodes maintain $2d$ neighbors. This limitation forces the system to use hashing to reduce the dimension. But, since hashing destroys the logical integrity of the data, such systems cannot support range queries over multi-dimensional data efficiently.

In this paper, we first overview the related works briefly, followed by the principal ideas and structures of RAQ. Query handling methods are discussed in sections 5 and 6, followed by the algorithms for joining and leaving a node.

2   Related Works

Supporting range queries in QDS systems has been the subject of several recent works. In SWAM [3], for example, the key space is partitioned according to *Voronoi Tesselation*. By this property, and using links based on *Small-World Phenomenon* [8, 10], SWAM resolves k-nearest-neighbor search and range queries via $O(\log n + R)$ message passing, where R is the size of the answer. But, the number of links of each node grows exponentially with the dimension size [14].

Prefix Hash Tree is a solution proposed by Rantasamy *et. al* [12] to face the problem of hashing used in DHTs that destroys the integrity of the data. Their approach is based on distributed *trie*. Given a query, this system attempts to recognize the longest prefix of the query that appears as a trie-node. The complexity of this operation is $O(\log \log d \times \log n)$, where d is the size of the discrete domain.

Recently, some authors have addressed the problem from the load balancing viewpoint. Karger and Ruhl [7], offer an approach to overcome the known limitation of hash functions, by designing an algorithm to maintain load balance in not-uniform distributed key space. Aspens *et al.* [1] have proposed a similar solution based on *skip graphs*.

Gao and Steenkiste [5], present a QDS which relies on a logical tree data structure, the Range Search Tree (RST), to support range queries in one dimensional space. In this system, nodes in RST are registered in groups. To handle the range queries, queries are decomposed into a small number of sub-queries where the cost depends on the load factor of the data and query capacity of the nodes in the network.

3   Partition Tree

We have n points in our d-dimensional space. *Partition Tree*, $\mathcal{P}_d$, is the main data structure used in RAQ. Similar to the data structure used in [11], $\mathcal{P}_d$ partitions the d-dimensional space, so that in the final level, each region has only one point. Each internal vertex of the tree corresponds to a region in space, and the root represents the whole space. Each pair of the sibling vertices divide their parent region into two parts, and each leaf represents an undivided region called a *cell*, each corresponds to one single point in that region. Figures 1 portrays the partitioning of $\mathcal{P}_2$.

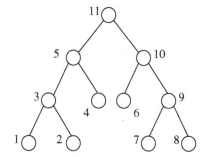
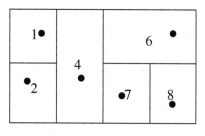

Fig. 1. The partition tree, $\mathcal{P}_2$, on the left corresponds to the points on the right

Each vertex x is assigned a *label* to specify the region space of x. We define $x_{label} = ((p_1, d_1), (p_2, d_2), \cdots, (p_{r(x)}, d_{r(x)}))$ where:

$r(x)$: The distance of x from the root of the tree.

p_i: The plane equation that partitions the current region into two parts.

d_i: Determines one side of the plane p_i.

$parent(x)_{label} = ((p_1, d_1), \cdots, (p_{r(x)-1}, d_{r(x)-1}))$

$sibling(x)_{label} = ((p_1, d_1), \cdots, (p_{r(x)}, \bar{d}_{r(x)}))$, where $\bar{d}_i$ is the opposite side of d_i. $root_{label} = \lambda$, i.e. the empty string.

We treat the labels as strings. The expression $l_1 \sqsubset l_2$ means that l_1 is a prefix of l_2, $|l|$ represents the size of l (i.e. the number of (p_i, d_i) pairs) and $+$ is the concatenation operator. Also, $[l] \stackrel{\text{def}}{=} \{x \in V | l \sqsubset x_{label}\}$ where l is a label and V is the vertex set of the partition tree. Obviously, for a vertex x, $|x_{label}| = r(x)$.

4   Design Principles of RAQ

RAQ is a structure on the nodes of a network. Each node maps to one point in the d dimensional search space. A partition tree $\mathcal{P}tree$ is constructed over the points and thus each node corresponds to a single cell. We say that a node owns its cell and is responsible for providing data to the queries targeting any point in that cell. Since there is a one-to-one map between nodes and the leaf points of the partition tree, we use them interchangeably. So, for example, we assume having labels for each node.

Moreover, each node has several *links* to other nodes of the network. Each link is basically the addressing information of the target node which can be its IP address. The links are established based on the partition tree information and the following rule.

Connection Rule: Consider node x and its label $x_{label} = ((p_1, d_1), (p_2, d_2), \cdots, (p_k, d_k))$. The connection rule of node x implies that x is connected to exactly one node in each of these $|x_{label}|$ sets:

$$[((p_1, \bar{d}_1))] \, , \, [((p_1, d_1), (p_2, \bar{d}_2))] \, , \, \cdots \, , \, [((p_1, d_1), (p_2, d_2), \cdots, (p_k, \bar{d}_k))]$$

For example, in figure 1, node 1 is connected to one node in each of these sets: $\{2\}, \{4\}, \{6, 7, 8\}$. We will show that the *join and leave* mechanisms guarantees the maintenance of connection rule over the network.

Lemma 1. *An arbitrary chosen vertex has link to $O(\log n)$ other nodes in the network.*

It is important to note that the partition tree is not directly maintained by the elements of RAQ; given the coordinates and the labels of the leaves, all information of the partition tree can be uniquely obtained, and these are the only data which are maintained by the nodes of the network. In fact, the partition tree is the *virtual data structure* of RAQ.

It is obvious that $\mathcal{P}tree$ is a balanced tree with the height of $O(\log n)$ when it is first constructed. We will argue that this property holds even in the dynamic environment where the nodes join and leaves the network.

5   Exact Match Query

In RAQ, exact-match queries are of the form $Exact - Query(target, metadata)$. The value of $target$ is the coordinate of the point that is searched for and $metadata$ contains the data to be used after the query reaches the target. Note that the queries aim to reach the target and the responses vary in the different cases. As mentioned, the target of a query is a node whose region contains the query target point.

We say, a point p matches a label l at level k, if k is the greatest value of i such that the subspace induced by a vertex x with $x_{label} = ((p_1, d_1), (p_2, d_2), \cdots, (p_i, d_i))$ contains p and $x_{label} \sqsubseteq l$. In other words, say l represents a leaf y in $\mathcal{P}tree$. Then, x is the lowest common ancestor of y and the node containing p.

Lemma 2. *Suppose node x receives a query targeting point p and p matches x_{label} at level k. If $k = |x_{label}|$ then the cell of x contains p. Otherwise, x has a link to a node y so that y_{label} matches p at a level greater than k.*

Proof. Let $x_{label} = ((p_1, d_1), \cdots, (p_{r(x)}, d_{r(x)}))$. If $k = |x_{label}|$, then, obviously, x contains p. If not, from the connection rule, we know that x is linked to a node y with $((p_1, d_1), \cdots, (p_{k+1}, \bar{d}_{k+1})) \sqsubseteq y_{label}$. Therefore, according to the definition, p matches y_{label} at a level not less than $k + 1$. □

Now, the algorithm for exact match routing becomes clear. Once the query Q is received by a node x, if x contains the target point, then we have done. Otherwise, x sends the query via a link to a node y with a label that matches the target point at a higher level. This will continue until the query reaches the target.

Theorem 1. *The exact match query resolves via $O(\log n)$ message passing.*

Proof. Suppose the target of query Q is the node x. From lemma 2, Q will reach to x in at most $|x_{label}|$ steps and $|x_{label}| = O(\log n)$. So, the routing operation is performed by $O(\log n)$ message passing. □

6   Range Query

We assume that a range query Q is of the form $Range\text{-}Query(label, pivot, func, d_1, d_2, metadata)$ where $label$ implies that Q must be sent only to the nodes x so that $label \sqsubseteq x_{label}$, we denote the label of Q by Q_{label}. The initial value of Q_{label} is set to empty string. The value of $pivot$ is the coordinate of the point that the distances are measured from, and $func$ is the distance function.

The above range query means that Q should be sent to every node in the network with the distance of $d_1 \leq d \leq d_2$ from $pivot$. $func$ can be any distance function $\mathcal{F}$ with the following characteristic: Given a point p, a hyper-cubic subspace S and a distance d, let $A = \{x | x \in S \text{ and } \mathcal{F}(x, p) < d\}$. The problem of whether or not A is empty must be computable. For example, $\mathcal{F}$ can be $L_p\text{-}norm$ function, in which case the answer space of Q is $\{x | d_1 \leq (\sum (p_i - x_i)^p)^{1/p} \leq d_2\}$.

To handle the range queries, we use *split the space, duplicate the query* method, or split-duplicate for short. Suppose that node x receives a range query Q and $x_{label} = ((p_1, d_1), \cdots, (p_{r(x)}, d_{r(x)}))$. Obviously, $Q_{label} \sqsubseteq x_{label}$. If $Q_{label} = x_{label}$, then x itself will give the appropriate response. Otherwise, we iterate the following sequence of operations:

1. Duplicate Q and name the results as Q_1 and Q_2. Set $Q_{1_{label}} = Q_{label} + (p_{|l|+1}, d_{|l|+1})$ and $Q_{2_{label}} = Q_{label} + (p_{|l|+1}, \bar{d}_{|l|+1})$.
2. If the answer space of Q has intersection with the subspace induced by $Q_{2_{label}}$, then send Q_2 via the link to node y where $Q_{2_{label}} \sqsubseteq y_{label}$. Note that by the connection rule, y exits.
3. Iterate split-duplicate operation subsequently on Q_1, while the split subspace has intersection with the answer space.

Lemma 3. *If node x receives range query Q, then Q will be routed to all nodes y where $Q_{label} \sqsubseteq y_{label}$, and the intersection of the cell of y and answer space of Q is not empty.*

Proof. We prove this by backward induction on $|Q_{label}|$. If $|Q_{label}| > |x_{label}|$, then obviously $Q_{label} \not\sqsubseteq x_{label}$ thus the induction basis holds.

Suppose that x receives the query and the induction hypothesis holds for $k > |x_{label}|$. If $Q_{label} = x_{label}$ then x is the only target of the query and we are done. Otherwise, two new queries are generated by the algorithm, while the second query Q_2 is sent to its adjacent node via an appropriate link, if the subspace induced by $Q_{2_{label}}$ has nonempty intersection with the answer space. The size of the labels of these new queries is increased by one. Thus, by the induction hypothesis, Q will be routed to all nodes z with $Q_{label} \sqsubseteq z_{label}$ where the cell of z has a nonempty intersection with the answer space. The union of the induced spaces of these labels covers the whole space of Q_{label}. The claim is therefore correct. □

Theorem 2. *RAQ resolves range queries in at most $O(\log n)$ communication steps. In other words, a query will be received by a target node by crossing $O(\log n)$ intermediate hops.*

Proof. As we mentioned, the basic-queries enters the system by initializing its label to an empty string. By lemma 3, the range-query will be received by all nodes whose cells have nonempty intersections with the query answer space. In each communication step, the size of the query label is increased at least by 1. Thus, when a node receives a query, the distance to the source must be $O(\log n)$, or equivalently the size of the label. □

7   Joining and Leaving

In this section, we describe the joining and leaving mechanism and demonstrate the validity of our claim in section 4 that the partition tree remains balanced all the time.

7.1   Joining Mechanism

Suppose that node x wants to join to the network. x chooses a fairly random point p, in the space and finds y one of the active nodes in the network. Several mechanism can be adopted for the arriving node to find an active node; we assume that RAQ uses the same mechanism as in CAN [11].

Sending an exact match query by y to find the node z whose cell contains p. z divides its cell into two parts, with one containing the corresponding point of z and the other includes p. We assume that x possess the cell containing p. This is just a simple insertion into the partition tree. This is done by updating the labels of x and z. Since we are not directly maintaining this tree, this update is sufficient.

The connections are now updated to follow the given connection rule: x chooses one random point in each of the subspaces induced by the labels specified by the connection rule. For each of these points, say r, z routes an exact match query to find the node that owns r. Consequently, x establishes a connection link to this node.

Theorem 3. *Join operation is done via $O(\log^2 n)$ message passing.*

Proof. The arriving node finds its region by an exact match query. By lemma 1, the arriving node has to create $O(\log n)$ connections. Establishing each connection is done by a exact-match lookup, via $O(\log n)$ message passing. Therefore, the whole operation is completed by $O(\log^2 n)$ message passing.   □

7.2   Leaving Mechanism

Let x be the node that wants to leave the network. After x leaves the network, its sibling in $\mathcal{P}tree$ will maintain the region once belonged to x. From $\mathcal{P}tree$ viewpoint, leaving is just a simple deletion of a leaf in a binary tree, so z_{label} and thus $\mathcal{P}tree$ will be updated easily. The difficult part is updating the connection links of the nodes that have links to x. To handle this issue *Departure links* or for short *dlinks*, are defined below.

In RAQ, node b maintains *addressing information* of a, or a *dlink* to a, when node a establishes a connection link to node b. When b decides to leave the network, it sends a message to each of the nodes referred to by its dlinks. In the following, we denote *d-degree* of b as the number its dlinks.

Theorem 4. *The expected value of d-degree is $O(\log n)$*

Proof. Here we argue the validity of our claim. From the mechanism described above to establish a connection link, and from the dynamic structure of the network where the nodes frequently join to and leave the network, we can fairly conclude that the probability that a node v has a link to a node u is equal to the probability that u has a link to v. We avoid discussing the uncomplicated details of this claim, due to the lack of space. Accordingly, $E[v_{degree}] = E[v_{d-degree}]$ for an arbitrary node v in the network.   □

Corollary 1. *By lemma 1 and theorem 4, each nodes of the network maintains the addressing information of $O(\log n)$ nodes of the network.*

Consider the time when x is to leave the network. x sends a departure message to all of its nodes on its dlink. As mentioned, every connection in RAQ is a link to a subspace, each of the nodes that receives this departure message, chooses a new random point, say p, in the corresponding subspace and sends an exact match query via x to establish a new connection link to the node that possesses p. After these operations, x will peacefully leaves the network and the connection rule of RAQ is maintained.

Theorem 5. *The leave operation is done via $O(\log^2 n)$ message passing.*

Proof. According to theorem 4, $O(\log n)$ links must be updated. Each update is performed by $O(\log n)$ message passing, thus the total number of message passing is $O(\log^2 n)$.   □

According to the discussion, arriving nodes are distributed all over the space. Thus, the partition grows uniformly and remains balanced. Uniform distribution of the nodes also implies that the nodes leave the network randomly in the entire space. We can thus conclude that the claimed proposition about the balancing of the partition tree is valid.

8   Conclusion

In this paper we presented RAQ, a range-queriable distributed data structure for peer-to-peer networks to organize the multidimensional data it holds, and to efficiently support exact and range queries on the data. Our structure is easy to implement and use $O(\log n)$ memory space for each of its n nodes. The exact match query can be performed, as in other works, by $O(\log n)$ message passings. The main contribution of this paper is that the structures broadcast the range query to the target nodes within at most $O(\log n)$ link traversing steps. We showed that all properties of RAQ can be maintained when nodes join the network or leave it.

We are currently working on other extensions of the RAQ model, including its probabilistic model to reduce the complexity of the degree of the nodes in the network. We also intend to validate our results through experimental evaluation with real data. Other ideas can be to design a fault tolerant model to handle different faults such as the situation the nodes abruptly leave the network or abstain to handle the queries temporarily. Load balancing is another important property of the RAQ to look at. In this case, we are going to study the situations that the data points are not uniformly distributed in the search space; also, the computing power of the nodes of network are different. Further works on these topics are underway.

References

1. J. Aspnes, y. Kirschz, A. Krishnamurthy. Load balancing and locality in range-queriable data structures. in Prod, PODC the twenty-third annual ACM symposium on Principles of distributed computing
2. J. Aspnes and G. Shah. Skip Graphs. In Proceedings of Symposium on Discrete Algorithms, 2003.
3. F. Banaei-Kashani and C. Shahabi, SWAM: Small-World Access Model. In Proc. CIKM 2004, Thirteenth Conference on Information and Knowledge Management CIKM 2004, Nov 2004.
4. Gnutella, http://gnutella.wego.com
5. J. Gao and P. Steenkiste, An Adaptive Protocol for Efficient Support of Range Queries in DHT-based Systems. Proc. 12th IEEE International Conference on Network Protocols (ICNP'04), Berlin, Germany, Oct. 2004.
6. N. J. A. Harvey, M. B. Jones, S. Saroiu, M. Theimer, and A. Wolman. SkipNet: A Scalable Overlay Network with Practical Locality Properties. In Proc. of Fourth USENIX Symposium on Internet Technologies and Systems, 2003.
7. D. R. Karger and M. Ruhl. Simple Efficient Load Balancing Algorithms for Peer-to-Peer Systems. In ACM Symposium on Parallelism in Algorithms and Architectures, June 2004.
8. J. Kleinberg. The small-world phenomenon: An algorithmic perspective. In Proc. 32nd ACM Symposium on Theory of Computing (STOC 2000), pages 163-170, 2000.
9. D. Malkhi, M. Naor, and D. Ratajczak. Viceroy: A scalable and dynamic emulation of the butter y. In Proc 21st ACM Symposium on Principles of Distributed Computing (PODC 2002), pages 183-192, 2002.
10. H. Nazerzade. Making Querical Data Networks Navigable. In Proc. ICI 2004 International Conference on Informatic, Sep 2004.
11. S. Ratnasamy, P. Francis, M. Handley, and R. M. Karp. A scalable content-addressable network. In Proc. ACM SIGCOMM 2001, pages 161-172, 2001.
12. S. Ratnasamy, J. Hellerstein, and S. Shenker. Range Queries over DHTs. Technical Report IRB-TR-03-009, Intel Research, 2003.
13. A. I. T. Rowstron and P. Druschel. Pastry: Scalable, decentralized object location, and routing for large-scale peer-to-peer systems. In IFIP/ACM International Conference on Distributed Systems Platforms (Middleware 2001), pages 329-350, 2001.
14. R. Seidel. Exact upper bounds for the number of faces in d-dimensional Voronoi diagrams, DIMACS Series, volume 4. American Mathematical Society, 1991.
15. I. Stoica, R. Morris, D. Karger, M. F. Kaashoek, and H. Balakrishnan. Chord: A scalable peer-to-peer lookup service for internet applications. In Proc. ACM SIGCOMM 2001, pages 149-160, 2001.

On Some Weighted Satisfiability and Graph Problems

Stefan Porschen

Institut für Informatik, Universität zu Köln, D-50969 Köln, Germany
porschen@informatik.uni-koeln.de

Abstract. In the first part of this paper we address several weighted satisfiability problems. Among others, we provide linear time algorithms solving the optimization problems MINV(MAXV)-NAESAT and MINV (MAXV)-XSAT for 2CNF formulas and arbitrary real weights assigned to the variables. In a second part we consider the relationship between the problems maximum weight independent set (MAX-IS) in a graph and the problem XSAT. We show that the counting problem #XSAT can be solved in time $O(2^{0.40567n})$ thereby significantly improving on a bound $O(2^{0.81131n})$ provided in [4].

Keywords: (weighted) exact satisfiability, not-all-equal satisfiability, maximum weight independent set, counting problem, exact algorithm, NP-completeness.

1 Introduction and Notation

The classical satisfiability problem (SAT) is a prominent problem, namely one of the first problems that have proven to be NP-complete [3]. Till nowadays SAT plays a fundamental role in computational complexity theory and in the theory of designing exact algorithms. SAT also has a wide range of applications occuring via reduction from the corresponding abstract problem kernel to SAT. This is helpful due to the fact that meanwhile several powerful solvers for SAT have been developed (cf. e.g [7] and refs. therein). Weighted satisfiability is a natural generalization of SAT and has also important applications (e.g. in the area of code generation [1,6]).

In the present paper we focus on some satisfiability problems for propositional formulas in conjunctive normal form (CNF)and graph problems related to them. In a first part, we show that XSAT and NAESAT (defined below) can be solved in linear time for 2CNF formulas when arbitrary weights are assigned to the variables. Further we show that XSAT can be solved in polynomial time for monotone formulas in which each variable occurs at most twice and has an arbitrary weight. In a second part we consider the connection between the problems XSAT and maximum weight independent set for graphs. Specifically, we provide an exact algorithm for counting all XSAT models of an arbitrary CNF formula. This algorithm runs in time $O(2^{0.40567n})$ and thus significantly improves an algorithm in [4] with running time $O(2^{0.81131n})$, where n is the number of variables.

M. Bieliková et al. (Eds.): SOFSEM 2005, LNCS 3381, pp. 278–287, 2005.

To fix the notation, let $V = \{x_1, \ldots, x_n\}$ be a set of propositional variables, where $x_i \in \{0, 1\}$. A *literal* is a variable x or its negation $\overline{x} := \neg x$ (negated variable). The *complement* of a literal l is denoted as $\overline{l}$. A *clause* c is the disjunction of different literals and thus is represented as a literal set. A CNF formula C is a conjunction of different clauses and is thus represented as a clause set. For short we use throughout the term "formula" meaning a clause set as defined. For a given formula C, clause c, by $V(C), V(c)$ we denote the set of variables contained in C, c, respectively. Similarly given a literal l, $V(l)$ denotes the underlying variable. We distinguish between the length $\|C\|$ of a formula C and the number $|C|$ of its clauses. Let CNF denote the set of all formulas. For $k \in \mathbb{N}$ let kCNF denote the set of formulas C such that $|c| \leq k$ for each $c \in C$. For $r \in \mathbb{N}$, let CNF(r) denote the set of formulas in which each variable occurs in at most r different positions either negated or unnegated. CNF$_+$ denotes the set of *positive monotone* formulas, i.e., each clause contains only variables, no negated variables.

The satisfiability problem (SAT) asks in its *decision* version, whether an input formula C is *satisfiable*, i.e., whether C has a *model*, which is a truth assignment $t : V(C) \rightarrow \{0, 1\}$ assigning at least one literal in each clause of C to 1. In its *search* version SAT means to find a model t if the input formula is satisfiable. There are some interesting variants of SAT, namely *exact satisfiability* (XSAT) and *not-all-equal satisfiability* (NAESAT). XSAT means to find a truth assignment that assigns exactly one literal of each clause of C to 1, called an *XSAT model*. NAESAT searches for a truth assignment that assigns at least one literal in each clause of C to 1 and at least one literal to 0, called a *NAESAT model*. The decision versions of XSAT and NAESAT are defined analogously. Both XSAT and NAESAT are known to be NP-complete [10]. The empty set also is a formula: $\varnothing \in$ CNF which is satisfiable w.r.t. all variants. However, a formula C containing the empty clause ($\varnothing \in C$) is never satisfiable.

When weights are assigned to the variables we obtain the weighted variants of the defined problems: Given $C \in$ CNF and $w : V(C) \rightarrow \mathbb{R}$, problem MINV-SAT asks whether C is satisfiable and if, searches for a *minimum model* for C, i.e., a model t of minimal variable weight $w(t) := \sum_{x \in t^{-1}(1)} w(x)$ among all models of C. The maximum version MAXV-SAT is defined analogously. Also MINV(MAXV)-XSAT and MINV(MAXV)-NAESAT are defined analogously. We shall make use of some graph concepts assigned to formulas. For a monotone formula $C \in$ CNF$_+$, we have two corresponding graph concepts: First, the *formula graph* $G_{V(C)}$ with vertex set $V(C)$. Two vertices are joined by an edge if there is a clause in C containing the corresponding variables. Second, the *clause graph* G_C with vertex set C. Two vertices are joined by an edge if the corresponding clauses have a non-empty intersection. Observe that, given a monotone formula C, each graph concept above can be computed in time $O(\|C\|^2 |V(C)|)$ or better. Finally, to an arbitrary formula $C \in$ CNF, we can assign its bipartite *incidence graph* I_C with vertex set $V(C) \cup C$. For a variable x occuring in clause c (regardless whether negated or unnegated) an edge $x - c$ is introduced. We call a formula $C \in$ CNF *connected* if its incidence graph I_C is connected. Observe that the incidence graph of a formula C can be computed in time $O(\|C\|)$.

2   Linear Time Algorithms Solving Optimum Variable-Weight XSAT and NAESAT for 2CNF

Clearly, MINV-SAT, in it decision version, is NP-hard because it is a generalization of SAT. But MINV-SAT remains NP-complete even when restricted to monotone instances $C \in \mathrm{CNF}_+$ and a constant weight function. Even for monotone 2CNF formulas, MINV-SAT remains NP-complete, which immediately follows from a reduction of the VERTEX COVER problem which is NP-complete (see e.g. [5]). The VERTEX COVER problem gets as input a graph G and searches for a subset X of vertices of smallest cardinality such that every edge is incident to at least one vertex in X. For the reduction consider G as the formula graph of a formula $C \in 2\mathrm{CNF}$ and interpret X as the set of all variables assigned to 1. By exchanging the roles of the truth values 0,1, the same reduction shows that also MAXV-SAT is NP-hard for the class 2CNF and constant weight functions. Hence, both problems remain surely NP-hard for the general case of arbitrary real weights, even if restricted to monotone formulas. In [9] an algorithm is provided establishing that MINV-SAT (resp. MAXV-SAT) is exactly solvable in time $O(2^{0.5248 \cdot |V(C)|}$, for input instances (C, w) where $C \in 2\mathrm{CNF}$ and $w : V(C) \to \mathbb{R}_+$ (resp. $w : V(C) \to \mathbb{R}_-$).

However, the next result tells us that MINV-XSAT and MINV-NAESAT are solvable in linear time, for the class 2CNF, and arbitrary weights assigned to the variables.

Theorem 1. *For $C \in 2\mathrm{CNF}$ and $w : V(C) \to \mathbb{R}$, MINV(MAXV)-XSAT resp. MINV(MAXV)-NAESAT can be solved in linear time $O(\|C\|)$. Moreover, if C is connected, then there are at most two solutions in either case.*

PROOF. First observe that MINV-XSAT and MINV-NAESAT are essentially the same for 2CNF formulas except for the different role played by unit clauses: A formula C containing a unit clause $\{l\}$ is not not-all-equal satisfiable whereas the literal l must be assigned to 1 for exactly satisfying C. These decisions are independent of weights and can be made in linear time. In what follows we assume that each $c \in C$ has length exactly 2. Exactly satisfying C is then synonymous to not-all-equal satisfying C (it follows that the same is true when a minimum weight solution is searched for). Hence it suffices to verify that MINV-XSAT can be solved in linear time for such an input instance.

To that end we first compute the connected components of C via its incidence graph I_C. Clearly these components are pairwise variable disjoint and can be checked for XSAT independently. The connected components of I_C and hence of C can be computed in linear time $O(\|C\|)$.

Before continuing let us mention a simple fact concerning clauses of the form $\{x, \overline{x}\}$. Clearly, a variable x that occurs exclusively in such a clause, can be assigned independently of the remaining variables, called *relevant* variables, as to minimize (resp. maximize) the total weight.

Let $C = \{c_1, \ldots, c_k\}$ be a connected formula. We claim that if $C \in \mathrm{XSAT}$ then there exist exactly two different exact models for C which can be computed in linear time $O(\|C\|)$. Indeed, the necessity to exactly satisfy the first clause $c_1 = \{l_1, l_2\}$ yields the constraint $l_1 = \bar{l}_2$. Let $C[l_1 \leftarrow \bar{l}_2]$ denote the formula obtained from C by substituting each occurence of $l_1, \bar{l}_1$ by $l_2, \bar{l}_2$ and deleting duplicate clauses. Then we have $C \in$ XSAT iff $C[l_1 \leftarrow \bar{l}_2] \in$ XSAT. Observe that, since C is connected, l_2 must be a relevant variable and hence is determined in a subsequent step. Recursively, proceed in that way, by storing the corresponding constraints until either a contradiction occurs, then C is not exact satisfiable, or the empty formula occurs, then $C \in$ XSAT. The constraints yield a chain of equalities in which each relevant variable in C occurs exactly once either negated or unnegated. That is, fixing one variable determines uniquely the truth values of all other relevant variables in order to exactly satisfy the whole formula. Hence, there are exactly two distinct models for each connected formula C. As an example consider the connected formula $C = \{\{x, y\}, \{z, \bar{x}\}, \{y, z\}, \{u, \bar{v}\}, \{z, u\}\}$ The described procedure from left to right yields contradiction-free dependencies as follows: $z = \bar{u} = \bar{v} = \bar{y} = x$. The only two exact models are thus provided by $z = x = 0, u = v = y = 1$ and $z = x = 1, u = v = y = 0$.

Computing an optimal exact model for each connected component C_i of an input formula C runs in linear time $O(\|C_i\|)$ (observe that the optimal solution needs not to be unique, in general). With regard to the input formula C these are disjoint partial exact truth assignments whose union yields the unique optimal XSAT model of C. The time needed is $O(\sum_i \|C_i\|) = O(\|C\|)$ as has been claimed.   □

3   Solving MINV-XSAT for $\mathrm{CNF}_+(2)$ in Polynomial Time

Recall that for a formula $C \in \mathrm{CNF}(2)$ holds by definition that each clause $c \in C$ can have arbitrary length but each variable $x \in V(C)$ occurs at most twice in C. In the following we provide the algorithmic outline for solving MINV-XSAT restricted monotone instances in $\mathrm{CNF}_+(2)$ and arbitrary variable weights in polynomial time. To that end we construct a transformation from MINV-XSAT to the MINIMUM PERFECT MATCHING problem (MIN-PM) which as input gets a graph and arbitrary real weights assigned to its edges. MIN-PM asks for a perfect matching of minimum weight. Recall that a perfect matching is a subset P of pairwise non-adjacent edges such that *every* vertex of G is incident to (exactly) one edge in P. We construct a slight variant of the clause graph of C, called the *matching graph* G_M, which also incorporates variables that occur only once in C. Such variables do not occur in any intersection of clauses, and thus are not in correspondence to any edge of the clause graph. The matching graph is constructed in two steps:

1) If there is no clause in C containing a unique variable, then $G_M := G_C$, i.e., the clause graph of C. Label each edge of G_C by the variable with the smallest weight occuring in the intersection of the corresponding clauses.

2) If there is any clause in C containing a unique variable, then define G_M as follows: Construct two copies of G_C and join both copies by introducing an additional edge between each two vertices in either copy that contain at least one

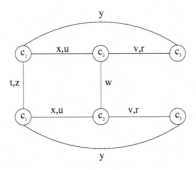

Fig. 1. Matching graph G_M corresponding to C

unique variable (both vertices clearly correspond to one and the same clause). Label each additional edge by the (unique) variable of the smallest weight in the clause corresponding to its incident vertices.

Example: For the formula $C = \{c_1, c_2, c_3\}$, where $c_1 = \{t, u, x, y, z\}$, $c_2 = \{r, u, v, w, x\}$, $c_3 = \{r, v, y\}$ the corresponding matching graph G_M is shown in Figure 1, where the labels are due to the variables in the clause intersections.

It is not hard to see that a minimum weight perfect matching in G directly corresponds to a minimum XSAT model of C, where the matching edges define exactly the variables that must be assigned to 1. Clearly, for $C \in \mathrm{CNF}_+(2)$ the matching graph is constructable in time $O(|C|^2|V(C)|)$. Since each variable that is not unique, occurs at most in two clauses it can be obtained only in the intersection of two clauses. Hence, there can occur no contradiction by selecting as edge label the minimum weight variable in each intersection. Because MIN-PM is solvable in time $O(|V|^2|E|)$ for $G = (V, E)$ and arbitrary real weights assigned to the edges [2], we immediately obtain:

Theorem 2. *For $C \in \mathrm{CNF}_+(2)$ and $w : V(C) \to \mathbb{R}$, MINV-XSAT is solvable in time $O(|C|^2|V(C)|)$.*   □

4   Reduction from Monotone XSAT to MAX-IS

The problem MAXIMUM INDEPENDENT SET (MAX-IS) gets as input a triple (G, w, k) where $G = (V, E)$ is a simple graph, $w : V(C) \to \mathbb{R}$ is a vertex weight function and $k \in \mathbb{R}$. It has to be decided, whether there is an independent set I in G, i.e., a set of pairwise non-adjacent vertices, that has total weight at least k. Observe that MAX-IS is NP-complete even when all weights are equal one which follows from the VERTEX COVER problem, because a minimum cardinality vertex cover in G is the complement of a maximum cardinality independent set. In its search version MAX-IS searches for an independent set of maximum weight. In this section we show that XSAT for arbitrary monotone CNF formulas can be considered as a specific version of MAX-IS by presenting a polynomial time reduction from XSAT to MAX-IS. This reduction will be useful to obtain a fast algorithm for #XSAT as provided in the next section. For $C \in \mathrm{CNF}_+$, let $G := G_{V(C)}$ be the formula graph corresponding to C. Recall that each variable $x \in V(C)$ constitutes a vertex and that two vertices are joined by an edge if the corresponding variables occur together in a clause. Since we have ony unnegated literals, for $x \in V(C)$, we have $C(x) = \{c \in C | x \in c\}$ which is the subformula of all clauses containing x. As weight function we define $w : V(C) \to \mathbb{N}$, with $w(x) = |C(x)|$, hence the weights are determined by the structure of the formula. As usual w extends to subsets X of $V(C)$ by $w(X) := \sum_{x \in X} w(x)$. The next lemma makes precise the reduction from monotone XSAT to MAX-IS:

Lemma 1. *For $C \in \mathrm{CNF}_+$ we have $C \in \mathrm{XSAT}$ if and only if G contains an independent set of weight $|C|$. Such a set then necessarily is a maximum weight independent set in G.*

PROOF. Clearly, if X is an independent set in G then all corresponding variables appear in different clauses, since by definition all variables in the same clause are joined pairwise by edges. Thus, an independent set X of total weight $|C|$ yields an exact model for C when setting exactly the variables in X to true, because all clauses are hitted then. To show the converse direction, assume that $C \in \mathrm{XSAT}$ and let t be an XSAT model of C, with $X := t^{-1}(1)$. Then all variables in X are contained in different clauses thus provide an independent set in G which obviously has total weight $|C|$.   □

5   Counting All XSAT Models

In [4] Dahlöf and Jonsson stated an upper bound of $O(2^{0.81131n})$ for computing the number of all XSAT models (#XSAT) in an instance $C \in \mathrm{CNF}$ where n is the number of variables. That bound is based on an algorithm for counting all maximum weight independent sets in a simple graph [4]. For convenience we refer to the latter algorithm as to the D-J-Algorithm. In this section we show that this bound can be reduced significantly, namely to $O(2^{0.4057n})$. The main idea is as follows: If the given instance $C \in \mathrm{CNF}$ for #XSAT is monotone, we proceed as in the last section and use the D-J-Algorithm [4] for counting all maximum weight independent sets in the formula graph $G_{V(C)}$, which are all independent sets of weight $|C|$. Due to Lemma 1 this number is equal that of all XSAT models of C. If C is not monotone we transform it (in polynomial time) into a monotone XSAT-equivalent formula C'; thereby a unique *model multiplicator N* is determined such that N times the number of solutions of C' equals the number of solutions of C. First, let us prove several lemmas that provide all tools for transforming an arbitrary formula into a monotone one such that the number of all models can be determined. Let $\mathrm{X}(C)$ denote the set of all (total) truth assignments $t : V(C) \to \{0, 1\}$ that exactly satisfy $C \in \mathrm{CNF}$. Since for the empty formula $\varnothing$ holds $V(\varnothing) = \varnothing$, we have $|\mathrm{X}(\varnothing)| = 2^0 = 1$.

If a clause contains more than one complemented pairs, then it can never be exactly satisfied, hence the formula containing such a clause has 0 XSAT models. However, clauses containing exactly one complemented pair can be removed from the formula by the so-called "complemented pair (cp)" rule as stated in the following lemma:

Lemma 2. *For $C \in \mathrm{CNF}$, let $c \in C$ contain exactly one complemented pair: $x, \overline{x} \in c$. Let $\phi_{\mathrm{cp}}(C)$ be the formula obtained from C by removing c and assigning all literals to 0 that occur in $c' := c - \{x, \overline{x}\}$ and finally removing all duplicate clauses. We have $V(\phi_{\mathrm{cp}}(C)) = V(C) - V(c)$ and:*
(i) $|\mathrm{X}(C)| = 2|\mathrm{X}(\phi_{\mathrm{cp}}(C))|$ if $x \notin V(\phi_{\mathrm{cp}}(C))$ and
   $|\mathrm{X}(C)| = |\mathrm{X}(\phi_{\mathrm{cp}}(C))|$ if $x \in V(\phi_{\mathrm{cp}}(C))$,
(ii) $\phi_{\mathrm{cp}}(C)$ can be obtained from C in time $O(\|C\|)$.

PROOF. Obviously $V(\phi_{\mathrm{cp}}(C)) = V(C) - V(c)$, because removing of duplicate clauses can remove no other variable. For (i) first observe that surely $C \in \mathrm{XSAT}$ iff $\phi_{\mathrm{cp}}(C) \in \mathrm{XSAT}$, since c is always exactly satisfiable only through x and all other literals in c are fixed to 0. Clearly, every model $t \in \mathrm{X}(C)$ fixes all variables in $V(C)$, thus defines by restriction to $V(C) - V(c)$ a unique model $t' \in \phi_{\mathrm{cp}}(C)$. On the other hand, if $t' \in \phi_{\mathrm{cp}}(C)$ and $x \notin V(\phi_{\mathrm{cp}}(C))$ then there are exactly two extensions t_0, t_1 of t' to $V(C)$ such that $t_0, t_1 \in \mathrm{X}(C)$. Both assign all variables in $V(c) - \{x\}$ to 0, but $t_i(x) = i$, for $i \in \{0, 1\}$. In the remaining case $x \in V(\phi_{\mathrm{cp}}(C))$ the extension of t' to $V(C)$ is unique since the value of x is fixed also. Thus we have (i). Part (ii) is obvious.   □

In the following, we call a formula *cp-free* if none of its clauses contains a complemented pair of variables. The transformation in the next lemma is called the "negation rule ($\neg$)" rule:

Lemma 3. *Let $C \in \mathrm{CNF}$ be a cp-free formula, let $x \in V(C)$ be a variable that exclusively occurs negated in C. Let $\phi_{\neg}(C)$ be the formula obtained from C by replacing each occurence of $\overline{x}$ by x. Then:*
(i) $|\mathrm{X}(C)| = |\mathrm{X}(\phi_{\neg}(C))|$,
(ii) $\phi_{\neg}(C)$ can be computed from C in time $O(\|C\|)$.

PROOF. The proof is obvious.   □

Next we state a transformation called "simple resolution (sr)" rule which in a slightly different form was used in [8]. Given a formula C and a literal l, we denote by $C(l) = \{c \in C : l \in C\}$ all clauses containing l.

Lemma 4. *Let $C \in \mathrm{CNF}$ be complement pair free formula and let $c_i = \{x\} \cup u$, $c_j = \{\overline{x}\} \cup v \in C$ where $x \in V(C)$ and u, v are literal sets. Let $\phi_{\mathrm{sr}}(C)$ be the formula obtained from C as follows:*
(1) Replace every clause $c \in C : x \in c$ by the clause $c - \{x\} \cup v$,
(2) Replace every clause $c \in C : \overline{x} \in c$ by the clause $c - \{\overline{x}\} \cup u$,
(3) Set all literals in $u \cap v$ to 0,
(4) remove all duplicate clauses from the current clause set.
Then we have $V(\phi_{\mathrm{sr}}(C)) = V(C) - \{x\} - V(u \cap v)$, $|\phi_{\mathrm{sr}}(C)| \leq |C| - 1$ and:
(i) $|\mathrm{X}(C)| = |\mathrm{X}(\phi_{\mathrm{sr}}(C))|$,
(ii) $\phi_{\mathrm{sr}}(C)$ can be obtained from C in time $O(\|C\|^2)$.

PROOF. Obviously x and $V(u \cap v)$ are removed from $V(C)$ by ϕ_{sr}. This holds due to the fact that neither u nor v can contain x or $\overline{x}$, because C is assumed to be cp-free and clauses are assumed to be duplicate-free. On the other hand, no other variable than x and those in $V(u \cap v)$ are removed by ϕ_{sr} since only duplicate clauses are removed. Thus we have $V(\phi_{\mathrm{sr}}(C)) = V(C) - \{x\} - V(u \cap v)$. Moreover, c_i and c_j are transformed by ϕ_{sr} into the same clause $u \oplus v$ (denoting the symmetric difference), hence because of (4) we have $|\phi_{\mathrm{sr}}(C)| \leq |C| - 1$.

Addressing (i), we first show that $C \in \mathrm{XSAT}$ iff $\phi_{\mathrm{sr}}(C) \in \mathrm{XSAT}$. To that end consider the following decomposition of C: $C = C(x) \cup C(\overline{x}) \cup C(u \cap v) \cup C_R$ where $C(u \cap v)$ denotes all clauses in C containing a literal in $u \cap v$, and C_R is the set of all remaining clauses in C that are not contained in any of the first three sets. Because C is assumed to be cp-free we have $C(x) \cap C(\overline{x}) = \varnothing$ and therefore the decomposition above is, in fact, a disjoint union. Now, assume that $C \in \mathrm{XSAT}$ and let $t \in \mathrm{X}(C)$, then we claim that the truth assignment t' defined as t restricted to $V(C) - \{x\} - V(u \cap v)$ exactly solves $\phi_{\mathrm{sr}}(C)$. To see this, let $c' \in \phi_{\mathrm{sr}}(C)$ be an arbitrary clause, then there is a unique clause $\hat{c} \in C$ such that $c' = \phi_{\mathrm{sr}}(\hat{c})$. Thus we have four cases. Case (a): $\hat{c} \in C_R$, then t' exactly satisfies c', since we have $\phi_{\mathrm{sr}}(C_R) = C_R$ and t' operates on C_R the same as t.

Case (b): $\hat{c} \in C(u \cap v)$, then c' is obtained from $\hat{c}$ by setting all literals in it to 0 that also occur in $u \cap v$. Cleary exactly one literal, say y in $\hat{c}$ is set to 1 by t but that literal cannot be contained in $u \cap v$ because otherwise c_i or c_j has two different literals set to 1, namely y and x or $\overline{x}$. Thus $t(y) = t'(y) = 1$ thus t' exactly satisfies c'.
Case (c): $\hat{c} \in C(x)$ then because of (3) $c' := (\hat{c} - \{x\}) \oplus v$. Clearly, $\hat{c}$ is either exactly satisfied because of $t(x) = 1$ in which case for exactly one literal, say $y \in v$ must also hold $t(y) = 1$ because otherwise $c_j \in C$ would not be satisfied by t. Because $y \notin \hat{c}$, we have that in this case $t'(y) = 1$ and therefore t' exactly satisfies c'. Or $\hat{c}$ is exactly satisfied by $t(z) = 1$ for exactly one literal $z \in \hat{c}$ different from x, then especially $t(x) = 0$. Observe that $z \notin v$ because otherwise c_j contains two literals set to 1 by t, namely $\overline{x}, z$.
Case (d) proceeds completely analogously by exchanging the roles of $x \leftrightarrow \overline{x}$, $v \leftrightarrow u$, and $c_j \leftrightarrow c_i$. Conversely, if $\phi_{\mathrm{sr}}(C) \in \mathrm{XSAT}$ and $t' \in \mathrm{X}(\phi_{\mathrm{sr}}(C))$ then $t \in \mathrm{X}(C)$ is uniquely defined by setting x accordingly in order to exactly satisfy c_i and c_j simultaneously. This is possible since all literals in $u \cap v$ are already fixed to 0.

Hence, we have proven that $C \in \mathrm{XSAT}$ if and only if $\phi_{\mathrm{sr}}(C)$. Assume that $C \in \mathrm{XSAT}$ and let $t \in \mathrm{X}(C)$ then from the proof above follows that $t' \in \mathrm{X}(\phi_{\mathrm{sr}}(C))$ is uniquely determined by t. Conversely, if $\phi_{\mathrm{sr}}(C) \in \mathrm{XSAT}$ and $t' \in \mathrm{X}(\phi_{\mathrm{sr}}(C))$ then also from the proof above follows that $t \in \mathrm{X}(C)$ is uniquely determined, since all variables in $V(C) - V(\phi_{\mathrm{sr}}(C))$ are fixed by the transformation. In summary we have constructed a one-one correspondence between $\mathrm{X}(C)$ and $\mathrm{X}(\phi_{\mathrm{sr}}(C))$, hence part (i) is implied. Part (ii) is obvious.   □

Now we are ready to present Procedure Monotonize which gets as input a non-monotone CNF C formula and calls itself recursively until C is monotone thereby it computes the model multiplicator for C:

Procedure Monotonize(C, N)
Input: $C \in$ CNF
Output: monotone formula C; model multiplicator N
begin
(1) $N \leftarrow 1$
(2) **if** $\varnothing \in C$ **then return** $N \leftarrow 0$
(3) **if** $\exists c \in C$ containing ≥ 2 complemented pairs **then return** $N \leftarrow 0$
(4) **if** $\exists c \in C$ containing 1 complemented pair $\{x, \overline{x}\}$ **then**   (∗ cp-rule ∗)
      $C \leftarrow \phi_{\mathrm{cp}}(C)$
      if $x \notin V(C)$ **then** $N \leftarrow 2 \cdot N$
      Monotonize(C, N)
(5) **if** $\exists x \in V(C)$ occuring only negated in C **then**   (∗ ¬-rule ∗)
      $C \leftarrow \phi_{\neg}(C)$; Monotonize$(C, N)$
(6) **if** $\exists c_i, c_j \in C, x \in V(C) : x \in c_i, \overline{x} \in c_j$ **then**   (∗ sr-rule ∗)
      $C \leftarrow \phi_{\mathrm{sr}}(C)$; Monotonize$(C, N)$
(7) **return** C, N
end

Theorem 3. *For $C \in$ CNF, Procedure Monotonize, in time $O(|C| \|C\|^2)$, correctly computes a monotone formula $C' \in$ CNF and a multiplicator N such that $|\mathrm{X}(C)| = N \cdot |\mathrm{X}(C')|$.*

PROOF. Correctness of steps (1) to (3) is obvious. Correctness of steps (4) to (6) follows by Lemmas 2 to 4, and by the fact that the current formula is cp-free when step (5) is executed first. Thus the current formula C returned in step (7) is monotone and the returned multiplicator value N is correct. Addressing the running time observe that each statement of Procedure Monotonize is executable at worst in quadratic time in the length of the formula. In each of the steps (4), (6), and (7) at least one clause is removed from the current formula. Step (5) is called at most $|V(C)|$ times. Thus there are at most $O(|V(C)| + |C|) = O(\|C\|)$ recursive calls of the procedure. Thus we have a polynomial running time bounded by $O(\|C\|^3)$.   □

Algorithm #XSAT$(C, |X(C)|)$
Input: $C \in$ CNF
Output: number of all XSAT-models $|X(C)|$ of C
begin
$N \leftarrow 1$
if C is not monotone **then**
(1)   compute Monotonize(C, N)
(2)   **if** $N = 0$ **then return** $|X(C)| = 0$
(3)   **if** $C = \emptyset$ **then return** $|X(C)| \leftarrow N$
(4) compute weighted formula graph $(G_{V(C)}, w)$ with $\forall x \in V(C) : w(x) = |C(x)|$
(5) by the D-J-Algorithm compute the number M of all independent sets of
      weight $|C|$ in $(G_{V(C)}, w)$
(6) **return** $|X(C)| \leftarrow N \cdot M$
end

Theorem 4. *Algorithm* #XSAT *correctly computes the number of all* XSAT *models of an arbitrary instance* $C \in$ CNF *in time* $O(2^{0.40567 \cdot |V(C)|})$.

PROOF. Theorem 3 establishes the correctness of statement (1). Clearly, if the multiplicator is 0 then $C \notin$ XSAT and the number of models is 0, hence (2) is correct. Correctness of (3) follows because the empty formula has only 1 XSAT model. In (4) the formula graph and the variable weights are computed for the current formula C. Hence in (5) the number of all maximum weight independent sets and therefore the number of all XSAT models is computed for the current formula. Correctness of step (5) follows by [4], Theorem 3.1. The running time of step (5) is, according to [4], Theorem 4.1, bounded by $O(1.3247^{|V(C)|}) = O(2^{0.40567 \cdot |V(C)|})$, and clearly dominates all other steps of the algorithm.   □

References

1. A. V. Aho, M. Ganapathi, and S. W. Tjiang, Code Generation Using Tree Matching and Dynamic Programming, ACM Trans. Programming Languages and Systems, 11 (1989) 491-516.
2. D. Applegate and W. Cook, Solving large-scale matching problems, in: D. S. Johnson, C. C. McGeoch (Eds.), Algorithms for Network Flows and Matching Theory, American Mathematical Society, pp. 557-576, 1993.
3. S. A. Cook, The Complexity of Theorem Proving Procedures, in: Proceedings of the 3rd ACM symposium on Theory of Computing, pp. 151-158, 1971.
4. V. Dahllöf, P. Jonsson, An Algorithm for Counting Maximum Weighted Independent Sets and its Applications, in: Proceedings of the 13th ACM-SIAM symposium on Discrete Algorithms, pp. 292-298, 2002.
5. M. R. Garey and D. S. Johnson, Computers and Intractability: A Guide to the Theory of NP-Completeness, W. H. Freeman and Company, San Francisco, 1979.
6. S. Liao, K. Kreutzer, S. W. Tjiang, and S. Devadas, A New Viewpoint on Code Generation for Directed Acyclic Graphs, ACM Trans. Design Automation of Electronic Systems, 3 (1998) 51-75.
7. D. Le Berre and L. Simon, The Essentials of the SAT 2003 Competition, in: E. Giunchiglia, A. Tacchella (Eds.), Proceedings of the 6th International Conference on Theory and Applications of Satisfiability Testing (SAT'03), Lecture Notes in Computer Science, Vol. 2919, pp. 172-187, Springer-Verlag, Berlin, 2004; and references therein.
8. B. Monien, E. Speckenmeyer and O. Vornberger, Upper Bounds for Covering Problems, Methods of Operations Research 43 (1981) 419-431.
9. S. Porschen and E. Speckenmeyer, Satisfiability Problems for Mixed Horn Formulas, in: H. Kleine Büning, X. Zhao, (Eds.), Proceedings of the Guangzhou Symposium on Satisfiability and its Applications, 2004, Guangzhou, China 2004, pp. 106–113.
10. T. J. Schaefer, The complexity of satisfiability problems, in: Proceedings of the 10th ACM Symposium on Theory of Computing, pp. 216-226, 1978.

On the Security and Composability of the One Time Pad

Dominik Raub, Rainer Steinwandt*, and Jörn Müller-Quade**

IAKS, Arbeitsgruppe Systemsicherheit, Prof. Dr. Th. Beth
Fakultät für Informatik, Universität Karlsruhe (TH),
Am Fasanengarten 5, 76131 Karlsruhe, Germany

Abstract. Motivated by a potentially flawed deployment of the one time pad in a recent quantum cryptographic application securing a bank transfer [1], we show how to implement a statistically secure system for message passing, that is, a channel with negligible failure rate secure against unbounded adversaries, using a *one time pad* based cryptosystem. We prove the security of our system in the framework put forward by Backes, Pfitzmann, and Waidner [2].

1 Introduction

It is well known that the *one time pad* (OTP) is perfectly concealing. Therefore OTP based encryption is the obvious choice when dealing with unbounded adversaries. However, the OTP on its own does not suffice to implement *secure message passing* (SMP), as it is "malleable" in the sense that plaintext bits can be flipped by flipping the corresponding ciphertext bit.

Recently, a bank transfer of EUR 3000 was secured by quantum cryptography [1], i.e., a quantum key agreement scheme was used to establish a shared secret and a one time pad encrypted money transfer form was sent. However, in the experiment the integrity of the message was not secured which can have devastating consequences (cf. [3–Section 1.4]): Say, the bank transfer form itself contains no authentication mechanism and there is a known position where the amount of money is specified in digits. Then an adversary can undetectedly flip bits at these positions, changing the specified amount of money. Hence the security of a bank transfer as described in [1] cannot be concluded from the security of the (authenticated!) quantum key agreement protocol alone.

Therefore, to implement SMP, the OTP needs to be combined with some kind of authentication scheme, to ensure non-malleability and of course authenticity. In this work we give a SMP protocol that achieves statistical security and prove its security against unbounded adversaries in the formal framework developed by Backes, Pfitzmann, and Waidner [2] (one could also think of applying Canetti's framework [4]). For being able to deal with an unbounded adversary, we will use an authentication scheme described by Stinson [5–Chapter 10.3], but note that

---

* work partially supported by DFG-project ANTI-BQP.
** work partially funded by the project PROSECCO of the IST-FET programme of the EC, and performed in cooperation with SECOQC.

M. Bieliková et al. (Eds.): SOFSEM 2005, LNCS 3381, pp. 288–297, 2005.

any other statistically secure and composable authentication scheme serves our purpose as well. Secure message passing in the presence of a computationally bounded adversary is treated by Canetti and Krawczyk in [6].

There are a number of other issues with the OTP, e.g., since the OTP is a stream cipher, synchronization needs to be maintained. If the adversary suppresses a message, the subsequent messages should still be readable. Therefore the current position in the encryption key needs to be transmitted (unencrypted but authenticated) along with the ciphertext. Finally, for reasons of applicability we choose the alphabet $\Sigma = \mathbb{F}_2$ for all subsequent discussions.

2   The Ideal Model

In a first step we need to define precisely, what secure message passing is supposed to mean. To this end we specify an ideal functionality TH, that obviously ensures the secrecy and authenticity of our messages, passes all information leaked through necessary imperfections of the SMP system to the adversary and offers the adversary a well defined interface to perform all actions the message passing system cannot prevent (i.e. reordering or suppressing messages). So it becomes obvious what the system does *not* conceal and *not* prevent. In defining the ideal model for SMP, we largely follow [7]. Since the handling of multiple sessions can be derived from the handling of individual sessions by means of the composition theorem, we can restrict ourselves to considering only a single session as is done in [6]. We do not need session identifiers, since multiple sessions can be distinguished by different port names. Also, differing from [7], we will for now only investigate unidirectional message transmission, since arbitrary networks can be composed from secure unidirectional point to point connections (this is analogous to [6]). There is a further reason, why it is sensible to first restrict oneself to a unidirectional connection:

Caveat 1. *As we investigate stream ciphers (more specifically the one time pad) in an asynchronous framework, any two participants need a* separate key *for each* unidirectional connection.

Rational. If two parties were to use the same key string for a bidirectional connection it would be hard to ensure, that a specific subsequence of the key is not used twice (e.g. simultaneously by each partner sending out some message). But this would undermine the security of the protocol, since by calculating the sum of two such ciphertexts the key can be cancelled. The resulting sum of two plaintexts may be deciphered using statistical methods.   □

Now let $s \in \mathbb{N}_{>0}$ and L be a non-zero polynomial with coefficients in $\mathbb{N}$, where s denotes the maximum number of messages the sender may send, and L the maximum message length as a function of the security parameter k.

All machines given throughout this work will be initialized to a state corresponding to the security parameter k in accordance with the framework [7] and the adversary is always the master scheduler, all buffers not explicitly scheduled by another machine are scheduled by the adversary.

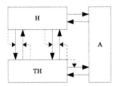

Fig. 1. Ideal Model: SMP

Since we only admit two participants, the sender and the receiver, a single dishonest party can, due to the nature of the message transmission task, already disclose all relevant information to the adversary A. Hence we require nothing in case one (or both) parties are corrupted and turn control over to the adversary—all messages are passed directly to the adversary who may also send arbitrary messages to the honest users ("environment") H. Therefore we only need to discuss the case where the set of honest participants $\mathcal{H}$ is the set of all participants $\mathcal{M}$, i.e. $\mathcal{H} = \mathcal{M} = \{1, 2\}$, where 1 is the sender and 2 the receiver. We now define the ideal system for secure message transmission as $Sys_{s,L}^{\text{secmsg,ideal}} = \{(\{\text{TH}\}, S)\}$ where $\text{ports}(\text{TH}) = \{\text{in}_u?, \text{out}_u!, \text{out}_u^{\triangleleft}! | u \in \mathcal{H}\} \cup \{\text{in}_{\text{sim}}?, \text{out}_{\text{sim}}!, \text{out}_{\text{sim}}^{\triangleleft}!\}$ and the specified ports are given by $S^c := \{\text{in}_u!, \text{out}_u? | u \in \mathcal{H}\}$. TH maintains data structures $init_1, init_2, key_1, key_2 \in \{0, 1\}$, $sc \in \{0, \ldots, s\}$ initialized to 0, and a list $deliver$ initially empty. $init_u$ stores if user u has initiated key exchange, key_u stores if user u would have received his set of keys (both encryption and authentication keys) in the real model, $deliver$ holds the messages due for delivery until the adversary schedules them. The state-transition function of TH is given by the rules below. Inputs not treated explicitly are ignored. If an input triggers a non-empty output, we say the machine accepts it.

(init) **to TH via** $\text{in}_u?$	(send, m) **to TH via** $\text{in}_1?$
- if $init_u = 1$ then abort (User u has already initiated key exch.) - $init_u := 1$ (User u has initiated key exchange.) - output (init, u) at $\text{out}_{\text{sim}}!$ and schedule $\text{out}_{\text{sim}}!$	- if $key_1 \neq 1$ then abort (No key yet, hence no encryption possible.) - if $sc \geq s$ or $\text{len}(m) > L(k)$ then abort (Too many messages sent already or message too long.) - $deliver[sc] := (sc, m)$ - $sc := sc + 1$
(initok, u) **to TH via** $\text{in}_{\text{sim}}?$	- output (busy, $sc - 1$, $\text{len}(m)$) at $\text{out}_{\text{sim}}!$ and schedule $\text{out}_{\text{sim}}!$
- if $init_1 = 0$ or $init_2 = 0$ or $key_u = 1$ then abort (One user has not initiated key exchange yet or user u has already received a key.) - $key_u := 1$ (The adversary has not disrupted key distrib. to user u.) - output (initialized) at $\text{out}_u!$ and schedule $\text{out}_u!$	(select, i) **to TH via** $\text{in}_{\text{sim}}?$ - if $key_2 \neq 1$ then abort (User 2 has not received a key yet, thus decryption not possible.) - if $0 \leq i < sc(= \text{size}(deliver))$ then output (receive, $deliver[i]$) at $\text{out}_2!$ and schedule $\text{out}_2!$

3   The Hybrid Model

We now define a hybrid model that uses an actual encryption algorithm, but still relies on an ideal authentication subsystem to deliver messages. We will prove this hybrid model to be perfectly as secure as the ideal model in the black box simulatability sense [7]. As encryption primitive we will use the one time pad, but we will attempt to give a general formulation, so that the one time pad can easily be replaced with a different stream cipher. Of course, no more than computational security can be expected then.

The hybrid real model is sketched in Figure 2. The two machines $M_{1,\text{enc}}$, $M_{2,\text{enc}}$ handle encryption and decryption respectively. Between the two machines we still have an authenticated channel, implemented by the ideal authentication functionality $F_{s,L_{\text{auth}}}^{\text{auth,ideal}}$. The maximal message length L_{auth} for the authentication subsystem is given as polynomial over $\mathbb{N}$ in k defined as $L_{\text{auth}}(k) := L(k) + s \cdot k \geq L(k) + \lceil \log_2(s \cdot k) \rceil$. The authentication subsystem has to handle messages that are $\lceil \log_2(s \cdot k) \rceil$ longer than the messages handled by the encryption machines, because the current position within the one time pad key used for encryption has to be authenticated with the original message. With $\mathcal{H} = \{1, 2\}$ as above, the hybrid system is formally given as

$$
\begin{aligned}
Sys_{s,L}^{\text{secmsg,hybrid}} = \{(\{M_{u,\text{enc}}, & F_{s,L_{\text{auth}}}^{\text{auth,ideal}}, \text{FKE}_{\text{enc}} | u \in \mathcal{H}\}, S) \text{ where} \\
\text{ports}(M_{u,\text{enc}}) = \{&\text{in}_u?, \text{out}_u!, \text{out}_u^{\triangleleft}!, \text{in}_{u,\text{auth}}!, \text{out}_{u,\text{auth}}?, \text{in}_{u,\text{auth}}^{\triangleleft}!, \qquad (1) \\
\text{in}_{u,\text{FKE}_{\text{enc}}}!, & \text{in}_{u,\text{FKE}_{\text{enc}}}^{\triangleleft}!, \text{out}_{u,\text{FKE}_{\text{enc}}}?\} \qquad \text{where } u \in \mathcal{H}
\end{aligned}
$$

$$
\text{ports}(F_{s,L_{\text{auth}}}^{\text{auth,ideal}}) = \{\text{in}_{u,\text{auth}}?, \text{out}_{u,\text{auth}}!, \text{out}_{u,\text{auth}}^{\triangleleft}! | u \in \mathcal{H}\} \qquad (2)
$$

$$
\cup \{\text{in}_{\text{sim,auth}}?, \text{out}_{\text{sim,auth}}!, \text{out}_{\text{sim,auth}}^{\triangleleft}!\}
$$

$$
\begin{aligned}
\text{ports}(\text{FKE}_{\text{enc}}) = \{&\text{in}_{u,\text{FKE}_{\text{enc}}}?, \text{out}_{u,\text{FKE}_{\text{enc}}}!, \text{out}_{u,\text{FKE}_{\text{enc}}}^{\triangleleft}!, \\
&\text{in}_{\text{sim,FKE}_{\text{enc}}}?, \text{out}_{\text{sim,FKE}_{\text{enc}}}!, \text{out}_{\text{sim,FKE}_{\text{enc}}}^{\triangleleft}! | u \in \mathcal{H}\} \qquad (3)
\end{aligned}
$$

and the specified ports are given by $S^c := \{\text{in}_u!, \text{out}_u? | u \in \mathcal{H}\}$. The machines in $Sys_{s,L}^{\text{secmsg,hybrid}}$ maintain the following data structures (where $u \in \mathcal{H}$):

FKE_{enc}: $init_u, distributed_u \in \{0, 1\}$, $key \in \Sigma^*$

$M_{u,\text{enc}}$: $init \in \{0, 1\}$, $enckey \in \Sigma^*$, $keypos \in \{0, \dots, s \cdot L(k)\}$, $sc \in \{0, \dots, s\}$

$F_{s,L_{\text{auth}}}^{\text{auth,ideal}}$: $init_1, init_2, key_1, key_2 \in \{0, 1\}$, $sc \in \{0, \dots, s\}$,

$\quad deliver \in (\{0, \dots, s\}, \Sigma^k)^*$

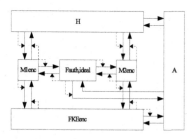

Fig. 2. Hybrid Model: SMP

where all variables are initialized to 0, the empty list [] or the empty string ε as applicable. When operating on strings in Σ^* or lists, we let $x[a : b]$ denote the substring (in case x is a string) or sublist (in case x is a list) from (and including) position a up to (but not including) position b. The state-transition functions of the machines are given by the rules in the box below. Inputs not treated explicitly are ignored. If an input triggers a non-empty output, we say the machine accepts it. (Note that the rules are ordered by the machines they belong to and may be invoked by rules listed further down.)

(init) to $M_{u,\text{enc}}$ via in_u?

- output (init) at $\text{in}_{u,\text{auth}}!$ and schedule $\text{in}_{u,\text{auth}}!$

(initialized) to $M_{u,\text{enc}}$ via $\text{out}_{u,\text{auth}}$?

- $i := s \cdot L(k)$
- output (generate, i) at $\text{in}_{u,\text{FKE}_{\text{enc}}}!$ and schedule $\text{in}_{u,\text{FKE}_{\text{enc}}}!$

(key, key) to $M_{u,\text{enc}}$ via $\text{out}_{u,\text{FKE}_{\text{enc}}}$?

- $enckey := key$
- $init := 1$
- output (initialized) at $\text{out}_u!$ and schedule $\text{out}_u!$

(send, m) to $M_{1,\text{enc}}$ via in_1?

- if $init \neq 1$ then abort (No key yet, hence no encryption possible.)
- if $sc \geq s$ or $\text{len}(m) > L(k)$ then abort (Too many messages sent already or message too long.)
- $sc := sc + 1$
- $c := (keypos, m \oplus enckey[keypos : keypos + \text{len}(m)])$ (For transmission we encode $keypos$ as $\lceil \log_2(s \cdot k) \rceil$ bit string.)
- $keypos := keypos + \text{len}(m)$
- output (cipher, c) at $\text{in}_{1,\text{auth}}!$ and schedule $\text{in}_{1,\text{auth}}!$

(cipher, (cnt, ($keypos$, c))) to $M_{2,\text{enc}}$ via $\text{out}_{2,\text{auth}}$?

- if $initauth \neq 1$ or $initenc \neq 1$ then abort (No key yet, hence no decryption possible.)
- if $\text{len}(c) > L(k)$ or $keypos + \text{len}(c) > s \cdot L(k)$ then abort (Message too long or $keypos$ out of range.)
- $m := c \oplus enckey[keypos : keypos + \text{len}(c)]$
- output (receive, cnt, m) at $\text{out}_2!$ and schedule $\text{out}_2!$

(init) to $F_{s,L_{\text{auth}}}^{\text{auth,ideal}}$ via $\text{in}_{u,\text{auth}}$?

- if $init_u = 1$ then abort (User u has already initiated key exchange.)
- $init_u := 1$ (User u has initiated key exchange.)
- output (init, u) at $\text{out}_{\text{sim,auth}}!$ and schedule $\text{out}_{\text{sim,auth}}!$

(initok, u) to $F_{s,L_{\text{auth}}}^{\text{auth,ideal}}$ via $\text{in}_{\text{sim,auth}}$?

- if $init_1 = 0$ or $init_2 = 0$ or $key_u = 1$ then abort (One user has not initiated key exchange yet or user u has already received a key.)
- $key_u := 1$
  (The adversary has not disrupted the distribution of keys to user u.)
- output (initialized) at $\text{out}_{u,\text{auth}}!$ and schedule $\text{out}_{u,\text{auth}}!$

(cipher, c) to $F_{s,L_{\text{auth}}}^{\text{auth,ideal}}$ via $\text{in}_{1,\text{auth}}$?

- if $key_1 \neq 1$ then abort (No key yet, hence no authentication possible.)
- if $sc \geq s$ or $\text{len}(m) > L_{\text{auth}}(k)$ then abort (Too many messages sent already or message too long.)
- $deliver[sc] := (sc, c)$
  (For transmission we encode sc as $\lceil \log_2 s \rceil$ bit string.)
- $sc := sc + 1$
- output (busy, $sc - 1$, c) at $\text{out}_{\text{sim,auth}}!$ and schedule $\text{out}_{\text{sim,auth}}!$

(select, i) to $F_{s,L}^{\text{auth,ideal}}$ via $\text{in}_{\text{sim,auth}}$?

- if $key_2 \neq 1$ then abort (User 2 has not received a key yet, thus authentication not possible.)
- if $0 \leq i < sc$ (= $\text{size}(deliver)$) then output (cipher, $deliver[i]$) at $\text{out}_{2,\text{auth}}!$ and schedule $\text{out}_{2,\text{auth}}!$

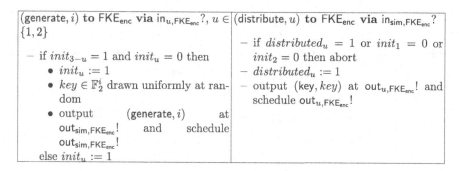

(generate, i) to $\mathsf{FKE_{enc}}$ via $\mathsf{in}_{u,\mathsf{FKE_{enc}}}$?, $u \in \{1,2\}$	(distribute, u) to $\mathsf{FKE_{enc}}$ via $\mathsf{in}_{\mathsf{sim},\mathsf{FKE_{enc}}}$?

− if $init_{3-u} = 1$ and $init_u = 0$ then
   • $init_u := 1$
   • $key \in \mathbb{F}_2^i$ drawn uniformly at random
   • output (generate, i) at $\mathsf{out}_{\mathsf{sim},\mathsf{FKE_{enc}}}$! and schedule $\mathsf{out}_{\mathsf{sim},\mathsf{FKE_{enc}}}$!
 else $init_u := 1$
	− if $distributed_u = 1$ or $init_1 = 0$ or $init_2 = 0$ then abort
	 − $distributed_u := 1$
	 − output (key, key) at $\mathsf{out}_{u,\mathsf{FKE_{enc}}}$! and schedule $\mathsf{out}_{u,\mathsf{FKE_{enc}}}$!

Caveat 2. *The position in the key sequence must be included with the message unencrypted. It should be authenticated with the message, unless tampering with the key sequence position ensures (with overwhelming probability) that authentication will fail.*

Rational. If the key sequence position is not included, the adversary suppressing one single message results in loss of synchronization. Thus decryption will fail for all subsequent messages. If the key position is not authenticated, the adversary may modify it without being noticed. If the authentication is then not guaranteed to fail, the receiver will be under the impression he is receiving nonsensical messages from the sender.   □

The next theorem (proven in the full version [8]) states that the ideal model (cf. Figure 1) and the hybrid real model (cf. Figure 2) are black box perfectly indistinguishable according to the definition set forth in [7]. That implies that the hybrid real model is perfectly at least as secure as the ideal model.

Theorem 1. *The hybrid real system $Sys_{s,L}^{\mathsf{secmsg,hybrid}}$ as in Figure 2 is black box perfectly at least as secure as the ideal system $Sys_{s,L}^{\mathsf{secmsg,ideal}}$ as in Figure 1: $Sys_{s,L}^{\mathsf{secmsg,hybrid}} \geq_{\mathsf{sec}}^{f,\mathsf{perf}} Sys_{s,L}^{\mathsf{secmsg,ideal}}$ where the valid mapping f between the systems is obvious from the machine names.*

4 The Real Model

The real model defined in this section describes the actual usable protocol for statistically secure message passing. There are two machines M_1 and M_2 for sender and receiver respectively. Each machine M_u decomposes into two submachines $\mathsf{M}_{u,\mathsf{enc}}$ and $\mathsf{M}_{u,\mathsf{auth}}$ that handle encryption and authentication. The machines $\mathsf{M}_{u,\mathsf{auth}}$ and $\mathsf{FKE_{auth}}$ constitute the real authentication subsystem $Sys_{s,L}^{\mathsf{auth,real}}$ that replaces the ideal authentication subsystem $Sys_{s,L}^{\mathsf{auth,ideal}}$. As such the real system is mostly identical to the hybrid-real model, only replacing the ideal authentication system with the real one, and thus allowing for a proof of security by composition. The real authentication system is based on a statistically secure authentication procedure described in [5–Chapter 10.3]. It views the message m as number in $\mathbb{F}_p$ and uses an affine authentication function, computing the authentication tag as $t(m, k_1, k_2) := (m \cdot k_1 + k_2) \mod p$ where $k_1, k_2 \in \mathbb{F}_p$ is the current pair of keys. A more detailed discussion of this scheme and its security will be given in the next section. With $\mathcal{H} = \{1, 2\}$ the real system is formally given as $Sys_{s,L}^{\text{secmsg,real}} = \{(\{M_{u,\text{enc}}, M_{u,\text{auth}}, \text{FKE}_{\text{auth}}, \text{FKE}_{\text{enc}} | u \in \mathcal{H}\}, S)\}$ where

$$\text{ports}(M_{u,\text{enc}}) = \{\text{in}_u?, \text{out}_u!, \text{out}_u^\triangleleft!, \text{in}_{u,\text{auth}}!, \text{out}_{u,\text{auth}}?, \text{in}_{u,\text{auth}}^\triangleleft!, \tag{4}$$

$$\text{in}_{u,\text{FKE}_{\text{enc}}}!, \text{in}_{u,\text{FKE}_{\text{enc}}}^\triangleleft!, \text{out}_{u,\text{FKE}_{\text{enc}}}?\} \qquad \text{where } u \in \mathcal{H}$$

$$\text{ports}(M_{1,\text{auth}}) = \{\text{in}_{1,\text{auth}}?, \text{out}_{1,\text{auth}}!, \text{out}_{1,\text{auth}}^\triangleleft!, \text{netout}!, \tag{5}$$

$$\text{in}_{1,\text{FKE}_{\text{auth}}}!, \text{in}_{1,\text{FKE}_{\text{auth}}}^\triangleleft!, \text{out}_{1,\text{FKE}_{\text{auth}}}?\}$$

$$\text{ports}(M_{2,\text{auth}}) = \{\text{in}_{2,\text{auth}}?, \text{out}_{2,\text{auth}}!, \text{out}_{2,\text{auth}}^\triangleleft!, \text{netin}?, \tag{6}$$

$$\text{in}_{2,\text{FKE}_{\text{auth}}}!, \text{in}_{2,\text{FKE}_{\text{auth}}}^\triangleleft!, \text{out}_{2,\text{FKE}_{\text{auth}}}?\}$$

$$\text{ports}(\text{FKE}_{\text{auth}}) = \{\text{in}_{u,\text{FKE}_{\text{auth}}}?, \text{out}_{u,\text{FKE}_{\text{auth}}}!, \text{out}_{u,\text{FKE}_{\text{auth}}}^\triangleleft!, \tag{7}$$

$$\text{in}_{\text{sim},\text{FKE}_{\text{auth}}}?, \text{out}_{\text{sim},\text{FKE}_{\text{auth}}}!, \text{out}_{\text{sim},\text{FKE}_{\text{auth}}}^\triangleleft! | u \in \mathcal{H}\}$$

$$\text{ports}(\text{FKE}_{\text{enc}}) = \{\text{in}_{u,\text{FKE}_{\text{enc}}}?, \text{out}_{u,\text{FKE}_{\text{enc}}}!, \text{out}_{u,\text{FKE}_{\text{enc}}}^\triangleleft!, \tag{8}$$

$$\text{in}_{\text{sim},\text{FKE}_{\text{enc}}}?, \text{out}_{\text{sim},\text{FKE}_{\text{enc}}}!, \text{out}_{\text{sim},\text{FKE}_{\text{enc}}}^\triangleleft! | u \in \mathcal{H}\}$$

and the specified ports are given by $S^c := \{\text{in}_u!, \text{out}_u? | u \in \mathcal{H}\}$. The machines in $Sys_{s,L}^{\text{secmsg,real}}$ maintain the following data structures (where $u \in \mathcal{H}$ and $v \in \{\text{enc}, \text{auth}\}$):

FKE_{enc}: $init_u, distributed_u \in \{0, 1\}, key \in \Sigma^*$
FKE_{auth}: $init_u, distributed_u \in \{0, 1\}, key \in \Sigma^*, p \in \{0, \ldots, 2^{k+1} - 1\}$
$M_{u,\text{enc}}$: $init \in \{0, 1\}, enckey \in \Sigma^*, keypos \in \{0, \ldots, s \cdot L(k)\}, sc \in \{0, \ldots, s\}$
$M_{u,\text{auth}}$: $initauth \in \{0, 1\}, authkey \in \Sigma^*, sc \in \{0, \ldots, s\}, p \in \{0, \ldots, 2^{k+1} - 1\}$

where all variables are initialized to $0, [], \varepsilon$ as applicable. Again, all differences to the hybrid real model are confined to the authentication subsystem. The state-transition functions of all machines are given as in the hybrid real model, except for the machines that belong to the real authentication subsystem. For those the state transition functions are described below. Again, inputs not treated explicitly are ignored.

(generate, i, j) to FKE_{auth} via $\text{in}_{u,\text{FKE}_{\text{auth}}}?$, $u \in \{1, 2\}$	(distribute, u) to FKE_{auth} via $\text{in}_{\text{sim},\text{FKE}_{\text{auth}}}?$

— if $init_{3-u} = 1$ and $init_u = 0$ then
 • choose $p \in \{2^{j-1} \leq q < 2^j : q \text{ prime}\}$ arbitrary
 • $init_u := 1, key := []$
 • while size$(key) < 2i$
    * $a \in \mathbb{F}_2^j$ drawn uniformly at random
    * if $a < p$ then
      $key := key\|a$
 • output (generate, i, p) at $\text{out}_{\text{sim},\text{FKE}_{\text{auth}}}!$ and schedule $\text{out}_{\text{sim},\text{FKE}_{\text{auth}}}!$
 else $init_u := 1$
	— if $distributed_u = 1$ or $init_1 = 0$ or $init_2 = 0$ then abort
	 — $distributed_u := 1$
	 — output (key, key, p) at $\text{out}_{u,\text{FKE}_{\text{auth}}}!$ and schedule $\text{out}_{u,\text{FKE}_{\text{auth}}}!$

(init) **to** $M_{u,\mathrm{auth}}$ **via** $in_{u,\mathrm{auth}}$?	(key, key, $modulus$) **to** $M_{u,\mathrm{auth}}$ **via** $out_{u,\mathrm{FKE_{auth}}}$?
$- \; j := \max\{k+1, L_{\mathrm{auth}}(k)+2\}$ $-$ output (generate, s, j) at $in_{u,\mathrm{FKE_{auth}}}$! and schedule $in_{u,\mathrm{FKE_{auth}}}$!	$- \; p := modulus$ $- \; authkey := key$ $- \; initauth := 1$ $-$ output (initialized) at $out_{u,\mathrm{auth}}$!
(cipher, c) **to** $M_{1,\mathrm{auth}}$ **via** $in_{1,\mathrm{auth}}$?	and schedule $out_{u,\mathrm{auth}}$!
$-$ if $initauth \neq 1$ then abort (No key yet, hence no authentication possible.)	(packet, pkt) **to** $M_{2,\mathrm{auth}}$ **via** netin?
$-$ if $sc \geq s$ or $\mathrm{len}(c) > L_{\mathrm{auth}}(k)$ then abort (Too many messages sent already or message too long.) $- \; m := 1\|c$ (Prefix the bitstring c with a leading one.) $- \; tag := (m \cdot key[2 \cdot sc] + key[2 \cdot sc+1])$ mod p (Calculating the authentication tag we view m as number in $\mathbb{Z}$.) $- \; pkt := (sc, m, tag)$ $- \; sc := sc + 1$ $-$ output (packet, pkt) at netout!	$-$ if $initauth \neq 1$ then abort (No key yet.) $- \; (pos, m, tag) := pkt$ (As pos, tag are bitstrings of fixed length, segmentation is easy) $-$ if $m[0] \neq 1$ or $pos \geq s$ or $\mathrm{len}(m) - 1 > L_{\mathrm{auth}}(k)$ then abort (Leading bit wrong, too many messages sent or message too long.) $-$ if $tag \neq (m \cdot key[2 \cdot pos] + key[2 \cdot pos + 1])$ mod p then abort (Authentication failed.) $-$ output (cipher, pos, $m[1 : \mathrm{len}(m)]$) at $out_{2,\mathrm{auth}}$! and schedule $out_{2,\mathrm{auth}}$!

Caveat 3. *Our authentication scheme does not protect leading zeros, since m and $0\|m$ correspond to the same number in $\mathbb{F}_p$. Therefore, we make sure, that every message starts with a one.*

Caveat 4. *We need to include the message sequence number with the authenticated message, since loss of synchronization would otherwise prevent us from authenticating messages after the adversary has suppressed one. The message sequence number need not be authenticated, since modification of the sequence number will just lead to failing authentication.*

It remains to show that $Sys_{s,L}^{\mathrm{secmsg,real}}$ is black box statistically as secure as $Sys_{s,L}^{\mathrm{secmsg,hybrid}}$. This is done by composition. We will prove that the real authentication subsystem $Sys_{s,L}^{\mathrm{auth,real}}$ is black box statistically as secure as the ideal authentication subsystem $Sys_{s,L}^{\mathrm{auth,ideal}}$ utilized in $Sys_{s,L}^{\mathrm{secmsg,real}}$. The statistical black box security of $Sys_{s,L}^{\mathrm{secmsg,ideal}}$ then follows from the perfect black box security of $Sys_{s,L}^{\mathrm{secmsg,hybrid}}$ using the composition theorem from [7].

5   Security of the Authentication Subsystem

The ideal authentication subsystem $Sys^{\text{auth,ideal}}_{s,L_{\text{auth}}} = \{(\{\mathsf{F}^{\text{auth,ideal}}_{s,L_{\text{auth}}}\}, S)\}$, utilizing the ideal authentication functionality (structure) $F^{\text{auth,ideal}}_{s,L_{\text{auth}}} = (\{\mathsf{F}^{\text{auth,ideal}}_{s,L_{\text{auth}}}\}, S)$ is depicted in Figure 3. The real authentication subsystem

$$Sys^{\text{auth,real}}_{s,L_{\text{auth}}} = \{(\{\mathsf{M}_{u,\text{auth}}, \mathsf{FKE}_{\text{auth}} | u \in \mathcal{H}\}, S)\}$$

utilizing the real authentication functionality $F^{\text{auth,real}}_{s,L_{\text{auth}}} = (\{\mathsf{M}_{u,\text{auth}}, \mathsf{FKE}_{\text{auth}} | u \in \mathcal{H}\}, S)$ is shown in Figure 4. All machine definitions and the trust model are as given above and the specified ports are given by $S^c := \{\text{in}_{u,\text{auth}}!, \text{out}_{u,\text{auth}}? | u \in \mathcal{H}\}$ where of course $\mathcal{H} = \{1, 2\}$. As shown in the full version [8] we have

Theorem 2. *The real authentication subsystem $Sys^{\text{auth,real}}_{s,L_{\text{auth}}}$ as in Figure 4 is black box statistically at least as secure as the ideal authentication subsystem $Sys^{\text{auth,ideal}}_{s,L_{\text{auth}}}$ as in Figure 3: $Sys^{\text{auth,real}}_{s,L_{\text{auth}}} \geq^{f,\text{ExpSmall}}_{\text{sec}} Sys^{\text{auth,ideal}}_{s,L_{\text{auth}}}$ where the valid mapping f between the systems is obvious from the machine names.*

Caveat 5. *The statistical security of the authentication scheme given here is only guaranteed, as long as the message m interpreted as a natural number is bounded by the modulus p.*

Rational. If we allowed messages $m \geq p$ the adversary could easily introduce a forged message $(m + p) \mod p$. This would go unnoticed, as $t(m + p, k_1, k_2) = ((m + p)k_1 + k_2) \mod p = (mk_1 + k_2) \mod p = t(m, k_1, k_2)$ □

Our system takes this into account by limiting the message length to at most $L_{\text{auth}}(k) + 1$ bits (including the leading one) and choosing p as $L_{\text{auth}}(k) + 2$ bit prime (or larger).

The composition theorem of [7–Theorem 4.1] is applicable to the systems $Sys^{\text{secmsg,hybrid}}_{s,L}$ and $Sys^{\text{secmsg,real}}_{s,L}$ with the respective subsystems $Sys^{\text{auth,ideal}}_{s,L_{\text{auth}}}$ and $Sys^{\text{auth,real}}_{s,L_{\text{auth}}}$, since each system is composed of only one single structure and because the consistency condition on the ports is clearly fulfilled. Thus we have

Theorem 3. *The real system for secure message passing as given above is black box statist. as secure as the ideal system: $Sys^{\text{secmsg,real}}_{s,L} \geq^{f,\text{ExpSmall}}_{\text{sec}} Sys^{\text{secmsg,ideal}}_{s,L}$ where the valid mapping f between the systems is obvious.*

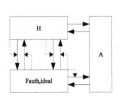

Fig. 3. Ideal Model: Authentication

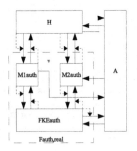

Fig. 4. Real Model: Authentication

Note that the theorem above still holds if we replace the authentication system with any other statistically secure and composable authentication system.

6   Conclusion

We have seen, that it is feasible, but not trivial, to use the one time pad to construct a statistically secure message passing system. In particular we note, that it is impossible to obtain a perfectly secure system (i.e. perfectly indistinguishable from the ideal system), because authentication can at best be statistically secure. The proof we presented is modular in the sense that it admits any choice of statistically secure authentication system. As indicated in Appendix A of the full version [8] it is also easily extensible to computationally secure ciphers and authentication systems, thus providing a framework for statements about stream ciphers in the model of [2].

Acknowledgements. We thank Dennis Hofheinz and Dominique Unruh for helpful comments and discussions.

References

1. Poppe, A., Fedrizzi, A., Loruenser, T., Maurhardt, O., Ursin, R., Boehm, H.R., Peev, M., Suda, M., Kurtsiefer, C., Weinfurter, H., Jennewein, T., Zeilinger, A.: Practical Quantum Key Distribution with Polarization-Entangled Photons. lanl.-arXiv.org e-Print archive, quant-ph/0404115 (2004)
2. Backes, M., Pfitzmann, B., Waidner, M.: Secure Asynchronous Reactive Systems. Cryptology ePrint Archive, Report 2004/082 (2004) http://eprint.iacr.org/2004/082/.
3. Boyd, C., Mathuria, A.: Protocols for Authentication and Key Establishment. Information Security and Cryptography. Springer (2003)
4. Canetti, R.: Universally Composable Security: A New Paradigm for Cryptographic Protocols. In: 42nd Annual Symposium on Foundations of Computer Science, FOCS 2001, IEEE Computer Society (2001) Full version available at Cryptology ePrint Archive, Report 2000/067; http://eprint.iacr.org/2000/067/.
5. Stinson, D.R.: Cryptography – Theory and Practice. CRC Press, Boca Raton, FL, USA (1995)
6. Canetti, R., Krawczyk, H.: Universally Composable Notions of Key Exchange and Secure Channels. Cryptology ePrint Archive, Report 2002/059 (2002) http://eprint.iacr.org/2002/059/. Extended version of [9].
7. Pfitzmann, B., Waidner, M.: A Model for Asynchronous Reactive Systems and its Application to Secure Message Transmission. Cryptology ePrint Archive, Report 2000/066 (2000) http://eprint.iacr.org/2000/066/.
8. Raub, D., Steinwandt, R., Müller-Quade, J.: On the Security and Composability of the One Time Pad. Cryptology ePrint Archive, Report 2004/113 (2004) http://eprint.iacr.org/2004/113/.
9. Canetti, R., Krawczyk, H.: Universally Composable Notions of Key Exchange and Secure Channels. In Knudsen, L., ed.: Advances in Cryptology – EUROCRYPY 2002. Volume 2332 of Lecture Notes in Computer Science., Springer (2002) 337–351

Lower Bounds on the OBDD Size of Graphs of Some Popular Functions*

Daniel Sawitzki**

University of Dortmund, Computer Science 2,
D-44221 Dortmund, Germany
daniel.sawitzki@cs.uni-dortmund.de

Abstract. Ordered binary decision diagrams (OBDDs) are a data structure for Boolean functions which supports many useful operations. It finds many applications in logic design, CAD, model checking, and symbolic graph algorithms. Nevertheless, many simple functions are known to have exponential OBDD size w. r. t. their number of variables. In order to investigate the limits of symbolic graph algorithms which work on OBDD-represented graph instances, it is useful to have simply-structured graphs whose OBDD representation has exponential size. Therefore, we consider fundamental arithmetic and storage access functions with exponential OBDD size and transfer these results to the graphs of these functions. Concretely, lower bounds for the graphs of integer multiplication, indirect storage access, and the hidden weighted bit function are presented. Finally, an exemplary application of the result for multiplication to the analysis of a symbolic all-pairs shortest-paths algorithm is sketched.

1   Introduction

The representation of Boolean functions by branching programs has been extensively studied both in complexity theory and logic design and verification. Lower bounds on the branching program size imply lower bounds on the space complexity of computations. Moreover, tradeoff results for the depth and size of branching programs imply time–space tradeoffs on sequential machines (see, e. g., [1]). Therefore, there are many lower bound results on the size of restricted types of branching programs for at best simple and important functions like arithmetic functions and storage access functions.

On the other hand, in the practical area of logic design and verification, there is the need of succinct representations for Boolean functions which allow efficient algorithms for functional manipulation. In this context, *oblivious read-once branching programs* [2, 3, 4] (also called *ordered binary decision diagrams*

---

*   An extended version of this paper is available via http://ls2-www.cs.uni-dortmund.de/˜sawitzki/LBotOSoGoSPF_Extended.pdf.
** Supported by the Deutsche Forschungsgemeinschaft (DFG) as part of the Research Cluster "Algorithms on Large and Complex Networks" (1126).

M. Bieliková et al. (Eds.): SOFSEM 2005, LNCS 3381, pp. 298–309, 2005.

(*OBDDs*)) have proved to be very useful for the implicit representation of state transition graphs and their symbolic manipulation. The research in this practical area is limited to some application-related problems and experimental analyses.

Recently, a new research branch has emerged which is concerned with the theoretical design and analysis of *symbolic algorithms* for classical graph problems on OBDD-represented graph instances. The input of these specialized heuristic algorithms consists of one or more OBDDs which represent the input graph instance in an *implicit* way avoiding an explicit enumeration of nodes and edges. For example, a directed graph $G = (V, E)$ with $V = \{v_0, \ldots, v_{2^n-1}\}$ can be represented by its *characteristic* Boolean function $\chi_G \colon \{0,1\}^{2n} \to \{0,1\}$ with $\chi_G(x,y) = 1 :\Leftrightarrow (v_{|x|}, v_{|y|}) \in E$ for the binary values $|x|$, $|y|$ of binary node number encodings $x, y \in \{0,1\}^n$.

Symbolic algorithms have to solve problems on G by efficient functional operations offered by the OBDD data structure. Until now, symbolic methods for flow maximization [5, 6], topological sorting [7], shortest paths computation [8, 9], and component analysis [10, 11] have been presented. Most papers justify the new OBDD-based approaches by an analysis of the number of executed OBDD operations [12, 10, 11, 13] or by experimental results [5, 14, 15, 16, 17]. Newer research also tries to analyze the over-all runtime of symbolic methods, which includes the analysis of all OBDD sizes occurring during the algorithm. In general, even basic problems like reachability analysis are PSPACE-hard on OBDD-represented graphs [18]. So analyses must investigate input instances with special properties that enable sublinear runtimes w. r. t. the explicit graph size.

OBDDs during the run of a symbolic graph algorithm represent intermediate results and, therefore, typically not well-structured functions. In order to prove that an OBDD-based algorithm needs exponential time w. r. t. input and output size we have to estimate the size of these intermediate OBDDs. Although lower bound techniques for OBDDs are well-known [19, 4, 20], it is not easy to apply them in such situations. Intermediate OBDDs often check whether some condition is fulfilled. We formalize this type of function by defining the *symbolic graph* function of a vector of Boolean functions. In the following, the class of Boolean functions $f \colon \{0,1\}^n \to \{0,1\}$, $n \in \mathbb{N}$, is denoted by B_n.

Definition 1. *Let $f = (f_0, \ldots, f_{m-1})$ be a vector of m Boolean functions $f_i \in B_n$ for $n, m \in \mathbb{N}$, $0 \le i \le m - 1$. The function $f\text{-}GRAPH \in B_{n+m}$ defined by*

$$f\text{-}GRAPH(x_0, \ldots, x_{n-1}, y_0, \ldots, y_{m-1}) = \bigwedge_{i=0}^{m-1} [f_i(x) = y_i]$$

is called the symbolic graph *of f.*

The contribution of this paper is to transfer existing lower bounds for some fundamental functions to their corresponding symbolic graphs. These can then be applied in the construction of worst-case inputs for symbolic algorithms yielding exponential running times.

Section 2 introduces general branching programs for Boolean functions as well as two restricted types that are of particular interest in this paper. In order to disprove the reasonable assumption that lower bounds for single output bits directly carry over to the symbolic graph of a vector of functions, Sect. 3 investigates the OBDD size of a specially constructed storage access function.

After these preliminaries, Sect. 4 gives a survey of the results of this paper. Exponential lower bounds on the OBDD size of the graphs of integer multiplication, indirect storage access, and the hidden weighted bit function are presented. We also mention an implication on squaring. Section 5 contains the proof of the result for multiplication. Then, Sect. 6 gives an exemplary application of the result for multiplication to the analysis of a symbolic all-pairs shortest-paths algorithm. Finally, Sect. 7 gives conclusions on the work and mentions open problems.

2   Branching Programs for Boolean Functions

We denote the value of a binary string $x = x_{n-1} \ldots x_0 \in \{0,1\}^n$ by $|x| := \sum_{i=0}^{n-1} x_i 2^i$. On the other hand, we denote by (a) the binary string with value $a = |(a)| \in \mathbb{N}$. Let $X := \{x_0, \ldots, x_{n-1}\}$ be a set of n Boolean variables.

Definition 2. Branching programs (BPs).

(a) *A* branching program *P* defined on the variable set X is a directed acyclic graph with two kinds of nodes: Inner nodes and sink nodes. Each inner node v is labeled with a variable $x_i =: \text{label}(v) \in X$ and left by two edges called 0- and 1-edge. Each sink w is labeled with a Boolean constant $c =: \text{label}(w) \in \{0,1\}$ and has no outgoing edge. A special node s is marked as source node.

(b) *Each assignment* $a = (a_0, \ldots, a_{n-1}) \in \{0,1\}^n$ to the variables in X defines a unique computation path p_a in P from s to a sink t_a by leaving inner nodes v labeled with $x_i = \text{label}(v)$ via their a_i-edge. The function $f \in B_n$ represented by P is defined by $f(a) := \text{label}(t_a)$.

(c) *The* size $\text{size}(P)$ of P is defined as its number of nodes.

We consider two restrictions of branching programs. Let $\pi \in X^*$ be a sequence of variables from X.

Definition 3. *A BP P is called π-oblivious if the sequence $\pi(p)$ of variables visited on any path p from the source node to a sink is a subsequence of π.*

Definition 4. Oblivious read-once branching programs *or* ordered binary decision diagrams (OBDDs).

(a) *A π-oblivious BP P is called π-oblivious read-once branching program or ordered binary decision diagram (π-OBDD) if π contains every variable from X at most once. Then, π is called the* variable order *of P.*

(b) *The size of a minimal OBDD for a Boolean function $f \in B_n$ is denoted by* $\text{OBDD}(f)$. *The size of a minimal π-OBDD for f is denoted by $\pi\text{-OBDD}(f)$.*

Every Boolean function $f \in B_n$ has a unique minimal-size π-OBDD $P_{\min}$ for any variable order π. It is $\pi\text{-OBDD}(f) \le (2 + o(1))2^n/n$.

The book of Wegener [4] gives a comprehensive survey on the topic of branching programs.

3   Lower Bounds Do Not Necessarily Carry over

Consider a vector $f = (f_0, \ldots, f_{m-1})$ of m Boolean functions $f_i \in B_n$, $0 \le i \le m - 1$. One could conjecture that any lower bound on the OBDD size for some f_i is also a lower bound on the OBDD size for the symbolic graph f-$GRAPH$ of f. This pleasant property would made it obsolete to transfer lower bounds for f to f-$GRAPH$ explicitly. Unfortunately, this is not the case as the following example shows.

We define generalized versions of the well-known functions DSA and ISA which will be combined to construct the counterexample function vector FSA.

Definition 5. *Let* $n = 2^m$, $m \in \mathbb{N}$, *and* $i \in \mathbb{N}_0$. *The* shifted direct storage access (SDSA) *function* $SDSA_{n,i} \in B_{n+m}$ *is defined by*

$$SDSA_{n,i}(x, y) := x_{\alpha(y)} \ , \qquad \text{where } \alpha(y) := (|y| + i) \bmod n$$

and $x \in \{0, 1\}^n$, $y \in \{0, 1\}^m$.

That is, y controls which bit of x to output. Note that $SDSA_{n,0} = DSA_n$ (see [4–Def. 4.3.1]).

Definition 6. *Let* $n = 2^m$, $m \in \mathbb{N}$, *and* $i \in \mathbb{N}_0$. *The* shifted indirect storage access (SISA) *function* $SISA_{n,i} \in B_{n+m}$ *is defined by*

$$SISA_{n,i}(x, y) := SDSA_{n,i}(x, SDSA_{n,0}(x, y), \ldots, SDSA_{n,m-1}(x, y))$$

for $x \in \{0, 1\}^n$, $y \in \{0, 1\}^m$.

That is, we output the bit $x_{(\alpha(x,y)+i) \bmod n}$ for the indirect address $\alpha(x, y) := |x_{|y| \bmod n} \cdots x_{(|y|+m-1) \bmod n}|$. Note that $SISA_{n,0} = ISA_n$ (see [4–Def. 2.2.5]).

We now combine SDSA and SISA to obtain a counterexample to the conjecture stated above.

Definition 7. *Let* $n = 2^m$ *and* $m \in \mathbb{N}$. *The* full storage access (FSA) *vector* $FSA_n = (FSA_{n,0}, \ldots, FSA_{n,2m-1})$ *of Boolean functions* $FSA_{n,i} \in B_{n+m}$, $0 \le i \le 2m - 1$, *is defined by*

$$FSA_{n,i}(x, y) := \begin{cases} SDSA_{n,i}(x, y) & , \ 0 \le i \le m - 1 \\ SISA_{n,i-m}(x, y) & , \ m \le i \le 2m - 1 \end{cases}$$

for $x \in \{0, 1\}^n$ *and* $y \in \{0, 1\}^m$.

The OBDD size of FSA is dominated by the hard function SISA.

Proposition 1. $\text{OBDD}(FSA_{n,m}) \ge 2^{\lfloor n/\log n \rfloor - 1}$.

Proof. The OBDD size of $FSA_{n,m} = SISA_{n,0} = ISA_n$ is bounded below by $2^{\lfloor n/\log n \rfloor - 1}$. (See [19] or [4–Theorem 4.3.3].)   □

Let FSA-$GRAPH_n$ be the symbolic graph of FSA_n. While the inclusion of SDSA into FSA_n did not influence the lower bound on its OBDD size (which is dominated by $SISA_{n,0}$), it *does simplify* the OBDD representation of FSA-$GRAPH_n$:

Proposition 2. $\text{OBDD}(FSA\text{-}GRAPH_n) = \mathcal{O}(n^3 \cdot \log n)$.

Proof. Let each $FSA_{n,i}$ be defined on the variables $x \in \{0,1\}^n$ and $y \in \{0,1\}^m$ for $n = 2^m$. Moreover, let $z \in \{0,1\}^{2m}$ be variables corresponding to the $2m$ results of the functions $FSA_{n,0}, \ldots, FSA_{n,2m-1}$.

We choose $\pi = (y_0, \ldots, y_{m-1}, z_0, \ldots, z_{2m-1}, x_0, \ldots, x_{n-1})$ as variable order. The following π-OBDD P for $FSA\text{-}GRAPH_n$ has size $\mathcal{O}(n^3 \cdot \log n)$. On the first $3m$ layers, P consists of the complete binary tree on the variables y and z having $\mathcal{O}(n^3)$ nodes. At each of its n^3 leaves $v_{y,z}$, y and z are already fixed. It remains to test if $z_i = SDSA_{n,i}(x,y)$ for all $0 \le i \le m-1$ and $z_j = SISA_{n,j-m}(x,y)$ for all $m \le j \le 2m-1$. This is done by connecting a chain of at most $2m = \mathcal{O}(\log n)$ nodes testing x-variables to each leaf $v_{y,z}$ implying size $\mathcal{O}(n^3 \cdot \log n)$ for P. $\square$

Nevertheless, the next section will reveal that the exponential lower bounds for three fundamental functions actually *do* carry over to the symbolic graph scenario.

4   Survey of Results

In this section, we will present exponential lower bounds for symbolic graphs of fundamental functions. Due to space limitations, only the result for integer multiplication will be proved in Sect. 5. For the remaining proofs, the reader is referred to the extended version of this paper. At first, we consider two arithmetic functions.

4.1   Integer Multiplication and Squaring

Integer multiplication is one of the most important and difficult functions in logic design. All data structures considered so far whose algorithmic properties allow efficient circuit verification have exponential size for this hard function (see, e.g., [4]). Only for the special case of Wallace-tree like multipliers, a polynomial formal verification method using multiplicative binary moment diagrams (*BMDs) has been presented by Keim et al. [21].

Definition 8. *The integer multiplication (MUL) vector $MUL_n = (MUL_{n,0}, \ldots, MUL_{n,2n-1})$ of Boolean functions $MUL_{n,i} \in B_{2n}$, $0 \le i \le 2n-1$, is defined by*

$$MUL_{n,i}(x,y) = (|x| \cdot |y|)_i$$

for $x, y \in \{0,1\}^n$.

Especially, the OBDD size of $MUL_{n,n-1}$ is exponential w.r.t. n.

Theorem 1 (Woelfel [20]). $\text{OBDD}(MUL_{n,n-1}) \ge 2^{n/2}/61$.

On the other hand, multiplication is simply-structured, and its graph is easy to encode into characteristic functions; Sect. 6 gives a corresponding application example in symbolic algorithm analysis.

So let $MUL\text{-}GRAPH_n$ be the symbolic graph of MUL_n. Jukna [22] presents an exponential lower bound on the size of nondeterministic read-k-times BPs for the symbolic graph of a subvector $(MUL_{n,i})_{i \in I}$ for indices $I \subset \{0, \ldots, n-1\}$, $|I| \leq \sqrt{n}$. Despite the title of Jukna's work, this is *neither* a lower bound for the graph of multiplication *nor* does it imply the following theorem and corollary.

Theorem 2. *Any π-oblivious BP for $MUL\text{-}GRAPH_n$ whose variable sequence π contains each variable at most k times has a size of at least $2^{n/((4k-1)(2^{8k}))-1}$.*

An OBDD is an oblivious branching program reading each variable at most once. Choosing $k = 1$ in Theorem 2, we obtain the following corollary.

Corollary 1. $\mathrm{OBDD}(MUL\text{-}GRAPH_n) \geq 2^{n/768-1}$.

Wegener [23] presents a read-once projection from integer multiplication to squaring (SQU) which is so far the only way to prove exponential OBDD sizes for the latter function. It can also be used to prove a lower bound for a restricted version of the symbolic graph of SQU which verifies only a subset of the result variables.

Corollary 2. *Let $n := 3m + 2$ for $m \in \mathbb{N}$ and $SQU\text{-}GRAPH_n^* \in B_{8m+4}$ be defined by $SQU\text{-}GRAPH_n(x, y) := \bigwedge_{i=2m+3}^{4m+2} \left[(|x|^2)_i = y_i \right]$. It is $\mathrm{OBDD}(SQU\text{-}GRAPH_n^*) \geq 2^{m/768-1} \geq 2^{n/2304-2}$.*

4.2 Storage Access Functions

In Sect. 3 Def. 6, we defined a generalized version called SISA of the well-known ISA function. Breitbart, Hunt III, and Rosenkrantz [19] present an exponential lower bound of $2^{\lfloor n/\log n \rfloor -1}$ for the OBDD size of ISA. Let $SISA\text{-}GRAPH_{n,w} \in B_{n+m+w}$ be the symbolic graph of the vector $(SISA_{n,0}, \ldots, SISA_{n,w-1})$ for $n = 2^m$ and $1 \leq w \leq n$. Also $SISA\text{-}GRAPH_{n,w}$ has superpolynomial OBDD size for $w = o(n/\log^2 n)$. Due to the storage access character of SISA, this restriction is not too prohibitive.

Theorem 3. $\mathrm{OBDD}(SISA\text{-}GRAPH_{n,w}) \geq 2^{n/(w \cdot \log n)-4}$.

We now consider a generalization of one further important and fundamental storage access function.

Definition 9. *Let $n \in \mathbb{N}$ and $i \in \mathbb{N}_0$. The shifted hidden weighted bit (SHWB) function $SHWB_{n,i} \in B_n$ is defined by*

$$SHWB_{n,i}(x) := x_{\alpha(x)}, \qquad \text{where } \alpha(x) := \left(\sum_{j=1}^{n} x_j + i \right) \bmod n ,$$

$x = (x_1, \ldots, x_n) \in \{0,1\}^n$, and $x_0 := 0$.

That is, the number of ones in x plus the shifting parameter i determines the address $\alpha(x)$ of the output bit $x_{\alpha(x)}$. Note that $SHWB_{n,0} = HWB_n$ (see [4– Def. 1.1.3]).

Wegener [4–Theorem 4.10.2] presents an exponential lower bound of $\Omega\left(2^{n/5}\right)$ on the OBDD size of the HWB function. Let $SHWB\text{-}GRAPH_{n,w} \in B_{n+w}$ be the symbolic graph of the vector $(SHWB_{n,0}, \ldots, SHWB_{n,w-1})$ for $n \in \mathbb{N}$ and $1 \leq w \leq n$. As for SISA, a certain restriction on the number w of verified result variables suffices to retain a superpolynomial lower bound.

Theorem 4. $\mathrm{OBDD}(SHWB\text{-}GRAPH_{n,w}) = \Omega\left(2^{n/(11w)}\right)$.

That is, the OBDD size of $SHWB\text{-}GRAPH_{n,w}$ is superpolynomial w.r.t. n if $w = o(n/\log n)$. Due to the storage access character of SHWB, this restriction is not too prohibitive. Moreover, for $n = 11m$ and $m \in \mathbb{N}$ the lower bound is $2^{\lfloor n/(11w)\rfloor - 1}$.

5   Proof of Theorem 2

We now prove Theorem 2 on the oblivious branching program size of $MUL\text{-}GRAPH_n \in B_{4n}$. We use techniques from Gergov's proof of Theorem 2 in [24].

Let the Boolean variables $X := \{x_0, \ldots, x_{n-1}\}$ and $Y := \{y_0, \ldots, y_{n-1}\}$ denote the factor variables of $MUL\text{-}GRAPH_n$ and $Z := \{z_0, \ldots, z_{2n-1}\}$ its result variables. That is, $MUL\text{-}GRAPH_n(x, y, z) = 1 \Leftrightarrow |x| \cdot |y| = |z|$ for $x = x_{n-1} \ldots x_0$, $y = y_{n-1} \ldots y_0$, and $z = z_{2n-1} \ldots z_0$. Let P be a π-oblivious BP for $MUL\text{-}GRAPH_n$ whose variable sequence π contains each variable at most k times.

First, we show a lower bound for a restricted symbolic graph of MUL_n. Therefore, let ρ be some arbitrary sequence that contains each variable from X, Y, and Z exactly $2k =: \ell$ times and which has π as subsequence.

Definition 10. *For two disjoint subsets S and T of X, an interval $(\rho_i, \ldots, \rho_j)$ of ρ is called* link *if $\rho_{i+1}, \ldots, \rho_{j-1} \notin S \cup T$ and either $\rho_i \in T \wedge \rho_j \in S$ or $\rho_i \in S \wedge \rho_j \in T$.*

Lemma 1 (Alon and Maass [25]). *Let $M = (m_1, \ldots, m_{n\ell})$ be a sequence in which each element $m_i \in X$ appears exactly ℓ times. Suppose $X_1 \dot\cup X_2$ is a partition of X into two disjoint non-empty sets. Then, there are two subsets $S \subseteq X_1$ and $T \subseteq X_2$ with $|S| \geq |X_1|/2^{2\ell-1}$, $|T| \geq |X_2|/2^{2\ell-1}$, and such that the number of links between S and T in M is bounded above by $2 \cdot \ell - 1$.*

Let $X_1 := \{x_0, \ldots, x_{\lfloor n/2 \rfloor}\}$ and $X_2 := \{x_{\lfloor n/2 \rfloor+1}, \ldots, x_{n-1}\}$. Due to Lemma 1, there are subsets $S \subseteq X_1$ and $T \subseteq X_2$ such that $|S|, |T| \geq (n/2 - 1)/2^{2\ell-1} > n/2^{2\ell} - 1$ and there are no more than $2 \cdot \ell - 1$ links between S and T in ρ.

Since $D := \{(x_i, x_j) \mid x_i \in S, \ x_j \in T\}$ contains at least $\left(n/2^{2\ell} - 1\right)^2$ pairs, there is some index set $I \subseteq \{0, \ldots, n-1\}$ and distance parameter $d \in \{1, \ldots, n-1\}$ such that $D' := \{(x_i, x_{i+d}) \mid i \in I\} \subseteq D$ contains at least $\left(n/2^{2\ell} - 1\right)^2/n \geq n/2^{4\ell} - 1$ pairs and $\max I < \min I + d$.

Let $MUL\text{-}GRAPH_n^* := \bigwedge_{i=\min I+d}^{\max I+d} [MUL_{n,i}(x,y) = z_i]$. Moreover, we consider the subfunction f_n of $MUL\text{-}GRAPH_n^*$ defined by the following partial variable assignments:

$$x_i := \begin{cases} 1 & \text{if } \begin{aligned} &(i = \min I) \vee (i = \min I + d) \\ &\vee [(\min I \leq i \leq \max I) \wedge (i \notin I)] \end{aligned} \\ 0 & \text{if } \begin{aligned} &(i = \max I) \vee (i = \max I + d) \\ &\vee [((i < \min I) \vee (\max I < i)) \wedge (i - d \notin I)] \end{aligned} \end{cases} \quad , \tag{1}$$

$$y_j := \begin{cases} 1 & \text{if } (j = 0) \vee (j = d) \\ 0 & \text{else} \end{cases} \quad , \tag{2}$$

$$z_r := \begin{cases} 1 & (r = \max I + d) \\ 0 & \text{else} \end{cases} \quad . \tag{3}$$

Lemma 2. *Any ρ-oblivious BP Q for f_n has a size of at least $2^{n/((2\ell-1)(2^{4\ell}))-1}$.*

Proof. In (1), we replace all variables between $\min I$ and $\max I$ by 1 which do not take part in D' as well as $x_{\min I}$ and $x_{\min I+d}$. All x-variables lying outside the interval $[\min I, \max I]$ and not taking part in D' are replaced by 0 as well as $x_{\max I}$ and $x_{\max I+d}$. In (2), the y-variables are chosen such that we sum up $|x|$ and $|x| \cdot 2^d$. The result is checked against $|z| = 2^{\max I+d}$ (see (3)). (See Fig. 1.)

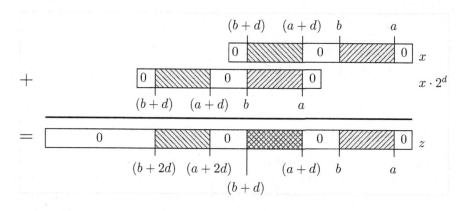

Fig. 1. Illustration of the effect of variable replacements (1), (2), and (3) for $a := \min I$ and $b := \max I$. $MUL\text{-}GRAPH_n^*$ verifies only the result bits $z_{a+d}, \ldots, z_{b+d}$ which correspond to the sum of $x_{b+d} \ldots x_{a+d}$ and $x_b \ldots x_a$.

The function f_n depends only on the x-variables being part of pairs in $D' \setminus \{(x_{\min I}, x_{\min I+d}), (x_{\max I}, x_{\max I+d})\}$; it can be easily seen that it computes 1 if and only if $x_i \neq x_{i+d}$ for all $i \in I \setminus \{\min I, \max I\} =: I'$. We now use methods from communication complexity (see, e.g., [26]).

Claim. The deterministic communication complexity of f_n w.r.t. the variable set partition $\{x_i \mid i \in I'\} \dot\cup \{x_{i+d} \mid i \in I'\}$ is at least $|I'| = |I| - 2$.

Let C be the $2^{|I'|} \times 2^{|I'|}$ communication matrix of f_n. It is $C(i, j) = 1$ iff (i) is bit-inverse to (j). Hence, C is a permutation matrix and has rank $2^{|I'|}$. The value $\log(\text{rank}(C)) = |I'|$ is known to be a lower bound on the deterministic communication complexity of a Boolean function.

Due to the construction of I', any ρ-oblivious BP Q for f_n can be partitioned into at most $2 \cdot \ell$ parts $S_1, T_1, \ldots, S_\ell, T_\ell$ such that S_i (T_i) contains only variables from S (T). Therefore, Q yields a communication protocol of length $(2 \cdot \ell - 1) \cdot \log(\text{width}(Q))$, where the width of Q is the maximum number of nodes labeled with the same variable. (For the trivial construction see, e. g., [4–Sect. 7.5].) Due to the lower bound of $|I'| = |I| - 2 \geq n/2^{4\ell} - 3$ on the communication complexity of f_n, Q must have at least width (and, therefore, size) $2^{n/((2\ell-1)(2^{4\ell}))-3/(2\ell-1)} \geq 2^{n/((2\ell-1)(2^{4\ell}))-1}$.   □

We are now able to show the lower bound on the size of P.

Proof (Theorem 2). The lower bound of Lemma 2 does also hold for P because we can construct an oblivious BP P' for f_n from P without enlarging it: At first, we apply variable replacements (1), (2), and (3). Then, we have to get rid of the z-variables $z_0, \ldots, z_{\min I + d - 1}, z_{\max I + d + 1}, \ldots, z_{2n-1}$. In satisfying inputs of f_n, it holds $z_i = x_i$ for $i \in I$ and $z_j = x_{j-d}$ for $j - 2d \in I$ (see Fig. 1). In order to force these variable pairs to be equal, we replace node labels z_i, $i \in I$, by x_i as well as labels z_j, $j - 2d \in I$, by x_{j-d}. The remaining z-variables are replaced by 0. The resulting BP P' represents f_n.

We added no more than k nodes for each z-variable, and P' is π'-oblivious for some variable sequence π' containing each variable $2k = \ell$ times and having π as subsequence. Lemma 2 implies the lower bound of $2^{n/((2\ell-1)(2^{4\ell}))-1} = 2^{n/((4k-1)(2^{8k}))-1}$ for P' and P.   □

6   An Application to Symbolic Algorithm Analysis

Finally, we sketch an application of the lower bound for the graph of integer multiplication to the analysis of a symbolic algorithm for the all-pairs shortest-paths problem in OBDD-represented weighted graphs. Input is the OBDD of a graph $G = (V, E, c)$, $c\colon E \to \mathbb{N}$, while the output OBDD represents the shortest path distances $\text{dist}\colon V^2 \to \mathbb{N}_0$. The algorithm presented in [9] (called $\mathcal{A}$ in the following) has polylogarithmic runtime $\mathcal{O}(\log^3(|V| \cdot c^{\max}))$, $c^{\max} := \{c(e) \mid e \in E\}$, if both input and output OBDD have constant width while not skipping any variable tests. However, the method works only for strictly positive edge weights $c(e) \in \mathbb{N}_{>0}$.

Another approach mentioned in [9] works also for weights $c(e) = 0$. It iteratively generates OBDDs for functions S_k with $S_k(x, y, d) = 1$ iff there is a shortest path from node $|x|$ to node $|y|$ of length $|d|$ visiting no more than 2^k edges. We call this algorithm $\mathcal{B}$. In order to justify algorithm $\mathcal{A}$, it can be shown that $\mathcal{B}$ has *not* necessarily polylog. runtime on constant width OBDDs. Similar to [27–Sect. 7], we construct an input graph $G_n = (V_n, E_n, c_n)$ such that both

G_n and the algorithm output dist_n have constant width OBDDs, while some intermediate function S_k generated by $\mathcal{B}$ has only exponential OBDDs.

G_n is the union of subgraphs $G_n^{i,j}$ with $2^{n-1} + 1 \leq i, j \leq 2^n - 1$ sharing only one special node s. Each subgraph $G_n^{i,j}$ is a path $(s, u_0, \ldots, u_{2^n-j-1}, v_{2^n-j}, \ldots, v_{2^n-1}) =: p_{i,j}$ with edge weights $c_n(\cdot, u.) := 1$, $c_n(v., \cdot) := i$, and $c_n(u_{2^n-j-1}, v_{2^n-j}) := i + j - 2^n$. Moreover, there are nodes $w_{2^n-j}, \ldots, w_{2^n-1}$ connected by *shortcut edges* (u_{2^n-j-1}, w_{2^n-j}), (w_{2^n-j}, v_{2^n-j}), $(v_\ell, w_{\ell+1})$, and $(w_{\ell+1}, v_{\ell+1})$ for $2^n - j \leq \ell \leq 2^n - 2$ with weight 1. Note that shortcut edges bridge all edges whose weight is larger than 1. Figure 2 shows subgraph $G_3^{6,5}$.

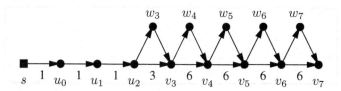

Fig. 2. $G_3^{6,5}$. Edges incident to w-nodes have weight 1

It can be easily verified that the shortest path from s to v_{2^n-1} in $G_n^{i,j}$ uses all shortcut edges and has length $2^n + j$. Using an appropriate binary node encoding, these shortest path lengths have OBDDs with constant width. This can also shown for all other node pairs in G_n and the OBDD for G_n itself (analogous to [27–Sect. 7.1]). That is, the preconditions are fulfilled and $\mathcal{A}$ has polylog. runtime on G_n.

On the other hand, the intermediate function S_n generated by $\mathcal{B}$ on G_n must not cover the shortest s–v_{2^n-1}-path in $G_n^{i,j}$ because only 2^n edges are allowed. Therefore, it has to represent the direct path $p_{i,j}$ that uses no shortcut edges, which has length $(2^n - j) + (i + j - 2^n) + (j-1)i = i \cdot j$. The OBDD representation of the corresponding characteristic function inherently contains the symbolic graph of integer multiplication. In this way, it is possible to show an exponential lower bound on the OBDD size of S_n w.r.t. its number of $\Theta(\log |V_n| + \log c_n^{\max})$ Boolean variables. This implies that $\mathcal{B}$, while allowing zero weights, has not the same convenient runtime properties as $\mathcal{A}$. A detailed discussion of this application can be found in the extended version of this paper.

So Theorem 1 has been used to show limits of the symbolic APSP-algorithm $\mathcal{B}$ and to justify the restriction of $\mathcal{A}$ to strictly positive edge weights.

7 Conclusions and Open Problems

Exponential lower bounds on the OBDD size of the fundamental and hard functions MUL, SQU, ISA, and HWB carry over to the symbolic graph scenario. This has not to be the case in general as seen for the counterexample function vector FSA. Graphs of such simply-structured functions can then be used to show limits of symbolic graph algorithms, which has been done exemplarily for a symbolic all-pairs shortest-paths algorithm.

Corollary 2 covers only a restricted version of the symbolic graph of squaring. The OBDD size of its unrestricted graph is an open question yet; experiments suggest that it is also exponential. Moreover, it is of interest whether exponential lower bounds for integer multiplication w. r. t. more general types of branching programs carry over to the symbolic graph scenario, too (see, e. g., [28, 29, 30]).

Acknowledgments. Thanks to André Gronemeier, Martin Sauerhoff, Detlef Sieling, and Ingo Wegener for proofreading and helpful discussions.

References

1. Borodin, A., Cook, S.: A time–space tradeoff for sorting on a general sequential model of computation. SIAM Journal on Computing **11** (1982) 287–297
2. Hachtel, G., Somenzi, F.: Logic Synthesis and Verification Algorithms. Kluwer Academic Publishers, Boston (1996)
3. McMillan, K.: Symbolic Model Checking. Kluwer Academic Publishers, Boston (1994)
4. Wegener, I.: Branching Programs and Binary Decision Diagrams. SIAM, Philadelphia (2000)
5. Hachtel, G., Somenzi, F.: A symbolic algorithm for maximum flow in 0–1 networks. Formal Methods in System Design **10** (1997) 207–219
6. Sawitzki, D.: Implicit flow maximization by iterative squaring. In: SOFSEM'04. Volume 2932 of LNCS., Springer (2004) 301–313
7. Woelfel, P.: Symbolic topological sorting with OBDDs. In: MFCS'03. Volume 2747 of LNCS., Springer (2003) 671–680
8. Sawitzki, D.: Experimental studies of symbolic shortest-path algorithms. In: WEA'04. Volume 3059 of LNCS., Springer (2004) 482–497
9. Sawitzki, D.: A symbolic approach to the all-pairs shortest-paths problem. To appear inWG'04 (2004)
10. Gentilini, R., Piazza, C., Policriti, A.: Computing strongly connected components in a linear number of symbolic steps. In: SODA'03. (2003) 573–582
11. Gentilini, R., Policriti, A.: Biconnectivity on symbolically represented graphs: A linear solution. In: ISAAC'03. Volume 2906 of LNCS., Springer (2003) 554–564
12. Bloem, R., Gabow, H., Somenzi, F.: An algorithm for strongly connected component analysis in $n \log n$ symbolic steps. In: FMCAD'00. Volume 1954 of LNCS., Springer (2000) 37–54
13. Ravi, K., Bloem, R., Somenzi, F.: A comparative study of symbolic algorithms for the computation of fair cycles. In: FMCAD'00. Volume 1954 of LNCS., Springer (2000) 143–160
14. Hojati, R., Touati, H., Kurshan, R., Brayton, R.: Efficient ω-regular language containment. In: CAV'93. Volume 663 of LNCS., Springer (1993) 396–409
15. Jin, H., Kuehlmann, A., Somenzi, F.: Fine-grain conjunction scheduling for symbolic reachability analysis. In: TACAS'02. Volume 2280 of LNCS., Springer (2002) 312–326
16. Moon, I., Kukula, J., Ravi, K., Somenzi, F.: To split or to conjoin: The question in image computation. In: DAC'00, ACM Press (2000) 23–28

17. Xie, A., Beerel, P.: Implicit enumeration of strongly connected components. In: ICCAD'99, ACM Press (1999) 37–40

18. Feigenbaum, J., Kannan, S., Vardi, M., Viswanathan, M.: Complexity of problems on graphs represented as OBDDs. Chicago Journal of Theoretical Computer Science **1999** (1999) 1–25

19. Breitbart, Y., III, H.H., Rosenkrantz, D.: On the size of binary decision diagrams representing Boolean functions. Theoretical Computer Science **145** (1995) 45–69

20. Woelfel, P.: New bounds on the OBDD-size of integer multiplication via universal hashing. In: STACS'01. Volume 2010 of LNCS., Springer (2001) 563–574

21. Keim, M., Drechsler, R., Becker, B., Martin, M., Molitor, P.: Polynomial formal verification of multipliers. Formal Methods in System Design **22** (2003) 39–58

22. Jukna, S.: The graph of integer multiplication is hard for read-k-times networks. Technical Report 95–10, Universität Trier (1995)

23. Wegener, I.: Optimal lower bounds on the depth of polynomial-size threshold circuits for some arithmetic functions. Information Processing Letters **46** (1993) 85–87

24. Gergov, J.: Time-space tradeoffs for integer multiplication on various types of input oblivious sequential machines. Information Processing Letters **51** (1994) 265–269

25. Alon, N., Maass, W.: Meanders and their applications in lower bound arguments. Journal of Computer and System Sciences **37** (1988) 118–129

26. Hromkovič, J.: Communication Complexity and Parallel Computing. Springer, Berlin Heidelberg New-York (1997)

27. Sawitzki, D.: On graphs with characteristic bounded-width functions. Technical report, Universität Dortmund (2004) Available via http://ls2-www.cs.uni-dortmund.de/~sawitzki/OGwCBWF.pdf.

28. Ablayev, F., Karpinski, M.: A lower bound for integer multiplication on randomized ordered read-once branching programs. Information and Computation **186** (2003) 78–89

29. Ponzio, S.: A lower bound for integer multiplication with read-once branching programs. SIAM Journal on Computing **28** (1998) 798–815

30. Woelfel, P.: On the complexity of integer multiplication in branching programs with multiple tests and in read-once branching programs with limited nondeterminism. In: CCC'02, IEEE Press (2002) 80–89

XML-Based Declarative Access Control

Robert Steele, William Gardner, Tharam S. Dillon, and Abdelkarim Erradi

Faculty of Information Technology, University of Technology, Sydney
PO Box 123, Broadway NSW 2007, Australia
{rsteele, wgardner, tharam, karim}@it.uts.edu.au

Abstract. XML, a self-describing and semi-structured data format, is becoming a standard to represent and exchange data between applications across the Web. XML repositories are also starting to be used either to store data or as an interoperability layer for legacy applications and data sources. The widespread use of XML highlights the need for flexible access control models for XML documents to protect sensitive and valuable information from unauthorised access. This paper presents a novel declarative access control model and elaborates how this model allows the expression of access control rules in XML. The paper further introduces the operational semantics of the model by describing the Xplorer engine which supports search-browse-navigate activities on XML repositories. Xplorer takes as inputs XML-based data schema, instance data and access control rules to auto-generate an access control-enabled Web application in accordance with these rules.

1 Introduction

eXtensible Markup Language (XML) [1] is a widely accepted standard for structuring data by adding metadata to elements using self-descriptive tags. It is a fundamental part of the future Web and eBusiness. This creates a need for new ways to securely and effectively search and browse semi-structured content by using more expressive mechanism than just keywords. Despite good results provided by the current keyword-oriented search, searching semi-structured content with such tools suffers from limited query interface, limited customisation to individual users and often returns inaccurate query results. This is because keyword-based search does not make use of the semantics and the relationships embedded in the content. This can result in a user having to filter through numerous mostly irrelevant documents to find the right information. In addition semi-structured data potentially supports sophisticated and fine-grain differentiation of access capabilities between different users.

Our research attempts to address these problems by empowering the information seeker with intuitive auto-generated multi-field search interface and also to provide a system that can support fine-grained differentiation of access capabilities between users. One of the critical issues related to search of XML content is the protection of valuable and sensitive information (e.g. Electronic Health Records - EHR) against unauthorised access; hence there is a strong need for defining powerful access control models for XML documents.

M. Bieliková et al. (Eds.): SOFSEM 2005, LNCS 3381, pp. 310–319, 2005.

XML is increasingly used for encoding all kinds of documents such as product catalogues, digital libraries and Electronic Health Records (EHR). This growth led to increasing interest in XML document access control. Our aim is to develop extensible models and tools for defining and configuring fine-grained access control policies for XML documents. These policies will be enforced by an access control aware framework to auto-generate a multi-field user interface to search XML repositories. The proposed framework dramatically reduces UI development and maintenance time as the interface and the access constraints are not hard-coded in the application and this also enables flexible customisation at low cost.

This paper provides a presentation of the features of the proposed access control model and interpreting engine, named Xplorer [2] system by showing how the techniques could be used for secure navigation and search of large EHR as an example. The rest of the paper is organised as follows. Section 2 presents a brief survey of related works. The system architecture is introduced in Section 3 and the declarative access control model introduced in Section 4. Section 5 describes the proposed access control model for auto-generated interface through a sample prototype. The last section concludes and gives an outlook for future work.

2   Related Work

There are three related areas in relation to the scope of this paper; 1) generation of the interface to XML data; 2) using a graphical user interface to pose queries and 3) XML access control. There are multiple existing works dealing with 1) and 2), graphical interfaces for displaying and querying data repositories such as Odeview [3, 4], Pesto [5] and BBQ [6]. Odeview and Pesto deal with graphical interfaces for browsing and querying object oriented databases not XML-based data. BBQ does address the searching and accessing of XML-based data and provides a new underlying query language XMAS. Other work in this area includes XQForms and QURSED [7, 8]. These works has looked at simpler schemas and at developer tools rather than complete auto-generation. This paper presents a more sophisticated approach to take into account of fine-grained data access constraints in the process of providing an auto-generated user interfaces for the manipulation of XML data. This is an extension to our previous work [9-11] to auto-generate UI from XML Schema.

Regarding XML security, the literature offers several approaches to define and enforce access rights on XML documents. Kudo and Hada [12] proposed XML Access Control Language (XACL). XACL is used to specify an object-subject-action-condition oriented access control policy. It supports flexible provisional authorisation to a document based on whether certain conditions are met; e.g., the subject is allowed access to confidential information, but the access must be logged. Bertino et al. [13, 14] defined Author-X system as a suite of tools focusing on access-control enforcement and security administration for XML documents. Damiani et al. [15, 16] also specify a language for encoding access restrictions at the DTDs/schemas level or for individual XML documents. Gabillon and Bruno [17] implement access control by converting their "authorization sheet" to an XSLT document that can then extract a view of the accessible part of the corresponding XML document.

XACML (eXtensible Access Control Markup Language) [18] is an OASIS standard based on work including that of Kudo, Damiani, and Bertino. It standardises access request/response format, architecture of the policy enforcement framework, etc., but it does not address deriving access control rules from the existing policy base. These approaches are based on XPath [19], XSL [20] and custom constructs that were developed to specify access conditions. Goel et al. [21] developed an XQuery-based [22] approach for deriving fine-grained access control rules from schema-level rules, document content, or rules on other documents. Miklau et al. [23] proposed cryptography technique to ensure that published data is only visible to authorised users. Recently a security views technique has been proposed in [24], it provides each group an XML view consisting of only the information that the users are authorised to access but the proposed technique only supports DTDs.

Our work extends ideas from both XML access control research and the research into user interface generation to provide an access control framework for viewer applications for semi-structured data. While a fine-grained policy definition is available, the access control policies are well keep separated from the actual underlying data. Our work differs from previous work in that it provides an access control model including the semantics of the access control privileges in terms of their interpretation by a *generic* semi-structured data viewer application.

3 System Architecture

As shown in Figure 1, the Xplorer system will rely on XML schemas of the data as well as access control rules encoded using the proposed constraints vocabulary. The framework will support searching XML repositories without requiring end-users to use complex XML query languages. The security enforcement module will enforce access control rules either by refining the XQuery to run against the XML repository or by generating and applying XML transformations via dynamically generated XSLT from the access control rules. In this paper we will focus on the XML security features of the framework.

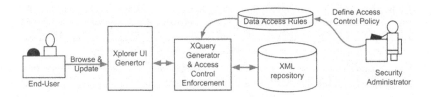

Fig. 1. Xplorer architecture

4 Declarative Access Control Model

After reviewing the literature about XML access control we have come to the conclusion that there is still a need for a clear and simple language to declaratively encode the semantics of an access control policy. The existing approaches require costly runtime security checks as access to each element requested by the user requires a request to be made to query the security policy. This will result in a poor performance and scalability.

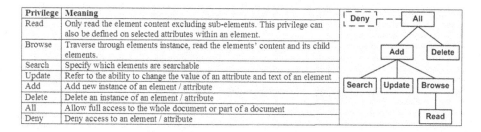

Privilege	Meaning
Read	Only read the element content excluding sub-elements. This privilege can also be defined on selected attributes within an element.
Browse	Traverse through elements instance, read the elements' content and its child elements.
Search	Specify which elements are searchable
Update	Refer to the ability to change the value of an attribute and text of an element
Add	Add new instance of an element / attribute
Delete	Delete an instance of an element / attribute
All	Allow full access to the whole document or part of a document
Deny	Deny access to an element / attribute

Fig. 2. Permissions and their semantics

The Role Based Access Control model is shown in Figure 3 and the set of possible privileges is shown in Figure 2. This model allows a role to be attached to a user and amongst other objects allows the adding of a sensitivity value/attribute to elements of an XML data schema. An access control policy, consisting of a set of declarative rules, defines the roles' privileges to various sensitivity levels and sections of the XML document. Finally the access control rules are dynamically taken into account by Xplorer [2] to restrict access to the parts of the document/ repositories the user (of a particular role) is allowed to access. The semantic relationships between privilege types are shown in the diagram on Figure 2.

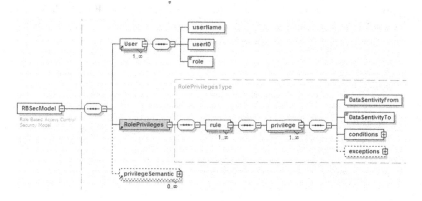

Fig. 3. Role based access control security model

The access control enforcement module will apply the access control policy by first retrieving the access control rules that apply to the user, based on the role that is assigned to the user. There is a configuration parameter to instruct the framework to cache the access control for a specific period in order to avoid repetitive access control queries. The system will then replace any references to session variables (e.g. current user Id) with the actual values from the session object. If any of the access control rules apply to the requested data then the system will refine the auto-generated query to ensure that the query only requests the data the user has access to. The query results will be intercepted by the UI generator to provide a user friendly view of the results.

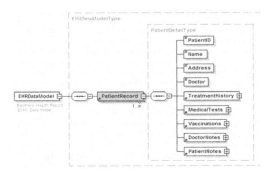

Fig. 4. eHealth Record data model

To illustrate this in this paper, we will use the example of EHR data, such as the one defined in Figure 4. Suppose that a hospital wants to impose the following security policy on the e-Health records, with 2 types of role defined in the security policy, namely Doctor and Patient, according to the following:

- The doctor can access the records of all patients but only the `MedicalTest` results of the patients he/she takes care of. For `PatientNote` the doctor can view and search but not add or update.
- A patient is only allowed to access his/her own record, which will allow the patient to view all information except `DoctorNote` and `MedicalTest`. However, the only editing privilege of a patient is for `PatientNote`.

```
<RolePrivileges>
    <rule RoleId="Doctor">
        <privilege PrivilegeType="All">
            <DataSentivityFrom>1</DataSentivityFrom>
            <DataSentivityTo>4</DataSentivityTo>
            <conditions>
                <predicate type="equal" scope="//MedicalTest">
                    <parameter type="xpath">//PatientRecord/Doctor</parameter>
                    <parameter type="variable">Server.GetValue(Session.UserId)</parameter>
                </predicate>
            </conditions>
            <exceptions>
                <exception type="Search">//PatientNotes</exception>
                <exception type="Browse">//PatientNotes</exception>
            </exceptions>
        </privilege>
    </rule>
    <rule RoleId="Patient">
        <privilege PrivilegeType="Browse">
            <DataSentivityFrom>1</DataSentivityFrom>
            <DataSentivityTo>2</DataSentivityTo>
            <conditions>
                <predicate type="equal" scope="//PatientRecord">
                    <parameter type="xpath">//PatientRecord/PatientID</parameter>
                    <parameter type="variable">Server.GetValue(Session.UserId)</parameter>
                </predicate>
            </conditions>
            <exceptions>
                <exception type="Add">//PatientNotes</exception>
            </exceptions>
        </privilege>
    </rule>
</RolePrivileges>
```

Fig. 5. Example of declarative access control rules

The above constraints can be easily encoded using the proposed method by firstly adding the data sensitivity level into the existing data model schema. We have decided four sensitivity levels, represented by the integer value of 1 to 4, where level 1 denote elements that are unclassified to level 4 that represent elements that are consider highly classified. In the case of this example, the data sensitivity rating of 4 was assigned to `DoctorNote` and `MedicalTest`.

With the data sensitivity level in place, the access control policy could then be defined based on the role of the user, and the access privilege allowed for the defined role type. The access policy for the eHealth records example is encoded via declarative access control rules as shown in Figure 5.

5   Interpreting Access Control Rules

The Xplorer provides the user interface for the searching, viewing and manipulation of semi-structured data. The UI generator constructs the user interface based on the semantics defined in the data model schema. Taking into account the access control policy in the role based model, Xplorer presents the appropriate data to user. Its interface at any one time is in one of three modes (Figure 6):

- Search
- View (element instance)
- Update (element instance)

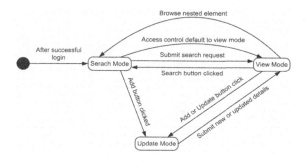

Fig. 6. State transitions for Xplorer

In the example of the eHealth records, Xplorer will provide the doctor (Figure 7) an user interface that has elements with sensitivity level between 1 and 4 and with the options to search and edit most of the data, while the patient will have a more limited access (Figure 8), but still be able to update patient note.

The Search mode (e.g. Figure 7a) is generated from the data model XML Schema so as to only display search fields (i.e. just the simpleType elements) corresponding to the elements from the schema for which the user has search privilege under the provided access control policy. In the Search mode the user will have the ability to either switch to View mode or Update mode. When in Search mode, in addition to the search text fields and hyperlinks to nested XML elements two buttons will be displayed; the Add button and the Find button. The Add button adds a new instance of the type of the current element. Clicking the Add button take Xplorer into the Update mode. Clicking the Find button submits the search request and takes Xplorer into the View mode.

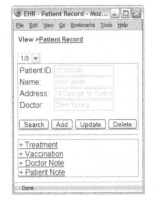

a) As the doctor has access to all elements and various permissions the default interface will prompt for search parameters for any of the simpleType XML elements. Complex elements appear as hyperlinks.

b) Patient records view for "John". Note that Medical Test link is unavailable because this patient is not under the care of the currently authenticated doctor. When the Update button is clicked screen c) is shown, Patient Note link will bring screen d), Treatment link will bring screen f).

c) Screen to update the Patient Detail elements.

d) Search screen for Patient Note, note that the Add button is unavailable because the Doctor does not have access to do this for Patient Note.

e) Patient Note search results in View Mode. Add, Update and Delete buttons are unavailable because a Doctor does not have such access.

f) Since there is only one instance of Treatment the system defaults to View Mode when Treatment link is clicked.

Fig. 7. Xplorer example screens showing the interface provided to the Doctor

The Search mode is the default behaviour if the element that the user gets into has more than one instance otherwise the View mode will be provided. For example for the patient browsing a Health Record using Xplorer (see Figure 8a) he/she will get a view of his/her records without the ability to search other records. In this case, links to the appropriate sub elements are also presented.

The View mode (e.g. Figure 7b) is generated from the XML instance elements returned by a search request. In this mode all the elements are read-only with values for simpleType elements displayed as labels and instance element of complexType are displayed once again as hyperlinks so that the user can opt to drilldown into various parts of the currently selected element. Depending on the current user access policy, the View mode would provide a Search button to go back to the Search mode, an Add button that takes Xplorer to Update mode to add a sibling instance of the current element, an Update button to switch to the Update mode to edit the values of the current element and a Delete button to delete the current element and its child elements.

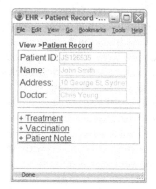

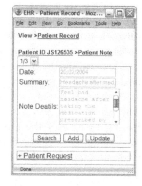

a) A Patient can only view his/her own records without the ability to search nor add or update. Medical Test, and Doctor Note are not shown because patient only access to elements at data sensitivity 1-2.

b) Treatment record of the currently logged in patient.

c) Patient Note is the only part of the EHR that the patient can search, add or update.

Fig. 8. Xplorer example screens showing the interface examples provided to the Patient

The Update mode (e.g. Figure 7c) is used to update values of an instance. This instance could be a new instance being added or an exiting instance being modified. The interface is generated from the schema of the current element to be edited or created. In this mode all the elements to which the user has write privilege will be rendered as editable text boxes. Once the update is done (Submit clicked) the user is returned to the View mode.

6 Discussion

The access control model comprises the key notions of access control rules, data sensitivity levels, roles, and privilege. The data sensitivity level of each element node that requires access control is declared in the data schema. Sensitivity level that is declared on the high level of the element tree will cascade down to the sub elements. However, a local declaration on any sub level can be used to override the cascaded sensitivity level. By attaching the sensitivity level to the data schema it is possible to change access behaviour for all instance data of an element without changing all of the instance documents – just a change to the schema is required. In addition a new data element can easily be added to the data model, and the need to modify the access control policy in the security model every time when a new element is added is greatly reduced. The privilege classification and their semantic relation, coupling with the use of data sensitivity level provide a simple yet powerful framework for the definition of access control rules.

Combining the auto-generated search interface framework with the role based control policy, the Xplorer interface allow user to search, view, update, and navigate through the XML repositories, by presenting simpleType element on the current level with the appropriate UI elements, while sub levels are show as hyperlinks. This presents a generic technique to search and browse XML data that takes into account access control rules. The propose framework could be utilized without modification to access and update business records as well as it can be used for EHRs. The notion of a simple user interface for the XML browse and search technique is also well suited to mobile clients as it can be used to produce a simple interface to incrementally browse the data.

A vocabulary was defined to encode the access rules outside the raw instance data. This allows the access control policies, roles and rules to be altered easily without having to modify the data schema, data instances or other business logic. Just by modifying the access control policy rules, new roles can be simply defined or existing roles altered with a high level of fine-grained expressive power.

7 Conclusion

In this paper, we presented a declarative access control model and Xplorer, an engine to interpret access control rules and provide secure searching and browsing of XML repositories. First we defined an access control model, consisting of a set of privileges and an access control rule schema, which provides powerful expressiveness to encode access rules to semi-structured content. Secondly, we described the Xplorer engine to take into account the access control rules. The auto-generated interface saves on development effort and eases maintainability. Our future work will extend the framework to define access rules and generate search interfaces for RDF/OWL-based ontologies. We will also study access control to XML repositories using XML views [25, 26].

References

[1] W3C-XML, *Extensible Markup Language (XML)*. 2004.
[2] R. Steele, W. Gardner and T.S. Dillon. *Xplorer: A Generic Search and Navigation Application for Semi-structured Data Repositories*. 3rd International Conference on Communications, Internet, and Information Technology (CIIT 2004).
[3] R. Agrawal, N.H. Gehani and J. Srinivasan. *OdeView: the graphical interface to Ode*. In Proceedings of the International Conference on Management of Data. 1990.
[4] S. Dar, N.H. Gehani, H.V. Jagadish and J. Srinivasan, *Queries in an Object-Oriented Graphical Interface*. Journal of Visual Languages and Computing, 1995. 6(1): p. 27-52.

[5] M. Carey, L. Haas, V. Maganty and J. Williams. *PESTO: an integrated query/browser for object databases.* In Proceedings of the International Conference on Very Large Databases (VLDB). 1996.

[6] K.D. Munroe and Y. Papakonstantinou. *BBQ: A Visual Interface for Integrated Browsing and Querying of XML.* In Proceedings of the International Conference on Very Large Databases (VLDB). 2000.

[7] M. Petropoulos, V. Vassalos and Y. Papakonstantinou. *XML query forms (XQForms): declarative specification of XML query interfaces.* In Proceedings of the International conference on World Wide Web. 2001.

[8] P. Mukhopadhyay and Y. Papakonstantinou. *Mixing querying and navigation in MIX.* In Proceedings of the 18th International Conference on Data Engineering. 2002.

[9] R. Steele and T. Dillon. Ontology Driven System for Mobile Device Access of Electronic Health Records. In Proceedings of the. 2004.

[10] R. Steele, Y. Ventsov and T. Dillon. *Object-Oriented Database-based Architecture for Mobile Enterprise Applications.* In Proceedings of the IEEE ITCC04. 2004.

[11] R. Steele, Y. Ventsov and T.S. Dillon. *XML Schema-based Discovery and Invocation of Mobile Services.* In Proceedings of the IEEE International Conference on e-Technology, e-Commerce and e-Service, EEE'04. 2004.

[12] M. Kudo and S. Hada. *XML document security based on provisional authorization.* In Proceedings of the 7th ACM conference on Computer and communications security (CCS). 2000.

[13] E. Bertino, S. Castano and E. Ferrari, *Securing XML documents with Author-X.* Internet Computing, IEEE, 2001. **5**(3): p. 21-31.

[14] E. Bertino and E. Ferrari, *Secure and selective dissemination of XML documents.* ACM Transactions on Information and System Security (TISSEC), 2002. **5**(3): p. 290-331.

[15] E. Damiani, P. Samarati, S. De Capitani di Vimercati and S. Paraboschi, *Controlling access to XML documents.* Internet Computing, IEEE, 2001. **5**(6): p. 18-28.

[16] E. Damiani, S.D.C.d. Vimercati, S. Paraboschi and P. Samarati, *A fine-grained access control system for XML documents.* ACM Transactions on Information and System Security (TISSEC), 2002. **5**(2): p. 169 - 202.

[17] A. Gabillon and E. Bruno. *Regulating access to XML documents.* In Proceedings of the the 15th Annual Conference on Database Security. 2001.

[18] OASIS, eXtensible Access Control Markup Language (XACML) version 1.0. 2003.

[19] W3C-XPath, XML Path Language (XPath) Version 1.0. 1999.

[20] W3C-XSL, Extensible Stylesheet Language (XSL). 2003.

[21] S.K. Goel, C. Clifton and A. Rosenthal. *Derived access control specification for XML.* In Proceedings of the Workshop On XML Security. 2003.

[22] W3C-XQuery, XQuery 1.0: An XML Query Language. 2004.

[23] G. Miklau and D. Suciu. *Controlling access to published data using cryptography.* In Proceedings of the VLDB. 2003.

[24] W. Fan, C.-Y. Chan and M. Garofalakis. *Secure XML querying with security views.* In Proceedings of the ACM SIGMOD International Conference on Management of Data, SIGMOD 2004. 2004.

[25] R. Rajugan, E. Chang, T.S. Dillon and L. Feng. *XML Views: Part I.* In Proceedings of the 14th International Conference on Database & Expert Systems Applications, DEXA 2003. 2003.

[26] V. Nassis, R. Rajugan, T.S. Dillon and W. Rahayu. *Conceptual Design of XML Document Warehouses.* In Proceedings of the 6th International Conference on Data Warehousing and Knowledge Discovery, DaWaK 2004. 2004.

VCD: A Visual Formalism for Specification of Heterogeneous Software Architectures*

David Šafránek and Jiří Šimša

Department of Computer Science, Faculty of Informatics,
Masaryk University Brno, Czech Republic
{xsafran1, xsimsa}@fi.muni.cz

Abstract. A visual formalism called Visual Coordination Diagrams (VCD) for high-level design of heterogeneous systems is presented in this paper. The language is based on a state-transition operational semantics, which allows application of formal methods to software design. Formal definition of VCD is included in the paper. Moreover, an example of use of the language is also given.

1   Introduction

The importance of visual modeling languages such as UML [1] is very significant in the domain of software engineering. The desired properties of such an universal visual design language are heterogeneity, hierarchy and component-based structure. Additionally, to be able to analyze the software design using formal methods, some unambiguous formal semantics is required. Unfortunately, there is no formal semantics of UML [2].

In this paper we present Visual Coordination Diagrams (VCD) – a *visual formalism* for specification of component-based distributed systems, based on the idea of GCCS [3] and its extensions [4]. The VCD formalism can be viewed as static architecture diagrams for specification of connections among components. The key property of VCD is its two-level *heterogeneity*. The first level of this heterogeneity is based on the possibility of combination of various coordination models (both synchronous and asynchronous) in a particular specification. The second level of the heterogeneity is the variability of specification of behavioral aspects. This can be done in various notations which have to be, in some well-defined sense, compatible with the supported coordination models.

The work on VCD is practically motivated by the formal verification project Liberouter [5]. In this project, we have to deal with formal modeling of a complex system composed of heterogeneous SW/HW units [6].

---

* This work has been supported by the Grant Agency of Czech Republic grant No. 201/03/0509, the FP5 project No. IST-2001-32603 and the CESNET activity Programmable hardware.

M. Bieliková et al. (Eds.): SOFSEM 2005, LNCS 3381, pp. 320–329, 2005.

1.1   Background and Related Work

There is a group of visual languages for concurrent systems in which the classical state transition diagrams have been extended to fulfill the needs of design of complex systems. Combining the concept of geometric inclusion with the concept of hi-graphs, the hierarchy of states has been added, leading to Harel's Statecharts [7]. The complexity of the syntactic richness of Statecharts has shown that reaching a compositional formal semantics for such a powerful language is impracticable. Various sub-dialects of Statecharts have been defined to achieve required semantic properties [8]. The concept of Statecharts was also incorporated in UML [9], [10].

Another group of visual languages is based on the concept of message flow graphs. They are employed to visually describe partial message passing interaction among concurrent processes. The high level message flow diagrams called Message Sequence Charts are based on this concept [11]. This notation does not support hierarchical design. For its simple nature, it is widely used in telecommunication industry and it is also a part of UML.

VCD extends and generalizes ideas of the work on Graphical calculus of communicating systems (GCCS) [3] and its synchronous extension SGCCS [4], which adopt the process algebraic approach as the underlying semantic model. In these languages, a very tight relation to the underlying process algebraic semantic model limits the heterogeneity of both coordination and behavioral layers. I.e., it is difficult to incorporate Statechart-like formalisms into GCCS. In VCD we try to overcome these inconveniences by using of a more general semantic model.

There is another architectural language, which is, similarly to VCD, based on the idea of GCCS. It is called Architectural Interaction Diagrams (AID) [12]. VCD and AID both achieve some level of heterogeneity by avoiding the tight relation with the CCS process algebra [13]. One of the significant differences between these two formalisms is in the underlying semantic model. AID is aimed to be used for specification of interactive systems while in VCD the interactive aspects can be additionally mixed with reactivity. At the behavioral layer, VCD supports more expressive formalisms than AID, and thus allows more heterogeneity at this level.

Similarly to some of the classical textual architecture description languages (ADLs) like Wright [14] or UniCON [15], VCD are based on the idea of taking connectors and computational components as different elements of system architecture. Moreover, this concept is further refined in VCD. In contrast to Wright, where the semantics of connectors is defined in terms of CSP processes, which are based on handshake-style coordination, in VCD more complex coordination mechanisms, e.g. multi-cast, can be modeled more conveniently. It is mainly due to the fact of semantic modeling of any coordination event, e.g. a broad-cast communication, as one atomic transition of a connector model. In the domain of ADLs, there are some languages which support dynamic changes of system structure, i.e. Darwin [16] or SOFA [17]. Unlike these languages, VCD does not support dynamism. VCD is aimed to be a simple visual formalism for hierarchical description of coordination of system components.

The main reason for developing VCD is our belief in importance of building a formal framework for coordination of various kinds of Statecharts and other visual formalisms for specification of component behavior. We would like to establish a simple syntactic visual notation with suitable underlying formal semantics. The chosen semantic model is based on composition of local transition systems, which represent particular components, resulting in one global transition system formally representing the whole architecture description.

2   Overview of VCD

VCD is aimed to be a formal language for specification of communication relationships in component-based systems. An example of a simple VCD is depicted in Fig. 1. Basic elements of the VCD formalism are component *interfaces*. Each interface contains input and output *ports*. Interfaces are organized in so called *networks* in which they can be connected by *links* to *buses*.

Buses represent connectors of components. They are used for specification of various types of coordination mechanisms. Different types of buses can be mixed together in a particular network. Consequently, systems with heterogeneous coordination mechanisms can be effectively specified using a single uniform formalism. In VCD there is a concept of *bus classes*, which allows to specify generic templates for various coordination media.

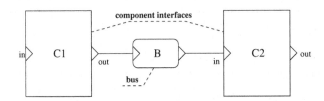

Fig. 1. A network of components *C1* and *C2* connected to bus *B*

The key concept of VCD is in its hierarchical network structure, which unfolds the *coordination layer*. This is achieved by the possibility of taking networks as components of other networks (higher-level networks). The relation between a network and its enclosing interface is defined by a *gate*. Gate maps ports of the lower-level network to ports of the enclosing interface in the higher-level network. An example of a network hierarchy is given in Fig. 2.

At the bottom-most level, behavior of system components has to be specified explicitly. There is no direct visual notation for behavioral specification in VCD. Instead of that, the formal semantic framework of so called VCD *leaves* is defined. It is called *behavioral layer*. The behavioral layer is based on the semantic model given by the notion of input/output labeled transition system (LTS) with sets of input and output actions taken as labels. This allows any language with

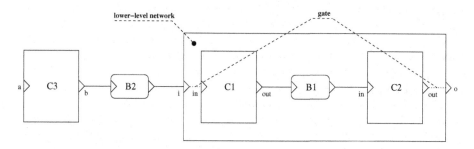

Fig. 2. Network hierarchy

semantics defined in the domain of LTS to be used for behavioral specification of system components. This property makes VCD heterogeneous also at the behavioral layer. Heterogeneity at this level is achieved with respect to the set of semantically compatible, but notationally different languages, which can be incorporated to VCD for the purpose of behavioral description. As examples of supported languages variants of Statecharts or Petri-Nets can be mentioned.

Semantics of VCD is based on a state transition model. By traversing the network hierarchy, it relies on a formal mechanism of combining component state transition models into one resulting state transition model of the top-most network. This is done with respect to the communication relationships specified by buses. Semantics of a particular bus class represents behavior of a specific communication media.

3   Syntax and Semantics of VCD

In this section, the formal syntax of VCD is defined and its semantics given.

3.1   Syntax

VCD networks are formally represented as VCD *terms*. Before capturing them formally we will build some basic notation.

Ports and Interfaces. The most basic elements of the coordination layer are *interfaces* with *ports*. We fix $\mathcal{W}$ a countable set of *write ports* and $\mathcal{R}$ a countable set of *read ports*. Interface I is defined as a pair consisting of a finite set of input ports and a finite set of output ports — $I = \langle W, R \rangle$, $W \subseteq \mathcal{W}$, $R \subseteq \mathcal{R}$, $W \cap R = \emptyset$. We mark projections $I^W = W$ the *write-interface*, $I^R = R$ the *read-interface*, respectively.

Buses and Bus Classes. The key construct of the coordination layer is *bus*. As it has been mentioned in the previous section, buses represent coordination mechanisms. Particular types of coordination mechanisms are represented as *bus* *classes*, which are formally defined as input/output labeled transition systems (I/O LTS).

Definition 1. Bus class $\mathcal{B}$ *is a tuple* $\langle Q, T, q_0 \rangle$ *where*

- Q *is a finite set of states*,
- $q_0 \in Q$ *an initial state*,
- $T \subseteq Q \times 2^{\mathcal{W}} \times 2^{\mathcal{R}} \times Q$ *a (countable) transition relation*.

Any bus class can be instantiated as a particular bus and used for specification of concrete connections among components in a network. The bus interface is determined by the set of links which connect the bus to the ports of surrounding components. The finiteness of the bus interface puts a finite bound to the transition relation of a bus instance. Formal definition of bus instance, given by its interface and its class, is the following.

Definition 2. Bus instance B *of a bus class* $\mathcal{B}$ *is a tuple* $B = \langle I, \mathcal{B} \rangle$, *where* I *is an interface and* $\mathcal{B}$ *a bus class*.

The interface of the bus instance B will be denoted as $I(B)$.

Gates, Networks and Leaves. Now we are going to define terms which formally represent VCD network diagrams. In the network depicted in Fig. 2 there are dashed lines which connect ports of subsystem interfaces to ports of the surrounding network interface. Later on in this subsection, these dashed links will be formalized as the notion of *gate*.

Definition 3. *A VCD term is:*

1. VCD leaf – *behavioral model specified in any LTS-compatible language*
2. VCD network $N = \langle \bar{C}, \bar{M}, L \rangle$, *where*
   (a) $\bar{C} = \langle C_1, \ldots, C_n \rangle$ – *vector of components*
   (b) $\forall i : C_i = \langle S_i, I_i, G_i \rangle$
      - $S_i \ldots$ *VCD term*
      - $I_i \ldots$ *interface*
      - $G_i \ldots$ *gate (see definition below)*
   (c) $\bar{M} = \langle M_1, \ldots, M_k \rangle$ – *vector of busses*
   (d) $\forall j : M_j = \langle I_j, \mathcal{B}_j \rangle$
      - $I_j \ldots$ *interface of a bus* M_j
      - $\mathcal{B}_j \ldots$ *class of a bus* M_j
   (e) $L \subseteq (\{1, \ldots, n\} \times (\mathcal{W} \cup \mathcal{R})) \times (\{1, \ldots, k\} \times (\mathcal{W} \cup \mathcal{R}))$ *a set of* links *satisfying:*
      if $\langle \langle i, p_1 \rangle, \langle j, p_2 \rangle \rangle \in L$ *then:*
      i. $p_1 \in \mathcal{W} \Leftrightarrow p_2 \in \mathcal{R}$
      ii. $p_1 \in I_i^W \cup I_i^R$
      iii. $p_2 \in I^W(M_j) \cup I^R(M_j)$
      iv. $\langle \langle l, p_1' \rangle, \langle j, p_2 \rangle \rangle \in L \Leftrightarrow l = i \wedge p_1' = p_1$
      v. $\langle \langle i, p_1 \rangle, \langle l, p_2' \rangle \rangle \in L \Leftrightarrow l = j \wedge p_2' = p_2$

The set of all VCD terms will be denoted by $\mathcal{S}$.

To formalize the feature of embedding a network into a higher-level network, we set up a function ϵ_R (ϵ_W) which for any VCD network returns a set of all its read (write) ports which have no connection to any bus. We call such ports *free ports*. To overcome ambiguity of port names in the context of a network, we index all the component interfaces in the scope of a particular network, and mark each port with the index of its interface.

Definition 4. *Let $N = \langle \bar{C}, \bar{M}, L \rangle$ be a network.*

- $\epsilon_W(N) = \{\langle i, w \rangle \mid w \in I_i^W \wedge \forall j, w' : \langle \langle i, w \rangle, \langle j, w' \rangle \rangle \notin L\}$
- $\epsilon_R(N) = \{\langle i, r \rangle \mid r \in I_i^R \wedge \forall j, r' : \langle \langle i, r \rangle, \langle j, r' \rangle \rangle \notin L\}$

We define *interface* of network N as a pair $I(N) = \langle \epsilon_W(N), \epsilon_R(N) \rangle$.

Gate is formally defined as a partial function relating ports of a particular component interface to free ports of the network which is nested in that interface. In the case when the nested structure is a leaf, the gate maps interface ports to eponymous actions of the nested process.

Definition 5. *Let I be an interface.*

1. *Let S be a VCD leaf encapsulated in the interface I. Let $ports(S) \subseteq \mathcal{W} \cup \mathcal{R}$ be a set of all actions of S. We define a gate of the leaf S as the identity function $G : I^W \cup I^R \to ports(S)$, $\forall x \in I^W \cup I^R . G(x) = x$.*
2. *Let $S = \langle \langle \langle S_1, I_1, G_1 \rangle, ..., \langle S_n, I_n, G_n \rangle \rangle, \langle M_1, ..., M_k \rangle, L \rangle$ be a VCD network embedded in interface I. We define a gate of the network S as the partial function $G : I^W \cup I^R \to I(S)$ satisfying:*
   - $\forall w \in I^W . G(w) = \langle i, w' \rangle \wedge \langle i, w' \rangle \in \epsilon_W(S)$
   - $\forall r \in I^R . G(r) = \langle i, r' \rangle \wedge \langle i, r' \rangle \in \epsilon_R(S)$

3.2   Semantics

In this subsection, the definition of the structural operational semantics of VCD terms is sketched. Its precise definition is presented in the full version of this paper [18].

As a semantic domain a class $\mathcal{L}$ of input/output labeled transition systems (I/O LTS) with sets of input and output actions in transition labels is used. Formally, the semantics is defined as a mapping $\psi : \mathcal{S} \to \mathcal{L}$ which assigns an I/O LTS to each VCD term.

First of all, we define the notion of I/O LTS, which makes the semantic domain for both the behavioral and the coordination layer.

Definition 6. *An I/O LTS is a tuple $\langle Q, T, q_0 \rangle$ where*

- *Q is a finite set of states,*
- *$q_0 \in Q$ an initial state,*
- *$T \subseteq Q \times 2^{\mathcal{R}} \times 2^{\mathcal{W}} \times Q$ a transition relation.*

At the behavioral layer, the state transition semantics captures the dynamics of atomic components. As VCD does not include any predefined syntactic construct for the behavioral layer, this I/O LTS is the structure in which the formalisms for behavioral description have to be encoded.

At the coordination layer, the semantics of a VCD network is defined as a global I/O LTS which composes transitions of local I/O LTSs representing the semantics of network components. This composition is realized with respect to the coordination model given by the specific bus classes instantiated in the network. States of the global I/O LTS are represented as *network configurations*. They respect the hierarchical structure of network terms. The formal definition of a network configuration is the following.

Definition 7. *Let* $N = \langle \langle C_1, ..., C_n \rangle, \langle M_1, ..., M_k \rangle, L \rangle$ *be a network. We define its configuration* $\langle \bar{s}, \bar{b} \rangle$ *as a vector of component and bus states* $\langle \langle s_1, ..., s_n \rangle, \langle b_1, ..., b_n \rangle \rangle$ *where* $\forall i \in \{1, ..., n\}. s_i$ *is a state of a component* C_i *and* $\forall j \in \{1, ..., k\}. b_j$ *is a state of a bus* M_j.

A network configuration contains a vector of current states of components and a vector of current states of buses. Such network configurations determine states of the resulting I/O LTS. Transitions of the global I/O LTS are defined with respect to the network hierarchy using Plotkin-style inference rules.

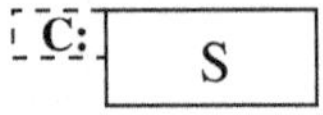

Fig. 3. Subsystem S embedded in a component C

In figure 3, there is a scheme how the component C is built by embedding of the subsystem S into a component interface. The subsystem S can be either a leaf or a network. Transition system of the component C is derived from the transition system of the embedded subsystem with respect to ports in the interface. Actions of S which have no ports in the interface of component C are hidden. The structure of the relevant inference rule for the situation when S is a leaf is the following.

$$\frac{\text{transition of S:} \quad q \xrightarrow{\Gamma}_{\Delta} q' \quad (\Gamma \subseteq ports(S) \cap \mathcal{R} \text{ and } \Delta \subseteq ports(S) \cap \mathcal{W})}{\text{transition of C:} \quad q \xrightarrow{I^R \cap \Gamma}_{I^W \cap \Delta} q'}$$

A similar inference rule captures the case when S is a network. This case is more complicated because of possible ambiguity of port names in the network scope. In appendix, the inference rule for this case is precised.

In figure 4, there is a scheme of a network with n components arbitrarily connected to m buses. For simplification, the links are not depicted. To define a global transition system for the network N, transition systems of the components and buses have to be composed. There are two different situations:

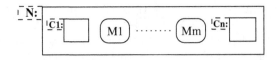

Fig. 4. Components $C_1 ... C_n$ and buses $M_1 ... M_m$ embedded in a network N

– Stand-alone components — their transitions are interleaved.
– Components connected to buses — their transitions are interleaved or synchronized w.r.t. semantics of instantiated bus classes.

The inference rules for both of the situations above are defined in [18].

4   An Example of Architectural Specification in VCD

In this section we will demonstrate on a very simple example how the VCD formalism, especially the concept of bus classes, can be used for specification of a distributed software architecture.

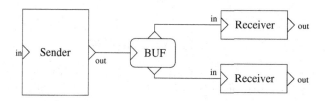

Fig. 5. Example of a concrete VCD network

In distributed software architectures, components of systems interact most typically in asynchronous way. One of the coordination mechanisms which captures this flavor of interaction is asynchronous message passing. In figure 5 there is a simple VCD network which specifies a distributed system with three components. It can be taken as a part of the specification of some communication protocol. There are one sender and two receiver components in the system. The intended behavior of the sender is to pass the output information to the communication media and continue some inner computation. On the other side, the behavior of any receiver is to take the information from the media asynchronously with computations of the sender.

To model this kind of interaction in the VCD framework, we establish a class $\mathcal{B}_{amp}$ of asynchronous message passing buses. It can be formally defined as the I/O LTS $\mathcal{B}_{amp} = \langle Q, T, q_0 \rangle$ where:

- $Q = \{q_w \mid w \in \mathcal{W}\} \cup q_0$
- T is defined by disjunction of the following expressions:

    1. $\forall w \in \mathcal{W}. \langle q_0, \{w\}, \emptyset, q_w \rangle \in T$
    2. $\forall q_x \in Q. \langle q_x, \emptyset, \{x\}, q_0 \rangle \in T$
    3. $\forall q_x \in Q. \langle q_x, \emptyset, \emptyset, q_x \rangle \in T$

The first expression defines the reaction of the bus to incoming write-actions. In the second expression, the interaction with receiver components is solved. The last expression adds the empty self-transitions, which allow the interactions to be potentially asynchronous.

The countable transition relation, which is the part of the bus class defined above, is made finite by the process of instantiation. In the example depicted in figure 5, the bus class $\mathcal{B}_{amp}$ is instantiated and placed in the context of three components. Thus, the number of transitions is bounded by the number of links which interconnect the bus with the surrounding components. In figure 6, there is the resulting transition system which represents this bus instance.

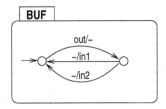

Fig. 6. Instance of asynchronous message-passing bus class

More complex types of bus classes modeling both synchronous and asynchronous coordination models can be defined following the scenario presented above. Together with the possibility of instancing different bus classes in the context of one particular network, this example demonstrates the heterogeneity of the VCD coordination layer.

5   Conclusions and Future Work

In this paper we have presented the formalism VCD for hierarchical specification of heterogeneous system architectures. The key concept of the language are buses which represent coordination models used in system architectures.

We see the main contributions of our work in three ways. First of all, the component-based character of VCD together with its hierarchical structure based on precise operational semantics allows to join the traditional design methods with the formal methods known from the theory of process algebras. On the other hand, the both syntactical and semantical separation of modeling the coordination aspects from modeling the behavioral aspects makes it possible to define a static communication infrastructure of a system independently of modeling the behavioral parts. Finally, heterogeneity supported in both behavioral and coordination layers of the language allows not only mixing of various coordination models in one specification, but also using of different models for behavioral description of components. For example, it is possible to put components defined as Statecharts together with components defined as Petri Nets and specify coordination relations among them using the constructs of the VCD coordination layer.

We are implementing a graphical tool [18] which allows VCD diagrams to be simply created and modified. In our future work, we would like to add the typed value-passing support to VCD. We would like to bring the notion of equivalences known from process algebraic theories and adapt them to VCD. We also aim to connect the editor of VCD with the distributed verification environment DiVinE [19].

References

1. OMG: Unified Modeling Language. Version 2.0. OMG (2003)
2. Harel, D., Rumpe, B.: Modeling Languages: Syntax, Semantics and All That Stuff. Technical Report MSC00-16, Weizman Institute of Science (2000)
3. Cleaveland, R., Du, X., Smolka, S.A.: GCCS: A Graphical Coordination Language for System Specification. In: Proceedings of COORD'00, LNCS, Springer Verlag (2000)
4. Šafránek, D.: SGCCS: A Graphical Language for Real-Time Coordination. In: Proceedings of FOCLASA'02. Volume 68 of ENTCS., Elsevier Science (2002)
5. Antoš, D., Fučík, O., Novotný, J.: Project of IPv6 Router with FPGA Hardware Accelerator. In: Proceeding of 13th International Conference on Field-Programmable Logic and Applications. Volume 2778., LNCS, Springer-Verlag (2003) 964–967
6. Jan, J.H., Kratochvíla, T., Řehák, V., Šafránek, D., Šimecek, P.: How to Formalize FPGA Hardware Design. Technical Report 4/2004, CESNET z.s.p.o. (2004)
7. Harel, D.: Statecharts: A Visual Formalism for Complex Systems. Technical report, The Weizmann Institute (1987)
8. Maggiolo-Schettini, A., Peron, A., Tini, S.: A Comparison of Statecharts Step Semantics. Theoretical Computer Science **290** (2003)
9. von der Beeck, M.: Formalization of UML-Statecharts. In: Proceedings of UML 2001. LNCS, Springer-Verlag (2001)
10. Harel, D., Kugler, H.: The Rhapsody Semantics of Statecharts (or, On the Executable Core of the UML). In: Proc. of 3rd Int. Workshop on Integration of Software Specification Techniques for Applications in Engineering. Volume 3147., LNCS, Springer-Verlag (2004) 325–354
11. Leue, S.: Methods and Semantics for Telecommunications Systems Engineering. PhD thesis, University of Berne (1994)
12. Ray, A., Cleaveland, R.: Architectural Interaction Diagrams: AIDs for System Modeling. In: Proc. of ICSE 2003, IEEE (2003)
13. Milner, R.: Communication and Concurrency. Prentice-Hall (1989)
14. Allen, R., Garlan, D.: A formal basis for architectural connection. ACM Trans. Softw. Eng. Methodol. **6** (1997) 213–249
15. Shaw, M., DeLine, R., Klein, D.V., Ross, T.L., Young, D.M., Zelesnik, G.: Abstractions for Software Architecture and Tools to Support Them. IEEE Trans. Softw. Eng. **21** (1995) 314–335
16. Magee, J., Kramer, J.: Dynamic Structure in Software Architectures. SIGSOFT Softw. Eng. Notes **21** (1996) 3–14
17. Plášil, F., Bálek, D., Janeček, R.: SOFA/DCUP: Architecture for Component Trading and Dynamic Updating. In: Proceedings of the International Conference on Configurable Distributed Systems, IEEE Computer Society (1998) 43
18. Šafránek, D., Šimša, J.: Visual Formalism for Specification of Heterogeneous Software Architectures. Technical Report FIMU-RS-2004-11, Faculty of Informatics, Masaryk University Brno (2004)
19. ParaDiSe Lab, Masaryk University Brno: DiVinE project home page. (2004)

Cost-Constrained Minimum-Delay Multicasting (Extended Abstract)

Satoshi Tayu, Turki Ghazi Al-Mutairi, and Shuichi Ueno

Department of Communications and Integrated Systems,
Tokyo Institute of Technology, Tokyo 152–8552–S3–57, Japan
{tayu, ueno}@lab.ss.titech.ac.jp

Abstract. We consider a problem of cost-constrained minimum-delay multicasting in a network, which is to find a Steiner tree spanning the source and destination nodes such that the maximum total delay along a path from the source node to a destination node is minimized, while the sum of link costs in the tree is bounded by a constant. The problem is $\mathcal{NP}$-hard even if the network is series-parallel. We present a fully polynomial time approximation scheme for the problem if the network is series-parallel.

1 Introduction

The multicasting is the simultaneous transmission of data from a source node to multiple destination nodes in a network. The multicasting involves the generation of a multicast tree, which is a Steiner tree spanning the source and destination nodes. The performance of multicasting is determined by both the cost of the multicast tree and the maximum delay between the source node and a destination node in the tree. Therefore, constructing efficient multicasting is formulated as a bicriteria Steiner tree problem.

In connection with the problem, the following problem has been considered in the literature [1, 2, 3, 4, 5]. The delay-constrained minimum cost multicast tree problem (DCMCMT) is to construct a multicast tree such that the cost of the tree is minimized while the delay between the source node and a destination node in the tree is bounded by a constant integer. DCMCMT is $\mathcal{NP}$-hard since it reduces to the Steiner tree problem, which is well-known to be $\mathcal{NP}$-hard. Chen and Xue proposed a fully polynomial time approximation scheme (FPTAS) for DCMCMT if the number of destination nodes is bounded by a constant [1], while many heuristic algorithms have been proposed in [2, 3, 4, 5]. We present a pseudo-polynomial time algorithm for DCMCMT if the network is series-parallel.

We also consider the following problem, which is another variant of the problem of constructing efficient multicasting. The cost-constrained minimum delay multicast tree problem (CCMDMT) is to construct a multicast tree such that the maximum delay between the source node and a destination node in the tree is minimized while the cost of the tree is bounded by a constant integer. CCMDMT is $\mathcal{NP}$-hard since it reduces to the cost-constrained shortest

M. Bieliková et al. (Eds.): SOFSEM 2005, LNCS 3381, pp. 330–339, 2005.

path problem (CCSP) which is known to be $\mathcal{NP}$-hard [6]. In fact, CCMDMT is $\mathcal{NP}$-hard even for series-parallel networks, since CCSP is $\mathcal{NP}$-hard for series-parallel networks as mentioned by Chen and Xue [7]. We present in this paper a pseudo-polynomial time algorithm and an FPTAS for CCMDMT if the network is series-parallel. This paper is the first to consider CCMDMT, as far as the authors know.

Due to space limitations most proofs are omitted in the extended abstract and will appear in the final version of the paper.

2   Problems

We consider a connected graph G with vertex set $V(G)$ and edge set $E(G)$. Each edge e is assigned a cost $\gamma(e)$ and a delay $\delta(e)$ which are assumed to be non-negative integers. The cost of a subgraph H of G, denoted by $\gamma(H)$, is defined as $\gamma(H) = \sum_{e \in E(H)} \gamma(e)$. The delay of a path P in G, denoted by $\delta(P)$, is defined as $\delta(P) = \sum_{e \in E(P)} \delta(e)$. A vertex s is designated as the source and a set D of vertices is designated as the destinations. A tree T is called a multicast tree if $\{s\} \cup D \subseteq V(T)$. The delay of a multicast tree T, denoted by $\delta(T)$, is defined as $\delta(T) = \max\{\delta(P(s,d)) | d \in D, P(s,d): (s,d)\text{-path in } T\}$. Let Γ and Δ be positive integers. The delay-constrained minimum cost multicast tree problem (DCMCMT) is to construct a multicast tree T such that $\delta(T) \leq \Delta$ and $\gamma(T)$ is minimized, while the cost-constrained minimum delay multicast tree problem (CCMDMT) is to construct a multicast tree T such that $\gamma(T) \leq \Gamma$ and $\delta(T)$ is minimized.

3   Pseudo-Polynomial Time Algorithms

A graph is said to be *series-parallel* if it contains no subdivision of K_4 as a subgraph. A maximal series-parallel graph is called a 2-*tree*. The 2-trees can be defined recursively as follows: (1) K_2 is a 2-tree on two vertices; (2) Given a 2-tree on n vertices ($n \geq 2$), a graph obtained from G by adding a new vertex adjacent to the ends of an edge of G is a 2-tree on $n+1$ vertices. A 2-tree on $n \geq 2$ vertices has $2n - 3$ edges by definition.

In this section, we will show an $\mathcal{O}(n\Delta^3)$ time algorithm and an $\mathcal{O}(n^4 \delta_{\max}^3)$ time algorithm to solve DCMCMT and CCMDMT, respectively, for a series-parallel graph G with n vertices, where $\delta_{\max} = \max\{\delta(e) | e \in E(G)\}$. We use methods similar to those used in [7]. We first augment a connected series-parallel graph with n vertices to a 2-tree on n vertices using a linear time algorithm presented in [8]. Each added edge has infinite cost and delay so that the added edges are never chosen in an optimal multicast tree. We next find an optimal multicast tree in the 2-tree. The algorithms are based on the dynamic programming.

3.1   Preliminaries

Let G be a 2-tree and $C_3(G)$ be the set of triangles of G. A tree T_G is defined as follows: $V(T_G) = E(G) \cup C_3(G)$; for any $e \in E(G)$ and $\nabla \in C_3(G)$, $(e, \nabla) \in$

$E(T_G)$ if and only if $e \in E(\nabla)$. T_G thus defined is indeed a tree since G is a 2-tree. T_G is considered as a rooted tree with root r, where r is an edge incident to s in G.

Let $\mathcal{C}(p)$ be the set of all children of $p \in E(G)$ in T_G. Notice that a child of p is a triangle in G. Let $\mathcal{D}(\nabla)$ be the set of triangles which are descendants of $\nabla \in C_3(G)$ in T_G. For $\mathcal{C}'(p) \subseteq \mathcal{C}(p)$, $G[p, \mathcal{C}'(p)]$ is a subgraph of G induced by the edges of triangles in $\bigcup_{\nabla \in \mathcal{C}'(p)} \mathcal{D}(\nabla)$ together with edge p.

Let $\prec$ be a partial order on $V(G)$ satisfying the following conditions:

- $s \prec v$ for all $v \in V(G)$;
- If ∇ is a triangle with $V(\nabla) = \{x, y, z\}$, and edge (x, z) is the parent of ∇ with $x \prec z$, then $x \prec y$ and $y \prec z$.

Such an order can be constructed recursively from the root of T_G as follows: First, we define $s \prec v$ for edge $r = (s, v)$. For every edge $p = (x, z)$ with $x \prec z$, if p has a child triangle C, we define $x \prec y$ and $y \prec z$ for vertex $y \in V(C) \setminus \{x, z\}$. We continue this process until $\prec$ is defined on every pair of endvertices of an edge. Then the transitive reflexive closure of $\prec$ is the desired partial order.

For any edge $p = (x, y)$ with $x \prec y$ and $\mathcal{C}'(p) \subseteq \mathcal{C}(p)$, $H_{\bullet\bullet}^{[p,\mathcal{C}'(p)]}$, $H_{\bullet\circ}^{[p,\mathcal{C}'(p)]}$, $H_{\circ\bullet}^{[p,\mathcal{C}'(p)]}$, and $H_{\bullet\bullet}^{[p,\mathcal{C}'(p)]}$ are subgraphs of $G[p, \mathcal{C}'(p)]$ such that each subgraph contains the vertices (destinations) in $D \cap V(G[p, \mathcal{C}'(p)])$ and;

- $H_{\bullet\bullet}^{[p,\mathcal{C}'(p)]}$ is a tree including both x and y,
- $H_{\bullet\circ}^{[p,\mathcal{C}'(p)]}$ is a tree with $x \in V(H_{\bullet\circ}^{[p,\mathcal{C}'(p)]})$ and $y \notin V(H_{\bullet\circ}^{[p,\mathcal{C}'(p)]})$,
- $H_{\circ\bullet}^{[p,\mathcal{C}'(p)]}$ is a tree with $x \notin V(H_{\circ\bullet}^{[p,\mathcal{C}'(p)]})$ and $y \in V(H_{\circ\bullet}^{[p,\mathcal{C}'(p)]})$,
- $H_{\bullet\bullet}^{[p,\mathcal{C}'(p)]}$ consists of vertex-disjoint two trees $T_x^{[p,\mathcal{C}'(p)]}$ and $T_y^{[p,\mathcal{C}'(p)]}$ such that $x \in V(T_x^{[p,\mathcal{C}'(p)]})$ and $y \in V(T_y^{[p,\mathcal{C}'(p)]})$.

Finally, let $\mathbb{S}_\Delta = \{-\infty, 0, 1, \ldots, \Delta\}$.

3.2   Functions

Let $p = (x, y)$ be an edge with $x \prec y$ and $\mathcal{C}'(p) \subseteq \mathcal{C}(p)$.

$W_{\bullet\bullet}(p, \mathcal{C}'(p); \xi_x, \xi_{xy})$ is the minimum cost of a tree $H_{\bullet\bullet}^{[p,\mathcal{C}'(p)]}$ in $G[p, \mathcal{C}'(p)]$ such that $\max\{\delta(x, d) | d \in D \cap V(G[p, \mathcal{C}'(p)])\} \le \xi_x$ and $\delta(x, y) \le \xi_{xy}$, where $\delta(u, v)$ is the delay of (u, v)-path in tree $H_{\bullet\bullet}^{[p,\mathcal{C}'(p)]}$, and $\xi_x, \xi_{xy} \in \mathbb{S}_\Delta$.

$\overline{W}_{\bullet\bullet}(p, \mathcal{C}'(p); \xi_y, \xi_{xy})$ is the minimum cost of a tree $H_{\bullet\bullet}^{[p,\mathcal{C}'(p)]}$ in $G[p, \mathcal{C}'(p)]$ such that $\max\{\delta(y, d) | d \in D \cap V(G[p, \mathcal{C}'(p)])\} \le \xi_y$ and $\delta(x, y) \le \xi_{xy}$, where $\delta(u, v)$ is the delay of (u, v)-path in tree $H_{\bullet\bullet}^{[p,\mathcal{C}'(p)]}$, and $\xi_y, \xi_{xy} \in \mathbb{S}_\Delta$.

$W_{\bullet\circ}(p, \mathcal{C}'(p); \xi_x)$ is the minimum cost of a tree $H_{\bullet\circ}^{[p,\mathcal{C}'(p)]}$ in $G[p, \mathcal{C}'(p)]$ such that $\max\{\delta(x, d) | d \in D \cap V(G[p, \mathcal{C}'(p)])\} \le \xi_x$, where $\delta(x, d)$ is the delay of (x, d)-path in tree $H_{\bullet\circ}^{[p,\mathcal{C}'(p)]}$, and $\xi_x \in \mathbb{S}_\Delta$.

$W_{\circ\bullet}(p, \mathcal{C}'(p); \xi_y)$ is the minimum cost of a tree $H_{\circ\bullet}^{[p,\mathcal{C}'(p)]}$ in $G[p, \mathcal{C}'(p)]$ such that $\max\{\delta(y, d) | d \in D \cap V(G[p, \mathcal{C}'(p)])\} \le \xi_y$, where $\delta(x, d)$ is the delay of (x, d)-path in tree $H_{\circ\bullet}^{[p,\mathcal{C}'(p)]}$, and $\xi_y \in \mathbb{S}_\Delta$.

$W_{\bullet\bullet}(p, \mathcal{C}'(p); \xi_x, \xi_y)$ is the minimum cost of a forest $H_{\bullet\bullet}^{[p,\mathcal{C}'(p)]}$ in $G[p, \mathcal{C}'(p)]$ such that $\max\{\delta(x,d)|d \in D \cap V(T_x^{[p,\mathcal{C}'(p)]})\} \leq \xi_x$, and $\max\{\delta(y,d)|d \in D \cap V(T_y^{[p,\mathcal{C}'(p)]})\} \leq \xi_y$, where $\delta(x,d)$ is the delay of (x,d)-path in tree $T_x^{[p,\mathcal{C}'(p)]}$ and $\delta(y,d)$ is the delay of (y,d)-path in tree $T_y^{[p,\mathcal{C}'(p)]}$, and $\xi_x, \xi_y \in \mathbb{S}_\Delta$.

$N_{\circ\circ}(p)$ is defined to be 0 if $G[p, \mathcal{C}(p)]$ has no destination and ∞ otherwise.

For an edge $p = (x, y) \in E(G)$ with $x \prec y$ and $\mathcal{C}'(p) \subseteq \mathcal{C}(p)$, the table $\mathcal{W}(p, \mathcal{C}'(p))$ for p and $\mathcal{C}'(p)$ is the list of values of $W_{\bullet\bullet}(p, \mathcal{C}'(p); \xi_x, \xi_{xy})$, $\overline{W}_{\bullet\bullet}(p, \mathcal{C}'(p); \xi_y, \xi_{xy})$, $W_{\bullet\circ}(p, \mathcal{C}'(p); \xi_x)$, $W_{\circ\bullet}(p, \mathcal{C}'(p); \xi_y)$, and $W_{\bullet\bullet}(p, \mathcal{C}'(p); \xi_x, \xi_y)$ for every $\xi_x, \xi_y, \xi_{xy} \in \mathbb{S}_\Delta$.

The following is immediate from the definition of functions above.

Lemma 1. *For any* $\xi \in \{0, 1, \ldots, \Delta\}$, $\min\{W_{\bullet\circ}(r, \mathcal{C}(r); \xi), W_{\bullet\bullet}(r, \mathcal{C}(r); \xi, \Delta)\}$ *is the minimum cost of a multicast tree* T *of* G *with* $\delta(T) \leq \xi$, *where* $r = (s, y) \in E(G)$ *is the root of* T_G. $\qquad\square$

3.3   Basic Algorithm BA$(G, s, D, \gamma, \delta, \Delta)$

We describe in this subsection a basic algorithm BA$(G, s, D, \gamma, \delta, \Delta)$ which computes $\mathcal{W}(r, \mathcal{C}(r))$ for a 2-tree G with n vertices in $\mathcal{O}(n\Delta^3)$ time.

BA$(G, s, D, \gamma, \delta, \Delta)$ first computes T_G and chooses an edge incident with s in G as the root of T_G.

Then, BA$(G, s, D, \gamma, \delta, \Delta)$ recursively computes the functions defined in Section 3.2. We distinguish three cases.

Case 1 : $\mathcal{C}'(p) = \emptyset$.

For every $p = (x, y) \in E(G)$ with $x \prec y$, and $\xi_x, \xi_y, \xi_{xy} \in \mathbb{S}_\Delta$,

$$W_{\bullet\bullet}(p, \emptyset; \xi_x, \xi_{xy}) = \begin{cases} \gamma(p) \text{ if the following conditions are satisfied:} \\ \quad \text{(i) if } y \in D \text{ then } \xi_x \geq \delta(p); \quad \text{(ii) if } x \in D \text{ then } \xi_x \geq 0; \\ \quad \text{(iii) } \xi_{xy} \geq \delta(p), \\ \infty \text{ otherwise.} \end{cases}$$

$$\overline{W}_{\bullet\bullet}(p, \emptyset; \xi_y, \xi_{xy}) = \begin{cases} \gamma(p) \text{ if the following conditions are satisfied:} \\ \quad \text{(i) if } y \in D \text{ then } \xi_y \geq 0; \quad \text{(ii) if } x \in D \text{ then } \xi_y \geq \delta(p); \\ \quad \text{(iii) } \xi_{xy} \geq \delta(p), \\ \infty \text{ otherwise.} \end{cases}$$

$$W_{\bullet\circ}(p, \emptyset; \xi_x) = \begin{cases} 0 \text{ if } y \notin D, \text{ and if } x \in D \text{ then } \xi_x \geq 0, \\ \infty \text{ otherwise.} \end{cases}$$

$$W_{\circ\bullet}(p, \emptyset; \xi_y) = \begin{cases} 0 \text{ if } x \notin D, \text{ and if } y \in D \text{ then } \xi_y \geq 0, \\ \infty \text{ otherwise.} \end{cases}$$

$$W_{\bullet\bullet}(p, \emptyset; \xi_x, \xi_y) = \begin{cases} 0 \text{ if the following conditions are satisfied:} \\ \quad \text{(i) if } x \in D \text{ then } \xi_x \geq 0, \quad \text{(ii) if } y \in D \text{ then } \xi_y \geq 0, \\ \infty \text{ otherwise.} \end{cases}$$

Case 2 : $\mathcal{C}'(p) = \{\nabla\}$ for some $\nabla \in \mathcal{C}(p)$.

For every $p = (x, z) \in E(G)$ with $\mathcal{C}(p) \neq \emptyset$ and $x \prec z$, for every $\nabla \in \mathcal{C}(p)$ with

$V(\nabla) = \{x, y, z\}$, $E(\nabla) = \{p = (x, z), q = (x, y), t = (y, z)\}$, and $x \prec y \prec z$, and for every $\xi_x, \xi_y, \xi_{xy} \in \mathbb{S}_\Delta$, the functions are computed as follows:

$$W_{\bullet\bullet}(p, \{\nabla\}; \xi_x, \xi_{xz}) =$$

$$\min \left\{ \begin{array}{l} \min \left\{\gamma(p) + W_{\bullet\circ}(q, \mathcal{C}(q); \xi_x') + W_{\circ\bullet}(t, \mathcal{C}(t); \xi_z'') \mid \xi_{xz} \geq \delta(p),\right. \\ \left. \quad \xi_x \geq \max\{\xi_x', \delta(p) + \xi_z''\}, (\xi_x', \xi_z'') \in \mathbb{S}_\Delta^2\right\}, \\[4pt] \min \left\{\gamma(p) + W_{\bullet\bullet}(q, \mathcal{C}(q); \xi_x', \xi_{xy}') + W_{\bullet\bullet}(t, \mathcal{C}(t); \xi_y'', \xi_z'') \mid \xi_{xz} \geq \delta(p),\right. \\ \left. \quad \xi_x \geq \max\{\xi_x', \xi_{xy}' + \xi_y'', \delta(p) + \xi_z''\}, (\xi_x', \xi_{xy}', \xi_y'', \xi_z'') \in \mathbb{S}_\Delta^4\right\}, \\[4pt] \min \left\{\gamma(p) + W_{\bullet\bullet}(q, \mathcal{C}(q); \xi_x', \xi_y') + \overline{W}_{\bullet\bullet}(t, \mathcal{C}(t); \xi_z'', \xi_{yz}'') \mid \xi_{xz} \geq \delta(p),\right. \\ \left. \quad \xi_x \geq \max\{\xi_x', \delta(p) + \xi_z'', \delta(p) + \xi_{yz}'' + \xi_y'\}, (\xi_x', \xi_y', \xi_z'', \xi_{yz}'') \in \mathbb{S}_\Delta^4\right\}, \\[4pt] \min \left\{W_{\bullet\bullet}(q, \mathcal{C}(q); \xi_x', \xi_{xy}') + W_{\bullet\bullet}(t, \mathcal{C}(t); \xi_y'' \xi_{yz}'') \mid \xi_{xz} \geq \xi_{xy}' + \xi_{yz}'',\right. \\ \left. \quad \xi_x \geq \max\{\xi_x', \xi_{xy}' + \xi_y''\}, (\xi_x', \xi_{xy}', \xi_y'', \xi_{yz}'') \in \mathbb{S}_\Delta^4\right\} \end{array} \right\}, \quad (1)$$

$$\overline{W}_{\bullet\bullet}(p, \{\nabla\}; \xi_z, \xi_{xz}) =$$

$$\min \left\{ \begin{array}{l} \min \left\{\gamma(p) + W_{\bullet\circ}(q, \mathcal{C}(q); \xi_x') + W_{\circ\bullet}(t, \mathcal{C}(t); \xi_z'') \mid \xi_{xz} \geq \delta(p),\right. \\ \left. \quad \xi_x \geq \max\{\xi_x', \delta(p) + \xi_z''\}, (\xi_x', \xi_z'') \in \mathbb{S}_\Delta^2\right\}, \\[4pt] \min \left\{\gamma(p) + \overline{W}_{\bullet\bullet}(q, \mathcal{C}(q); \xi_y', \xi_{xy}') + W_{\bullet\bullet}(t, \mathcal{C}(t); \xi_y'', \xi_z'') \mid \xi_{xz} \geq \delta(p),\right. \\ \left. \quad \xi_x \geq \max\{\xi_x', \xi_{xy}' + \xi_y'', \delta(p) + \xi_z''\}, (\xi_y', \xi_{xy}', \xi_y'', \xi_z'') \in \mathbb{S}_\Delta^4\right\}, \\[4pt] \min \left\{\gamma(p) + W_{\bullet\bullet}(q, \mathcal{C}(q); \xi_x', \xi_y') + \overline{W}_{\bullet\bullet}(t, \mathcal{C}(t); \xi_z'', \xi_{yz}'') \mid \xi_{xz} \geq \delta(p),\right. \\ \left. \quad \xi_x \geq \max\{\xi_x', \delta(p) + \xi_z'', \delta(p) + \xi_{yz}'' + \xi_y'\}, (\xi_y', \xi_y'', \xi_z'', \xi_{yz}'') \in \mathbb{S}_\Delta^4\right\}, \\[4pt] \min \left\{\overline{W}_{\bullet\bullet}(q, \mathcal{C}(q); \xi_y', \xi_{xy}') + \overline{W}_{\bullet\bullet}(t, \mathcal{C}(t); \xi_z'', \xi_{yz}'') \mid \xi_{xz} \geq \xi_{xy}' + \xi_{yz}'',\right. \\ \left. \quad \xi_z \geq \max\{\xi_y' + \xi_{yz}'', \xi_z''\}, (\xi_y', \xi_{xy}', \xi_z'', \xi_{yz}'') \in \mathbb{S}_\Delta^4\right\} \end{array} \right\}, \quad (2)$$

$$W_{\bullet\circ}(p, \{\nabla\}; \xi_x) = \min \left\{ \begin{array}{l} \min \left\{W_{\bullet\circ}(q, \mathcal{C}(q); \xi_x') + N_{\circ\circ}(t) \mid \xi_x \geq \xi_x',\right. \\ \left. \quad \xi_x' \in \mathbb{S}_\Delta\right\}, \\ \min \left\{W_{\bullet\bullet}(q, \mathcal{C}(q); \xi_x', \xi_{xy}') + W_{\bullet\circ}(t, \mathcal{C}(t); \xi_y'') \mid\right. \\ \left. \quad \xi_x \geq \max\{\xi_x', \xi_{xy}' + \xi_y''\}, (\xi_x', \xi_{xy}', \xi_y'') \in \mathbb{S}_\Delta^3\right\} \end{array} \right\}, \quad (3)$$

$$W_{\circ\bullet}(p, \{\nabla\}; \xi_z) = \min \left\{ \begin{array}{l} \min \left\{N_{\circ\circ}(q) + W_{\circ\bullet}(t, \mathcal{C}(t); \xi_z'') \mid \xi_z \geq \xi_z'',\right. \\ \left. \quad \xi_z'' \in \mathbb{S}_\Delta\right\}, \\ \min \left\{W_{\circ\bullet}(q, \mathcal{C}(q); \xi_y') + \overline{W}_{\bullet\bullet}(t, \mathcal{C}(t); \xi_z'', \xi_{yz}'') \mid\right. \\ \left. \quad \xi_z \geq \max\{\xi_z'', \xi_{yz}'' + \xi_y'\}, (\xi_y', \xi_z'', \xi_{yz}'') \in \mathbb{S}_\Delta^3\right\} \end{array} \right\}, \quad (4)$$

$$W_{\bullet\bullet}(p, \{\nabla\}; \xi_x, \xi_z) =$$

$$\min \left\{ \begin{array}{l} \min \left\{W_{\bullet\circ}(q, \mathcal{C}(q); \xi_x') + W_{\circ\bullet}(t, \mathcal{C}(t); \xi_z'') \mid \xi_x \geq \xi_x', \xi_z \geq \xi_z'',\right. \\ \left. \quad (\xi_x', \xi_z'') \in \mathbb{S}_\Delta^2\right\}, \\[4pt] \min \left\{W_{\bullet\bullet}(q, \mathcal{C}(q); \xi_x', \xi_{xy}') + W_{\bullet\bullet}(t, \mathcal{C}(t); \xi_y'', \xi_z'') \mid\right. \\ \left. \quad \xi_x \geq \max\{\xi_x', \xi_{xy}' + \xi_y''\}, \xi_z \geq \xi_z'', (\xi_x', \xi_{xy}', \xi_y'', \xi_z'') \in \mathbb{S}_\Delta^4\right\}, \\[4pt] \min \left\{W_{\bullet\bullet}(q, \mathcal{C}(q); \xi_x', \xi_y') + \overline{W}_{\bullet\bullet}(t, \mathcal{C}(t); \xi_z'', \xi_{yz}'') \mid \xi_x \geq \xi_x',\right. \\ \left. \quad \xi_z \geq \max\{\xi_z'', \xi_{yz}'' + \xi_y'\}, (\xi_x', \xi_{xy}', \xi_z'', \xi_{yz}'') \in \mathbb{S}_\Delta^4\right\} \end{array} \right\}. \quad (5)$$

Case 3 : $\mathcal{C}'(p) = \mathcal{C}''(p) \cup \{\nabla\}$ for some $\mathcal{C}''(p) \subseteq \mathcal{C}(p)$ and $\nabla \in \mathcal{C}(p) - \mathcal{C}''(p)$.

For every $p = (x, y) \in E(G)$, $\mathcal{C}'(p) \subseteq \mathcal{C}(p)$, $\nabla \in \mathcal{C}'(p)$, and $\xi_x, \xi_y, \xi_{xy} \in \mathbb{S}_\Delta$, the functions are computed as follows:

$$
W_{\bullet\bullet}(p, \mathcal{C}'(p); \xi_x, \xi_{xy}) = \\
\min \left\{
\begin{array}{l}
\min \left\{ W_{\bullet\bullet}(p, \mathcal{C}''(p); \xi_x', \xi_{xy}') + W_{\bullet\bullet}(p, \{\nabla\}; \xi_x'', \xi_y'') \mid \xi_{xy} \geq \xi_{xy}', \right. \\
\quad \left. \xi_x \geq \max\{\xi_x', \xi_x'', \xi_y'' + \xi_{xy}''\}, (\xi_x', \xi_{xy}', \xi_x'', \xi_y'') \in \mathbb{S}_\Delta^4 \right\}, \\
\min \left\{ W_{\bullet\bullet}(p, \mathcal{C}''(p); \xi_x', \xi_y') + W_{\bullet\bullet}(p, \{\nabla\}; \xi_x'', \xi_{xy}'') \mid \xi_{xy} \geq \xi_{xy}'', \right. \\
\quad \left. \xi_x \geq \max\{\xi_x', \xi_x'', \xi_y' + \xi_{xy}''\}, (\xi_x', \xi_y', \xi_x'', \xi_{xy}'') \in \mathbb{S}_\Delta^4 \right\}
\end{array}
\right\}, \quad (6)
$$

$$
\overline{W}_{\bullet\bullet}(p, \mathcal{C}'(p); \xi_y, \xi_{xy}) = \\
\min \left\{
\begin{array}{l}
\min \left\{ \overline{W}_{\bullet\bullet}(p, \mathcal{C}''(p); \xi_y', \xi_{xy}') + W_{\bullet\bullet}(p, \{\nabla\}; \xi_x'', \xi_y'') \mid \xi_{xy} \geq \xi_{xy}', \right. \\
\quad \left. \xi_y \geq \max\{\xi_y', \xi_y'', \xi_x'' + \xi_{xy}'\}, (\xi_y', \xi_{xy}', \xi_x'', \xi_y'') \in \mathbb{S}_\Delta^4 \right\}, \\
\min \left\{ W_{\bullet\bullet}(p, \mathcal{C}''(p); \xi_x', \xi_y') + \overline{W}_{\bullet\bullet}(p, \{\nabla\}; \xi_y'', \xi_{xy}'') \mid \xi_{xy} \geq \xi_{xy}'', \right. \\
\quad \left. \xi_y \geq \max\{\xi_y', \xi_y'', \xi_x' + \xi_{xy}''\}, (\xi_y', \xi_y', \xi_x'', \xi_{xy}'') \in \mathbb{S}_\Delta^4 \right\}
\end{array}
\right\}, \quad (7)
$$

$$
W_{\bullet\circ}(p, \mathcal{C}'(p); \xi_x) = W_{\bullet\circ}(p, \mathcal{C}''(p); \xi_x) + W_{\bullet\circ}(p, \{\nabla\}; \xi_x), \quad (8)
$$

$$
W_{\circ\bullet}(p, \mathcal{C}'(p); \xi_y) = W_{\circ\bullet}(p, \mathcal{C}''(p); \xi_y) + W_{\circ\bullet}(p, \{\nabla\}; \xi_y), \quad (9)
$$

$$
W_{\bullet\bullet}(p, \mathcal{C}'(p); \xi_x, \xi_y) = W_{\bullet\bullet}(p, \mathcal{C}''(p); \xi_x, \xi_y) + W_{\bullet\bullet}(p, \{\nabla\}; \xi_x, \xi_y). \quad (10)
$$

The computation of the tables for functions proceeds as follows. We first compute $\mathcal{W}(p, \mathcal{C}(p)) = \mathcal{W}(p, \emptyset)$ for every leaf p of T_G as in Case 1 above.

For every triangle ∇ with parent p and children q and t, $\mathcal{W}(p, \{\nabla\})$ is computed using tables $\mathcal{W}(q, \mathcal{C}(q))$ and $\mathcal{W}(t, \mathcal{C}(t))$ as in Case 2.

For every $p \in E(G)$ with $\mathcal{C}(p) = \{\nabla_1, \nabla_2, \ldots, \nabla_{|\mathcal{C}(p)|}\}$, $\mathcal{W}(p, \mathcal{C}(p))$ is computed as follows. Let $\mathcal{C}^{(i)}(p) = \{\nabla_1, \nabla_2, \ldots, \nabla_i\}$ for $1 \leq i \leq |\mathcal{C}(p)|$. $\mathcal{W}(p, \mathcal{C}^{(i)}(p))$ is computed using $\mathcal{W}(p, \mathcal{C}^{(i-1)}(p))$ and $\mathcal{W}(p, \{\nabla_i\})$ as in Case 3 for $2 \leq i \leq |\mathcal{C}(p)|$.

Finally, BA$(G, s, D, \gamma, \delta, \Delta)$ outputs $\mathcal{W}(r, \mathcal{C}(r))$.

3.4   Analysis of BA$(G, s, D, \gamma, \delta, \Delta)$

We use the following lemmas to prove Theorem 1 below. The proofs of the lemmas are omitted in the extended abstract.

Lemma 2. BA$(G, s, D, \gamma, \delta, \Delta)$ *computes* $\mathcal{W}(r, \mathcal{C}(r))$, *correctly.*   □

Lemma 3. $\mathcal{W}(p, \emptyset)$ *is computed in* $\mathcal{O}(\Delta^2)$ *time for any leaf* p *of* T_G. □

Lemma 4. *Let* ∇ *be a triangle with parent* p *and children* q *and* t. *Given* $\mathcal{W}(q, \mathcal{C}(q))$ *and* $\mathcal{W}(t, \mathcal{C}(t))$, $\mathcal{W}(p, \{\nabla\})$ *is computed in* $\mathcal{O}(\Delta^3)$ *time.*   □

Lemma 5. *Let* $p \in E(G)$, $\mathcal{C}(p) = \{\nabla_1, \nabla_2, \ldots, \nabla_{|\mathcal{C}(p)|}\}$, *and* $\mathcal{C}^{(i)}(p) = \{\nabla_1, \nabla_2, \ldots, \nabla_i\}$ *for* $1 \leq i \leq |\mathcal{C}(p)|$. *Given* $\mathcal{W}(p, \mathcal{C}^{(i-1)}(p))$ *and* $\mathcal{W}(p, \{\nabla_i\})$, $\mathcal{W}(p, \mathcal{C}^{(i)}(p))$ *is computed in* $\mathcal{O}(\Delta^3)$ *time for* $2 \leq i \leq |\mathcal{C}(p)|$.   □

Theorem 1. *For a 2-tree G on n vertices, $\mathrm{BA}(G, s, D, \gamma, \delta, \Delta)$ computes $\mathcal{W}(r, \mathcal{C}(r))$ in $\mathcal{O}(n\Delta^3)$ time.*

Proof. The tables $\mathcal{W}(p, \emptyset)$ for all leaves p can be computed in $\mathcal{O}(n\Delta^2)$ time by Lemma 3. Since the number of triangles is $\mathcal{O}(n)$, the tables $\mathcal{W}(p, \{\nabla\})$ for all triangles ∇ can be computed in $\mathcal{O}(n\Delta^3)$ time by Lemma 4. By Lemma 5, $\mathcal{W}(p, \mathcal{C}(p))$ can be computed in $\mathcal{O}(|\mathcal{C}(p)|\Delta^3)$ time. Since $\sum_{p \in E(G)} |\mathcal{C}(p)| = \mathcal{O}(n)$, the tables $\mathcal{W}(p, \mathcal{C}(p))$ for all edges p can be computed in $\mathcal{O}(n\Delta^3)$ time. It follows that $\mathrm{BA}(G, s, D, \gamma, \delta, \Delta)$ computes $\mathcal{W}(r, \mathcal{C}(r))$ in $\mathcal{O}(n\Delta^3)$ time by Lemma 2.   □

By Lemma 1 and Theorem 1, $\mathrm{BA}(G, s, D, \gamma, \delta, \Delta)$ computes the minimum cost of a multicast tree with delay at most ξ for any $\xi \in \{0, 1, \ldots, \Delta\}$. If we perform some bookkeeping operations such as recording how the minimum was achieved during the computation of the tables for functions, we can construct a delay-constrained minimum cost multicast tree in the same time complexity. Thus, we have the following.

Corollary 1. *Given a 2-tree G on n vertices, s, D, γ, δ, Δ, and an integer ξ, $0 \le \xi \le \Delta$, a minimum cost multicast tree T with $\delta(T) \le \xi$ can be constructed in $\mathcal{O}(n\Delta^3)$ time.*   □

We denote by $\mathrm{MT}(G, s, D, \gamma, \delta, \Delta, \xi)$ such an $\mathcal{O}(n\Delta^3)$ time algorithm constructing a minimum cost multicast tree T with $\delta(T) \le \xi$ for a given 2-tree G, s, D, γ, δ, Δ, and an integer ξ, $0 \le \xi \le \Delta$.

3.5   Pseudo-Polynomial Time Algorithm for DCMCMT

Given a connected series-parallel graph G' with cost and delay functions γ' and δ', we denote by $\mathrm{EXT}(G', \delta', \gamma')$ a linear time procedure for augmenting G' to a 2-tree G with $V(G) = V(G')$ [8], and extending γ' and δ' to γ and δ, respectively, by defining $\gamma(e) = \infty$ and $\delta(e) = \infty$ for each $e \in E(G) - E(G')$, and $\gamma(e) = \gamma'(e)$ and $\delta(e) = \delta'(e)$ for each $e \in E(G')$. Then, it is easy to see that Algorithm 1 shown in Fig. 1. solves DCMCMT for series-parallel graphs, and we have the following by Theorem 1.

Theorem 2. *For a series-parallel graph G with n vertices and a positive integer Δ, Algorithm 1 solves DCMCMT in $\mathcal{O}(n\Delta^3)$ time.*   □

Input a series-parallel graph G', $s \in V(G')$, $D \subseteq V(G')$,
     $\gamma' : E(G') \to \mathbb{N}$, $\delta' : E(G') \to \mathbb{N}$,
     $\Delta \in \mathbb{Z}^+$.
Output a minimum cost multicast tree T with delay at most Δ.

begin
     $\mathrm{EXT}(G', \gamma', \delta')$;
     $\mathrm{BA}(G, s, D, \gamma, \delta, \Delta)$;
     $\mathrm{MT}(G, s, D, \gamma, \delta, \Delta, \Delta)$;
     if $\gamma(T) < \infty$
          return T;
     else
          return "NO";
     endif
end

Fig. 1. Algorithm 1

3.6   Pseudo-Polynomial Time Algorithm for CCMDMT

Given a cost bound Γ and the table $\mathcal{W}(r, \mathcal{C}(r))$ for functions, we denote by $\mathrm{MIN\_DELAY}(\Gamma, \mathcal{W}(r, \mathcal{C}(r)))$ a linear time procedure for computing the minimum ξ satisfying $\min\{W_{\bullet\bullet}(r, \mathcal{C}(r); \xi, \Delta), W_{\bullet\circ}(r, \mathcal{C}(r); \xi)\} \leq \Gamma$ if exists. It returns ∞ if there exists no such ξ.

Since the number of edges of multicast tree is at most $n - 1$, the maximum delay of a multicast tree is at most $(n - 1)\delta_{\max}$, where $\delta_{\max} = \max_{e \in E(G')} \delta'(e)$. Thus, it is easy to see that Algorithm 2 shown in Fig. 2 is a pseudo-polynomial time algorithm for CCMDMT, and we have the following by Theorem 1.

Theorem 3. *For a series-parallel graph G with n vertices and a non-negative integer Γ, Algorithm 2 solves CCMDMT in $\mathcal{O}(n^4 \delta_{\max}{}^3)$ time if $\delta_{\max} \geq 1$.*   □

Input a series-parallel graph G',
  $s \in V(G')$, $D \subseteq V(G')$,
  $\gamma' : E(G') \to \mathbb{N}$, $\delta' : E(G') \to \mathbb{N}$,
  $\Gamma \in \mathbb{Z}^+$.

Output a minimum delay multicast tree T with cost at most Γ.

begin
  $\delta_{\max} := \max_{e \in E(G')} \delta(e)$;
  $\Delta' := (n - 1)\delta_{\max}$;
  $\mathrm{EXT}(G', \gamma', \delta')$;
  $\mathrm{BA}(G, s, D, \gamma, \delta, \Delta')$;
  $\mathrm{MIN\_DELAY}(\Gamma, \mathcal{W}(r, \mathcal{C}(r)))$;
  if $\xi < \infty$
    $\mathrm{MT}(G, s, D, \gamma, \delta, \Delta, \xi)$;
    return T;
  else
    return "NO";
  endif
end

Fig. 2. Algorithm 2

4   FPTAS for CCMDMT

We use standard techniques [7, 6, 9, 10] to turn $\mathrm{BA}(G, s, D, \gamma, \delta, \Delta)$ into an FP-TAS for CCMDMT. We show in Section 4.1 a pair of upper and lower bounds U and L for the minimum delay of a cost constrained multicast tree such that $U/L \leq n - 1$. For any $\varepsilon > 0$, we show in Section 4.2 a $(1 + \varepsilon)$-approximation algorithm for CCMDMT. The algorithm runs in $\mathcal{O}(n^7/\varepsilon^3)$ time, provided that we have a pair of upper and lower bounds U and L for the delay of a cost constrained multicast tree such that $U/L = \mathcal{O}(n)$. It follows that we have an FPTAS for CCMDMT.

4.1   Upper and Lower Bounds for Minimum Delay

We use a technique similar to [7]. Let $\nu_1 < \nu_2 < \cdots < \nu_k$ be different edge delays, and γ_j be the cost function defined as $\gamma_j(e) = \gamma(e)$ if $\delta(e) \leq \nu_j$, and $\gamma_j(e) = \infty$ otherwise. Let T_j be a minimum cost multicast tree of G for γ_j, and J be the minimum j such that $\gamma_j(T_j) \leq \Gamma$.

By the definition of J, the minimum delay of a cost constrained multicast tree is at least ν_J and at most $(n - 1)\nu_J$. Since such J and also T_J can be computed in $\mathcal{O}(n \log n)$ time [7], we have the following.

Theorem 4. *A pair of upper and lower bounds U and L for the minimum delay of a cost constrained multicast tree satisfying $U/L = n - 1$ can be computed in $\mathcal{O}(n \log n)$ time. Moreover, a multicast tree T_J with cost at most Γ and delay at most U can also be computed in $\mathcal{O}(n \log n)$ time.* $\qquad\square$

Given a 2-tree G with source s and destinations D, cost and delay functions γ and δ, and a positive integer Γ, we denote by $\textsc{Comp\_UL}(G, s, D, \gamma, \delta, \Gamma)$ an $\mathcal{O}(n \log n)$ time procedure for computing upper and lower bounds U and L with $U/L \leq n - 1$.

4.2   FPTAS for CCMDMT

For any $\alpha > 0$, let δ_α be a delay function defined as $\delta_\alpha(e) = \lfloor \alpha \delta(e) \rfloor$ for any $e \in E(G)$. Let T_α be a minimum delay multicast tree with cost at most Γ for δ_α and $\mathrm{OPT}(\delta_\alpha) = \delta_\alpha(T_\alpha)$. Notice that T_1 is a minimum delay multicast tree with cost at most Γ for $\delta = \delta_1$. We denote by P_α a maximum delay path in T_α for δ_α.

By the definition of δ_α, we have $\delta(e) \geq \delta_\alpha(e)/\alpha$ and

$$\delta(e) < (\delta_\alpha(e) + 1)/\alpha \qquad (11)$$

for any $e \in E(G)$. If we denote by P_1' a maximum delay path of T_1 for δ_α,

$$\mathrm{OPT}(\delta) = \sum_{e \in E(P_1)} \delta(e) \geq \sum_{e \in E(P_1')} \delta(e) \geq \sum_{e \in E(P_1')} \delta_\alpha(e)/\alpha \geq \sum_{e \in E(P_\alpha)} \delta_\alpha(e)/\alpha,$$

where the second inequality follows from $\delta(e) \geq \delta_\alpha(e)/\alpha$. Thus, we have

$$\mathrm{OPT}(\delta) \geq \delta_\alpha(T_\alpha)/\alpha$$
$$= \mathrm{OPT}(\delta_\alpha)/\alpha. \qquad (12)$$

Moreover, if we set $\alpha = (n-1)/\varepsilon L$, and denote by P_α' a maximum delay path in T_α for δ, we have

$$\delta(T_\alpha) = \sum_{e \in E(P_\alpha')} \delta(e)$$
$$< \frac{1}{\alpha} \sum_{e \in E(P_\alpha')} (\delta_\alpha(e) + 1) \qquad (13)$$
$$\leq \frac{1}{\alpha}|E(P_\alpha')| + \frac{1}{\alpha} \sum_{e \in E(P_\alpha')} \delta_\alpha(e)$$
$$\leq \frac{n-1}{\alpha} + \frac{1}{\alpha} \sum_{e \in E(P_\alpha)} \delta_\alpha(e)$$
$$= \varepsilon L + \frac{1}{\alpha}\mathrm{OPT}(\delta_\alpha)$$
$$\leq \varepsilon L + \mathrm{OPT}(\delta) \qquad (14)$$
$$\leq (1 + \varepsilon)\mathrm{OPT}(\delta),$$

Input a series-parallel graph G', $s \in V(G')$, $D \subseteq V(G')$, $\gamma' : E(G') \to \mathbb{N}$, $\delta' : E(G') \to \mathbb{N}$, $\Gamma \in \mathbb{Z}^+$, $\varepsilon > 0$.

Output a multicast tree T with cost at most Γ and delay at most $(1 + \varepsilon)\mathrm{OPT}(\delta')$.

begin
    $\textsc{Ext}(G', \gamma', \delta')$;
    $\textsc{Comp\_UL}(G, s, D, \gamma, \delta, \Gamma)$;
    $\alpha := (n-1)/\varepsilon L$;
    $\delta_\alpha(e) := \lfloor \alpha\delta(e) \rfloor \ \forall e \in E(G)$;
    $\Delta_\alpha := \alpha U$;
    $\textsc{Ba}(G, s, D, \gamma, \delta_\alpha, \Delta_\alpha)$;
    $\textsc{Min\_Delay}(\Gamma, \mathcal{W}(r, \mathcal{C}(r)))$;
    $\textsc{Mt}(G, s, D, \gamma, \delta_\alpha, \Delta_\alpha, \xi)$;
    return T;
end

Fig. 3. Algorithm 3

where inequality (13) and (14) follow from (11) and (12), respectively.

Thus, we conclude that Algorithm 3 shown in Fig. 3 is an FPTAS for CCMDMT. Since $\Delta_\alpha = (n-1)U/\varepsilon L = \mathcal{O}(n^2/\varepsilon)$, we have the following by Theorem 2.

Theorem 5. *For a series-parallel graph G with n vertices and a non-negative integer Δ, Algorithm 3 computes a $(1+\varepsilon)$-approximate solution for CCMDMT in $\mathcal{O}(n \log n + n^7/\varepsilon^3)$ time.*   □

Finally, it should be noted that our method to obtain FPTAS for CCMDMT cannot apply to DCMCMT in a straightforward way, since Δ can be exponentially large.

References

1. Chan, G., Xue, G.: k-pair delay constrained minimum cost routing in undirected networks. ACM-SIAM Symposium on Discrete Algorithms (2001) 230–231
2. Kompella, V., Pasquale, J., Polyzos, G.: Multicast routing for multimedia communication. IEEE/ACM Transactions on Networking **1** (1993) 286–292
3. Parsa, M., Zhu, Q., G.-L.-Aceves, J.J.: An iterative algorithm for delay-constrained minimum-cost multicast. IEEE/ACM Trans. Networking **6** (2001) 213–219
4. Sriram, R., Manimaran, G., Murthy, C.: Algorithms for delay-constrained low-cost multicast tree construction. Computer Communications **21** (1998) 1693–1706
5. Youssef, H., A.-Mulhem, A., Sait, S., Tahir, M.: QoS-driven multicast tree generation using tabu search. Computer Communications **25** (2002) 1140–1149
6. Hassin, R.: Approximation schemes for the restricted shortest path problem. Mathematics of Operations Research **17** (1992) 36–42
7. Chen, G., Xue, G.: A PTAS for weight constrained steiner trees in series-parallel graphs. Theoretical Computer Science **304** (2003) 237–247
8. Wald, J., Colbourn, C.: Steiner trees, partial 2-trees and minimum IFI networks. Networks **13** (1983) 159–167
9. Lorenz, D., Raz, D.: A simple efficient approximation scheme for the restricted path problem. Operations Research Letters **28** (2001) 213–219
10. Warburton, A.: Approximation of pareto optima in multiple-objective shortest path problems. Operations Research **35** (1987) 70–79

Ontology-Based Inconsistency Management of Software Requirements Specifications

Xuefeng Zhu[1,2] and Zhi Jin[1,3]

[1] Institute of Computing Technology, Chinese Academy of Sciences, P.O.Box 2704
Beijing 100080, People's Republic of China
[2] Graduate School of the Chinese Academy of Sciences, Beijing 100080, People's
Republic of China
[3] Academy of Mathematics and System Sciences, Chinese Academy of Sciences,
Beijing 100080, People's Republic of China
zhuxuefeng@ict.an.cn

Abstract. Management of requirements inconsistency is key to the development of trustworthy software systems. But at present, although there are a lot of work on this topic, most of them are limited in treating inconsistency at the syntactic level. We still lack a systematical method for managing requirements inconsistency at the semantic level.

This paper first proposes a requirements refinement model, which suggests that interactions between software agents and their ambiences are essential to capture the semantics of requirements. We suppose that the real effect of these interactions is to make the states of entities in the ambiences changed. So, we explicitly represent requirements of a software agent as a set of state transition diagrams, each of which is for one entity in the ambiences. We argue that, based on this model, the mechanism to deal with the inconsistency at the semantic level. A domain ontology is used as an infrastructure to detect, diagnose and resolve the inconsistency.

1 Introduction

Inconsistency is inevitable in software requirements specifications. How to handle inconsistency in requirements specifications is key to the development of trustworthy software systems[1]. But at present, although there are a lot of work[1][2] on inconsistency management, we still lack of systematical method for managing requirements inconsistency at the semantic level.

How to formulate the requirements inconsistency? Based on B.Nuseibeh's original definition[3], G.Spanoudakis gave a more formal one[2]: "Assume a set of software model $S_1, \ldots, S_n$, the set of overlap relations between them $O_a(S_i, S_j)$ (i, j=1,...,n), a domain theory D, and a consistency rule CR, $S_1, \ldots, S_n$ will be said to be inconsistent with CR given the overlaps between them as expressed by the sets $O_a(S_i, S_j)$ and the domain theory D if it can be shown that the rule CR is not satisfied by the models."

M. Bieliková et al. (Eds.): SOFSEM 2005, LNCS 3381, pp. 340–349, 2005.

As for inconsistency management, B.Nuseibeh also proposed a framework[1], which has been widely accepted. Centering to this framework is the set of consistency checking rules, which express properties that software models should hold. Main activities in this framework include detection, diagnosis, treatment,handle, etc. . .

Following this framework, lots of work, mainly on inconsistency detection, have been done. Firstly, as we know, the prerequisite of requirements inconsistency is that there exists denotational overlaps between requirements statements. At present, the detection methods of denotational overlaps fall into four kinds ranked by their automation and intelligence. They are the human inspection of M.Jackson[4], the shared ontology based method of Z.Jin[5], the representation convention of V.Lamsweerde[6], and the similarity analysis of G.Spanoudakis[7]. Secondly, inconsistency detection methods differ from each other because of the difference requirements representation schemes. V.Lamsweerde[6] used the theorem provers to check consistency of their logic-based specifications. C.Heimeyer[8] used the model checker to find inconsistency in their automata-based specifications. S.Clarke[9] checked inconsistency of their graph-based specifications in terms of special properties or semantics of the graph. V.Lamsweerde[6] adopted human collaboration technique to assist inconsistency detection. And finally, V.Lamsweerde[6] solved inconsistency by using heuristic rules.

Managing the inconsistency in requirements is key to the development of the trustworthy software systems. But at present, although there are a lot of work on this topic, most of them are

limited in treating inconsistency at syntactic level. We still lack a promising method for managing requirements inconsistency at semantic level. This paper first proposes a requirements refinement model, which suggests that interactions between software agents and their ambiences are essential to capture the semantics of requirements. We suppose that the real effect of these interactions is to make the states of entities in the ambiences changed. We argue that, based on this model, the proposed mechanism to deal with the inconsistency at the semantic level. A domain ontology then is used as infrastructure to detect, diagnose and resolve inconsistency.

The paper is structured as follows. In section 2, a requirements refinement process is proposed which act as a reference model for requirements engineering. In section 3, a state transition model is proposed as a schema to represent interactions between software agents and their ambiences. In section 4, an innovative framework based on above two points is proposed to handle inconsistency at semantic level. Section 5 summarizes our work. An example from London Ambulance Service[10] has been used to validate our method.

2   Process for Refining Requirements

To further grasp the essence of requirements inconsistency, we need begin with the process of requirements capture and requirements refinement. Suppose we take software system as an agent and real world entities outside this agent as its ambience, then external requirement is observations that agent perceives related to it and relations among these observations. Internal structure which corresponding to external requirement consists of a group of subsystems and constraints among these subsystems, these subsystems act as realization of external requirement. Thus system requirements can be represented as a tuple which consists of external requirements and internal requirements.

As we can see, software requirements are based on external requirements. However, in terms of the development method and the decomposition of the system, we can get its internal structure. Requirements which has internal structure are called decomposable, otherwise they are called atomic. Refinements process of decomposable software requirements can be represent as a requirements refinement tree. There we can give the structural definition of software requirements:

Definition 1. *External requirements of a software system S can be represented as a triple ExReq=< Amb, Obv, Rel >*

1. $Amb = \{amb_1, amb_2, \cdots, amb_n\}$ is a set of all ambiences that interact with S, $amb_i (i = 1, 2 \cdots, n)$ can be an entity, a human being or another software system of the real world;

2. $Obv \subseteq (\{S\} \times Ph \rightarrow Amb) \cup (Amb \times Ph \rightarrow \{S\})$ is a set of observations which occur between ambiences and S, $Ph = \{ph_1, ph_2, \cdots, ph_m\}$ is a set of phenomena. Any $obv \in Obv$ indicates that S generate a phenomenon to one of its ambience or one ambience of S generates a phenomenon to S;

3. $Rel \in < (Obv, Obv) \cup \Re(Ph, Ph), <$ indicates only a caused sequence between two observations, $\Re$ indicate not only a caused sequence but also a special domain relation that two observations should satisfy.

$\Re$ in above definition represents $Input \rightarrow Output$ functional abstraction of the software system. We need not qualify properties of phenomena. They can be any perceivable phenomena between the real world and the software system. This definition reveals that requirements stimulate the context of the expected software system in its future ambience. However, for a complex software system, such requirements description may be insufficient for implementing. For this kind of systems, the process of requirements refinement is stepwise decomposition of requirements, until we get a set of requirements, each of which can be implemented directly.

Definition 2. *Suppose the external requirements of system S is ExReq =<Amb, Obv, Rel> and S is decomposable, then InReq=<Sub, Ass, InObv, InRel> is the internal structure of S, in which*

1. $Sub = \{S_1, S_2, \cdots, S_n\}$ is the set of all subsystems of S;

2. $Ass \subseteq Obv(S) \rightarrow \mathbb{P}Sub$ is the realization of external interaction of S. It shows that each external observation of S can be implemented by a group of subsystems;

3. $InObv \subseteq (Sub \times Ph \rightarrow Sub)$ are observations about interactions among subsystems of S. They are the internal observations of S;

4. $InRel \subseteq < (InObv, InObv) \cup \Re(Ph, Ph)$ contains the relations among the internal observations, $<$ and $\Re$ have the same meaning as above.

Decomposability can be seen as projection or decomposition. They are two basic principles of requirements engineering. Based on above definitions, we can draw the conclusion that construction process of requirements specification is the requirements refinement process. This refinement process can be represented as a finite tree $RefiningTree(S)$, its root is $< ExReq(S), InReq(S) >$, other nodes in the tree can unify be represent as $< ExReq(S_{i_1}, S_{i_2}, \cdots, S_{i_k}),$ $InReq(S_{i_1}, S_{i_2}, \cdots, S_{i_k}) > (i_1, i_2, \cdots, i_k = 1, 2, 3, \cdots)$. All nodes with $InReq(S_{i_1},$ $S_{i_2}, \cdots, S_{i_k}) = NULL$ are leaf node of $RefiningTree(S)$.

Definition 3. *Assume that T is a requirements refinement tree of a software system S. Every node in T has tangent plane. A set is called a tangent plane of T if it contains and only contains the following elements:*

1. If a is a leaf node in T then $TangentPlane(a) = \{a\}$;

2. If a is an AND node, and has sub node $a_1, a_2, \cdots, a_m$, then $TangentPlane(a_i) \in TangentPlane(a)$;

3. If a is an OR node, and has sub node $a_1, a_2, \cdots, a_m$, then there exists only one a_k, $TangentPlane(a_k) \in TangentPlane(a)$; $(1 \leq k \leq m, i = 1, 2, \cdots, m)$

From above definitions, we can give an abstract definition of requirements specification in the process of requirements refinement.

Definition 4. *Any tangent plane $TangentPlane(T)$ of requirements refinement tree T of software system S is a specification of S in certain stage of the requirements refinement process.*

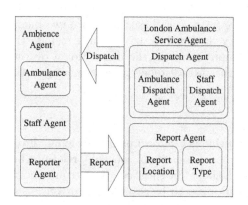 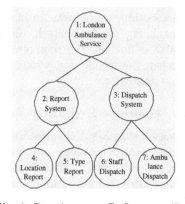

Fig. 1. Agent Hierarchical   **Fig. 2.** Requirements Refinement Tree

For example, figure 1 is the agent hierarchical structure of LAS system, and figure 2 is corresponding requirements refinement tree of figure 1. The top level agent in figure 1 is LAS agent, and the ambience agent which can be decomposed into Ambulance, Staff and Reporter. The interactions between LAS agent and its ambience include dispatch and report. Figure 2 shows the decomposition of LAS agent, in which node 3 is an OR node and the others are AND nodes. This refinement tree has tangent planes such as $\{1\}, \{2, 3\}, \{2, 6\}, \{3, 4, 5\}$, etc. ... Every tangent plane is a description of LAS of particular granularity.

3   State Transition Model

According to the requirements refinement process, software requirements consist of the external requirements which shows the system as an integrity and the internal structure which represents internal system decomposition.

The entities outside of the software system constitute its ambiences. We argue that the real effect of interactions between software system and its ambiences is to make the state of the entities in ambiences changed. So, we explicitly represent requirements of a software system as a set of state transition diagrams, each of which is for one entity in the ambiences. We can get the following definition for state transition of entity in ambience, which can then be used as a description schema for entity ontology.

Definition 5. *The state transition model which express the interactions between the software system and its ambience can be represented as $M =< S, E, T, S_0 >$*

1. $S = \{s_0, \ldots, s_n\}$ is the set of entity states in ambience;

2. $E = \{e_0, \ldots, e_m\}$ is the set of events related to entity state transition, ;

3. T is the set of the state transitions in ambience. For each $t \in T$, $t = < s_s, s_t, e_t >$, in which s_s is the source state of t, S_t is the target state of t, e_t is the event which causes the state transition.

4. $S_0 \subseteq S$ is the set of the initial states of entity.

4   Inconsistency Management

In this section, we propose an innovative framework based on above two model for managing requirements inconsistency at semantic level, which include the operable definition, the requirements representation, the denotational overlap detection, the inconsistency detection, the inconsistency measurement and the resolve strategies.

4.1   Operable Definition

M.Jackson suggests that software engineering is a subject of description[4]. From the viewpoint of ontology, any domain model consists of concepts, relations between the concepts and constraints which should be satisfied. Since requirements specification is a kind of model of the problem to be solved, it also consists of the above three elements. So, we can give the following definition for software requirements inconsistency:

Definition 6. *Assume that T is a requirements refinement tree of software system, TangentPlane(T)=$\{< ExReq(S_1), \ldots >, \ldots, < ExReq(S_n), \ldots >\}$ is a tangent plane of T, and CR is a consistency checking rule, then*

1. If there exists $ExReq(S_i)$, DS_1 and DS_2 are two homogeneous specification elements in it and have overlap relation o. If we can prove that CR is not satisfied by $ExReq(S_i)/\{DS/DS_1, DS/DS_2\}$, then we can say that at given overlap o, $ExReq(S_i)$ is inconsistent allowing for CR. Here DS is a new name in $ExReq(S_i)$ and $i = 1, 2, \cdots, n$;

2. If there exists $ExReq(S_i)$ and $ExReq(S_j)$, DS_1 and DS_2 are corresponding specification elements in them and have overlap relation o, if we can prove that CR can not be satisfied by $ExReq(S_i) \cup ExReq(S_j)/\{DS/DS_1, DS/DS_2\}$, then we can say that at given overlap relation o, $ExReq(S_i) \cup ExReq(S_j)$ is inconsistent allowing for CR. Here, DS is a new name in $ExReq(S_i) \cup ExReq(S_j)$ and $i,j = 1, 2, \cdots, n, i \neq j$;

3. If we can prove that CR can not be satisfied by $ExReq(S_1) \cup \cdots \cup ExReq(S_n)$, then we can say $ExReq(S_1) \cup \cdots \cup ExReq(S_n)$ is inconsistent allowing for CR.

4.2 Requirements Representation

Software requirements consist of the external requirements which represent the interactions between the software system and its ambiences, and the internal requirements which represent the internal structure of the system. We assume that entities in ambience have finite states. These entities can transit from one state to another triggered by events. So, we can represent software requirements as: $Requirements :=< Name, Ambience, Structure, Event >$.

Software requirements specification as above can be obtained by stepwise system decomposition. Ambience includes all relevant entities outside of this system. Structure relies on Ambience and the software system. Event include a set of event which cause entity transit from one state to another.

For example, reporter agent's requirements of reporting an accident to accident controller room can be represented as: <accident agent, {ambulance, driver}, {report system}, {reporter} >.

4.3 Denotational Overlap Detection

Denotational overlaps are prerequisite of requirements inconsistency. But multiple requirements descriptions usually can not be comparable. That is one of the main reasons why inconsistency is so difficult to deal with.

Shared ontology can be used as a translator among different requirements descriptions. Software requirements rely on its ambience, owing to the difference of its ambience, any requirements description can be seen as a positive projection of the system being described on shared ontology. Only when all description elements can be mapped onto shared ontology will it be possible that different description elements or different descriptions be comparable.

Given the requirements specification representation schema above, two specifications are said to have denotational overlap if and only if their ambiences have some common entities, and the event sequence can make these entities transit to common state. The following is an algorithm to detect the denotational overlap.

Algorithm 1. *Given requirements specifications spe_1 and spe_2 which have representation scheme of Agent-Ambience interaction:*

1. *Mapping spe_i and all its ambiences onto shared ontology, $spe_i; =$ project $(spe_i, SharedOntology)(i = 1, 2)$;*

2. *If $amb_1 \cap amb_2 = \phi$, then spe_1 and spe_2 have no denotational overlap;*

3. If $amb_1 \cap amb_2 \neq \phi$ and $entity_1 \cap entity_2 = \phi$, then spe_1 and spe_2 have no denotational overlap;

4. We assume $stateseq_i$ is state transition sequence of $entity_i(i = 1, 2)$, if $stateseq_1 \cap stateseq_2 = \phi$, then there is no denotational overlap, otherwise spe_1 and spe_2 have denotational overlap which is $< entity, stateseq_1 \cap stateseq_2 >$.

For example, driver and reporter requirements of dispatching ambulance have the same ambience {ambulance}, and event of driver and reporter contain a same state {<ambulance, ready>}, thus they have denotational overlap which is <ambulance, {<ambulance, ready>} > .

4.4   Inconsistency Detection

Given software requirements specifications and consistency checking rule CR, we call that specifications are inconsistent if entities in ambience transit to a contradictory state which CR states that it should not reach. Based on the state transition model and the requirements representation, we have the following definition for detecting requirements inconsistency:

Definition 7. *Given two requirements specifications spe_1 and spe_2. amb_1 and amb_2 are corresponding ambiences. $eventseq_1$ and $eventseq_2$ are corresponding event sequences. If $amb_1 \cap amb_2 \neq \phi$, then if $\exists entity \in (amb_1 \cap amb_2)$, such that the sets of states that $eventseq_1$ and $eventseq_2$ act upon this entity are $stateset_1$ and $stateset_2$, if $\exists state \in (stateset_1 \cap stateset_2), (state_1 \in stateset_1) \cap (state_2 \in stateset_2) \cap (state_1 \cap state_2 \vdash False)$, then we call spe_1 and spe_2 are inconsistent.*

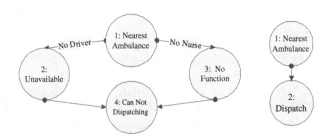

Fig. 3. State Transition Graphs

For example, Figure 3 contains two state transition graphs. The first one expresses the accident controller's view. It says that if an ambulance has no operator in it, then it should not dispatch to the accident spot. This rule can be represented as CTL formula: $AG(ambulance.state = nooperator \rightarrow AF(\neg ambulance.state = dispatch))$. On the other hand, the second graph captures the reporter's view. It says that under such condition, the nearest ambulance should be dispatched to that spot unconditionally. Thus we can see that the accident reporter's requirements is inconsistent with the domain rule.

4.5   Inconsistency Measurement

From the requirements refinement model, we can see that inconsistency management of software requirements can be equal to the inconsistency management of the tangent planes of requirements refinement tree. Inspired by irregular knowledge processing, we can give an abstract model for inconsistency reduction.

Definition 8. *If we take specifications in the requirements refinement tree as a formula set R, if*

$$\bigcap_{r \in R} r \vdash False \text{ and } SR = \{R' | \bigcap_{r \in R'} r \vdash True, \forall t \in R - R', \{t\} \cap (\bigcap_{r \in R'} r) \vdash False\},$$

then we call each $R' \in SR$ is a maximal consistency subset of R.

Suppose we assign different factor to formulas according to their importance, then SR'=$\{R' | R' \subseteq SR$ and $weight(R') = maximal\}$ contains all the consistency subsets that have maximal benefit, and (weight(R)-weight(R'))/weight(R) can be used to estimate severe degree of inconsistency.

Definition 9. *If we take specifications in the requirements refinement tree as a formula set R, if*

$$\bigcap_{r \in R} r \vdash False \text{ and } SR' = \{R' | \bigcap_{r \in R'} r \vdash False, \forall t \in R', (\bigcap_{r \in R' - \{t\}} r) \vdash True\},$$

then each $R' \in SR'$ is a minimal inconsistency subset of R.

Theorem 1. *Assume formula s belongs to k minimal inconsistency subsets of R, then number of minimal inconsistency subset in R-$\{s\}$ is $|SR'|$-k.*

Proof. According to above definition, number of minimal inconsistency subset in R is $|SR'|$. while (1)reduce a formula in R will not increase the number of minimal inconsistency subset; (2) if s not belongs to any minimal inconsistency subset $S \subseteq R$, then reduce s in R will not influence inconsistency of S. So the number of minimal inconsistency subset in R-$\{s\}$ is $|SR'|$-k.

Let s_1=s and k_1=k, if R-$\{s_1\}$ still is inconsistent, then we can take formula s_2 in one of the minimal inconsistency subsets of R-$\{s_1\}$, and that s_2 is contained in k_2 minimal inconsistent subsets,.... If after reducing m formulas, we can get formula set that no longer inconsistent, then we have $\sum_{i=1}^{m} k_i = |SR'|$.

Definition 10. *We call each formula in the minimal inconsistency set a conflicting point, if a formula contained in k minimal inconsistency sets, then we call this formula a k-folds conflicting point.*

Definition 11. *We call sequence S=$(s_1, s_2, \ldots, s_m)$ a conflicting point sequence of R if*

1. *s_1 is a conflicting point of R;*
2. *$\forall j < m$, we have already reduced $(s_1, s_2, \ldots, s_j)$, but s_{j+1} still is a conflicting point of R-$\{s_1, s_2, \ldots, s_j\}$.*

The severe degree of inconsistency caused by conflicting point s in R can be calculated as Impact(R,s)=(Weight(R)-Wight(R-$\{s\}$))/weight(R). The severe degree of inconsistency caused by conflicting point sequence S=$(s_1, s_2, \ldots, s_m)$ in R can be calculated as Impact(R,S)=(weight(R)-weight(R-$(s_1, s_2, \ldots, s_m)$))/weight(R). And if s_j is the j-th conflicting point of conflicting point sequence S=$(s_1, s_2, \ldots, s_m)$, then conditional inconsistency caused by s_j is Impact(R, $s_j/(s_1, s_2, \ldots, s_{j-1})$)= Impact(R, $\{s_1, s_2, \ldots, s_{j-1}\}$)-Impact(R, $\{s_1, s_2, \ldots, s_j\}$).

4.6   Inconsistency Resolution

The aim of inconsistency management is to eliminate the influence of inconsistency. However we need not eliminate inconsistency directly. According to the severe degree of inconsistency, there are two kinds of actions for solving inconsistency.

Pattern Matching. Most work on inconsistency resolution use heuristic rules to find the most appropriate actions that might be taken. These heuristic rules can be used in generalizing the inconsistency handling patterns. The following is a semantic matching algorithm:

Algorithm 2. *Given an inconsistent state transition graph G, a set of inconsistency handling pattern P=$\{P_1, P_2, \ldots, P_k\}$ and a threshold v:*

1. Assign zero to j, and assign v to min, min is minimal semantic distance and j indicate the number of pattern that most appropriate to G;
2. For i:=1 to k do $\{D_i := distance(G, P_i)$; if min> D_i then min:=D_i and j:=i$\}$, $distance(G_1, G_2)$ is a function that calculates the semantic distance of two graphs;
3. If $distance(G, P_j) < v$ then P_j is the pattern that most appropriate to handle G, otherwise there are no existing pattern that can handle G;

Requirements Distribution. As to the inconsistency that can not be solved by the pattern matching, we can distribute requirements according to ambiences. So that we can keep internal consistency of each requirements set. Referring to CYC program and microtheory architecture for knowledge organization, we can give the following algorithm for tolerating inconsistency:

Algorithm 3. *Given leaf requirements node N in requirements refinement tree T, $\{R_1, R_2, \ldots, R_n\}$ are formulas in N, we can get the following consistency subset N'.*

1. For i:=1 to n do $\{consistencychecking(R_i)$, if $\neg consistent(R_i)$ then we need partition R_i into subformulas that are consistent and add them in formula subset of N', number of formulas in N' is n'$\}$
2. Divide N' into subset $N_1', \ldots, N_m'(m < n)$, $\bigcup_{i=1}^{m} N_i' = N'$ and $N_i' \cap N_j' = \phi$, every N_i' is internal consistent (i,j=1,2,...,m; $i \neq j$), and m is minimal;
3. Generate m node which contain formula set $N_i'(i = 1, \ldots, m)$, and generate an AND node which pointing to these m sub nodes and replace node N with it.

5   Conclusions

Based on the requirements refinement model, this paper takes domain ontology as infrastructure to manage inconsistency at semantic level. The main contributions of this paper include the following points:(1)software requirements can be seen as interactions between software agents and their ambiences;(2)software requirements can be represented as a set of state transition diagrams of entities in the ambiences;(3)an ontology-based inconsistency management framework for handling inconsistency at semantic level.

Acknowledgement

This work is partly supported by the National Natural Science Key Foundation of China under Grant No.60233010, and Grant No.60496324, the National Key Research and Development Program under Grant No.2002CB312004, the Knowledge Innovation Program of the Chinese Academy of Sciences, and MADIS of the Chinese Academy of Sciences.

References

1. B.Nuseibeh, S.Easterbrook and A.Russo: Leveraging Inconsistency in Software Development. IEEE Computer, 2000, Vol.33, No.4, 24-29
2. G.Spanoudakis, A.Zisman: Inconsistency Management in Software Engineering: Survey and Open Research Issues. In S.Chang(ed), Handbook of Software Engineering and Knowledge Engineering, World Scientific Publishing Co. 2001, 329-380
3. B.Nuseibeh: To be and not to be: On managing inconsistency in software development. In Proceedings of the 8th International Workshop on Software Specification and Design, Germany, 1996, 164-169
4. M.Jackson: The meaning of Requirements. Annuals of Software Engineering, 1997, Vol.3, 5-21
5. Z.Jin, R.Lu and D.Bell: Automatically MultiParadigm Requirements Modeling and Analyzing: An Ontology-Based Approach. Science in China(Series F), 2003, Vol.46, No.4, 279-297
6. V.Lamsweerde, R.Darimont and E.Letier: Managing Conflict in Goal-Driven Requirements Engineering. IEEE Transactions on Software Engineering, 1998, Vol.24, No.11, 908-926
7. G.Spanoudakis and A.Finkelstein: Reconciling Requirements: A Method for Managing Interference, Inconsistency and Conflict. Annuals of Software Engineering, 1997, Vol.3, 433-457
8. C.Heitmeyer, R.Jeffords and D.Kiskis: Automated Consistency Checking Requirements Specifications. ACM Transactions on Software Engineering and Methodlogy, 1996, Vol.5, No.3, 231-261
9. S.Clarke, J.Murphy and M.Roantree: Composition of UML Design Models: A Tool to Support the Resolution of Conflicts. In Proceedings of the International Conference on Object-Oriented Information Systems, Springer Verlag, 1998, 464-479
10. A.Finkelstein and J.Dowell: A Comedy of Errors: The London Ambulance Service Case Study. In Proceedings of the 8th International Workshop on Software Specification and Design. IEEE Computer Society Press, 1996, 2-4

Suffix Tree Based Data Compression

Martin Senft

Faculty of Mathematics and Physics, Charles University,
Malostranské náměstí 25, 118 00 Praha, Czech Republic
senft@atrey.karlin.mff.cuni.cz

Abstract Suffix tree is a widely studied data structure with a range
of applications. Although originally developed for string matching algorithms, it has proved to be useful for implementation of various lossless
data compression methods. This interesting fact inspired us to explore
the possibilities of creating new data compression methods, based entirely on properties of suffix trees. Resulting from a thorough study of
existing suffix tree construction algorithms and their modifications, we
propose and evaluate several variants of suffix tree based compression.

1 Introduction

Suffix tree is a widely studied data structure with a range of applications. Although originally developed by Weiner [1] for text indexing in string matching
algorithms, it has also proved as useful for the data compression. The most
prominent lossless data compression algorithms, including the Burrows-Wheeler
Transform [2], PPM [3] and ZL77 [4] methods are based on different ideas, but
still have one thing in common: a suffix tree can be used for their implementation. This interesting fact inspired us to explore the possibilities of creating new
data compression methods, based entirely on properties of suffix trees. Resulting
from a thorough study of existing suffix tree construction algorithms [1, 5, 6, 7]
and their modifications [8, 9], we propose several variants of suffix tree based
compression and compare them to existing compression methods on a standard
text corpus.

This paper is organised as follows: The next section summarises some necessary notation and terminology, leading to the definition of a suffix tree. Section 3
reviews two classical suffix tree construction algorithms. Our main contribution
is in Sect. 4, where the idea of the Suffix Tree Compression (STC) is described
and developed to create a class of new methods. The next two sections are devoted to implementation and evaluation of two selected methods. We conclude
the paper with final remarks and suggestions for future development in Sect. 7.

2 Concepts and Notation

2.1 Notation

Alphabet is a nonempty finite set, its elements are *characters* and any sequence
δ of n characters is a *string* of *length* $|\delta| = n$. If $n = 0$, δ is an *empty string*,
denoted by λ. Σ^* is a set of all strings over alphabet Σ and $\Sigma^+ = \Sigma^* - \{\lambda\}$.

M. Bieliková et al. (Eds.): SOFSEM 2005, LNCS 3381, pp. 350–359, 2005.
© Springer-Verlag Berlin Heidelberg 2005

Let $\alpha, \beta \in \Sigma^*$, where $\alpha = a_1a_2\ldots a_m$ and $\beta = b_1b_2\ldots b_n$, then $\alpha\beta = a_1\ldots a_m b_1\ldots b_n$ denotes the *concatenation* of α and β. If $\delta = \alpha\beta\gamma$, then α, β and γ are called *prefix*, *factor* and *suffix* of string δ, respectively. The sets of all prefixes, factors and suffixes of a string δ are denoted in order by $\mathrm{Prefix}(\delta)$, $\mathrm{Factor}(\delta)$ and $\mathrm{Suffix}(\delta)$.

The i-th character of string δ is denoted by $\delta[i]$ for $1 \le i \le |\delta|$. The string $\delta[i]\delta[i+1]\ldots\delta[j]$, where $1 \le i \le j \le |\delta|$, is denoted by $\delta[i:j]$. It is convenient to have $\delta[i:j] = \lambda$ for $i > j$. Let $\alpha \in \Sigma^*$, if there exists i, j such that $\alpha = \delta[i:j]$, then α is said to *occur at position i* in δ. If $\alpha \in \mathrm{Factor}(\delta)$ occurs only once in δ, it is called *unique*. The set of all unique suffixes of δ is denoted by $\mathrm{UniqueSuffix}(\delta)$.

Further on, all characters and strings are taken from Σ and Σ^* and denoted by latin and greek characters, respectively.

2.2 Strings

Definition 1. *Let $\delta \in \Sigma^*$ and $\alpha \in \mathrm{Factor}(\delta)$, then α is* right branching *in δ if there exist characters $a \ne b$, such that $\alpha a, \alpha b \in \mathrm{Factor}(\delta)$. We denote the set of all strings that are right branching in δ by $\mathrm{Branch}^{\mathrm{R}}(\delta)$.*

Right branching factors of δ are strings that occur at least twice in δ and two of these occurrences are followed by two different characters.

Definition 2. *Let $\delta \in \Sigma^*$. For any $\alpha \in \mathrm{Factor}(\delta)$, we define $\overset{\delta}{\Longrightarrow}{\alpha}$ as follows:*

$$\overset{\delta}{\underset{\alpha}{\Longrightarrow}} = \begin{cases} \lambda & \alpha = \lambda \\ \alpha & \alpha \in \mathrm{Branch}^{\mathrm{R}}(\delta) \\ \alpha & \alpha \in \mathrm{UniqueSuffix}(\delta) \\ \overset{\delta}{\underset{\alpha a}{\Longrightarrow}}, \text{ where } a \in \Sigma \text{ and } \alpha a \in \mathrm{Factor}(\delta) & \text{otherwise} . \end{cases}$$

The set $\left\{ \alpha \;\middle|\; \overset{\delta}{\underset{\alpha}{\Longrightarrow}} = \alpha \right\}$ is denoted by $\mathrm{Proper}^{\mathrm{L}}(\delta)$.

Less formally, $\overset{\delta}{\underset{\alpha}{\Longrightarrow}}$ is the string obtained by appending characters to α until we get a unique suffix of δ or a right branching factor of δ.

2.3 Suffix Trees

Here we define the suffix tree data structure using the standard graph terminology (cf. [10]) with one simplification: vertices of a tree that are not leaves are called *nodes*. The suffix tree is defined as an *edge-labelled graph* (V, E), where V is a set of vertices and $E \subseteq V \times \Sigma^+ \times V$ is a set of edges. The second edge component is its *label*.

Definition 3. *Let $\delta \in \Sigma^*$. $\mathrm{STree}(\delta)$ is an edge-labelled graph (V, E), where*

$$V = \left\{ \alpha \;\middle|\; \alpha \in \mathrm{Proper}^{\mathrm{L}}(\delta) \right\},$$

$$E = \left\{ (\alpha, a\beta, \gamma) \;\middle|\; \alpha, \gamma \in V \text{ and } \overset{\delta}{\underset{\alpha a}{\Longrightarrow}} = \alpha a\beta = \gamma \right\}.$$

The edge $(\alpha, a\beta, \gamma)$ is called an a-edge from α to γ of length $|a\beta|$.

It is easy to see that STree(δ) is really a tree. We generally root it at λ.

Definition 4. *A location of $\alpha \in$ Factor(δ) in STree(δ) is a pair (β, γ), where β is an STree(δ) node and $\alpha = \beta\gamma$. $\text{Loc}_\delta(\alpha)$ is the set of all locations of α and the canonical location of α, denoted by $\text{CanonLoc}_\delta(\alpha)$, is its member with the shortest γ.*

Let $\text{CanonLoc}_\delta(\alpha) = (\beta, \gamma)$. If $\gamma = \lambda$, then we say that α is *located in the node β*. Otherwise, α is said to be *located on the edge* $(\beta, \gamma\eta, \varphi)$ at position γ. On the other hand, all nodes and positions on edges *represent* some string α and its length defines the *depth* of such location. Note that we can *canonise* any location of α to obtain $\text{CanonLoc}_\delta(\alpha)$ by following some edges down from β.

3   Suffix Tree Construction

This section describes two linear time suffix tree construction algorithms due to McCreight [5] and Ukkonen [6].

3.1   Ukkonen's Algorithm

Ukkonen's algorithm works on-line, e.g. it constructs STree(δa) directly from STree(δ). The sets of vertices of these trees are $\text{Proper}^{\text{L}}(\delta)$ and $\text{Proper}^{\text{L}}(\delta a)$, respectively, and contain λ, all right branching factors and all unique suffixes. Clearly, all right branching factors of δ are also right branching in δa, but many unique suffixes of δ may not be suffixes of δa at all. However, all changes to the set of vertices come from the suffixes of δ, which we divide into the following groups:

1. $\alpha \in$ UniqueSuffix(δ),
2. $\alpha \notin$ UniqueSuffix(δ) and $\alpha a \in$ UniqueSuffix(δa),
   (a) $\alpha \notin \text{Branch}^{\text{R}}(\delta)$,
   (b) $\alpha \in \text{Branch}^{\text{R}}(\delta)$,
3. $\alpha \notin$ UniqueSuffix(δ) and $\alpha a \notin$ UniqueSuffix(δa).

Note that all suffixes shorter (longer) than a right branching (unique) one, must also be right branching (unique). Consequently, suffixes are split into these groups according to their lengths. The longest suffix (i.e. δ) is unique, the shortest suffix (i.e. λ) is generally non-unique and right branching, and all suffixes in group 3 are right branching. The longest non-unique (repeated) suffix of δ (i.e. the longest suffix not in group 1) is denoted by LRS(δ). The following list summarises contribution of each group to $\text{Branch}^{\text{R}}(\delta a)$ and UniqueSuffix(δa):

1. $\alpha a \in$ UniqueSuffix(δa) and $\alpha \notin \text{Branch}^{\text{R}}(\delta a)$,
2. $\alpha a \in$ UniqueSuffix(δa) and $\alpha \in \text{Branch}^{\text{R}}(\delta a)$,
3. $\alpha a \notin$ UniqueSuffix(δa) and $\alpha \in \text{Branch}^{\text{R}}(\delta a)$.

We transform STree(δ) to STree(δa) by doing the following in each suffix group:

1. Suffixes in this group are leaves and every leaf α is replaced by a leaf αa and its incoming edge by an edge to αa with a appended to its label.
2. For each α in this group a new leaf αa and edge $(\alpha, a, \alpha \alpha)$ are created. In case 2a, the edge $(\beta, \gamma \varphi, \eta)$, where α is located, is split into edges (β, γ, α) and (α, φ, η) and node α is created.
3. Nothing to do.

Case 1 can be solved by an implementation trick of so called "open edges" (see [6] for details). Case 2 appears to be quite simple, however, we have to find all suffixes in this group somehow. To do this fast, we augment the suffix tree with auxiliary edges called *suffix links* (introduced by Weiner [1]) and one auxiliary node called *nil*, denoted by $\perp$. Modified suffix tree definition follows:

Definition 5. *Let $\delta \in \Sigma^*$. STree(δ) is an edge-labelled graph (V, E), such that*

$$V = \left\{ \alpha \mid \alpha \in \text{Proper}^L(\delta) \right\} \cup \{\perp\},$$

$$E = \left\{ (\alpha, a\beta, \gamma) \mid \alpha, \gamma \in V \text{ and } \overrightarrow{\alpha a} \stackrel{\delta}{=} \alpha a \beta = \gamma \right\} \cup$$

$$\cup \text{NilEdges} \cup \text{SuffixLinks}(\delta), \text{ where}$$

$$\text{NilEdges} = \{(\perp, a, \lambda) \mid a \in \Sigma\} \text{ and}$$

$$\text{SuffixLinks}(\delta) = \left\{ (a\alpha, \lambda, \alpha) \mid \alpha, a\alpha \in \text{Branch}^R(\delta) \right\} \cup \{(\lambda, \lambda, \perp)\} .$$

Members of NilEdges *and* SuffixLinks(δ) *are called* nil edges *and* suffix links, *respectively.*

It is convenient to denote CanonLoc$_\delta$($\alpha[2 \ldots |\alpha|]$), i.e. the canonical location of the longest suffix of α different from α, by SuffixLoc$_\delta$(α). Note that SuffixLoc$_\delta$(α) may be obtained from $(\beta, \gamma) = $ CanonLoc$_\delta$(α) easily by following the suffix link (β, λ, η) and then canonising location.

All we have to do to transform STree(δ) to STree(δa) is to follow the so called *active point* (AP) and make the necessary changes:

1. At the beginning, AP is equal to CanonLoc$_\delta$(LRS(δ)).
2. While AP $(= (\beta, \gamma))$ is not the location of the longest suffix in group 3
   (a) If the last step created a new node, add a suffix link from this node to the node AP is in or the node that will be created in 2b.
   (b) Make the changes required for this suffix in group 2.
   (c) Set AP to SuffixLoc$_\delta$($\beta\gamma$). This is called the *sideways movement* of AP.
3. Set AP to $(\beta, \gamma a)$ and canonise.

3.2   McCreight's Algorithm

McCreight's and Ukkonen's algorithms only group the same steps differently. Ukkonen's algorithm proceeds "character by character", while McCreight's algorithm builds the tree "unique suffix by unique suffix" and only the first and the last tree in the building process are guaranteed to be suffix trees. It reads characters and moves AP down as long as possible, then it makes the changes necessary to add a new leaf for the new unique suffix, moves AP sideways and reads characters again. This way it adds a new unique suffix in each step.

4   Suffix Tree Based Data Compression

While other authors [8, 9, 11] employ the suffix tree as a suitable data structure for implementation of existing compression methods, we explore the possibility of an opposite approach. Our goal is to design a brand new algorithm, based entirely on properties of suffix trees. We follow the strategy of constructing a suffix tree for the data to be compressed and storing the decisions made while doing so. The original data can be reconstructed later from that information.

4.1   The General Idea

When constructing $STree(\delta)$, the two algorithms of Sec. 3 follow the active point. AP's movement through the tree is given by the string δ, on the other hand, the movement of AP in the tree fully describes δ. It can be reconstructed as follows: For every single character of δ, AP is moved down either immediately or after a few sideways moves. So, we either already know the character and can write it down or just wait until AP is moved down again.

4.2   Describing the AP Movement

Now, when we know that AP movement alone suffices for the original string reconstruction, a way to save it is needed. There are basically two ways to describe AP movement: the absolute and the relative.

Absolute Description. Two of many possible ways are:

1. If the downward movement stopped in a node, the node id is saved. If it stopped on an edge, then the id of the second edge vertex together with another value describing the actual location on the edge is saved. This may be the distance (in characters) from the first edge vertex. This is similar, but not identical to Fiala's and Greene's [8] method C2.
2. Save the position of one of the unique suffixes that has a leaf in the subtree where AP's downward movement ended, incremented by the depth of the location where AP ended after the last sideways move. Also save the difference between depths of this location and the location where AP ended the downward movement. This creates a Ziv and Lempel [4] compression variant and suffix tree is not necessary for the decompression.

Relative Description. We look for inspiration at the construction algorithms and denote the Ukkonen based and McCreight based approaches STC_U and

STC_M, respectively. The Ukkonen's algorithm is "cautious" and proceeds one character at a time as it fears that unexpected character will follow. So we will be "cautious" as well and describe one decision made during AP movement at a time. On the other hand, the McCreight's algorithm is "optimistic" and proceeds one unique suffix at a time as it believes that it will take some time, before it reads an unexpected character. An "optimistic" description method may describe AP movement decisions in chunks that end in a new leaf creation.

STC_U. If AP is on an edge, there are just two possibilities: either it is moved down or sideways. However, when in a node, AP may move down on one of node's edges or move sideways. Things are a bit different in $\perp$, where for every $a \in \Sigma$ there is an a-edge leading from $\perp$ to λ and AP always moves down here. This method is similar to PPM [3] type methods. Note that the sideways movement corresponds to the *context escape*.

STC_M1. Save the chunk length and the edge choices made in nodes.

STC_M2. Save the edge choice when in a node. When on an edge, write down whether or not AP reached the second vertex. If it did not, note how far it moved from the current position on this edge.

STC_M3. differs from STC_M2 in case AP does not reach the second vertex. If this happens, it falls back to STC_U-like single decision description.

STC_M4. is an STC_M1 modification. When AP ends up in a node after a sideways movement, just the single decision made in that node is saved and only later choices are grouped.

All the STC_M methods behave like a dictionary compression algorithm with a PPM dictionary choice. However, when compared to the work of Hoang et.al. [12], STC_M dictionaries are more compact and higher level contexts are incorporated. Another similar method by Bloom [13] has only one phrase in each context dictionary and is based on hashing.

4.3   Example

To illustrate how STC algorithms work, we will use a simple example with string *cocoa* and method STC_M1. All 10 steps of STree(*cocoa*) construction are depicted in Fig. 1. Note that STC_M1 chunks are: (0), $(1, c)$, $(3, oc)$, (0), (0) and $(1, a)$, where the first component of each tuple is chunk length and the optional second component is list of edge choices made.

5   Implementation

To evaluate the proposed algorithms, we tested methods STC_U and STC_M1 paired with a combination of two well known arithmetic coding techniques: the Piecewise Integer Mapping [14] and Schindler's Byte Renormalisation [15]. Suffix tree implementation is similar to "sliding window" implementations by Larsson

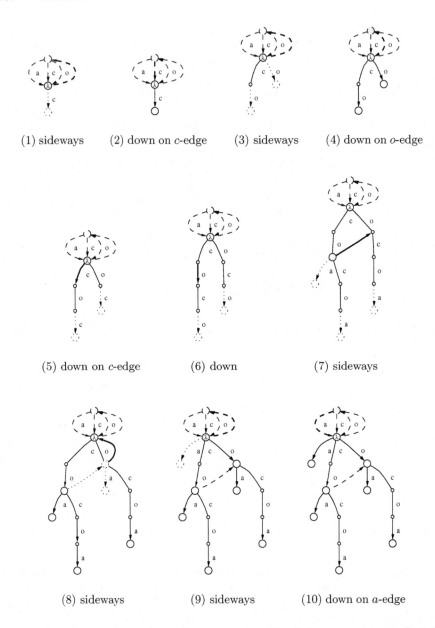

(1) sideways   (2) down on c-edge   (3) sideways   (4) down on o-edge

(5) down on c-edge   (6) down   (7) sideways

(8) sideways   (9) sideways   (10) down on a-edge

Fig. 1. Step by step STree (*cocoa*) construction. Big circles are nodes, small circles are locations on edges and lines that start in node go through zero or more edge locations and end with arrow in another node are edges. Solid objects are regular suffix tree nodes and edges, dashed objects are auxiliary and dotted objects were created in the particular step. Single characters along edges form edge labels. Bold arrows and subfigure labels describe AP movement

[9] and Fiala and Greene [8]. To estimate the probability of movement choice, we employ a very simple method using counters in nodes combined with AX estimation method [16] and *exclusion* technique [17]. For more details see [18].

6   Experiments

We have performed a series of experiments to compare the proposed methods with existing methods represented by BZip2 [19], CTW [20, 21], GZip [22]. STC_M1 and STC_U were run with the "sliding window" size of 2^{21} and other methods with their maximum compression settings turned on. All tests were run under Debian GNU/Linux 3.0 on Athlon 1.3GHz processor with 256MB DDR RAM. The source code was compiled using a GNU C++ version 2.95.4. More detailed tests can be found in [18].

The following three data sets were used: the standard Canterbury Corpus [23], its optional large part and the Linux 2.4.14 kernel archive [24]. Note that

Table 1. Canterbury Base Corpus compression results

Cant. Base	text 149K	play 122K	html 24K	csrc 11K	list 4K	excl 1006K	tech 417K	poem 471K	fax 501K	sprc 37K	man 4K	Avg. 250K
BZip2	2.27	2.53	2.48	2.18	2.76	1.01	2.02	2.42	**0.78**	2.70	3.33	2.23
CTW	**2.06**	**2.32**	**2.31**	**1.97**	**2.38**	**0.97**	**1.81**	**2.18**	0.79	**2.54**	**2.98**	**2.03**
GZip	2.85	3.12	2.60	2.25	2.68	1.63	2.71	3.23	0.82	2.67	3.32	2.53
STC_M1	2.88	3.24	2.91	2.49	2.94	2.36	2.65	3.20	1.21	3.14	3.62	2.79
STC_U	2.42	2.70	2.41	2.11	2.47	1.45	2.18	2.65	0.90	2.67	3.08	2.28

Table 2. Canterbury Large Corpus and Linux 2.4.14 results

Cant. Large	bible 3953KB com. bpc	com. KB/s	dec. KB/s	E.coli 4530KB com. bpc	com. KB/s	dec. KB/s	world 2415KB com. bpc	com. KB/s	dec. KB/s	Avg. 3632KB com. bpc
BZip2	1.671	1168	3790	2.157	**1091**	3109	1.583	1125	3594	1.804
CTW	**1.464**	13	13	**1.939**	6	6	**1.298**	22	22	**1.567**
GZip	2.325	**2653**	**47641**	2.240	474	**46224**	2.333	**4745**	**51392**	2.299
STC_M1	2.245	274	2	2.968	242	4	1.949	314	2	2.387
STC_U	1.798	266	239	2.665	258	238	1.513	298	262	1.992

Linux 2.4.14	linux-2.4.14.tar 123600K com. bpc	com. KB/s	dec. KB/s
BZip2	1.466	1172	5154
CTW	1.814	11	11
GZip	1.814	**3878**	**60707**
STC_M1	1.689	202	5
STC_U	**1.373**	192	176

there are many results available for the Canterbury Corpus at the corpus site [23]. The compression performance in bits per character (bpc) and speed in kilobytes per second are in tables 1 and 2. Note that method STC_U is much better than STC_M1 in both compression performance and decompression speed. STC_U is sometimes very close to the best compression performance of the other three methods and surprisingly even outperforms them on the Linux test, where most of the decisions are made on edges and its simple probability estimation works well. The speed tests have shown that when compressing the STC methods are significantly faster than CTW and also much slower than BZip and GZip. This changes a bit on decompression, where STC_M1 is the slowest method, probably due to a very large number of short chunks.

7   Conclusion

In this paper we have demonstrated the possibility of creating a suffix tree based compression methods. We have shown that many different methods can be simulated by STC variants, moreover, we have presented some new methods from the same family and surely other methods may be added.

Though the tested algorithms have shown promising results, there definitely is some room for improvement both in speed and compression. The slow part of our methods is the modelling as the coding part takes only about 3% of the total run time. The most time-consuming part is the edge list traversal algorithm. To help compression, we should use a better method for escape probability estimation and a clever method of probability estimation in higher-level contexts. Promising candidates for this are Shkarin's PPMII techniques [25].

While this paper concentrated on the use of the suffix tree, it is clear that related data structures like suffix trie, DAWG and CDAWG could be used as well (see for example [18, 26]). Especially DAWG and CDAWG look interesting, however, there are problems with the leftest character removal for these two structures (cf. [27]).

Acknowledgements. This research was supported in part by GAČR grant 201/03/0912.

References

1. Weiner, P.: Linear pattern matching algorithms. In: Proceedings of the 14th Annual IEEE Symposium on Switching and Automata Theory. (1973) 1–11
2. Burrows, M., Wheeler, D.J.: A block-sorting lossless data compression algorithm. Research Report 124, Digital Systems Research Center, Palo Alto, California, USA (1994)
3. Cleary, J.G., Witten, I.H.: Data compression using adaptive coding and partial string matching. IEEE Transactions on Communications **COM-32** (1984) 396–402
4. Ziv, J., Lempel, A.: A universal algorithm for sequential data compression. IEEE Transactions on Information Theory **IT-23** (1977) 337–343

5. McCreight, E.M.: A space-economical suffix tree construction algorithm. Journal of the Association for Computing Machinery **23** (1976) 262–272
6. Ukkonen, E.: On-line construction of suffix trees. Algorithmica **14** (1995) 249–260
7. Farach, M.: Optimal suffix tree construction with large alphabets. In: Proceedings of the 38th Annual IEEE Symposium on Foundations of Computer Science. (1997) 137–143
8. Fiala, E.R., Greene, D.H.: Data compression with finite windows. Communications of the Association for Computing Machinery **32** (1989) 490–505
9. Larsson, N.J.: Structures of String Matching and Data Compression. PhD thesis, Department of Computer Science, Lund University, Sweden (1999)
10. Harary, F.: Graph Theory. Addison-Wesley, New York (1969)
11. Balkenhol, B., Kurtz, S., Shtarkov, Y.M.: Modifications of the Burrows and Wheeler data compression algorithm. In Storer, J., Cohn, M., eds.: IEEE Data Compression Conference. (1999) 188–197
12. T.Hoang, D., Long, P.M., Vitter, J.S.: Dictionary selection using partial matching. Information Sciences **119** (1999) 57–72
13. Bloom, C.R.: LZP: A new data compression algorithm. In Storer, J., Cohn, M., eds.: IEEE Data Compression Conference. (1996) 425 http://www.cbloom.com/src/index_lz.html.
14. Stuiver, L., Moffat, A.: Piecewise integer mapping for arithmetic coding. In Storer, J., Cohn, M., eds.: IEEE Data Compression Conference. (1998) 3–12
15. Schindler, M.: Byte oriented arithmetic coding. In Storer, J., Cohn, M., eds.: IEEE Data Compression Conference. (1998) Poster presentation.
16. Moffat, A., Neal, R.M., Witten, I.H.: Arithmetic coding revisited. ACM Transactions on Information Systems **16** (1998) 256–294
17. Bell, T., Witten, I.H., Cleary, J.G.: Modelling for text compression. ACM Computing Surveys **21** (1989) 557–591
18. Senft, M.: Lossless data compression using suffix trees. Master's thesis, Faculty of Mathematics and Physics, Charles University, Prague, Czech Republic (2003) (in czech).
19. Seward, J.: BZip2. (http://sources.redhat.com/bzip2/)
20. Willems, F.M.J., Shtarkov, Y.M., Tjalkens, T.J.: The context-tree weighting method: Basic properties. IEEE Transactions on Information Theory **IT-41** (1995) 653–664
21. Franken, E., Peeters, M., Tjalkens, T.J.: CTW 0.1b2. (http://www.ele.tue.nl/ctw)
22. Gailly, J.L.: GZip. (http://www.gzip.org/)
23. Arnold, R., Bell, T.: A corpus for the evaluation of lossless compression algorithms. In Storer, J., Cohn, M., eds.: IEEE Data Compression Conference. (1997) 201–210 http://corpus.canterbury.ac.cz/.
24. Torvalds, L., et al.: Linux. (ftp://ftp.cz.kernel.org/pub/linux/)
25. Shkarin, D.: PPM: One step to practicality. In Storer, J., Cohn, M., eds.: IEEE Data Compression Conference. (2002) 202–211
26. Inenaga, S., et al.: On-line construction of compact directed acyclic word graph. In Amir, A., Landau, G.M., eds.: Proceedings of the 12th Annual Symposium on Combinatorial Pattern Matching (CMP '01). Volume 2089 of Lecture Notes in Computer Science., Springer-Verlag (2001) 169–180
27. Inenaga, S., Shinohara, A., Takeda, M., Arikawa, S.: Compact directed acyclic word graphs for a sliding window. Journal of Discrete Algorithms **2** (2004) 25–51 (special issue for SPIRE'02).

Tier Aspect Model Based on Updatable Views

Radosław Adamus[1] and Kazimierz Subieta[1, 2, 3]

[1] Computer Engineering Department, Technical University of Lodz, Lodz, Poland
[2] Institute of Computer Science PAS, Warsaw, Poland
[3] Polish-Japanese Institute of Information Technology, Warsaw, Poland
radamus@kis.p.lodz.pl, subieta@ipipan.waw.pl

Abstract. The tier aspect model addresses Aspect Oriented Programming (AOP) in the context of database applications. It is a new technique of separation of concerns through tiers implemented as virtual object-oriented updatable views. The code of an aspect related to particular objects is encapsulated within a separate view that overrides all generic operations performed on the objects. Aspect tiers form a chain of views. Such additional codes can do any extra action e.g. security, licensing, integrity constraints, monitoring and others. The idea is based on the Stack-Based Approach to object-oriented query languages and the novel idea of virtual updatable views.

1 Introduction

In this paper we focus attention on specific database features that in the context of software change become tangled aspects [2,4], i.e. they tend to be dispersed along many pieces of data or code, assuming typical relational, object-relational or object-oriented DBMSs. Our idea is to introduce to an object-oriented DBMS some new facilities that allow the programmer to focus such aspects in continuous code pieces.

As an example of a tangled database aspect consider a bank database whose *Account* objects contain the *Balance* attribute. Assume that some 5 years after launching the application the requirement to this attribute has been changed: any user that reads this attribute or makes any other operation must be recorded at a special log file. The *Balance* attribute is used in hundreds of places along the application code. We can suppose that references to *Balance* can be hidden within dynamic SQL statements, i.e. they are not explicitly seen from the application program code. This could make the task of discovering all places where the attribute *Balance* is used extremely difficult. Note that the trigger technique in this case is inapplicable because triggers cannot be fired on read events. In classical databases the only way to fulfill this requirement is long and costly adjustment of the code in those hundreds of places.

We propose to cope with such cases by means of virtual updatable database views. The idea is that each generic operation acting on virtual objects (including read operations) can be overloaded by a procedure which implements the mapping of the operation to operations on stored database objects. Regarding the above example, we propose to use the view named *Balance* that overloads the original *Balance* attribute. Within the view one can put any additional code that does the required action. Because the name of the view is the same as the name of the attribute all the bindings

M. Bieliková et al. (Eds.): SOFSEM 2005, LNCS 3381, pp. 360–363, 2005.

to *Balance* come to the view. Only the view code contains bindings to the original *Balance* attribute. The updating semantics of the original *Balance* attribute can be retained or modified according to new requirements.

2 General View of the Tier Aspect Model

The tier aspect model [1] allows the designers and programmer to isolate, name encapsulate and implement software requirements related to aspects that are usually tangled in classical database applications. The goal of the model is twofold:

- Conceptually independent requirements can be implemented as separable (adjustable, deletable) conceptual units of the software;
- A new requirement, as a separate conceptual unit, can be added on top of the existing implemented conceptual units related to a particular data type.

A tier is implemented as a virtual updatable database view. To our purposes we introduce to the concept the following original qualities that so far are not implemented and even not considered in the database literature:

- Full transparent updateability of virtual objects delivered by the view. In current DBMS-s updateability is severely limited to avoid updating anomalies.
- A language for view definitions should have full computational power.
- A view definition can access and update entities from the database, metabase, application environment and operating system environment (e.g. files).
- The view language should address a powerful object-oriented database model.

Database views with the above properties are being developed in the context of the Stack-Based Approach (SBA) to object-oriented query languages [5]. SBA introduces a query/programming language SBQL which is then used to define virtual updatable views [3]. For the tier model we use views in different configuration by assuming that a new view named A is defined on top of (virtual or stored) objects named A. After introducing the view all external bindings to objects A will come through the view, and only the view can access the original objects A. In this way one can introduce any new tier related to a new requirement on objects A. Any population of database objects (collections, attributes, methods, etc.) can be covered by a chain of virtual updateable overloading views, where each view implements some isolated requirement concerning access semantics of corresponding objects.

3 Implementation of the Tier Aspect Model

The tier aspect model influences the typical stack-based semantics of query languages. We assume that stored or virtual objects named *n* in the database can be overloaded by an updateable view that delivers virtual objects named *n*. Overloading means that after the view has been inserted all bindings of name *n* invoke the view rather than return references to objects named *n*. Access to the original objects named *n* is possible only inside the overloading view, through special syntax. A managerial name of a view [3] (independent from the name of virtual objects delivered by the view) allows the administrator to make managerial operations on the views, e.g.

delete a view, update it, or change its position in a chain. Virtual objects delivered by an overloading view can be overloaded by a next overloaded view, with the same rules, Fig.1.

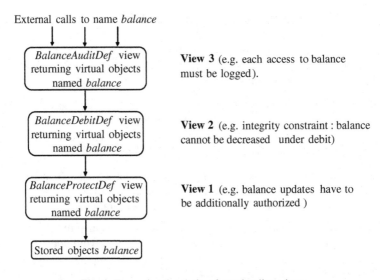

External calls to name *balance*

BalanceAuditDef view returning virtual objects named *balance*

View 3 (e.g. each access to balance must be logged).

BalanceDebitDef view returning virtual objects named *balance*

View 2 (e.g. integrity constraint : balance cannot be decreased   under debit)

BalanceProtectDef view returning virtual objects named *balance*

View 1 (e.g. balance updates have to be additionally authorized)

Stored objects *balance*

Fig. 1. Example of a chain of overloading views

A chain of overloading views has to be formed into a database structure. It should be possible to find the most outer view to which all external calls are to be bound. This should enable localizing the next view in the chain (the calling order). For full updating power it is also necessary to find a next view in the chain. In Fig.1 the most outer view is *BalanceAuditDef*.

4   Query Processing for Overloading Views

The view programmer needs language constructs that allow him/her to call the original (overloaded) object from the view body. On the syntax level of the query language the constructs introduce a keyword **original**. This syntax informs the binding mechanism that the binding must be special. The example below presents the definition of the *BalanceAuditDef* view defining the new semantic of the *Balance* objects (every read of the balance must be logged).

```
create overloading view BalanceAuditDef {
   virtual objects Balance { return original Balance as b}
   on_retrieve do { SaveLogFile(); return deref b }
   on_update(NewBalance) do { b:= NewBalance }
}
```

insert *BalanceAuditDef* **into** *AccountClass* **on top of chain** *Balance*; //administrative operation

Name *Balance* preceded by the keyword **original** requires binding to the previous view in the chain or to the original *Balance* object (if there is no more views in the chain). *Balance* not preceded by this keyword is treated as recursive call to the **virtual objects** *Balance* procedure placed in the first overloading view. There are no other binding possibilities in the model. In particular, it is not possible to call from a view another view in the chain. This assumption gives the opportunity to equip the administrative module with operation allowing deleting and changing the order of views in chains.

The environment stack mechanism should ensure that every call to name n preceded with the keyword **original** causes execution of a **virtual objects** procedure located in the view that is previous in the chain or, if no more views, binding to the original object. Because of the paper size limits we skip more detailed description.

5 Conclusion

In the paper we have proposed the tier aspect model as a new technique of separation of concerns through tiers implemented as virtual object-oriented updatable views. Thanks to full control over retrieve, update, insert and delete operations acting on virtual objects we are able to add new semantic to database objects. Our idea is that such views specifically overload original database objects; thus any new requirement can be put within a chain of such views. A view can be easily inserted in a chain, removed from it, changed or replaced after the requirements are changed.

Currently the prototype implementing updateable views is ready. We are adopting the prototype to our idea of the tier aspect model based on overloading views for the prototype object-oriented database server ODRA.

References

[1] R.Adamus, K.Subieta. Security Management Through Overloading Views. Proc. of International Conference on Ontologies, Databases and Applications of Semantics (ODBASE), 25 - 29 October 2004, Larnaca, Cyprus, Springer LNCS, to appear
[2] G.Kiczales, J.Lamping, A.Mendhekar, C.Maeda, C.Lopes, J.Loingtier, J.Irwin. Aspect-Oriented Programming. Proc. ECOOP Conf., Springer LNCS 1241, 220-242, 1997
[3] H.Kozankiewicz, J.Leszczyłowski, K. Subieta. Updateable XML Views. Proc. 7th ADBIS Conf., Springer LNCS 2798, 2003, 381-399
[4] A.Rashid. Aspect-Oriented Database Systems. Springer-Verlag, Berlin Heildelberg 2004.
[5] K.Subieta. Theory and Construction of Object-Oriented Query Languages. Editors of the Polish-Japanese Institute of Information Technology, 2004, 522 pages

Well-Founded Metamodeling for
Model-Driven Architecture

Liliana Favre

Universidad Nacional del Centro de la Provincia de Buenos Aires,
Comisión de Investigaciones Científicas de la Provincia de Buenos Aires,
Argentina
lfavre@exa.unicen.edu.ar

Abstract. The Model-Driven Architecture (MDA) is emerging as a technical framework to improve productivity, portability, interoperability , and evolution. Metamodeling plays a key role in MDA. A combination of formal specification techniques and metamodeling can help us to address Model-Driven developments. In this paper we describe a conceptual framework for MDA-based metamodeling that integrates UML/OCL and formal specifications. We present the specification language NEREUS for specifying UML metamodels. NEREUS can be viewed as an intermediate notation open to many other formal languages. A transformational system to translate UML/OCL to NEREUS was defined. We investigate the way to define metamodel mappings using UML/OCL and NEREUS.

1 Introduction

The Object Management Group (OMG) is facing a paradigm shift from object-oriented software development to model-centric development. A recent OMG initiative is the Model Driven Architecture (MDA) that is emerging as a technical framework for information integration and tools interoperation based on the separation of Platform Specific Models (PSMs) from Platform Independent Models (PIMs) [8].

One of the key challenges of the Model-Driven paradigm is to define, manage, and maintain traces and relationships between different models and model views in an automatic manner. MDA standard is still evolving and few tools provide support for it [11].

Metamodeling is a key facility in the new MDA paradigm. The metamodeling framework for the UML is based on an architecture with four layers: meta-metamodel, metamodel, model and user objects. The UML metamodel has traditionally been based on an abstract syntax consisting of UML class diagrams, well-formedness rules using the Object Constraint Language (OCL) and natural language [9]. Related metamodels and meta-metamodels such as MOF (Meta Object Facility), SPEM (Software Process Engineering Metamodel) and CWM (Common Warehouse Model) share common design philosophies [10]. To date several metamodeling approaches have been proposed to Model-Driven development [1], [2], [4], [7].

M. Bieliková et al. (Eds.): SOFSEM 2005, LNCS 3381, pp. 364–367, 2005.
© Springer-Verlag Berlin Heidelberg 2005

A combination of formal specification techniques and metamodeling can help us to address MDA. To enable automatic transformation of a model, we need the UML metamodel that is written in a well-defined language. In this direction, we propose the NEREUS language that is aligned with MDA. A transformational system to translate UML/ OCL to NEREUS was defined. Thus, the UML metamodel can be formalized, and its resulting models can be simulated and validated for consistency. We describe different correspondences that may hold between several models and metamodels that are expressed in terms of UML/OCL and NEREUS. NEREUS has a formal semantics in terms of a translation into the algebraic language CASL [3].

The structure of the rest of this article is as follows. Section 2 describes how to formalize metamodels. Section 3 presents the approach that is used for transforming models. Finally, Section 4 concludes and discusses further work.

2   Formalizing Metamodels

A combination of formal specification techniques and metamodeling can help us to specify model transformations independently of any particular technology. A formal specification clarifies the intended meaning of metamodels, helps to validate them and provides reference for implementation.

We propose the metamodeling language NEREUS that is aligned with metamodels based at least on the concepts of entity, associations and packages. NEREUS can be viewed as an intermediate notation open to many other formal languages. In particular, we define its semantics by giving a precise formal meaning to each of the constructions of the NEREUS in terms of the algebraic language CASL [3] that can be viewed as the centerpiece of a standardized family of specification languages.

NEREUS consists of several constructions to express classes, associations and packages. It is relation-centric, that is it expresses different kinds of relations (dependency, association, aggregation, composition) as primitives to develop specifications. It provides constructions to express clientship, inheritance and subtyping relations. NEREUS supports higher-order operations. Also, it is possible to specify any of the three levels of visibility for operations: public, protected and private. The axioms can be arbitrary first-order formulae.

NEREUS provides a taxonomy of constructor types that classifies associations according to kind (aggregation, composition, association, association class, qualified association), degree (unary, binary), navigability (unidirectional, bidirectional), connectivity (one-to-one, one-to-many, many-to-many). Associations describe interaction relations among objects as well as constraints over such related objects.

The UML metamodel can be described using a combination of OCL and UML. In this direction, we define a transformational system to transform UML/OCL class diagrams into NEREUS. The transformation process of OCL specifications into NEREUS is supported by a system of transformation rules. Analyzing OCL specifications we can derive axioms that will be included in the NEREUS specifications. This kind of transformations can be automated. A detailed description may be found at [5], [6].

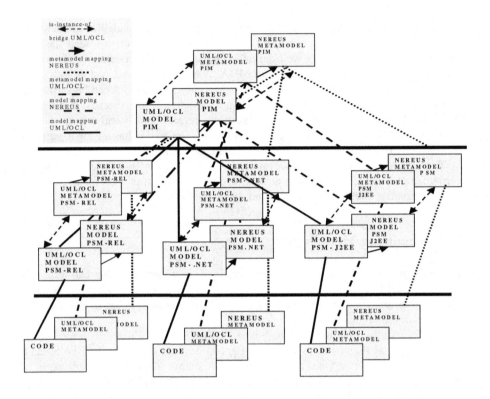

Fig. 1. Relations among models and metamodels

3 MDA Transformations

MDA proposes to use different models (PIM and PSMs) at different stages of the development and to employ automated transformations between them. The MDA process is divided into three main steps: construct a model with a high level of abstraction that is called Platform Independent Model (PIM); transform the PIM into one or more Platform Specific Models (PSMs), each one suited for different technologies, and transform the PSMs to code.

The PIMs, PSMs and code describe a system in different levels of abstraction. We propose to define PIMs and PSMs by integrating UML/OCL and NEREUS specifications. A PSM is tailored to specify different realizations of a PIM that are available in one specific technology, for example dot.NET, J2EE or relational.

The notion of mapping among models is crucial in the MDA approach. Fig. 1 shows a correspondence model. It defines various correspondences that may hold between several models and metamodels. Mappings are expressed in OCL and NEREUS. In the first case, a transformation between models/metamodels is a sequence of OCL transformations. A transformation rule is defined by its name, a source language, a target language, a source condition, a target condition and a mapping description in terms of rules written in OCL [9]. In NEREUS mappings are built by instantiating reusable schemes. Relations are employed to define which pairs the target and source models are connected. Transformations allow us to produce a trace between the elements of the target model and the elements of the source model. They can be automated and preserve the integrity among the different models.

4   Conclusions

In this paper we describe a conceptual framework for MDA-based metamodeling that integrates UML/OCL and formal specifications. The bases of our approach are the metamodeling language NEREUS and a transformational system to translate OCL to NEREUS. Mapping relations across models are also described.

We foresee to integrate our results in the existing UML tools and experiment with different platform such as dotNET and J2EE.

References

1. Akehurst, D., Kent, S.: A relational approach to defining transformations in a metamodel. In: Jezequel, J.M., Hussmann, H., Cook, S. (eds): Lecture Notes in Computer Science, Vol. 2460. Springer-Verlag, Berlin Heidelberg New York (2002) 243-258.
2. Bézivin, J., Farcet, N., Jézéquel, J., Langlois, B., Pollet, D.: Reflective Model Driven Engineering. In: Stevens, P., Whittle, J., Booch, G. (eds): Lecture Notes in Computer Science, Vol. 2863. Springer-Verlag, Berlin Heidelberg New York (2003) 175-189.
3. Bidoit, M., Mosses, P.: CASL User Manual- Introduction to Using the Common Algebraic Specification Language. Lecture Notes in Computer Science, Vol. 2900. Springer-Verlag, Berlin Heidelberg New York (2004).
4. Caplat, G., Sourrouille, J.: Model Mapping in MDA. In: Bezivin, J. France, R. (eds): Proc. UML'2002 Workshop in Software Model Engineering (WiSME 2002). http://www. metamodel.com/wisme-2002 (2002).
5. Favre, L.: A Formal Mapping between UML Static Models and Algebraic Specifications. In: Evans, A. France,R., Moreira, A., Rumpe, B. (eds): Practical UML-Based Rigorous Development Methods-Countering or Integrating the eXtremist, Lecture Notes in Informatics (P 7). SEW, GI Edition, Alemania (2001) 113-127.
6. Favre, L.: Foundations for MDA-based Forward Engineering. Journal of Object Technology (JOT). ETH Zurich, to appear (Jan/Feb 2005)
7. Haussmann, J.: Relations-Relating metamodels. In: Evans, A., Sammut, P. , Williams, J. (eds): Proc. Metamodeling for MDA. First International Workshop. York, UK (2003) 147-161.
8. MDA. The Model Driven Architecture: Object Management Group. www.omg.org/mda (2004).
9. OCL Specification. Versión 2.0.: Documento ptc/03-03-14. www.omg.org (2004).
10. OMG. Unified Modeling Language Specification, v1.5: Object Management Group. http://cgi.omg.org /cgi-bin/doc?ad/01-02-14 (2004).
11. UML Tools: www.objectsbydesign.com/tools/ (2004).

Stepwise Optimization Method for k-CNN Search for Location-Based Service

Jun Feng[1], Naoto Mukai[2], and Toyohide Watanabe[2]

[1] Hohai University, Nanjing, Jiangsu 210098 China
`fengjun-cn@vip.sina.com`
[2] Nagoya University, Nagoya, Aichi 464-8603 Japan

Abstract. In this paper, we focus on a typical situation of LBS which is to provide services for users in cars that move in a road network. For such kind of users, the problem of k-CNN search along a specific route on road network is to find out k nearest neighbor (k-NN) objects for any place on the route. k nearest neighbors are selected based on the path length from the route to the objects, and the continuous search for all the points on the route should be considered. A k-CNN search method is proposed by using an incremental k-NN search based on road network.

1 Introduction

The issue of how to provide location-based service (LBS) attracts many researchers. We focus on a typical situation of LBS which is to provide services for users in cars that move in a road network. For such kind of users, the problem of k continuous nearest neighbors (k-CNN) search along a specific route on road network is to find out k nearest neighbor objects for any place on the route, k nearest neighbors are selected based on the path length from the route to the objects. It is an instance of spatial distance semi-join problem [1]. We propose new strategies for efficiently processing k-CNN search on road network. The existing work for CNN search is almost presented from the computational geometry perspective [2] [3] [4]. CNN search was based on the straight-line distance between objects in these works. While researches on k-NN search have been conducted from the viewpoint of incremental spatial join [5, 6, 7] and distance browsing [1, 8]. The distance functions are all based on a distance metric for points, $dis(s, t)$, such as the Chessboard, Manhattan or Euler metric.

2 Preliminaries

A road network with nodes and links representing the cross-points and road segments can be regarded as a graph $G : G = (V, L)$, where V is a set of vertices $\{ v_1, v_2, ...v_n\}$, and L is a collection of edges $\{ l_1, l_2, ...l_m\}$, used to indicate the relationship between vertices. The predefined route from a start point v_s to an end point v_e is given by an array $Route(v_s, v_e)\{(v_{r1}, ..., v_{ri}, ..., v_{rn})|v_{r1} =$

M. Bieliková et al. (Eds.): SOFSEM 2005, LNCS 3381, pp. 368–371, 2005.

$v_s, v_{rn} = v_e, v_{ri} \in V$ A sub-route of $Route(v_s, v_e)$ is defined as $(v_{rl}, ..., v_{rj})$, which overlaps with $Route(v_s, v_e)$. If the target object set is $T = \{t_a, t_b, ...\}$ and $t_i \in T$ with a corresponding node $v_{ti} \in V$, the NN for v_{ri} on $Route(v_s, v_e)$ is t_i when the shortest path $Path_{Vri\_ti} = \{min(v_{ri}, ..., v_j, ...v_{ti})|v_j \in V, ti = t_a, t_b, ...\}$.

[Definition 1]. *Node $v_d \in V$ is called the divergence point between $Route(v_s, v_e)$ and $Path_{Vri\_ti}$ if: 1) The sub-route $(v_{ri}, ..., v_d)$ of $Path_{Vri\_ti}$ is also a sub-route of $Route(v_s, v_e)$; 2) The node following v_d along $Path_{Vri\_ti}$ is not on $Route(v_s, v_e)$.*

[Definition 2]. *Node $v_c \in V$ is called the computation point of $Route(v_s, v_e)$ if: 1) v_c is the start point of $Route(v_s, v_e)$; 2) v_c is on $Route(v_s, v_e)$ and is also a node on the route following a divergence point between $Route(v_s, v_e)$ and $Path_{Vri\_ti}$.*

To solve the CNN problem, there are two main issues: one is the selection of computation point on the route; and another is the computation of NN for the computation point. Here, we give two propositions on the road network for nearest object search, and the details of them is given in [9].

[Proposition1]. *For a source point S and a target object t, when the length of a path from S to t is r, any target object which is nearer to S than t can only be found inside a circle region, denoted as r-region, whose center is S and whose radius is r .* □

[Proposition 2]. *For two points S and t on the road network with straight-line distance d, the test of whether there is a path shorter than r from S to t can be based on a path search region, denoted as p-region, the sum of the straight-line distance between any nodes inside this region and S and that between this node and t is not longer than r.* □

3 k-CNN Search Method

The problem of k-CNN search which we address in this paper is to find k-NN's for any point along a specific route on a large road network. k-NN's are the first k target objects from the point on the route on the sort of the shortest path length.

3.1   Bounds of the Shortest Path Length

Observe Figure 1: in 2-CNN search, 2-NN's for the first computation point S are t and t_g. To find 2-NN for the next computation point c can take advantage of the previous computation, for example at least t and t_g can be regarded as 2-NN up to now, which are with the possible longest paths from c: $(Path_{cq} + Path_{qt})$ and $(Path_{cq} + Path_{qt_g})$. However, the real path length from c to them may be varied. This is because there may be some shorter paths, and the lower and

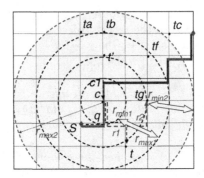

Fig. 1. There are lower and upper bounds of path length from the current computation point to the previous k-NN's

upper bounds of the path length from c to t are r_{min} and r_{max}, which can be decided by $r_{min} = |Path_{cq} - Path_{qt}|$; $r_{max} = Path_{cq} + Path_{qt}$. Those for t_g are r_{ming} and r_{maxg}. It means that t is a NN candidate for c with a possible path length varied from r_{min} to r_{max}. The value of $Path_{qt}$ has been computed in the previous NN search step for S, and the value of $Path_{cq}$ is the curve length between c and q. Though r_{maxg} is greater than r_{max}, it can to say the path length to t_g is longer than that to t.

Based on **Proposition 1**, if r-*region* is decided for C with the radius r_{maxg}, the target objects nearer than the up-to-now known 2-NN's can be found only inside this r-*region*.

3.2   Data Structures for k-CNN Search

The NN search process bases on the R-tree index and a priority queue *Queue* [10]. *Queue* is used to record the intermediate results: the candidate targets or the internal nodes of R-tree [11] inside r-*region*. The key used to order the elements on *Queue* is the straight-line distance of R-tree node and the path length computed for target object. In the process of searching k-NN's for a computation point, the priority queue can also be used. When an object with the computed path length turns out on the head of the queue, the first (or the nearest) neighbor is found. On the next time, the second (2-nearest) neighbor will be returned. However, in the process of CNN search, it assures the objects found on the head of queue are in order, but cannot assure that there are enough (here, k) objects in the queue. To solve this problem, another data structure, called k-queue, is adopted to keep the up-to-now k candidates and distances (or distance bounds) for k-NN search. The greatest upper bound of distance is regarded as a cutoff value for pruning nodes or objects: a node or an object with a longer path is not inserted into the priority queue, while there are at least k candidates kept in the priority queue. k-queue is defined as a fixed length queue, where 1) k is the number of nearest neighbors found for every point on the predefined route. It is decided at the beginning of k-CNN search, and be kept in the process of the search. 2) The elements inside the queue are triplets $< t, r_{min}, r_{max} >$, where t is a target object and r_{min} and r_{max} are the lower and upper bounds of the path length from the current computation point to t. If the real path length to t has been computed, then r_{min} and r_{max} are set as the same value. There are at most k elements in the queue. 3) The longest r_{max} in k-queue is regarded as a cutoff value, which is used to set r-*region* for pruning objects in the priority queue *Queue*.

4   Conclusion

From a viewpoint of decreasing the times of disk access, we proposed a method for k-CNN search along route on road network by using small search regions. Our method is based on the premise of that the distance from one place to another on the road network is the path length of them. The search regions can be decided based on the path length, and the filtering condition in the search via hierarchical data structures can take advantage of it. In ITS applications, there are applications, such as CNN, k-NN or k-CNN search, are based on the travel cost, sometimes dynamical values, from one place to another. Because the travel cost may not be in direct proportion to their path length or the straight-line distance, our method proposed in this paper cannot be used to solve this problem directly.

References

1. Hjaltson, G., Samet, H.: Incremental distance join algorithms for spatial databases. Proc.of ACM-SIGMOD (1998) 237–248
2. Tao, Y.F., Papadias, D., Shen, Q.M.: Continuous nearest neighbor search. Proc. of VLDB'02 (2002) 287–298
3. Song, Z.X., Roussopoulos, N.: K-nearest neighbor search for moving query point. Proc. of SSTD'01 (2001) 79–96
4. Bespamyatnikh, S., Snoeyink, J.: Queries with segments in voronoi diagrams. SODA (1999)
5. Jacox, E.H., Samet, H.: Iterative spatial join. ACM Trans. Database Syst. **28** (2003) 230–256
6. Lee, D.H., Kim, H.J.: An efficient technique for nearest-neighbor query processing on the spy-tec. IEEE Trans. on Knowledge and Data Engineering **15** (2003) 1472–1486
7. Shin, H., Moon, B., Lee, S.: Adaptive and incremental processing for distance join queries. IEEE Trans. on Knowledge and Data Engineering **15** (2003) 1561–1578
8. Hjaltson, G., Samet, H.: Distance browsing in spatial databases. ACM Transactions on Database Systems **24** (1999) 265–318
9. Feng, J., Mukai, N., Watanabe, T.: Multi-level transportation network adaptable to traffic navigation. Proc. of IEA/AIE 2004, (2004) 164–169
10. Feng, J., Watanabe, T.: A fast search method of nearest target object in road networks. Journal of the ISCIE **16** (2003) 484–491
11. Guttman, A.: R-trees: A dynamic index structure for spatial searching. Proc. of ACM SIGMOD'84 (1984) 47–57

An Approach for Integrating Analysis Patterns and Feature Diagrams into Model Driven Architecture

Roman Filkorn and Pavol Návrat

Institute of Informatics and Software Engineering,
Faculty of Informatics and Information Technologies,
Slovak University of Technology,
Ilkovicova 3, SK-84216 Bratislava, Slovakia
{filkorn, navrat}@fiit.stuba.sk

Abstract. Knowledge-capturing approaches cover a wide variety of software engineering activities, and this variety impedes to find relations and integration between them. One of movements towards knowledge reuse and integration is Model Driven Architecture, a model-centric approach based on platform-specific abstractions and model transformations. We attempt to integrate knowledge captured in Analysis Patterns extended with Feature Modelling, and we propose a method for the initial steps in MDA approach.

1 Introduction

Techniques capturing software engineering best practices knowledge in a concept of a pattern matured to the state when new opportunities arise. The reuse of proved solutions is applied into consecutive development activities. Possible integration platform is OMG's Model Driven Architecture (MDA) [5]. The competence to navigate in a space of patterns' inherent variability and the potential to integrate the knowledge from the various patterns is still open and a challenge to research.

2 Developing PIMs – Platform Independent Models

Analysis patterns constitute an attempt to reuse vertical domain knowledge, usable as a base for building a PIM. To be reusable, the patterns are represented in a highly abstract and general form. The patterns abstract and generalize platform specific details that have to be supplemented (concretised, specialized) in the process of pattern instantiation. The patterns cover a lot of variability inherent in the particular context and the problem they solve, and that variability needs to be more appropriately documented. It is the accurate selection and order of specializations and concretisations, along with proper variation and commonality localization that has impact on the successfulness of a software development process.

For the purpose of our study, we chose Fowler's Corporate Finance analysis patterns in their published form (in Fig.1, adapted from [1]). To systematically cope with the granularity of analysis patterns catalogues, the abstraction, the generality and the variability of analysis pattern, we formulated following steps:

M. Bieliková et al. (Eds.): SOFSEM 2005, LNCS 3381, pp. 372–375, 2005.

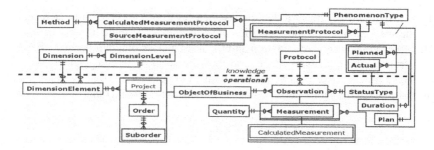

Fig. 1. Corporate Finance Analysis Patterns (adapted from [1])

1. Find an appropriate analysis patterns schema.
2. For each schema's pattern, decide if it shall be used, choose from alternatives.
3. For each selected pattern, abstract from details not indicated in the system context.
4. From the selected and modified patterns, develop a single model; according to the patterns definition, distinguish carefully the knowledge and the operational level.
5. Review the model; search for new relationships that could be abstracted.
6. Replace the points that play a significant role in the original model and should be substituted with specifics from the context of the problem that is being solved.
7. Review the model to identify variation points.

For each variation point, carry out feature modelling process; consider a single feature model that covers all principal concepts and their features.

The steps 7 and 8 introduce feature modelling process (see [2]) into the development of PIMs. The variation points are not explicitly documented in the analysis pattern documentation. Modelling variability is helpful for two levels: for the analysis pattern itself (results as an additional reusable knowledge), and for the concrete context (in specifying other aspects of the system). Omitting a discussion, we show an example in Fig. 2.

Representing multiple variation points in a single feature diagram can lead to an overwhelming set of diagram objects, lowering diagram's comprehensibility and clarity. We enriched out feature diagrams with a new notation for feature nodes that represent variability points but the diagram abstracts the details away.

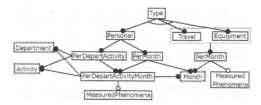

Fig. 2. Feature diagram with introduced notation

3 Transformation into PSMs – Platform Specific Models

In Table 1, we make an attempt in stating the marking method in a more formal way. The source models are the developed PIM and additional feature diagrams. The method marks and separates a data-persistence level and an object-oriented realization. The specification and the separation are better done in single transformation step relating the persistence level diagram and the design class diagram because they both source from a single analysis model and they are strongly interdependent. We view the marking process as a creative process with an opportunity to find and model additional concepts and relationships, therefore influencing both diagrams.

Table 1. Proposed transformation method

Source Platform Models	**Analysis patterns**, documenting the knowledge of the problem domain **Feature models**, capturing additional information about commonality and variability in the analysis model
Target Platform Models	**Persistent Data Class diagram**, defining persistent data entities and logical view relationships **Design Class diagram**, defining the framework for representing Domain Model in the application

Trans-formation Steps	**1. Mark the PIM According to the Level**
	     Attach the "level" mark (with values "operational", or "knowledge") to all concepts according to the PIM
	 2. Mark the PIM According to the Feature Models
	     Attach the "variation point" mark according to the identified variation points in the developed Feature Models
	 3. Mark the PIM Towards the Persistent Data Model
	     Select and attach marks on persistent constituents
	 4. Mark the PIM Towards the Design Diagram
	     Select and attach marks on concepts on design level
	 5. Perform the Transformations

In our ongoing research, we analyse and relate used modelling concepts and propose modelling alternatives for a mapping into both persistent level and object-realization model. We attempt to state a set of rules for marking processes in steps 3 and 4 of our method. A general outline of the marking process method is defined as follows:

1. If not empty, select first from the list of non-marked concepts, otherwise end
2. According to „level" mark, choose the *set of applicable marks*
3. From the set of applicable marks, *choose* a mark and attach it to the concept
4. Add the concept to the list of marked concepts
5. According to the *set of rules*, resolve all rules applicable to the new situation
6. Continue with step 1.

For both marking processes, we are working to specialize the three open points: a set of applicable marks, selection of an appropriate mark, and a set of rules. An example of a rule is in Fig. 3.

```
if   concept.hasMark("level", "operational") and
     concept.hasMark("variationpoint")
then concept.attachMark("enumeration")
```

Fig. 3. Example from the set of rules

It is essential to carefully choose the concepts that will have a persistent mapping with a correct mechanism. To quantify the variability of the concepts in the hierarchy, we use feature modelling technique. The concepts from the analysis model knowledge level represent valuable abstractions, even though their persistence might not be needed. It results in a modified PIM in the way that reduces or concretises the generalization hierarchies. The marking process towards the object-realization level of the analysis pattern is a promising area for reusing knowledge from design patterns. We are professed to iterative, incremental, and interactive method from [3].

The method is not automatable and requires human developer, who performs the steps and decides on designing questions not covered by it.

4 Conclusions and Future Work

We are working on a methodological framework for integration of various knowledge capturing pattern approaches into the context of Model Driven Development. One can anticipate that their tighter interconnection and integration would bring new qualities and impulses for both the knowledge capturing techniques and the quality of software development process.

We are considering switching to an UML-based representation of the knowledge captured in the patterns. We continue in specifying and concretising open areas, i.e. our proposed marking processes. We plan to develop a catalogue that will to some extent help in automating analysis, design and generation of a software product.

Acknowledgements: The work reported here was partially supported by Slovak Scientific Agency, grant No. VG 1/0162/03.

References

1. Fowler, M.: Analysis Patterns: Reusable Object Models. ISBN: 0201895420. Addison-Wesley Professional (1996)
2. Czarnecki, K., Eisenecker, U.: Generative Programming: Methods, Tools, and Applications. ISBN: 0201309777. Addison-Wesley Professional (2000)
3. Marko, V.: Template Based, Designer Driven Design Pattern Instantiation Support, In: Proc. ADBIS 2004 (LNCS 3255), Budapest, Springer Verlag (2004)
4. Filkorn, R., Návrat, P.: Modeling Optimization Components for XML Processing, In: M. Benes (ed.): Proc. of ISIM'04, Ostrava, ISBN 80-85988-99-2, Stefan MARQ (2004).
5. Miller, J., Mukerji, J. (eds.): MDA Guide, accessible: http://www.omg.org/docs/omg/03-06-01.pdf, OMG (2003)

Outerplanar Crossing Numbers of 3-Row Meshes, Halin Graphs and Complete p-Partite Graphs*

Radoslav Fulek[1], Hongmei He[2], Ondrej Sýkora[2], and Imrich Vrťo[3]

[1] Department of Computer Science, Comenius University, Mlynská dolina, 842 48
Bratislava, Slovak Republic
radkofulek@pobox.sk
[2] Department of Computer Science, Loughborough University, Loughborough,
Leicestershire LE11 3TU, United Kingdom
{h.he, o.sykora}@lboro.ac.uk
[3] Department of Informatics, Institute of Mathematics, Slovak Academy of Sciences,
Dúbravská 9, 841 04 Bratislava, Slovak Republic
vrto@savba.sk

Abstract. An outerplanar (also called circular, convex, one-page) drawing of an n-vertex graph G is a drawing in which the vertices are placed on a circle and each edge is drawn using one straight-line segment. We derive exact results for the minimal number of crossings in any outerplanar drawings of the following classes of graphs: 3-row meshes, Halin graphs and complete $p-$partite graphs with equal size partite sets.

1 Introduction

Let $G = (V, E)$ denote a graph and $deg(v)$ denote the degree of $v \in V$. A drawing of a graph G with vertices of the graph placed on a circle and the edges drawn as straight-line segments is called *outerplanar drawing* of G. The *outerplanar crossing number* $\nu_1(G)$ is defined as the minimum number of pairs of crossing edges over all outerplanar drawings of G [1] (to find the outerplanar crossing number is NP-hard problem [2]).

We use this notation for outerplanar crossing number in accordance with the k-page crossing number ν_k [4, 5]). There are other notations and terminologies used for this quantity as circular, convex and one-page crossing number.

Let $D(G)$ denote an outerplanar drawing of G and let $\nu_1(D(G))$ denote the number of crossings in this drawing.

The only known exact result for outerplanar crossing numbers is in [3]. It is shown for complete bipartite graphs and for example in the case that m divides n, it holds that $\nu_1(K_{m,n}) = \frac{1}{12}n(m-1)(2mn - 3m - n)$ and in the case $m = n, \nu_1(K_{n,n}) = n\binom{n}{3}$.

* This research was supported by the EPSRC grant GR/S76694/01 and by VEGA grant No. 2/3164/23.

M. Bieliková et al. (Eds.): SOFSEM 2005, LNCS 3381, pp. 376–379, 2005.

In this short contribution we give the propositions on outerplanar crossing numbers of the 3-row meshes, Halin graphs and complete p-partite, $p \geq 3$, graphs with equal sizes of the partite sets. One of the motivations for this kind of the research is to provide graphs which can be used to test heuristics for low crossing outerplanar drawings.

2   Meshes

Let P_m be an m-vertex path. Let $P_m \times P_n$ denote the $m \times n$ mesh graph. The graph $P_m \times P_n$ is a union of m paths of type P_n (called rows) and n paths of type P_m (called columns). Let us call the subgraph consisting of edges incident to the first (last) two column vertices the comb, respectively.

Theorem 1. *For any $n \geq 3$,*

$$\nu_1(P_3 \times P_n) = \begin{cases} 2n - 3, & \text{if } n \text{ is odd,} \\ 2n - 4, & \text{if } n \text{ is even.} \end{cases}$$

Proof. The order of vertices on the circle in an optimal drawing for even n is given by the Hamiltonian cycle. For odd n the order is same, except the last column, which follows the order of the last column in even case.

The lower bound is proved by induction on odd n. The cases $n = 3, 4, 5$ are proved by using a computer. By using a computer we also proved that in any drawing of $P_3 \times P_4$ every comb contains at least 4 crossings. Suppose that the claim holds for some odd $n \geq 5$. Consider an optimal outerplanar drawing of $P_3 \times P_{n+2}$. By deleting the edges of the 4-th,...,$(n+1)$-st column we get a graph homeomorphic to $P_3 \times P_4$. we conclude that the left comb of $P_3 \times P_4$ contains at least 4 crossing. It means that there are at least 4 crossings on the comb edges in the drawing of $P_3 \times P_{n+2}$. By deleting the comb edges we get a drawing $D(P_3 \times P_n)$. By the inductive assumption we have

$$\nu_1(P_3 \times P_{n+2}) \geq 4 + \nu_1(D(P_3 \times P_n)) \geq 4 + 2n - 3 = 2(n + 2) - 3.$$

Proof for even n is similar.

3   Outerplanar Crossing Numbers of Halin Graphs

A Halin graph G consists of a tree with m leaves connected by a cycle, with no vertices of degree 2 except the root.

Theorem 2. *For a Halin graph G, with a tree with m leaves:*

$$\nu_1(G) = m - 2 \tag{1}$$

Proof. Lower bound.
Given a drawing of G, number the leaves of G on circle starting from root in clockwise manner as $u_0, ...u_{m-1}$. The leaves of the tree of G divide the circle

into m intervals of internal tree vertices which are either empty or contain some vertices of the tree of G. Observe that any edge between vertices of two different intervals cross at least 2 edges of the cycle of G. Further observe that an edge incident with a leaf and a vertex from a non-neighbouring interval causes at least one crossing on an edge of the cycle. Now, consider edges incident with leaf vertices except u_0 and u_{m-1}. Any leaf edge except edges incident with u_0 and u_{m-1} must cause at least one crossing apart from the case when it is incident with a vertex in the neighbouring intervals. Denote the vertex v. For every interval there are at most two such leaves. As between the vertex v and the root exists a unique path in the tree, there have to be an edge in this path crossing the cycle at least twice. Thus we can assign to every leaf except u_0 and u_{m-1} at least one crossing which implies at least $m-2$ crossings in any outerplanar drawing of G.

Upper bound.

First we describe the construction of the order of the vertices. We assign a type $pre, in,$ or $post$ to every vertex of G in the following way: the root will be of type pre, the left most child of the root will be of type pre, the right most child of the root will be of type $post$, rest of the children of the root will be of type in. Type of $deg(v)-1$ children $a_1, ... a_{deg(v)-1}$ of a vertex v, except the root, is calculated as follows: if v is of the type in, then the children $a_1, ... a_{\lfloor \frac{deg(v)-1}{2} \rfloor -1}$ and $a_{\lfloor \frac{deg(v)-1}{2} \rfloor +2}, ... a_{deg(v)-1}$ are of type in, $a_{\lfloor \frac{deg(v)-1}{2} \rfloor}$ are of type $post$, $a_{\lfloor \frac{deg(v)-1}{2} \rfloor +1}$ are of type pre, if v is of type pre, $a_2, ... a_{deg(v)-1}$ are of type in, a_1 are of type pre, if v is of type $post$, then the children $a_1, ... a_{deg(v)-2}$ are of type in, $a_{deg(v)-1}$ are of type $post$. Define $p(v)$ the sequence of vertices of subtree with root v as follows (children of v are the same as above): if v is of type pre then $p(v) = v, p(a_1), p(a_2), ... p(a_{deg(v)-1})$. If v is of type $post$ then $p(v) = p(a_1), p(a_2), ... p(a_{(deg(v)-1)}), v$. If v is of type in then $p(v) = p(a_1), ... p(a_{\lfloor \frac{deg(v)-1}{2} \rfloor}), v, p(a_{\lfloor \frac{deg(v)-1}{2} \rfloor +1}), ... p(a_{deg(v)-1})$.

This drawing of the Halin graph contains exactly $m-2$ crossings. For lack of space, we skip the proof of this.

4   Complete p-Partite Graphs $K_n(p)$

In this section we prove an exact result for the outerplanar crossing number of complete p-partite graphs with equal sizes of the partite sets. Denote

$$K_n(p) = K_{\underbrace{n, n, ..., n}_{p}}.$$

Theorem 3. *For the complete p-partite graph with n vertices in each partite set*

$$\nu_1(K_n(p)) = n^4 \binom{p}{4} + \frac{1}{2}n^2(n-1)(2n-1)\binom{p}{3} + n\binom{n}{3}\binom{p}{2}.$$

Proof. First we use 2 known facts shown in [3].

$$\nu_1(K_{n,2n}) = \frac{1}{6}n^2(n-1)(4n-5) \tag{2}$$

$$\nu_1(K_{n,n}) = n\binom{n}{3} \tag{3}$$

Lower bound. In every drawing of $K_n(p)$ there are 3 types of crossings of edges: for $i = 2, 3, 4$, in the i-coloured crossing, the endvertices of the corresponding edges are coloured by i colours.

The number of the 2-coloured crossings is clearly $\binom{p}{2}\nu_1(K_{n,n})$.

The number of 3-coloured crossings is clearly at least $p\binom{p-1}{2}(\nu_1(K_{n,2n}) - 2\nu_1(K_{n,n}))$. We have p possibilities to choose the colour c_1 which appears twice among the endvertices of a crossing and $\binom{p-1}{2}$ possibilities to choose two distinct colours c_2 and c_3. Then we identify the colours c_2 and c_3, which gives the total number of $\nu_1(K_{n,2n})$ crossings. However this number contains the numbers of 2-coloured crossings given by the colours c_1, c_2 and c_1, c_3 which must be subtracted.

The number of the 4-coloured crossings is $\binom{p}{4}n^4$. Realize that any four vertices of distinct colours produce one 4-coloured crossing. Summing up the numbers of all 3 types of crossings and substituting (2) and (3) we get the lower bound.

Upper bound. Place the vertices of the partite sets evenly around a cycle, i.e., the vertices of every partite set form a regular n-gon. Then one can check, that the number of i-coloured crossings, for $i = 2, 3, 4$, is the same as in the lower bound proof.

References

1. Kainen, P.C., The book thickness of a graph II, *Congressus Numerantium*, 71 (1990), 121–132.
2. Masuda, S., Kashiwabara, T., Nakajima, K., Fujisawa, T., On the NP-completeness of a computer network layout problem, in: Proc. *IEEE Intl. Symposium on Circuits and Systems 1987*, IEEE Computer Society Press, Los Alamitos 1987, 292–295.
3. Riskin, A., On the outerplanar crossing numbers of $K_{m,n}$, *Bulletin ICA* 39 (2003), 7–15.
4. Shahrokhi, F., Sýkora, O., Székely, L.A., Vrťo, I., Crossing numbers: bounds and applications, in: *Intuitive Geometry*, Bolyai Society Mathematical Studies 6, (I. Bárány and K. Böröczky, eds.), Akadémia Kiadó, Budapest 1997, 179–206.
5. Shahrokhi, F., Sýkora, O., Székely, L.A., Vrťo, I., The book crossing number of graphs, *J. Graph Theory*, 21 (1996), 413–424.

Fast Bit-Vector Algorithms for Approximate String Matching Under Indel Distance

Heikki Hyyrö[1], Yoan Pinzon[2,*], and Ayumi Shinohara[1,3]

[1] PRESTO, Japan Science and Technology Agency (JST), Japan
helmu@cs.uta.fi
[2] Department of Computer Science, King's College, London, UK
pinzon@dcs.kcl.ac.uk
[3] Department of Informatics, Kyushu University 33, Fukuoka 812-8581, Japan
ayumi@i.kyushu-u.ac.jp

Abstract. The approximate string matching problem is to find all locations at which a query p of length m matches a substring of a text t of length n with at most k differences (insertions, deletions, substitutions). The fastest solutions in practice for this problem are the bit-parallel NFA simulation algorithms of Wu & Manber [4] and Baeza-Yates & Navarro [1], and the bit-parallel dynamic programming algorithm of Myers [3]. In this paper we present modified versions of these algorithms to deal with the restricted case where only insertions and deletions (called indel for short) are permitted. We also show test results with the algorithms.

1 IndelMYE Algorithm

The bit-parallel approximate string matching algorithm of Myers [3], MYE, is based on the classical dynamic programming approach where a $(m+1) \times (n+1)$ matrix D is computed using the well-known recurrence $D_{i,j} = \min \{D_{i-1,j-1} + \delta(p_i, t_j), D_{i-1,j}, D_{i,j-1}\}$, subject to the boundary condition $D_{0,j} = 0$ and $D_{i,0} = i$, where $\delta(p_i, t_j) = 0$ iff $p_i = t_j$, and 1 otherwise. The solution to the approximate string matching problem is all the locations j where $D_{m,j} \leq k$. MYE is based on the observation that the vertical and horizontal differences between adjacent cells in D (i.e. $D_{i,j} - D_{i-1,j}$ and $D_{i,j} - D_{i,j-1}$) have the value -1, 0, or 1, and the diagonal differences (i.e. $D_{i,j} - D_{i-1,j-1}$) have the value 0 or 1. This enables the algorithm, as presented in [2], to use the following length-m bit-vectors to represent the vertical, horizontal and diagonal differences:

— $Pv_i = 1$ iff $D_{i,j} - D_{i-1,j} = 1$, $Nv_i = 1$ iff $D_{i,j} - D_{i-1,j} = -1$
— $Ph_i = 1$ iff $D_{i,j} - D_{i,j-1} = 1$, $Nh_i = 1$ iff $D_{i,j} - D_{i,j-1} = -1$
— $Zd_i = 1$ iff $D_{i,j} = D_{i-1,j-1}$

The values of Pv and Nv are known for the initial case $j = 0$. The steps of the algorithm at text position j are as follows. First the new diagonal vector Zd'

---

* Part of this work was done while visiting Kyushu University. Supported by PRESTO, Japan Science and Technology Agency (JST).

M. Bielikóva et al. (Eds.): SOFSEM 2005, LNCS 3381, pp. 380–384, 2005.

is computed by using Pv, Nv and $M(t_j)$, where for each character λ, $M(\lambda)$ is a precomputed length-m match vector where $M(\lambda)_i = 1$ iff $p_i = \lambda$. Then the new horizontal vectors Ph' and Nh' are computed by using Zd', Pv and Nv. Finally the new vertical vectors Pv' and Nv' are computed by using Zd', Nh' and Ph'. The value of the dynamic programming cell $D_{m,j}$ is maintained during the process by using the horizontal delta vectors (the initial value is $D_{m,0} = m$). A match of the pattern with at most k errors is found whenever $D_{m,j} \leq k$.

The dynamic programming recurrence for indel distance is $D_{i,j} = $ (if $p_i = t_j$ then $D_{i-1,j-1}$ else $\min\{D_{i-1,j},\ D_{i,j-1}\}$), which makes also the case $D_{i,j} - D_{i-1,j-1} = 2$ possible. To help deal with this complication, we will use the following additional vertical and horizontal zero vectors.

— $Zv_i = 1$ iff $D_{i,j} - D_{i-1,j} = 0$, $Zh_i = 1$ iff $D_{i,j} - D_{i,j-1} = 0$

Naturally $Zv = \sim (Pv \mid Nv)$ and $Zh = \sim (Ph \mid Nh)$, where $\sim$ is the bit-wise complement operation. In the following we describe the steps of our algorithm for updating the bit-vectors at text position j, where $1 \leq i \leq m$ and $1 \leq j \leq n$.

***i.*)** The new diagonal vector Zd' is computed exactly as in MYE. That is, $Zd'=(((M(t_j)\ \&\ Pv) + Pv)\ ^\wedge\ Pv) \mid M(t_j) \mid Nh$.

***ii.*)** The new horizontal zero vector Zh' is computed by using Pv, Zv and Zd'. By inspecting the possible difference combinations, we get the formula $Zh'_i = (Zh'_{i-1}\ \&\ Pv_i\ \&\ (\sim Zd'_i)) \mid (Zv_i\ \&\ Zd'_i)$. We use a superscript to denote bit repetition (e.g. $001011 = 0^2101^2 = 0(01)^21$). The self-reference of Zh' implies that each set bit $Zh'_i=1$ in Zh' can be assigned into a distinct region $Zh'_{a..b}=1^{b-a+1}$, where $1 \leq a \leq i \leq b \leq m$, $Zv_a\ \&\ Zd'_a=1$ or $a=1$, $Zh'_{r-1}\ \&\ Pv_r\ \&\ (\sim Zd'_r)=1$ for $r \in a+1\ldots b$, and $Zh'_{b+1}\ \&\ Pv_{b+1}\ \&\ (\sim Zd'_{b+1})=0$. We also notice that the conditions $Zv_a\ \&\ Zd'_a=1$ and $Pv_r\ \&\ (\sim Zd'_r)=1$ for $r \in a+1\ldots b$ imply that $Zh'_r=1$ for $r \in a\ldots b$. If we shift the region $a+1\ldots b$ of set bits one step right to overlap the region $a\ldots b-1$ and perform an arithmetic addition with a set bit into position a, then the bits in the range $a\ldots b-1$ will change from 1 to 0 and the bit b from 0 to 1. These changed bits can be set to 1 by performing XOR. From this we derive the final formula: $Zh'=(((Zv\ \&\ Zd') \mid ((Pv\ \&\ (\sim Zd'))\ \&\ 0^{m-1}1)) + ((Pv\ \&\ (\sim Zd')) >> 1))\ ^\wedge\ ((Pv\ \&\ (\sim Zd')) >> 1)$.

***iii.*)** The new horizontal vector Nh' is computed as in MYE by setting $Nh' = Pv\ \&\ Zd'$, after which we can also compute $Ph'= \sim (Zh' \mid Nh')$.

***iv.*)** The new vertical vector Zv' is computed by using Zh', Ph', Zd' and Zv. We notice that $D_{i,j}=D_{i-1,j}$ iff either $D_{i-1,j}=D_{i-1,j-1}=D_{i,j}$, or $D_{i,j-1}=D_{i-1,j-1}$ and $D_{i-1,j}=D_{i-1,j-1} + 1$. In terms of the delta vectors this means that $Zv'_i=1$ iff $Zh'_{i-1}\ \&\ Zd'_i=1$ or $Ph'_{i-1}\ \&\ Zv_i\ \&\ (\sim Zd'_i)=1$. From this we get the formula $Zv'=(((Zh' << 1) \mid 0^{m-1}1)\ \&\ Zd') \mid ((Ph' << 1)\ \&\ Zv\ \&\ (\sim Zd'))$.

***v.*)** The new vertical vector Nv' is computed as in MYE by setting $Nv'=(Ph' << 1)\ \&\ Zd'$, after which we can also compute $Pv'= \sim (Zv' \mid Nv')$.

Fig. 1 (*upper left*) shows a high-level template for the bit-parallel algorithms, and Fig. 1 (*lower left*) shows the complete formula for computing the new difference

Algo-BitParallelSearch($p_1 \ldots p_m, t_1 \ldots t_n, k$)
▷ Preprocess bit-vectors
  Algo-PreprocessingPhase()
  For $j \in 1 \ldots n$ **Do**
    ▷ Update bit-vectors at text character j
      and check if a match was found
    Algo-UpdatingPhase()

IndelMYE-PreprocessingPhase
  For $\lambda \in \Sigma$ **Do**
    $M(\lambda) \leftarrow 0^m$
  For $i \in 1 \ldots m$ **Do**
    $M(p_i) \leftarrow M(p_i) \mid 0^{m-i}10^{i-1}$
  $Pv \leftarrow 1^m,\ Zv \leftarrow 0^m,\ Nv \leftarrow 0^m$
  $currDist \leftarrow m$

IndelMYE-UpdatingPhase
  $Zd' \leftarrow (((M(t_j)\&Pv) + Pv)\ {}^{\wedge}Pv) \mid M(t_j) \mid Nv$
  $X \leftarrow (Pv\ \& (\sim Zd')) >> 1$
  $Y \leftarrow (Zv\ \& Zd') \mid ((Pv\ \& (\sim Zd'))\ \& 0^{m-1}1)$
  $Zh' \leftarrow (X' + Y')\ {}^{\wedge}\ X'$
  $Nh' \leftarrow Pv\ \& Zd'$
  $Ph' \leftarrow \sim (Zh' \mid Nh')$
  $Zv' \leftarrow (((Zh' << 1) \mid 0^{m-1}1)\ \& Zd')$
      $\mid ((Ph' << 1)\ \& Zv\ \& (\sim Zd'))$
  $Nv' \leftarrow (Ph' << 1)\ \& Zd'$
  $Pv' \leftarrow \sim (Zv' \mid Nv')$
  If $Ph'\ \& 10^{m-1} \neq 0^m$ **Then**
    $currDist \leftarrow currDist + 1$
  If $Nh'\ \& 10^{m-1} \neq 0^m$ **Then**
    $currDist \leftarrow currDist - 1$
  If $currDist \leq k$ **Then**
    Report a match at position j

IndelWM-PreprocessingPhase
  For $\lambda \in \Sigma$ **Do** $\overline{M(\lambda)} \leftarrow 1^m$
  For $i \in 1 \ldots m$ **Do** $\overline{M(p_i)} \leftarrow \overline{M(p_i)} \mid 1^{m-i}01^{i-1}$
  For $d \in 0 \ldots k$ **Do** $\overline{R_d} \leftarrow 1^{m-d}0^d$

IndelWM-UpdatingPhase
  $\overline{R_0'} \leftarrow (\overline{R_0} << 1) \mid M(t_j)$
  For $d = 1$ **to** k **Do**
    $\overline{R_d'} \leftarrow ((\overline{R_d} << 1) \mid \overline{M(t_j)})$
        $\&\ \overline{R_{d-1}}\ \& (R'_{d-1} << 1)$
  If $\overline{R_k'}\ \& 10^{m-1} = 0^m$ **Then**
    Report a match at position j

IndelBYN-PreprocessingPhase
  For $\lambda \in \Sigma$ **Do** $Md(\lambda) \leftarrow (01^{k+1})^{m-k}$
  For $i \in 1 \ldots m - k - 1$ **Do**
    $Md(p_i) \leftarrow\ \sim ((0^{k+2}1)^{k+1} << (k+2)(m-k-i))$
        $\&\ Md(p_i)$
  For $i \in m - k \ldots m$ **Do**
    $Md(p_i) \leftarrow\ \sim ((0^{k+2}1)^{k+1} >> (k+2)(i-m+k))$
        $\&\ Md(p_i)$
  $Rd \leftarrow (01^{k+1})^{m-k}$

IndelBYN-UpdatingPhase
  $x \leftarrow (Rd >> (k+2)) \mid Md(t_j)$
  $Rd' \leftarrow ((Rd << (k+3)) \mid (0^{k+1}1)^{m-k-1}01^{k+1})$
      $\& (((x + (0^{k+1}1)^{m-k})\ {}^{\wedge}\ x) >> 1)$
      $\& (01^{k+1})^{m-k}$
  If $Rd'\ \& 0^{(k+2)(m-k-1)+1}10^k = 0^m$ **Then**
    Report a match at position j

Fig. 1. Bit-parallel approximate string matching under indel distance: a template (*upper left*), indelMYE (*lower left*), indelWM (*upper right*), and indelBYN (*lower right*)

vectors at text position j. The running time of indelMYE is $O(\lceil m/w \rceil n)$ as a vector of length m may be simulated in $O(\lceil m/w \rceil)$ time using $O(\lceil m/w \rceil)$ bit-vectors of length w. The cost of preprocessing is $O(\sigma \lceil m/w \rceil + m)$.

2   Bit-Parallel NFA Simulation Algorithms: IndelWM and IndelBYN

The bit-parallel approximate string matching algorithms of Wu & Manber [4] (WM) and Baeza-Yates & Navarro [1] (BYN) simulate a non-deterministic finite automaton (NFA), R, by using bit-vectors. R has $(k + 1)$ rows, numbered from 0 to k, and each row contains m states. Let $R_{d,i}$ denote the ith state on row d of R. $R_{d,i}$ is active after processing the jth text character iff $D_{i,j} \leq d$ in the corresponding dynamic programming matrix D, and so an approximate occurrence of the pattern is found when the state $R_{k,m}$ is active.

WM uses $(k + 1)\lceil m/w \rceil$ and BYN $\lceil (k + 2)(m - k)/w \rceil$ bit-vectors of length w to encode R. Both perform a constant number of operations per bit-vector at text position j. For reasons of space, we do not discuss here further details of these algorithms. We just note that the bit-vector update formulas for

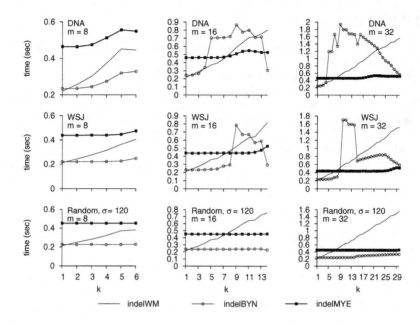

Fig. 2. The average time for searching in a $\approx$ 20 MB text. The first row is for DNA (a duplicated yeast genome), the second row for a sample of Wall Street Journal articles from TREC-collection, and the third row for random text with alphabet size $\sigma = 120$

both correspond to the dynamic programming recurrence, where each edit operation has a distinct part in the formula. Hence the modification for indel distance is straightforward: we only need to remove the part for substitution. Fig. 1 (*upper right*) shows indelWM and Fig. 1 (*lower right*) shows indelBYN: our versions of WM and BYN, respectively, that are modified for indel distance. The running time of indelWM is $O(k\lceil m/w\rceil n)$, and its time for preprocessing is $O(\sigma\lceil m/w\rceil + m)$. The running time of indelBYN is $O(\lceil (k+2)(m-k)/w\rceil n)$, and its time for preprocessing is $O(\sigma\lceil (k+2)(m-k)/w\rceil + m)$.

3 Experiments

We implemented and tested the three bit-parallel variants for approximate string matching under indel distance. IndelWM and IndelMYE were implemented fully by us, and IndelBYN was modified from a code by Baeza-Yates & Navarro. The computer was a 3.2Ghz AMD Athlon64 with word size w=32, 1.5 GB RAM, Windows XP, and MS Visual C++ 6.0 compiler using high optimization. The patterns were selected randomly from the text, and for each (m, k) we recorded the average time over searching 100 patterns. Fig. 2 shows the results.

References

1. R. Baeza-Yates and G. Navarro. Faster approximate string matching. *Algorithmica*, 23(2):127–158, 1999.
2. H. Hyyrö. Explaining and extending the bit-parallel approximate string matching algorithm of Myers. Technical Report A-2001-10, University of Tampere, 2001.
3. G. Myers. A fast bit-vector algorithm for approximate string matching based on dynamic progamming. *Journal of the ACM*, 46(3):395–415, 1999.
4. S. Wu and U. Manber. Fast text searching allowing errors. *Comm. of the ACM*, 35(10):83–91, October 1992.

Feature Selection by Reordering[*]

Marcel Jirina[1] and Marcel Jirina Jr.[2]

[1] Institute of Computer Science, Pod vodarenskou vezi 2,
182 07 Prague 8 – Liben, Czech Republic
marcel@cs.cas.cz
[2] Center of Applied Cybernetics, FEE, CTU Prague, Technicka 2,
166 27 Prague 6 – Dejvice, Czech Republic
jirina@fel.cvut.cz

Abstract. Feature selection serves for both reduction of the total amount of available data (removing of valueless data) and improvement of the whole behavior of a given induction algorithm (removing data that cause deterioration of the results). A method of proper selection of features for an inductive algorithm is discussed. The main idea consists in proper descending ordering of features according to a measure of new information contributing to previous valuable set of features. The measure is based on comparing of statistical distributions of individual features including mutual correlation. A mathematical theory of the approach is described. Results of the method applied to real-life data are shown.

1 Introduction

Feature selection is a process of data preparation for their consequential processing. Simply said, the feature selection filters out unnecessary variables. There are two aspects for feature selection. The first one is the time requirement for processing of large amount of data during learning [6]. The other one is finding that results of the induction algorithm (classification, recognition or also approximation and prediction) may be worse due to the presence of unnecessary features [2].

There exist two essentially different views, so called filter model and wrapper model [5]. In filter model features are selected independently on the induction algorithm. Wrapper models (methods) are tightly bound to an induction algorithm. Another approach can be more quantitative stating that each feature has some "weight" for its use by induction algorithm. There are lots of approaches trying to define and evaluate feature weights, usually without any relation to induction algorithm, e.g. [2], [6], [7].

2 Problem Formulation

The suggested method is based on a selection of relevant (appropriate) feature set from a given set. This can be achieved without the need of a metric on the feature

---

[*] The work was supported by Ministry of Education of the Czech Rep. under project LN00B096.

sets. In fact, a proper ordering of features or feature sets is sufficient. The measure for the ordering need not be necessarily a metrics in pure sense. It should give a tool for evaluating how much a particular feature brings new information to the set of features already selected.

3   The Method

The suggested method considers features for a classification task into two classes (0 and 1). The method for stating the measure of feature weight utilizes comparisons of statistical distributions of individual features and for each feature separately for each class. Comparison of distributions is derived from testing hypothesis whether two probability distributions are from the same source or not. The higher the probability that these distributions are different the higher is the influence of particular feature (variable) to proper classification. In fact, we do not evaluate correlation probability between a pair of features, but between subsets corresponding to the same class only.

After these probabilities are computed, the ordering of features is possible. The first feature should bring maximal information for good classification, the second one a little less including ("subtracting") also correlation with the first, the third again a little less including correlations with two preceding features etc.

The standard hypothesis testing is based on the following considerations: Given some hypothesis, e.g. two distributions are the same, or two variables are correlated. To this hypothesis some variable V is defined. Next, a probability p is computed from value of V often using some other information or assumptions. Then some level (threshold) P is chosen. If $p \geq P$ the hypothesis is assured, otherwise rejected. Sometimes instead of p the $1 - p$ is used and thus P and the test must be modified properly.

The logic used in this paper is based on somethig "dual" to the considerations above: Let $q = 1 - p$ be some probability (we call it the probability levels of rejection of hypothesis), $Q = 1 - P$ be some level. If $q < Q$ the hypothesis is assured, otherwise rejected. The larger q, the more likely the hypothesis is rejected (for the same level Q or P). In fact, the weights assigned to individual features are probability levels q related to rejection of hypotheses that distributions are the same or variables are correlated.

Let F_i be a feature. For the first ordering of individual features F_1, F_2, ... as to their influence on proper classification we use the probability levels of rejection p_{ii} of the hypothesis that probability distributions of feature the F_i for class 0 and for class 1 are the same. The p_{ii} is based on a suitable test, e.g. Kolmogorov - Smirnov test [8] or Cramér – von Mises test [1]. No correlation of features is considered for this first ordering.

To include influence of correlations let p_{ij0} and p_{ij1} be probability levels of rejection that distributions of variables for class 0 are correlated and that distributions of variables for class 1 are correlated, respectively. Taking all probability levels together, we have two triangular matrices, one for p_{ij0} and another for p_{ij1}, $i,j = 1, 2,$ ..., n. All results of pairwise distribution comparisons or correlations can be written in square matrix $n \times n$ as follows

$$
M = \begin{bmatrix}
p_{11} & p_{120} & \cdots & p_{1n0} \\
p_{211} & p_{22} & \cdots & p_{2n0} \\
\vdots & \vdots & \ddots & \vdots \\
p_{n11} & p_{n21} & \cdots & p_{nn}
\end{bmatrix}.
$$

In this matrix in diagonal entries are probability levels of rejection of hypothesis that that for feature of a given index and for class 0, and class 1 the distributions are the same. In the upper triangular part there are probability levels p_{ij0} for class 0, and in the bottom triangular part the probability levels p_{ij1} for class 1.

In the beginning the ordering of features is arbitrary. We now sort rows and columns in descending order according to diagonal elements p_{ii} of the matrix M. After it, first, we reassign indexes according to this ordering. The first feature now is a feature having the largest difference in distributions for both classes. The second feature has lesser difference in distributions for both classes and can be possibly somehow correlated to the preceding feature, etc.. Then, first, we state correlation coefficient for class 0 of variables 1 and 2, second, correlation coefficient for class 1 of variables 1 and 2 getting then probability levels p_{120} and p_{211} of rejection that distributions are correlated. The lesser these probabilities, the stronger correlation between features F_1 and F_2 exists.

Let us define independence level of feature F_i on preceding features F_k, $k < i$:

$$
\pi_i = p_{ii} \prod_{k=1}^{i-1} \sqrt{p_{ik0} p_{ki1}} \ . \tag{1}
$$

According to this formula the probability level of rejection that distributions for class 0 and 1 are the same is modified by measure of dependence on preceding variables. For calculation of probability levels p_{ik0} and p_{ki1} we use a standard approach [3] based on the fact that for small number of samples a nonzero correlation coefficient is found.

We associate these probability levels to corresponding rows and columns and again we sort rows and columns according to π_i in descending order. After it we again compute π_i according to (1) using new ordering and new indexing of variables. This step is repeated until no change in ordering occurs. Fast convergence of this process was found. By this procedure features are reordered from original arbitrary ordering to new ordering such that the first feature has the largest π_i, and the last the smallest π_i.

4 Results

The suggested method is demonstrated on a task of feature ordering of eight UCI MLR real-life databases: Heart, Vote, Splice, Spam, Shuttle, Ionosphere, German, and Adult [9]. Results after feature reordering are shown in Fig. 1.

5 Conclusion

We have presented a procedure for evaluating feature weights based on the idea that we need not evaluate subsets of features or build some metrics in the space of feature subsets. Instead of metrics some ordering would suffice. This is much weaker condition then metric. In fact, we need ordering of features from the viewpoint of the ability of feature to bring something new to the set of features already selected. If features are properly ordered we need not measure any distance. Knowledge that one feature is more important than the other should be sufficient. Having features already ordered, the question on proper feature set selection is reduced from combinatorial complexity to linear or at worst polynomial complexity – depending on the induction algorithm.

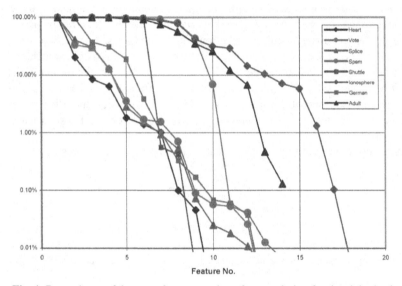

Fig. 1. Dependence of the π_i on feature number after reordering for the eight databases

References

[1] Csörgö, S., Faraway, J.J.: The Exact and Asymptotic Distributions of Cramér-von Mises Statistics. J.R. Statist. Soc. B vol. 58, No. 1 (1996) 221-234
[2] Dong, M., Kothari, R.: Feature subset selection using a new definition of classifiability. Pattern Recognition Letters 24 (2003) 1215–1225
[3] Hátle, J, Likeš, J.: Basics in probability and mathematical statistics (in Czech), SNTL/ALFA Praha (1974)
[4] Jirina, M., Jirina,M.jr.: Feature Selection By Reordering according to their Weights. Technical Report No. 919, Institute of Computer Science AS CR, November 2004, 9 pp. http://www.cs.cas.cz/people/homepages/jirina_marcel.shtml
[5] John, J.K., Kohavi, R., Pfleger, K.: Irrelevant features and the Subset Selection problem. In: Machine Learning: Proc. of the Eleventh Int. Conf. ed. Cohen,W., Hirsh,H., Morgan Kaufmann Publishers, San Francisco, Ca., USA (1994) 121-129

[6] Koller, D., Sahami, M.: Toward Optimal Feature Selection. Proc. of the Thirteenth Int. Conf. on Machine Learning. Morgan Kaufmann Publishers, San Francisco, Ca., USA, Morgan-Kaufman (1996) 284-292

[7] Last, M., Kandel, A., Maimon, O.: Information-theoretic algorithm for feature selection. Pattern Recognition Letters 22 (2001) 799-811

[8] Smirnov, N.: Table for estimating the goodness of fit of empirical distributions, Annals of Math. Statist. 19 (1948) 279-281

[9] UCI Repository of Machine Learning Databases. http//www.ics.uci.edu/~mlearn/ MLRepository.html

A Management Scheme for the Basic Types in High Level Languages

Fritz Mayer-Lindenberg

Technical University of Hamburg-Harburg (TUHH)
Schwarzenbergstrasse 95, D21073 Hamburg
mayer-lindenberg@tuhh.de

Abstract. This note describes a type concept for the basic data types handled in the applications of digital systems, numbers and bit fields, which are usually predefined in higher level languages and serve as the basis for application specific data types and classes. It has been implemented in a real-time language for parallel embedded systems and is motivated by the fact that such systems may have to use a multitude of application-specific number types. The type concept actually applies to programming languages for all classes of applications. It proposes the use of an abstract data type of numbers for which the various encodings of numbers provide implementations. This simple approach not only allows non-standard encoding types to be added as needed but also provides common formats for input and output and derived numeric data types that aren't bound to a specific encoding. Related to the handling of the basic data types is a conversion policy. For bit fields, conversions are substituted by word number changes of multi-word codes. Finally, the abstract number type can be used to simplify the related typing of functions that no longer need to specify the encoding of all their arguments and results.

1 Basic Data Types in Programming Languages

Abstractly speaking, a data type is a set M [1]. M may be defined through enumeration, through a set of operations and their properties (this holds for the set of natural numbers), or through other properties and constructions (e.g. in the case of the real numbers). Digital systems process bit fields (finite sequences of bits), and all abstract data types to be supported must first be encoded by special sets of bit fields.

An important special case is the encoding of numbers by bit fields. For the sake of efficiency, codes of a fixed word length are preferred. If C is the finite set of size n bit fields representing numbers through some encoding, a two-argument arithmetic operation would only be defined on a subset $D \subset C \times C$ and take values in C. For the purpose of error checking, it can be extended to all of $C \times C$ by mapping the complement of D to some extra error code 'F'.

For the current imperative and object-oriented languages [2] there is a de-facto standard for the provided data types. In C, these are 'integer' (16-bit or 32-bit twos complement numbers), 'char' (8-bit unsigned numbers), 'float' (32-bit IEEE floating point numbers), 'double' (same for 64-bit), and 'bool' (with the elements 'true', 'false'). 'integer' and 'float' are actually bit field types. The C integer type even provides the bit field 'and', 'or' and 'not' operations and substitutes the true integer

M. Bieliková et al. (Eds.): SOFSEM 2005, LNCS 3381, pp. 390–393, 2004.

operations (which are defined only partially) by modulo operations. Conversions between integers and floating point are inserted automatically. They round a floating point number to the closest integer (i.e. perform an invisible, non-trivial operation). Starting from the basic types, Cartesian product types can be constructed. Object-oriented languages use class definitions also including the operations to be performed on the Cartesian product types, and the notion of abstract data types for such. An abstract data type for the numbers isn't provided, however (except for some functional languages using integers of unlimited size).

The data type handling concept proposed in this work was introduced for the real-time language π-Nets developed at the TUHH [3]. It starts by extending the use of abstract data types to the base type of numbers to neatly distinguish between the abstract num-ber type and the bit field types implementing it. It supports application specific enco-dings of numbers on special processors, but has other advantages common to abstract data types as well that now get extended to the basic types.

2 The Number Type

The abstract number type NUM of π-Nets is equipped with the basic operations on numbers including the comparisons and some transcendental functions as well as input and output formats for NUM literals. Conceptually, NUM is the set of computable real numbers. NUM is used as a reference for the encoding types used by the processors. As a first benefit from using NUM, the formats of NUM literals are no longer related to any particular encoding. Output formats aren't related to the encoding types, as well. Integers can e.g. be output in an engineering format with a reduced number of places. The use of the NUM type allows the specification of numeric algorithms that can eventually be executed on every processor implementing some encoding type.

The π-Nets programming environment also provides a runtime environment and permits the execution of NUM algorithms (functions) on the host workstation using a selectable integer or floating point encoding. Thus the effect of choosing a particular encoding can be investigated without having to redefine the algorithm.

The standard type extension concepts can be applied to the NUM type in order to derive other types of interest. An abstract 'complex' data type useful for digital signal processing applications is obtained as the Cartesian product type NUM×NUM. It is predefined in π-Nets as a derived data type. Input and output formats for the complex type are couples of real numbers. The complex type is available to *every* processor providing a NUM encoding. In a similar fashion, various other useful types such as the 'quaternion' type used in computer graphics [4] can be defined without reference to a particular encoding.

3 Bit Field Types and Conversions

Besides the NUM type, sets of bit fields need to be provided as the canonical data types of the processors of the target, in particular, the sets of bit fields of a constant size of 16 or 32, and selected basic Boolean operations on these, the common 'and', 'or' and 'not' operations and some primitives for implementing binary arithmetic. All of these bit field types are considered to be subsets of the type BIN of all finite bit fields. There are BIN literals that have different formats from NUM literals and are used for input and output for every bit field type. It is therefore possible to specify the binary code of a number, or to directly display it. Symbolic constants replacing literals are only distinguished as being NUM or BIN constants.

The number encoding types are sets of bit fields, too, yet equipped with other opera-tions, namely those implementing the NUM operations as partially defined operations on the codes. The number encoding types don't provide the Boolean operations. NUM literals and constants are automatically encoded according to the selected type. Some integer types have implementations in terms of bit field types while others just have a behavioral definition. π-Nets e.g. provides a predefined non-standard 48-bit software floating point type defined in terms of basic 16-bit operations on bit fields. Consequently it is available on every processor executing these 16-bit operations. It uses a non-normalized floating point representation of numbers as described in [11]. Applications specific number types can be defined in terms of Boolean operation to define their behavior and to be able to used them in a simulation, but be implemented differently on a target processor.

It is natural and the choice for π-Nets to automatically convert between different number codes representing the same abstract number. No conversions occur between bit fields encoding numbers and raw bit fields. For bit field types not encoding numbers, a different, non-standard 'conversion' scheme is employed (the term 'con-version' doesn't really apply). A bit field presented as n words of size k is simply converted into n' words if size k' with n*k=n'*k' (with a zero extension, if necessary), leaving the total bit pattern unchanged. This faithful conversion scheme also applies to communications between processors, in particular of multiword number codes, and to input and output via ports of a reduced word size.

4 Function Types

Any choice of data types carries over to the storage objects for these, and to the functions constructed from the elementary operations on the data types. The type of a program function (with possible side effects) is defined through its input domain and its output range. Starting from NUM and BIN, functions will be either numerical functions on NUM^n with values in NUM^m for some n,m or bit field functions on BIN with values in BIN^m. There are hardly applications mapping NUM arguments to BIN elements of vice versa, except for the indexing of sequences of bit fields, or controlling BIN operations through NUM comparisons. Then, (n,m) and the common type T of its arguments and results define the type of a function. This approach implies that functions on encoded numbers use a consistent encoding for all of their arguments and results, and that functions have arbitrary yet fixed numbers of arguments and results. Functions of a number encoding type T may still call functions or read and write to variables of other number encoding types. Then conversions occur. A restricted use of mixed types is needed for indexing purposes only.

As already the elementary operations and the implicit conversions, functions may be incompletely defined. This is taken up systematically in π-Nets by letting every function of type (n,m,T) also have a Boolean result indicating whether its output is valid. Functions of type (n,0,T) only output this Boolean result and represent relations. The Boolean results are uniquely used for the control flow, a failing arithmetic operation causing e.g. a branch to an error handler. With this unification there remains no need for an extra type 'bool' for the results of comparisons.

5 Summary

The proposed, quite basic type concept has the special features of

- distinguishing abstract data from their encodings
- numeric algorithms not depending on particular codes
- supporting processors specialized in non-standard data types
- defining a conversion scheme and word size transformations
- using the same formats and constants for all number codes
- carrying over abstract numeric data types to all encodings
- extending into a simple type system for functions and objects

Advantages of the proposed type system are its flexibility at the level of number encodings and the support of a type to processor assignment for application specific systems. Its use of abstract data types doesn't involve runtime overheads and is suitable for simple, embedded processors.

References

[1] J. Gruska, Foundations of Computer Science, Thomson Computer Press 1997
[2] R.W. Sebesta, Concepts of Programming Languages, Benjamin 1989
[3] F. Mayer-Lindenberg, Dedicated Digital Processors, Wiley 2004
[4] R. Ablamowicz et al, Lectures on Clifford Algebras, Birkhäuser-Verlag 2004
[5] D. E. Knuth, The Art of Computer Programming, Addison-Wesley 1981

Bayesian Networks in Software Maintenance Management

Ana C.V. de Melo and Adilson de J. Sanchez

University of São Paulo (USP), Dep. of Computer Science, 05508 090, SP, Brazil
`acvm@ime.usp.br` `Adilson.Sanchez@itau.com.br`

Abstract. Managing software maintenance is rarely a precise task due to uncertainties concerned with resources and services descriptions. Even when a well-established maintenance process is followed, the risk of delaying tasks remains if the new services are not precisely described or when resources change during process execution. Also, the delay of a task at an early process stage may represent a different delay at the end of the process, depending on complexity or services reliability requirements. This paper presents a knowledge based representation (Bayesian Networks) for maintenance project delays based on specialists experience.

1 Introduction

The advances in technologies and the competitiveness of business products made computational support essential to have products in time to market. Computational development became part of business strategy because business rules are continuously changing and this can be rapidly acquired by computational systems. Changing business rules and its corresponding system implies in adapting existing systems to accomplish new rules, the so called adaptive maintenance [8]. In this scenario, software maintenance process is essential to business products competitiveness.

Despite of being widely studied and of interest to market, software maintenance is still a complex and costly task. It is pointed out as one of the most expensive tasks in software development and difficulties inherently to its execution is the main cause of software high costs. Having no maintenance plans or, even when plans are established, having the plan schedule not achieved lead to more costly maintenance projects [8]. Uncertainties in the maintenance process drives to unpredictable situations hard to be recovered during project execution [12, 13].

Due to uncertainties, project managers have to decide how to reconfigure plans as project schedule fails. Replanning is a hard task: What to do to finish a system in time? How much resource should be added in order to finish a delayed project in time? Can we implement the system in a shorter time without dropping its quality requirements? These and many other issues must be answered when a project is replanned and, most time, managers use their own experience (or feeling) to do that.

M. Bieliková et al. (Eds.): SOFSEM 2005, LNCS 3381, pp. 394–398, 2005.

In order to help in manage maintenance project delays, this paper presents a technique to represent specialists knowledge about project development regarding delays. The main goal is to be able to calculate the probability of a project being delayed during its execution. Managers can use this information to replan a project in order to finish it in time. The forthcoming sections are as follows: Section 2 presents a particular Bayesian Network for the delay propagation problem; and the last section is about limitations of the present work and the effective use of this technique in software development.

2   A Bayesian Network to Help Software Maintenance Management

Decision systems are concerned with a way of embedding knowledge in order to make a mimic of human decisions [2, 4, 9]. This involves how to represent knowledge and use such an information to make decisions. Software development and maintenance require a plan to define which tasks and in which order they must be performed [1]. The tasks execution involves people and resources and, because of that, uncertainties in the plan execution emerge. Due to those uncertainties, managers' experience is, in general, the main information available for replanning projects.

There is, nowadays, a set of techniques to represent and use knowledge as, for example, belief networks and evidence theory [5]. Belief networks are graphical structures to represent systems based on knowledge with a variety of theories. Bayesian networks [3, 10, 7, 6, 5, 11] is one of these theories based on probability theory. In Bayesian Networks, the probability distribution over the domain variable is calculated by Bayes Theorem [7, 5, 10] presented in Appendix A.

The Bayesian network in the present work has as main goal establishing a relationship between developers, software and tasks that can cause software delay. All this information is taken from specialists and their experience on past projects. The first step of building a Bayesian network is to identify software development phases as much as the factors that influence each phase task. For simplicity, only three macro-phases of maintenance development have been considered: definition, implementation and testing. In each of these phases, we may have a set of factors that suggest the maintenance risk as much as the probability of project delay.

The first step of building the Bayesian network is recognising the context in which risk management can be figured out: **Platform Expertise**- this variable quantifies how many years the professionals involved in have on system platform; **System Expertise**- concerned with the experience in the system itself; **System Documentation**- this variable denotes if there exists (or not) a good system documentation; **Maintenance Complexity**- it is related to how difficult is to make the service required.

Once we have defined the relevant variables for risk management, we must establish how they are related: how a variable can influence others when an evidence is given. **Maintenance Risk** is influenced by all variables above. So, if an evidence is given to one of the variables, the **Maintenance Risk** must be updated. Moreover, if evidences are given to more than one variable, all these must be considered to calculate the new **Maintenance Risk** probability. Its calculation is given by Equation in Appendix A.

To calculate probabilities over Bayesian network variables, we must first establish the probability vector for each variable, the so called quantification part of Bayesian networks (Figure 1). The probability related to each value means that, in a general case, the value occurs for the associated variable with such a frequency. For example, **Platform Expertise** ranges over three values interval: the first one denotes the developers experience from "0 to 1" years with 20% of probability, the second is from "1 to 3" with 40% of occurring, while the third ranges over "3 to 5" years of experience with probability of 40%. The probability assigned to each experience years interval comes from specialists information.

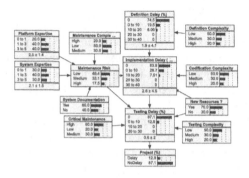

Fig. 1. Maintenance Risk and Delay Network

New variables together with their values and relations must be defined for the delay propagation network. We want to answer a question like " what is the real delay of maintenance projects during their executions? ". Again, the first step is finding out variables relevant to project delays that can have values measured. Since we work on possible delays of projects execution, the choice of variables is based on common tasks performed on maintenance processes: **Definition Delay**, **Implementation Delay** and **Testing Delay**. All of them are calculated from probability distribution factors that influence each phase. Here, a set of factors related to complexity, resources or whether the service involved is critical or not to the system being maintained have been defined: **Definition Complexity**, **Codification Complexity**, **Testing Complexity**, **New Resources?**, **Critical Maintenance**, and **Project** that denotes delay over the whole maintenance project.

The delay variables for each maintenance project phase, together with the factors above, constitute the whole elements of the delay propagation network. Since delay is realised after definition phase, maintenance risk can only influence the implementation and testing phases. As such, the network is made of the resulting **Maintenance Risk** variable influencing both **Implementation Delay** and **Testing Delay** variables.

3   Final Considerations

This paper presented a study on using Bayesian Belief Networks to help in managing maintenance projects. In general, maintenance projects are hard to follow their schedules and most of them have to be replanned during execution. Delaying any project task for a certain time doesn't mean that the whole project is late for the same period. Despite of that, most management tools are not able to calculate the real delays. In this project we have used Bayesian networks to store probability distribution of project delays based on project features to calculate a more accurate delay probabilities of maintenance projects.

The probability distribution for each environment variable embedded in the network has been obtained from specialists information. Also, variables chosen as "relevant" to determine project delays and maintenance risk are from specialists, regarding the network operational limitations. Acquiring data about maintenance risk may be extended to other elements, such as details on working environment. However, some of them are hard or even impossible to be measured. Because of this, they are not inserted as variable in our Bayesian network. A similar analysis can be done to the delay propagation network.

For this application domain, we could detect two main problems in using Bayesian Networks: information on factors probability comes mainly from specialists and the number of conditionally dependent variables must be controlled. The former can be amended by keeping information about systems development (maintenance). The last one is more related to how much a factor make influence over its conditionally dependent factor: there must be considered a combination of all dependent variables, and an accurate measure of influences is not trivial.

References

1. Project Management Instittute. *A Guide to the Project Management Body of Knowledge (PMBOK Guide)*. Project Management Institute, 2001.
2. P. Jackson. *Introduction to Expert Systems*. Addison-Wesley, Workingham, 1990.
3. F. Jensen. *An Introduction to Bayesian Networks*. Springer, 1996.
4. P. J. F. Lucas and Linda C. van der Gaag. *Principles of Expert Systems*. Addison-Wesley, Workingham, 1991.
5. J. Pearl. *Propabilistic Reasoning in Intelligent Systems. Networks of Pausible Inference*. Morgan Kaufmann, Palo Alto, CA, 1988.
6. J. Pearl. A probabilistic calculus of actions. In R. Lopez de Mantaras and D. Poole, editors, *Proceedings of UAI 94*, San Mateo, CA, 1994. Morgan Kaufmann.
7. J. Pearl and Stuart Russell. Bayesian networks. MIT Press, 2003.
8. Roger Pressman. *Software Engineering. A practitioner's approach*. McGraw–Hill, 2004.
9. Stuart Russell and Peter Norvig. *Artificial Intelligence: A Modern Approach*. Prentice Hall, 2003.
10. Linda C. van der Gaag. Bayesian belief networks: Odds and ends. *The Computer Journal*, 39(2):97–113, 1996.
11. A. D. Wooff, M. Goldstein, and P. A. F. Coolen. Bayesian graphical models for software testing. *IEEE Transactions on Software Engineering*, 28(5), 2002.

12. Hadar Ziv. Constructing bayesian-network models of software testing and mainte-nance uncertainties. In Mary J. Harrold and Giuseppe Visaggio, editors, *Proceedings of ICSM'97*, Bari, Italy, 1997. IEEE Computer Society.
13. Hadar Ziv, Debra J. Richardson, and René Klosch. The uncertainty principle in software engineering. Technical Report 96-33, University of California, Irvine, CA, USA, August 1996.

Appendix A

Bayes Theorem

$$P(H|e) = \frac{P(e|H).P(H)}{P(e)} \tag{1}$$

H represents a hypothesis and e an evidence. The formula above denotes the belief associated to hypothesis H as evidence e is given. Its calculation is obtained by multiplication of the *a priori* hypothesis probability $P(H)$ by the *a posteriori* probability $P(H|e)$.

$$P(e) = P(e|H).P(H) + P(e|\neg H).P(\neg H) \tag{2}$$

A Multiagent System Aiding Information Retrieval in Internet Using Consensus Methods

Ngoc Thanh Nguyen, Adam Blazowski, and Michal Malowiecki

Institute of Control and Systems Engineering,
Wroclaw University of Technology, Poland
thanh@pwr.wroc.pl

Abstract. This paper presents a multiagent system (called AGWI) which should assist users in information retrieval in Internet. In this system consensus methods are applied for reconciling inconsistency of answers generated by agents for the same query. The system has been created by means of platform IBM Aglets.

1 Introduction

For information retrieval in Internet, a very popular environment, there are still too few of such systems. Menczer [5] has designed and implemented a public Web intelligence called *Myspiders*, a multiagent system for information discovery in Internet. By the tool he has shown that augmenting search engines with adaptive intelligent search agents can lead to significant competitive advantages. In this paper we present a multiagent system (called AGWI [2]) assisting information retrieval in Internet. Our system differs from the one of Menczer in that it can solve conflicts of agents using consensus methods. The general idea of AGWI is based on using not one but several agents for a retrieval task. The advantage of this approach is that agents may use different resources and different search engines. Owing to this the answer will be more complete and relevant. However, different agents may generate different results (as rankings of URLs), thus they are in conflict. For conflict resolution we use consensus methods [1][3][6] which is based on selecting such URLs' ranking which has the minimal sum of distances to those given by agents. The consensus, after checking its consistency level, will be given to the user as the final answer. Each agent has its knowledge base and a tool for processing.

2 The Aims and Description of the Project

Aglets are a product of a IBM research program, the aim of which is relied on creating a platform for mobile agents. As the results of this program the authors have created in Open Source a packet named ASDK (Aglets Software Development Kit), owing to which a user knowing Java can create agents for his needs [4]. There has been arisen also project Aglets Community which has associated managers of free Aglets servers.

M. Bieliková et al. (Eds.): SOFSEM 2005, LNCS 3381, pp. 399–402, 2005.

The aim of this project is creating a multiagent system which should aid users in information retrieval in Internet. In this system consensus methods are applied for reconciling the differences in search results generated by agents. This approach enables to solve the following two problems often occurring in information retrieval processes in Internet:

- The first problem is related to low relevance degree of answers generated by an search engine. The reason of this problem is included in non-effective work of filters, and as the consequence, many documents non-related with user queries (e.g. advertising pages) may be by force pressed. Below we give an example.

- The second problem is based on giving to the user repeating URLs, i.e. URLs which are identical or very similar to each other. This phenomenon is a burden for the user because he must lose a lot of time and forward many screens for finding interested information.

There are two kinds of agents:

- Searching Agent (SA) and
- Managing Agent (MA).

The tasks of a searching agent consist of:

- selecting in its knowledge base the best search engine for given query,
- generating answer using chosen search engine,
- actualizing its knowledge base on the basis of the final result determined as the consensus of results generated by all searching agents. As the result of the actualization the used search engine obtains a weight referring to given query.

The tasks of a managing agent consist of:

- creating searching agents,
- determining consensus of answers generated by searching agents and checking consistency degree of the set of these answers,
- calculating the distances between the consensus and answers generated by searching agents and sending the results to these agents for actualizing their knowledge bases.

The general scheme of activity of m agents SA and one MA and their cooperation is presented in Figure 1.

The circles of work of agents SA and MA are described as follows:

Agent SA:

- Loading information about search engines;
- Receiving new query from agent MA;
- Loading the database about queries;
- Comparing new query with queries stored in the database;
- Choosing a search engine (if the database is empty then the choice is made in a random way) and running it for generating the answer;
- Eliminating repeating URLs and ordering them;
- Sending the answer (in form of a ranking of URLs) to the MA agent;
- After receiving the evaluation from the MA agent (this evaluation is calculated on the basis of the distance between its answer and the final answer determined by MA agent), and next actualizing its knowledge base;
- Reporting its readiness for the next search.

Agent MA:

- Gathering answers from agents SA (in form of URLs' rankings);
- Determining consensus *C* of the set of rankings;
- Calculating consistency level of the set of rankings;
- If the consistency level is low (the set of rankings is not consistent enough) then MA agent must wait for the decision of the user. The user can require repeating the search or indicate good answer among these given by SA agents;
- If the consistency level is high (the set of rankings is consistent) the MA agent calculates the weights of agents SA (i.e. it calculates elements of matrix *W*, at the beginning they are equal 1)
- These weights will be used by agent MA for calculating the next consensus.

In this system the following algorithms are used (their detail descriptions are given in report [2]):

- Algorithm for eliminating repeating URLs ;
- Algorithm for transferring a binary matrix into a ranking;
- Three algorithms for calculating 3 kinds of distances between queries;
- Algorithm for calculating distances between rankings;
- Algorithm for calculating consensus for a set of rankings based on the method *branch and bound*.
- Algorithm for calculating the consistency level of a set of rankings.

3   Description of the Program

To run the program it is needed to install server Aglets with environment JDK. The components needed for the systems are: JDK 1.1.6, Swing 1.0.1, JBCL 2.0 and IBM Aglets 1.1b3. For generating answer using a search engine each agent SA formulates the query in URL form and sends it to Internet. As the result the agent receives a HTML page which is next processed in order to generate proper markups HREF (by means of class HTMLDocument). During the system work it is possible to spy on actual state of agents MA and of agents SA and their histories. Each agent SA for the first query in its life draws a search engine from the above given list. A MA agent should take care of different search engines for different SA agents.

After each retrieval process and evaluation given by agent MA, a SA agent writes the query to its knowledge base with the accuracy degree (that is the distance between its answer and the consensus). For a next query an agent SA does not draw the search engine but determines it in the following way: The agent compares the query to queries stored in the base and chooses such query which is the nearest and has accuracy degree greater than 50%. The chosen search engine will be this which has been used for this query. Agents SA have 3 algorithms for comparing queries and they can draw one of them. Owing to this their query bases are not similar to each other.

After all SA agents have finished their retrieval processes agent MA determines the consistency level of answers generated by agents SA taking into account their weights. At the beginning the weights are equal 1 for all SA agents. If the set of answers is not consistent enough (the consistency level is too low) then agent MA shows the user all answers. The user can choose the answer of a SA agent which in his opinion is the best or require to renew the retrieval process. If the set of answers is consistent (the consistency level is high) then agent MA determines the consensus and calculates the weights for agents SA.

If a URL is repeated in many answers then it means that the relevance degree of this URL is large. If a URL occurs many times in an answer then an agent SA must eliminate the repetitions first. The way for eliminating of repeating URLs in the answer of one agent is based on comparison of this URL to other URLs, if they are similar in 80% than the system decides that they are identical. Owing to this the answers of SA agents are more "clear" before the consensus choice.

If a URL occurs in few answers or occurs on lower positions of the answers then in the consensus it should not occur at all, or if so, only on a lower position. The consensus determining algorithm allows to omit non-relevant URLs or at least to place them in the last positions of the final ranking. Concretely, assume that a URL (e.g. an advertising page) occurs at the first position of the answer generated by of an agent, but it does not occur in the rankings of other agents. Then, in dependence on the number of agents SA taking part in the retrieval this URL may not appear in the final ranking, and even if so, it should be at a very low position. The larger is the number of SA agents the lower is the position for this URL. So one can expect that the final ranking should be more relevant than each generated by an agent SA.

4   Conclusions

Presented in this paper multiagent system can be useful in information retrieval in Internet. The approach for using several search engines for the same query is not novel, but the worked out method for reconciling the results has not been proposed by other authors. The advantage of this method is that it does not need the information about the user (his preferences, profiles etc.). It accepts such assumption that if several experts (in this case search engines) solve the same problem, then in general the reconciled solution should be more credible than those proposed by the experts.

References

1. Barthelemy, J.P., Janowitz M.F.: A Formal Theory of Consensus. SIAM J. Discrete Math. **4** (1991) 305-322
2. Błażowski, A, Nguyen, N.T.: AGWI - Multiagent System Aiding Information Retrieval in Internet. Technical Reports of Dept. of Information Systems, Wroclaw Univ. of Tech., No. 32 (2002) (in Polish)
3. Day, W.H.E.: Consensus Methods as Tools for Data Analysis. In: Bock, H.H. (ed.): Classification and Related Methods for Data Analysis. North-Holland (1988) 312-324
4. Lange, D., Oshima, M.: Programming and Developing Java^{TM} Mobile Agents with Aglets, Longman (1998)
5. Menczer, F.: Complementing Search Engines with Online Web Mining Agents. Decision Support Systems **35** (2003) 195-212
6. Nguyen, N.T.: Consensus System for Solving Conflicts in Distributed Systems. Journal of Information Sciences **147** (2002) 91-122

Interval-Valued Data Structures and Their Application to e-Learning

Adam Niewiadomski

Technical University of Lodz, Institute of Computer Science,
ul.Wolczanska 215, 93-005, Lodz, Poland
aniewiadomski@ics.p.lodz.pl

Abstract. The paper is devoted to the problem of replacing crisp numbers with interval numbers in soft computations. The original concept of an interval-valued vector (IVV) is introduced, and the new extensions of classic similarity measures are proposed to handle IVV matching. Finally, the presented data structure and the matching methods are used in the process of an automated evaluation of tests in e-learning (distance learning within the Internet).

1   Introduction

Many of natural language (NL) and technical data are expressed with intervals instead of crisp numbers. In contemporary NL people frequently construct the sentences containing uncertainties of the interval e.g. *I drink coffee 3–4 times a day.* Analogously, in technical and engineering data, intervals also appear frequently, because of some tolerance in measuring real parameters, e.g. *Device powered with 220–230 V.* There also exist logical systems and calculi using intervals as truth values [2].

We see a great opportunity for applications of interval structures and operations in the process of automated evaluation of tests in e-learning; particularly, the scope is to model expert non-crisp marks with intervals and then to compare them to results obtained from intelligent marking algorithms whether the latter complete human intuitions.

2   Interval-Valued Vectors

We denote the set of all closed intervals in $\mathcal{R}$ as $Int(\mathcal{R})$. $Int(\mathcal{R})$ is the subset of $Pow(\mathcal{R})$ — powerset (set of all subsets) of $\mathcal{R}$. The closed interval $[\underline{a}, \overline{a}]$ is called *degenerated* iff $\underline{a} = \overline{a}$. Each degenerated interval $[\underline{a} = a, \overline{a} = a]$ may be treated as the real number a.

The structure called *an interval-valued vector* (IVV) is proposed: let $k \in \mathcal{N}$. Vector $V = [v_i], i \leq k$ built of k elements is an interval-valued vector in $(Int(\mathcal{R}))^k$, iff each v_i is an interval in $Int(\mathcal{R})$, i.e. $v_i = [\underline{v_i}, \overline{v_i}] \in Int(\mathcal{R}), \underline{v_i} \leq \overline{v_i}$, for each $i = 1, 2, ..., k$. Thus, vector V is of the form: $V = \{[\underline{v_1}, \overline{v_1}], [\underline{v_2}, \overline{v_2}], ...,$

M. Bieliková et al. (Eds.): SOFSEM 2005, LNCS 3381, pp. 403–407, 2005.

$[\underline{v}_k, \overline{v}_k]\}$. This extension is based on replacing elements being crisp numbers with elements being intervals in a vector. Each classic vector is a special case of an interval-valued vector where its each element is a degenerated interval $[\underline{v}, \overline{v}], \underline{v} = \overline{v} \in \mathcal{R}$. See the following example: let $k = 3$, and the sample interval-valued vector V in $Int([0,1])^3$ is of the form: $V = \{[0, 0.8], [0.2, 0.5], [0.9, 0.1]\}$.

3   Similarity Measures for IV Vectors

In this approach, we concentrate on reflexive and symmetrical fuzzy binary relations (the so-called *neighbourhood relations*, [1]) as models for similarity connections for interval-valued vectors. Three neighbourhood relations between numerical vectors are recalled in [9]: a) correlation coefficient, b) min-max method, and c) arithmetic average-minimum method. We propose the extended versions of these methods due to the interval character of vectors being compared. Let $A = \{a_1, a_2, ..., a_n\}, V = \{v_1, v_2, ..., v_n\}$ — interval-valued vectors in $(Int(\mathcal{R}))^n$. The following reflexive and symmetrical IV similarity measures between A and V are proposed

a) IV correlation coefficient — $r_{cc} = [\underline{r}_{cc}, \overline{r}_{cc}]$

$$\underline{r}_{cc}(A, V) = \frac{\sum_{i=1}^n |\underline{a}_i - \underline{av}(A)| \cdot |\underline{v}_i - \underline{av}(V)|}{\sqrt{\sum_{i=1}^n (\underline{a}_i - \underline{av}(A))^2 \cdot \sum_{i=1}^n (\underline{v}_i - \underline{av}(V))^2}} \tag{1}$$

$$\overline{r}_{cc}(A, V) = \frac{\sum_{i=1}^n |\overline{a}_i - \overline{av}(A)| \cdot |\overline{v}_i - \overline{av}(V)|}{\sqrt{\sum_{i=1}^n (\overline{a}_i - \overline{av}(A))^2 \cdot \sum_{i=1}^n (\overline{v}_i - \overline{av}(V))^2}} \tag{2}$$

where $av(A) = [\underline{av}(A), \overline{av}(A)] = \left[\frac{1}{n}\sum_{i=1}^n \underline{a}_i, \frac{1}{n}\sum_{i=1}^n \overline{a}_i\right]$ is the arithmetic average of the interval-valued vector A, and $av(V) = [\underline{av}(V), \overline{av}(V)]$ — analogously;

b) IV minimum-maximum method — $r_{mm} = [\underline{r}_{mm}, \overline{r}_{mm}]$

$$\underline{r}_{mm}(A, V) = \frac{\sum_{i=1}^n \min\{\underline{a}_i, \underline{v}_i\}}{\sum_{i=1}^n \max\{\underline{a}_i, \underline{v}_i\}} \tag{3}$$

and $\overline{r}_{mm}(A, V)$ — analogously;

c) IV arithmetic average-minimum method — $r_{am} = [\underline{r}_{am}, \overline{r}_{am}]$

$$\underline{r}_{am}(A, V) = \frac{\sum_{i=1}^n \min\{\underline{a}_i, \underline{v}_i\}}{\frac{1}{2}\sum_{i=1}^n (\underline{a}_i + \underline{v}_i)} \tag{4}$$

and $\overline{r}_{am}(A, V)$ — analogously.

Notice that the values indicating the similarity levels between IVVs are also intervals. Notice also that if all elements of compared vectors are degenerated intervals, the results of the a)–c) methods are degenerated intervals, too; in that case, formulae (1)–(4) are strictly relevant measures for classic vectors.

4   IV Evaluation and IV Marking in e-Learning

Contemporarily, evaluating and marking in e-learning must be done manually by a tutor, even if s/he works on modern Web platforms. Moreover, the process of manual marking should be repeated by a teacher at least as many times as the number of students at a course is. There exist the procedures of automated checking of choice tests, mostly based on Hamming distance, however, the results of the more complicated tests, as open questions, writing sentences, programming, image recognition, definitely lack of automated marking procedures. The original propositions are given for two subjects only: programming [6] [7] and German grammar [8]. The possible application of an intelligent marking procedure on an e-learning platform is depicted in Fig.1.

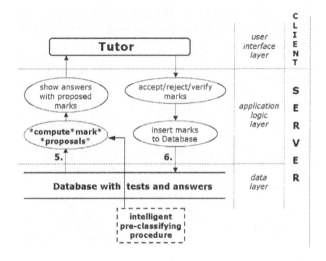

Fig. 1. The part of e-testing process supported by AI algorithms

In this paper, due to its title, the presentation of marking algorithms is not intended. Their thorough descriptions and the explanation of their connections with human intuitions are given in [6]–[8]. The computational power of these procedures in textual information searching and retrieving is described and exemplified in [5][10].

4.1   Experiment Construction

The aim of the experiment is to determine the relevancy of six intelligent marking algorithms with expert opinions, see: [8]. The association is measured by computing the similarity level between the vector of marks given by experts and the vector of analogous marks computed by a one of the algorithms. The test consists of 240 answers (correct, partially correct, and wrong) to ca 50 questions at German grammar (ordering sets of mixed words to correct sentences; see [8]). Each algorithm marks the answers with real numbers from $[0, 1]$. The marks given by the algorithms 1–6 are stored as the elements of vectors A_1, A_2,..., A_6, respectively.

The scale of marking for the experts is more natural: $\{0, 0.25, 0.5, 0.75, 1.0\}$. The marks collected from k experts are stored as the elements of vector $E = \{e_1,$ $e_2,..., e_{240}\}$, where $e_i = \frac{\sum_{j=1}^{k} e_{ij}}{k}$, and j is the number of the expert. The *clou* of the method is that experts sometimes use intervals instead of crisp marks (i.e. $[0.5, 0.75]$), so the data collected from them must be stored in an IV vector and evaluated with a one of procedures given in Section 3.

4.2   Evaluation

The similarity levels r_{cc}, r_{mm}, and r_{am} for the pairs of vectors (A_1, E), (A_2, E),..., (A_6, E) are computed with formulae (1)–(4). Notice that A_1,..., A_6 are degenerated IVVs, what simplifies the computations. The results of the comparisons are collected in Tab. 1

Table 1. IV similarity levels of A_1,..., A_6 to IVV E

Algorithm	$r_{cc}(A_n, E)$	$r_{mm}(A_n, E)$	$r_{am}(A_n, E)$
A_1	$[0.87, 0.89]$	$[0.72, 0.84]$	$[0.83, 0.91]$
A_2	$[0.81, 0.81]$	$[0.69, 0.86]$	$[0.82, 0.92]$
A_3	$[0.85, 0.86]$	$[0.70, 0.81]$	$[0.82, 0.90]$
A_4	$[0.86, 0.87]$	$[0.67, 0.74]$	$[0.80, 0.85]$
A_5	$[0.87, 0.87]$	$[0.72, 0.81]$	$[0.83, 0.89]$
A_6	$[0.81, 0.82]$	$[0.68, 0.88]$	$[0.81, 0.94]$
E	$[1.0, 1.0]$	$[1.0, 1.0]$	$[1.0, 1.0]$

4.3   Conclusions

Tab. 1 contains the IV similarity degrees of vectors A_1, A_2,..., A_6 to the expert marks E. As it is observed, Algorithms 1. and 5. provide the results which are the most relevant to the human marks — it is indicated by the highest lower bounds of similarity levels r_{cc}, r_{mm}, and r_{am} for pairs (A_1, E) and (A_5, E). The additional conclusion is that Algorithm 5. provides the results slightly more precisely than Algorithm 1., because of the smaller widths of the similarity coefficients for the pair (A_5, E), than for (A_1, E). Values 1.0 in the last row indicate, that the IVV E is identical with itself; it is implied by the reflexivity of neighbourhood fuzzy relations.

It should be emphasised, that the standard processing of interval data (i.e. reducing them to crisp numbers) would disable obtaining the interval-valued results, relevant to the intuitions of the experts, and, what is even more important, to their natural way of expressing opinions. The application of IVVs does not increase meaningfully the computational cost, and it provides better handling of uncertainties in automated e-testing process.

Currently, the authors are working on applications of *type-2 fuzzy logic systems* [3] in e-testing. There are strong premises that this newer approach enables the effective modeling the NL uncertainties appearing in more complicated tests.

References

1. Cross V.V., Sudkamp T.A.: Similarity and Compatibility in Fuzzy Set Theory. Physica-Verlag, c/o Springer-Verlag, 2002
2. Entemann C.W.: A fuzzy logic with interval truth values. Fuzzy Sets and Systems, **113** (2000) 162–183
3. Karnik N.N., Mendel J.M.: An Introduction to Type-2 Fuzzy Logic Systems. University of Southern California, Los Angeles, 1998
4. Moore R., Lodwick W.: Interval analysis and fuzzy set theory. Fuzzy Sets and Systems **135** (2003) 5–9
5. Niewiadomski A., Szczepaniak P.S.: Fuzzy Similarity in E-Commerce Domains. In: Segovia J., Szczepaniak P.S., Niedzwiedzinski M.: E-Commerce and Intelligent Methods. Springer-Verlag, 2002
6. Niewiadomski A., Grzybowski R.: Fuzzy measures of text similarity in automated evaluation of exams tests. Therotical and Applied Computer Science, **5** (2003) 193–200 (in Polish)
7. Niewiadomski A., Jedynak A., Grzybowski R.: Automated evaluation of e-tests. Proceedings of 4th Ukrainian-Polish Conference Environmental Mechanics, Methods of Computer Science And Simulations, Lviv, Ukraine (June 24-26, 2004) 133–140 (in Polish)
8. Niewiadomski A., Rybusinski B., Sakowski K., Grzybowski R.: The application of multivalued similarity relations to automated evaluation of grammar tests. In: Academy On-line — e-learning, methodics, technologies, management (to appear; in Polish)
9. Ross T.J.: Fuzzy Logic with Engineering Applications. McGraw Hill Inc., 1995.
10. Szczepaniak P.S., Niewiadomski A.: Internet Searched Based on Text Intuitionistic Fuzzy Similarity. In: Szczepaniak P.S., Segovia J., Kacprzyk J., Zadeh L.A. (Eds.): Intelligent Exploration of the Web. Physica Verlag, c/o Springer-Verlag, 2003

Boolean Functions with a Low Polynomial Degree and Quantum Query Algorithms

Raitis Ozols[1], Rūsiņš Freivalds[1,*], Jevgeņijs Ivanovs[1], Elīna Kalniņa[1],
Lelde Lāce[1], Masahiro Miyakawa[2], Hisayuki Tatsumi[2], and Daina Taimiņa[3]

[1] Institute of Mathematics and Computer Science, University of Latvia,
Raina bulv. 29, Riga, Latvia
Rusins.Freivalds@mii.lu.lv
[2] Tsukuba College of Technology, 4-3-15 Amakubo,
Tsukuba, Ibaraki, 305-0005 Japan
mamiyaka@cs.k.tsukuba-tech.ac.jp
[3] Department of Mathematics, Cornell University, 511 Malott Hall,
Ithaca, NY, 14853, U.S.A.
dtaimina@math.cornell.edu

Abstract. The complexity of quantum query algorithms computing
Boolean functions is strongly related to the degree of the algebraic polynomial representing this Boolean function. There are two related difficult open problems. First, Boolean functions are sought for which the
complexity of exact quantum query algorithms is essentially less than
the complexity of deterministic query algorithms for the same function.
Second, Boolean functions are sought for which the degree of the representing polynomial is essentially less than the complexity of deterministic
query algorithms. We present in this paper new techniques to solve the
second problem.

1 Introduction

In the query model, the input $x_1, \ldots, x_N$ is contained in a black box and can
be accessed by queries to the black box. In each query, we give i to the black
box and the black box outputs x_i. The goal is to solve the problem with the
minimum number of queries. The classical version of this model is known as
decision trees The quantum version of this model is described in [2].

Deutsch [3] constructed an unexpected quantum query algorithm computing the 2-argument Boolean function PARITY with 1 query only such that
this algorithm produces the correct result with probability 1. Such quantum
query algorithms are called exact quantum query algorithms. It is a well-known
open problem to construct exact quantum query algorithms with complexity

---

* Research supported by Grant No.01.0354 from the Latvian Council of Science and
by the European Commission, Contract IST-1999-11234 (QAIP).

M. Bieliková et al. (Eds.): SOFSEM 2005, LNCS 3381, pp. 408–412, 2005.

(number of queries) much smaller than the complexity of deterministic query algorithms for the same Boolean function. Deutsch's algorithm allows to save half of queries needed for deterministic query algorithms to compute PARITY.

Unfortunately, up to now this is the best proved advantage of exact quantum vs. deterministic query algorithms. On the other hand, the complexity of an exact quantum query algorithm $Q_E(f)$ can never be less than $\sqrt[3]{D(f)}$ for the same function f (Midrijānis [5], improving over Beals et al. [1]). However there is a huge gap between $D(f)/2$ and $\sqrt[3]{D(f)}$. Many researchers have tried to bridge the gap but with no success.

Every Boolean function can be expressed as an algebraic polynomial. For instance, $x_1 \vee x_2$ can be expressed as $x_1 + x_2 - x_1 x_2$, $x_1 \oplus x_2$ can be expressed as $x_1 + x_2 - 2x_1 x_2$, the function $x_1 \vee x_2 \vee x_3$ can be expressed as $x_1 + x_2 + x_3 - x_1 x_2 - x_1 x_3 - x_2 x_3 - x_1 x_2 x_3$. This representing polynomial is unique. The degree of the representing polynomial is called the degree of the Boolean function and it is denoted as *deg(f)*.

Theorem 1. *[1] For arbitrary Boolean function f , $deg(f) \leq D(f)$.*

Theorem 2. *[1] For arbitrary Boolean function f , $\frac{1}{2}deg(f) \leq Q_E(f)$.*

Hence to find a Boolean function for which the exact quantum query complexity is essentially smaller than the deterministic query complexity, we are to consider Boolean functions with small *deg(f)* and larger *D(f)*. However even this problem presents a big difficulty. We present in this paper new techniques to solve this problem but the concrete results are still moderate. We have improved the existing best upper bound but the open problem is still open.

2   Graphs and Functions

We write variables $x_1, x_2, \ldots x_7$ where $x_i \in \{0,1\}$ (Figure 1) in a circle and consider them as graph vertices. Let us compare variables x_1 and x_2. If $x_1 = x_2$ then we connect them with a continuous line . If $x_1 \neq x_2$ then we connect them with dashed line. After that we compare x_2 and x_3 and again connect them with the appropriate line. We continue until we get variables x_7 and x_1 which we again connect. This is the way how we get closed cyclic graph with "coloured" edges (Figure 2). Edges which are drawn with dashed lines let us call differences. Now we show that number of differences will always be an even number. Sum of

$$(x_1 - x_2)^2 + (x_2 - x_3)^2 + \ldots + (x_6 - x_7)^2 + (x_7 - x_1)^2$$

is always an even number 0, 2, 4 or 6. Hence the number of differences can only be 0, 2, 4 or 6.

Moreover, it is easy to understand that the number of differences in the graph can be expressed by a function

$$\varphi = (x_1 - x_2)^2 + (x_2 - x_3)^2 + \ldots + (x_6 - x_7)^2 + (x_7 - x_1)^2.$$

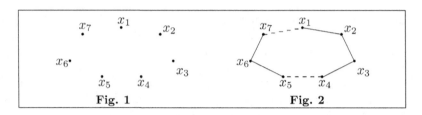

Hence for all $x_1, x_2, \ldots, x_7 \in \{0,1\}$, $\varphi \in \{0,2,4,6\}$. From here $\varphi - 3 \in \{-3,-1,1,3\}$ and $(\varphi - 3)^2 \in \{9,1,1,9\}$. $(\varphi - 3)^2 - 1 \in \{8,0,0,8\}$ and $f_0 = ((\varphi - 3)^2 - 1)/8 \in \{1,0,0,1\}$. Thus, the derived function $f_0 = f_0(x_1, x_2, \ldots x_7)$ is a Boolean function, because all of its variables $x_i \in \{0,1\}$, $f_0 \in \{0,1\}$. We have $deg(f_0) = deg(((\varphi - 3)^2 - 1)/8) = 4$. Obviously,

$$f_0 = \begin{cases} 0 \text{ if } d{=}2 \text{ or } d{=}4 \\ 1 \text{ if } d{=}0 \text{ or } d{=}6 \end{cases}, \text{(d is the number of differences)}.$$

Considering the case $x_1 = x_2 = \ldots = x_7$ we see that $D(f_0) = 7$.

The technique described above can be generalized:

1. We take any Boolean function $h(x,y)$ with two variables which has no fictive variables (for example, in case of function f_0 we consider function $h(x,y) = (x-y)^2 = x + y - 2xy$.

2. We consider a graph G with n vertices. For each vertice we assign a variable x_i. After then for each graph edge (x_i, x_j) we calculate $h(x_i, x_j)$ and the sum for all edges. Then we get a function $\varphi(x_1, x_2, \ldots, x_n)$:

$$\varphi(x_1, x_2, \ldots, x_n) = \sum_{\substack{x_i, x_j \in G}}^{1 \le i < j \le n} h(x_i, x_j)$$

where $(x_i, x_j) \in G$ means that the graph G contains an edge (x_i, x_j).

3. We find out what values can the function φ take. Then we define that for all x_i, $\varphi(x_1, x_2, \ldots, x_n) \in A$, where $A = \{a_1, a_2, \ldots, a_k\}$, $a_i \in N \cup \{0\}$.

4. We construct a polynomial $P = P(y)$ such that for $\forall y \in A$, $P(y) \in \{0,1\}$. We try to minimalize the degree of the polynomial.

5. We consider the function $f(x_1, x_2, \ldots, x_n) = P(\varphi(x_1, x_2, \ldots, x_n))$. This function is a Boolean function with $deg(f) \le deg(P)deg(\varphi) = 2deg(P) \le 2(|A|-1) = 2(k-1)$. If we manage to prove that $D(f)$ is sufficiently large or even $D(f) = n$ and k is small, then the function f is suitable for our purpose.

Now we wish to generalize the idea described above. We started from a particular graph. We used an algebraic polynomial (of low degree). However this polynomial brought us not Boolean values $\{0,1\}$ but values from a larger alphabet. In the example described above the polynomial was a second degree polynomial $\varphi = (x_1{}^2 - 2x_1x_2 + x_2{}^2) + (x_2{}^2 - 2x_2x_3 + x_3{}^2) + \ldots + (x_7{}^2 - 2x_7x_1 + x_1{}^2)$ and the values were $\{0,2,4,6\}$. Then we used another polynomial $\frac{1}{8}((d-3)^2 - 1)$ which transformed this set $\{0,2,4,6\}$ into Boolean values $\{0,1\}$. Again we were interested to have this polynomial of as low degree as possible.

Unfortunately, the graphs with more than 7 vertices do not possess an algebraic polynomial like our φ giving us no more than 4 possible values. However we can generalize this idea and start not with a graph but with an algebraic polynomial from variables $x_1, x_2, \ldots, x_N$.

Using this generalized idea, we now consider a polynomial $\varphi = (x_1 + x_2 - x_1 x_2 - x_1 x_3 - x_2 x_4 + x_3 x_4) + (x_5 + x_6 - x_5 x_6 - x_5 x_7 - x_6 x_8 + x_7 x_8) + (x_9 + x_{10} - x_9 x_{10} - x_9 x_{11} - x_{10} x_{12} + x_{11} x_{12})$ It is easy to check that this polynomial is a second degree polynomial taking values $\{0, 1, 2, 3\}$. Using the polynomial $\frac{1}{8}((2d-3)^2 - 1)$, we obtain Boolean values $\{0, 1\}$ as the values of the combined polynomial $\frac{1}{8}((2(x_1 + x_2 - x_1 x_2 - x_1 x_3 - x_2 x_4 + x_3 x_4) + (x_5 + x_6 - x_5 x_6 - x_5 x_7 - x_6 x_8 + x_7 x_8) + (x_9 + x_{10} - x_9 x_{10} - x_9 x_{11} - x_{10} x_{12} + x_{11} x_{12}) - 3)^2 - 1)$ of the degree 4. This polynomial represents a Boolean function f of 12 variables. We will immediately prove that the deterministic query complexity of this Boolean function is 9.

Theorem 3. *For f_n we have $D(f_n) = 9^n$ and $\deg(f_n) = 4^n$.*

Our main idea can be still generalized. We can use more values in the algebraic polynomial φ, provided we use a corresponding polynomial substituting $\frac{1}{8}((d-3)^2 - 1)$. There is a big temptation to start searching for low degree polynomials to transform many values into two Boolean values. However in all our polynomials φ we had the finite set values organized in an arithmetic progression. This is a serious restriction. Either we are to overcome this restriction or we are to use higher degree polynomials for the transformation.

We have found some rather low degree polynomials for such transformation. However they are somewhat exceptional polynomials obtained by exhaustive search.

Polynomial of Degree 4:

$$\frac{1}{24}x^4 - \frac{5}{24}x^3 + \frac{35}{24}x^2 - \frac{25}{12}x + 1$$

p(0)=1, p(1)=0, p(2)=0, p(3)=0, p(4)=0, p(5)=1

Polynomial of Degree 6:

$$\frac{1}{720}x^6 - \frac{7}{240}x^5 + \frac{35}{144}x^4 - \frac{49}{48}x^3 + \frac{203}{90}x^2 - \frac{49}{20}x + 1$$

p(0)=1, p(1)=0, p(2)=0, p(3)=0, p(4)=0, p(5)=0, p(6)=0, p(7)=1

Polynomial of Degree 8:

$$\frac{1}{40320}x^8 - \frac{1}{1120}x^7 + \frac{13}{960}x^6 - \frac{9}{80}x^5 + \frac{1069}{1920}x^4 - \frac{267}{160}x^3 + \frac{29531}{10080}x^2 - \frac{761}{280}x + 1$$

p(0)=1,p(1)=0,p(2)=0,p(3)=0,p(4)=0,p(5)=0,p(6)=0,p(7)=0,p(8)=0,p(9)=1

References

1. Robert Beals, Harry Buhrman, Richard Cleve, Michele Mosca, Ronald de Wolf: *Quantum Lower Bounds by Polynomials*. FOCS 1998: 352-361.
2. H. Buhrman and R. de Wolf. *Complexity Measures and Decision Tree Complexity : A Survey*. Theoretical Computer Science, v. 288(1): 21-43 (2002)
3. David Deutsch. *Quantum theory, the Church-Turing principle and the universal quantum computer*. Proceedings of the Royal Society, London, A400:97-117, (1985).
4. J. Gruska. *Quantum Computing*. McGraw-Hill, 1999.
5. Gatis Midrijānis. *Exact quantum query complexity for total Boolean functions*. http://arxiv.org/abs/quant-ph/0403168
6. M. Nielsen, I. Chuang. *Quantum Computation and Quantum Information*. Cambridge University Press, 2000

Representation of Extended RBAC Model Using UML Language

Aneta Poniszewska-Maranda[1], Gilles Goncalves[2], and Fred Hemery[3]

[1] Institute of Computer Science, Technical University of Lodz, Poland
anetap@ics.p.lodz.pl
[2] LGI2A - Universite d'Artois, Technoparc-Futura, Bethune, France
[3] IUT Bethune, Universite d'Artois, Bethune, France

Abstract. This paper presents an extension of the standard role-based access control (RBAC) model together with its representation using the Unified Modeling Language (UML). The presented model is developed for the role engineering in the security of information system.

The presented implementation of the RBAC model consists in role creation via defining appropriate permissions. The entire procedure is performed in two stages: defining the permissions assigned to a function and providing the definitions of functions assigned to a particular role.

1 Introduction

The role-based access control (RBAC) model was chosen to present and realize the access control in security processes of an information system. It allows the administrator to assign users to roles rather then users directly to permissions. The most important concept of RBAC is a role. It facilitates the access control management performed by the security administrator since users and permissions can be assigned to role. Each user has a utilization profile that defines his roles in the enterprise. The role is the set of permissions that allow execution of system functions and the access to the data used by these functions [1, 2].

The Unified Modeling Language (UML) [3] has been chosen for the representation of extended RBAC model because nowadays it is a standard tool, properly reflecting the description of the information system and its needs. In the recent years the UML has become a standard also for the object-oriented modeling in the field of the software engineering.

The first part of the paper describes our extension of the standard RBAC model, the second part the representation of the RBAC model concepts using UML, and the third part deals with the implementation of the RBAC model.

2 Extended RBAC Model

The RBAC model regulates the access of users to information basing on activities that users perform in a system [1, 2]. It requires identification of roles in this

M. Bieliková et al. (Eds.): SOFSEM 2005, LNCS 3381, pp. 413–417, 2005.

system. In course of study of this issue, the standard model was extended by addition of new elements [4] (Figure 1).

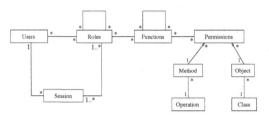

Fig. 1. Extension of the RBAC model

Each role defined in the RBAC model realizes a specific task in the enterprise process. A role contains many functions that the user can take. For each role it is possible to choose the system functions which are necessary for it. Thus a role can be presented as a set of functions that this role can take and realize. Each function can have one or more permissions, and consequently a function can be defined as a set of permissions. If an access to an object is required, then the necessary permissions can be assigned to the function to complete the desired job. All permissions can give the possibility to perform the role responsibilities.

In order to access data stored in an object a message has to be sent to this object. This message causes execution of a particular method on this object. Permission can be viewed as a possibility of execution a given method on an object in order to access data stored in this object. The same permission is defined for all instances of the object except the constraint specification.

3   Representation of the Extended RBAC Model Using the UML Concepts

UML [3] contains a suite of diagrams for requirements, analysis design and implementation of the development of systems. Out of nine types of diagrams defined by UML and representing different viewpoints of the modeling, two types, use case diagram and sequence diagram, are in the focus of attention of this study.

The use case diagram represents the functions of the system from the users' point of view. According to the definition of the UML meta-model, each use case diagram contains some use cases and each use case represents a collaboration, i.e. a scenario. To determine each collaboration an interaction diagram can be created. The interaction diagram represents the objects and messages exchanged in the use case. Two types of interaction diagrams, sequence diagram and collaboration diagram, are defined to capture how objects collaborate with each other to realize the use cases. The purpose of the presented study is to implement and realize the extended RBAC model with the use of UML. To this aim, the conceptions of the UML and RBAC model (Figure 2) should be first joined.

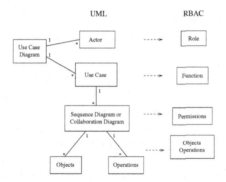

Fig. 2. UML concepts and their relationships with the RBAC model

Role-Actor. The UML supports the conception of the use case, on which the developed system is based. This approach is realized via use case diagram from which the concept of an actor very close to that of the role concept in the RBAC model is derived. In UML the actor defined as a role or a set of roles played by a person or by a group of people interacting with a system, represents a category of users that share the same functions or the same activities in an organization.

Function - Use Case. Each actor co-operates with one or more use cases representing the essential functions in a system. Use cases can be assigned to the functions and for each role the functions realized by this role to co-operate with the system can be chosen. This process is similar to that of connecting actors with their use cases. Thus, in the application realized with the use of the UML, a function can be replaced by a use case.

Methods and Objects. The methods of RBAC model are represented in UML by the methods executed in different types of diagrams, e.g. in a sequence diagram, while for the objects of RBAC model the UML object concept is used.

Permissions - Sequence Diagram. A use case contains a sequence of actions executed by an actor in a system. Although for the purpose of this study sequence diagrams have been chosen to present the use cases, the same results could be obtained with the use of collaboration diagrams. A sequence diagram represents an interaction of objects by means of a sequence of messages sent between these objects. In each sequence diagram there is one or more objects representing actors who in this way participle directly in the process of sending messages between the objects. This process is realized by the execution of methods on objects. A message sent to an object is a method. Access to the object is controlled with respect to the right of the method execution possessed by a subject over an object. Thus, for each use case it is necessary to specify permissions for the method execution. These permissions are assigned to the functions by the function-permission (F-P) relation.

Constraints. The concept of constraints of the RBAC model is connected directly with that of the constraint conception existing in the UML language.

Relations. The relations that occur between the elements of the RBAC model can be found in the use case diagrams and interaction diagrams. A use case diagram exploits four types of relations: communication relation between an actor and a use case (being a relation between a role and a function, R-F relation), generalization relation between actors (representing a heritage relation between roles, R-R relation), two types of relations: extension relation and utilization relation, both of them occurring between use cases (represent relations between functions in the RBAC model, F-F relation). The relations between functions and permissions (F-P relations) can be specified by means of connections between the use cases and the interaction diagrams that describe them.

4   Implementation of RBAC Model - Creation of Roles

The UML meta-model is applied to define the roles of RBAC models, the functions that are used by these roles to co-operate with the information system and the permissions needed to realize these functions. Owing to use case diagrams a list of actors co-operating with the information system and a list of use cases of this system are obtained. An analysis of these diagrams allows automatic specification of relations of the following types: R-R relation (with the use of generalization relation between the actors), R-F relation (association relation between the actors and the use cases) and F-F relation (generalization relation between the use cases). The description of a use case using the interaction diagrams presents activities needed to realize the functions of a system. Each activity is a definition of execution of a method on an object. Therefore the F-P relations can also be automatically managed.

Our definition of a set of roles in an information system with the use of UML diagrams contains two stages: assignment of a set of privileges (permissions) to the use case in order to define the function and assignment of a set of use cases (functions) in order to define the role.

5   Conclusions

The paper proposes a representation of the extension of the classic RBAC model using the UML language. It shows the relationships between the extended RBAC concepts and those of the UML language. The elaborated RBAC extension gives the possibility to manage the system and to make the process of the role creation simpler, particularly in the conception phase. The implementation of the RBAC model is also based on the UML. The creation of the user profile, performed in two stages is realized using the UML concepts.

References

1. Sandhu R.S., Coyne E.J., Feinstein H.L, Youman C.B.: Role-Based Access Control Models. IEEE Computer, 1996
2. Ferraiolo D., Sandhu R.S., Gavrila S., Kuhn D.R., Chandramouli R.: Proposed NIST Role-Based Access control. ACM TISSEC, Volume 4, Number 3, 2001
3. OMG Unified Modeling Language Specification. OMG, Reference Manual, 2003
4. Goncalves G., Hemery F., Poniszewska A.: Verification of Access Control Coherence in Information System during Modifications. 12th IEEE WETICE, Austria, 2003

A Methodology for Writing Class Contracts

Nele Smeets* and Eric Steegmans

Katholieke Universiteit Leuven, Department of Computer Science,
Celestijnenlaan 200A, 3001 Leuven, Belgium
{Nele.Smeets, Eric.Steegmans}@cs.kuleuven.ac.be

Abstract. One of the principles of Design by Contract is that contracts for software components must be written in a declarative way, using a formal, mathematically founded notation. When we apply the Design by Contract methodology in a naive and straightforward way, we risk ending up with unwanted duplication. In this paper, we describe a methodology for writing class contracts that avoids specification duplication and that gives rise to uniform class specifications with a clear and fixed structure.

1 Introduction

Design by Contract (DBC) [12] states that contracts for software components must be written in a declarative way, using a formal, mathematically founded notation. When we apply the DBC methodology in a naive and straightforward way, we risk ending up with unwanted duplication.

For example, consider the following Java code, describing a class of bounded lists. To save space, only a formal specification is given. When specification and implementation are similar, the implementation is left out too.

```
/** A class of bounded lists.
 * @invar   getElements () != null
 * @invar   getCapacity () > 0
 * @invar   getElements ().size () <= getCapacity ()
 */
public class BoundedList {
  private List elements;
  /** @basic */ public List getElements () { ... }
  private int capacity;
  /** @basic */
  public int getCapacity () { return capacity; }
  /** Set the capacity of this list to the given value.
   * @pre     capacity > 0
   * @pre     getElements ().size () <= capacity
   * @post    getCapacity () == capacity
   */
  public void setCapacity (int capacity) { ... }
  ...
}
```

---

* Research Assistant of the Fund for Scientific Research - Flanders (Belgium).

M. Bieliková et al. (Eds.): SOFSEM 2005, LNCS 3381, pp. 418–422, 2005.

Despite its simplicity, the class of bounded lists already gives an example of unwanted specification duplication. Indeed, part of the class invariant is repeated in the precondition of the `setCapacity` method to ensure that, after application of the method, the bounded list still satisfies its class invariant.

Duplicating constraints imposed on the characteristics of objects has a negative impact on software quality. Duplication of constraints hampers adaptability. Moreover, we risk ending up with an inconsistent class definition. Further, correctness is at stake, because there is a considerable risk of being incomplete, when writing the preconditions of a method. When we take inheritance into account, things get even more complicated. When the class invariant is strengthened in a subclass, the precondition of the `setCapacity` method must be strengthened accordingly. Since the Liskov substitution principle [11] expresses that preconditions can only be weakened, we are forced to use abstract preconditions [12]. Abstract preconditions will play an important role in our methodology.

Summarizing, a naive and straightforward application of the principles underlying DBC violates a basic principle of good software engineering: each fact must be worked out in one, and only one, place [14].

In this paper, we describe a methodology for writing class contracts. Our methodology avoids specification duplication and gives rise to uniform class specifications with a clear and fixed structure. We describe our methodology using the Java programming language, but it is also applicable to other object-oriented languages.

2   Basic Queries

In our methodology, a first important step in writing class contracts is choosing a set of *basic queries*. This way of working is also advocated in [13] and [15]. Basic queries are a minimal set of queries chosen in such a way that the entire state of an object can be inspected using basic queries only. In our example, there are two basic queries, indicated by the @basic-tag. Class contracts can be specified thoroughly in terms of the chosen basic queries. We also avoid the risk of writing circular assertions.

To describe the conditions imposed on the characteristics of the objects of a class, we introduce two new concepts: partial conditions (Sect. 3) and complete conditions (Sect. 4). Both are represented by Boolean queries. The concepts will be presented for a class X containing two basic queries $getv_1()$ and $getv_2()$ with return types V_1 and V_2.

3   Partial Conditions

A *partial condition* on a set of characteristics is a constraint involving all these characteristics. For example, the partial condition on the capacity of a bounded list expresses that the capacity should be positive. The partial condition on the capacity and the list elements expresses that the number of elements should not exceed the capacity of the list. The developer of a class is responsible for providing the partial conditions of the class. For each non-empty subset of characteristics, a partial condition must be introduced. A partial condition can be deterministic or non-deterministic.

In our methodology, the following Boolean queries are introduced to encapsulate the partial conditions on the characteristics of class X.

```
/** @return   ... */
public boolean partialConditionv₁(V₁ v₁) { ... }
/** @return   ... */
public boolean partialConditionv₂(V₂ v₂) { ... }
/** @pre      partialConditionv₁(v₁)
 *   @pre      partialConditionv₂(v₂)
 *   @return   ...
 */
public boolean partialConditionv₁v₂(V₁ v₁, V₂ v₂) { ... }
```

In general, evaluating `partialConditionv₁v₂(v₁, v₂)` is only meaningful when `partialConditionv₁(v₁)` and `partialConditionv₂(v₂)` evaluate to true. This explains the two preconditions of the `partialConditionv₁v₂` method.

4   Complete Conditions

A *complete condition* on a set of characteristics is the conjunction of all partial conditions concerning at least one of those characteristics. Thus, a complete condition specifies all conditions that must be satisfied by a set of characteristics. In our example, the complete condition on the capacity expresses that the capacity must be positive and that the current number of list elements must be smaller than or equal to the capacity.

Since all complete conditions have similar semantics, the complete conditions on proper subsets of characteristics are defined in terms of the complete condition on all characteristics. The complete condition *on all characteristics* is given by the following Boolean query. The query is made non-deterministic to support overriding in subclasses (see below).

```
/** @return   if ( not { partialConditionv₁(v₁) &&
 *                        partialConditionv₂(v₂) &&
 *                        partialConditionv₁v₂(v₁, v₂) } )
 *                then result == false
 */
public boolean completeConditionv₁v₂(V₁ v₁, V₂ v₂) { ... }
```

The complete conditions *on proper subsets of characteristics* are defined in terms of the complete condition on all characteristics `completeConditionv₁v₂`.

```
/** @return   completeConditionv₁v₂(v₁, getv₂()) */
public boolean completeConditionv₁(V₁ v₁) { ... }
/** @return   completeConditionv₁v₂(getv₁(), v₂) */
public boolean completeConditionv₂(V₂ v₂) { ... }
```

When a new characteristic is introduced in a subclass of X, represented by the basic query $getv_3$, the three complete conditions shown above are redefined in terms of the complete condition $completeConditionv_1v_2v_3$. It can be shown that these redefinitions satisfy the Liskov substitution principle [11].

5   Advantages and Disadvantages

In this section, we examine the influence of our methodology on the quality factors described in [12].

Our methodology has a positive impact on *correctness*, since it avoids duplication and helps in writing complete specifications.

It also has a good influence on *extendibility*. When a condition on a certain characteristic or set of characteristics changes, only local changes are required. The reason is that this knowledge is not spread around over the whole class, but is concentrated in one single Boolean query, namely the corresponding partial condition.

Using our methodology gives rise to well-documented classes with a transparent structure. In this way, it has a positive influence on *reusability*.

When writing a class containing n characteristics, $2^n - 1$ partial and complete conditions are needed. This exponential number of methods considerably increases the size of a class, thereby requiring more memory, so *efficiency* is negatively influenced.

Using our methodology, we abstract away from the code level, as prescribed by MDA (Model Driven Architecture) [5]. From the partial conditions, a proper tool can generate the complete conditions, the basic queries, the set methods and the class invariant. Since the basic structure of the class is generated, the developer can pay more attention to the other parts of the class, which has a positive influence on productivity and on the general quality of the code.

6   Related Work

The ideas underlying DBC go back to the work of pioneers in computer science such as Dijkstra [2] and Hoare [6]. DBC, as we know it today, was developed by Bertrand Meyer as a part of Eiffel [12]. Most object-oriented languages, including Java, C++ and C#, lack support for DBC. Nevertheless, there is a growing interest in DBC and several tools have been developed that provide support for DBC in Java ([1], [3], [4], [7], [8], [9]). The Java Modeling Language [10] and the Object Constraint Language [16] are examples of formal specification languages.

7   Conclusion

In this paper, we described a methodology for writing class contracts. Our methodology avoids specification duplication and gives rise to uniform class specifications with a clear and fixed structure. Using our methodology, the basic structure of a class can be generated automatically by a tool, allowing the developer to pay more attention to the other parts of the class.

References

1. Della Torre Cicalese, C., Rotenstreich, S.: Behavioral Specification of Distributed Software Component Interfaces. IEEE Computer, Vol. 32, No. 7 (1999) 46–53
2. Dijkstra, E.W.: A Discipline of Programming. Prentice Hall, ISBN 013215871X (1976)
3. Duncan, A., Hölzle, U.: Adding Contracts to Java with Handshake. http://www.cs.ucsb.edu/labs/oocsb/papers/TRCS98-32.pdf
4. Findler, R.B., Felleisen, M.: Contract Soundness for Object-Oriented Languages. http://www.ccs.neu.edu/scheme/pubs/oopsla01-ff.pdf
5. Frankel, D.S.: Model Driven Architecture. Applying MDA to Enterprise Computing. Wiley Publishing, ISBN 0-471-31920-1 (2003)
6. Hoare, C.A.R.: An axiomatic basis for computer programming. Communications of the ACM, Vol. 12, No. 10 (1969) 576–583
7. The Jass Page. Home Page. http://semantik.informatik.uni-oldenburg.de/~jass/
8. Karaorman, M., Hölzle, U., Bruno, J.: jContractor: A Reflective Java Library to Support Design By Contract. http://www.cs.ucsb.edu/labs/oocsb/papers/TRCS98-31.pdf
9. Kramer, R.: iContract – The Java Design by Contract Tool. TOOLS 26: Technology of Object-Oriented Languages and Systems, Los Alamitos, California (1998) 295–307
10. Leavens, G.T., Baker, A.L., Ruby, C.: Preliminary Design of JML: A Behavioral Interface Specification Language for Java. Department of Computer Science, Iowa State University, TR #98-06x (2004) ftp://ftp.cs.iastate.edu/pub/leavens/JML/prelimdesign.pdf
11. Liskov, B., Wing, J.: A behavioral notion of subtyping. ACM Transactions on Programming Languages and Systems, Vol. 16, No. 6 (1994) 1811–1841
12. Meyer, B.: Object-Oriented Software Construction, Second Edition. Prentice-Hall Inc, ISBN 0-13-629155-4 (1997)
13. Mitchell, R., McKim, J.: Design by Contract, by Example. Addison-Wesley, ISBN 0-201-63460-0 (2002)
14. Parnas, D.L.: On the Criteria to Be Used in Decomposing Systems into Modules. Communications of the ACM, Vol. 15, No. 12 (1972) 1053–1058.
15. Steegmans, E., Dockx, J.: Objectgericht programmeren met Java. Acco, ISBN 90-334-4535-2 (2002)
16. Warmer, J.B., Kleppe, A.G.: The Object Constraint Language, Precise Modeling with UML. Addison-Wesley, ISBN 0-201-37940-6 (1999)

Volumes of 3D Drawings of Homogenous Product Graphs (Extended Abstract)

Lubomir Torok*

Institute of Mathematics and Computer Science,
Slovak Academy of Sciences,
Severna 5, 974 01 Banska Bystrica, Slovak Republic
torok@savbb.sk

Abstract. 3-dimensional layout of graphs is a standard model for orthogonal graph drawing. Vertices are mapped into the 3D grid and edges are drawn as the grid edge disjoint paths. The main measure of the efficiency of the drawing is the volume which is motivated by the 3D VLSI design. In this paper we develop a general framework for efficient 3D drawing of product graphs in both 1 active layer and general model. As a consequence we obtain several optimal drawings of product graphs when the factor graphs represent typical networks like CCC, Butterfly, star graph, De Bruijn... This is an analogue of a similar work done by Fernandez and Efe [2] for 2D drawings using a different approach. On the other hand our results are generalizations of the optimal 3D drawings of hypercubes [9].

1 Preliminaries

Two models of 3D drawings are considered. One-active-layer (1-AL) model is a natural generalization of 2D layout, when vertices are placed in the basic plane and edges are routed in the volume above the basic plane in edge-disjoint manner. In general model there are no restrictions on vertices placement and edges are routed in edge-disjoint manner. The main measure of drawing effectivity is its volume.

The cartesian product is well-known operation defined on graphs. When applied, the cartesian product combines a set of "factor" graphs into a "product" graph. Several well-known networks are instances of product networks, including the grid, the hypercube and the torus. In this paper, we consider only homogenous products, i.e. the factor graphs are *isomorphic*.

The following theorem provides the general lower bounds for volumes of graph G in both models [9].

Theorem 1. *The optimal volume of 3-dimensional 1-active layer layout of any graph G with cutwidth $cw(G)$ satisfies*

---

* This research was supported by VEGA grants No. 2/3164/23 and No. 2/2060/23.

M. Bieliková et al. (Eds.): SOFSEM 2005, LNCS 3381, pp. 423–426, 2005.

$$VOL_{1-AL}(G) \geq cw(G)\sqrt{\Sigma_{v\in V} deg^2(v)}.$$

The optimal volume of 3-dimensional layout of any graph G with cutwidth $cw(G)$ satisfies

$$VOL(G) \geq \left(cw(G) - \sqrt{2cw(G)}\right)^{\frac{3}{2}}.$$

Definition 1. *[4] Let the routing ρ be defined as follows. For every two distinct vertices u, v of G there exist 2 paths, from u to v and from v to u. The number of paths of ρ is then $n(n-1)$. The edge forwarding index of (G, ρ) denoted by $\pi(G, \rho)$ is the maximum number of paths, specified by routing ρ going through any edge of G. More precisely :*

$$\pi(G, \rho) = max\{\pi_e(G, \rho) : e \in E(G)\}$$

and the edge-forwarding index of G is defined as

$$\pi(G) = min\{\pi(G, \rho) : \forall\rho\}.$$

The following lemma [2] provides the lower bound for the bisection width of product graph G^r. Since the bisection width is a lower bound for cutwidth, this lemma is useful for the approximation of the cutwidth of a product graph G^r.

Lemma 1. *If the edge forwarding index of factor graph G with n vertices is $\pi(G)$ then the bisection width of the product graph G^r satisfies*

$$bw(G^r) \geq \frac{n^{2r} - 1}{2\pi(G)n^{r-1}}.$$

2   Layout Volumes of Cartesian Product Graphs

In this section we consider the volumes of cartesian product graphs in 1-AL layer and general model. We provide the lower bounds and the constructions of the upper bounds for the layout volumes in both models.

The following lemma offers the relation between the cutwidths of cartesian product graphs and their factors.

Lemma 2. *Let G^r be the homogenous product graph with factor graph G, and let n be the number of vertices of G. Then*

$$cw(G^r) \leq \frac{n^r - 1}{n - 1}cw(G) = O\left(n^{r-1}cw(G)\right).$$

2.1   One-Active-Layer Model

Theorem 2. *Let G^r be the homogenous product graph with factor graph G and let r be divisible by 2. Then for the volume of the layout in 1-AL model we have the following bounds.*

$$V_{1-AL}(G^r) = \Omega\left(n^{\frac{r}{2}}r\Delta(G)cw(G^r)\right)$$

$$V_{1-AL}(G^r) = O\left(n^r r\Delta(G)cw(G^{\frac{r}{2}})\right)$$

Proof. For the upper bound we generalize the construction of 1-AL layout from [9]. The lower bound comes from Theorem 1.

Observation 1. *The construction of 1-AL layout from proof of Theorem 2 is asymptotically optimal if for the cutwidth of the product graph holds*

$$cw(G^r) = \Theta\left(n^{r-1}cw(G)\right).$$

2.2   General Model

Theorem 3. *Let G^r be the homogenous product graph with factor graph G and let r be divisible by 3. Then we have the following bounds for the volume of the layout in general model of graph G.*

$$VOL(G^r) = \Omega\left(cw^{\frac{3}{2}}(G^r)\right)$$

$$VOL(G^r) = O\left(n^r(\Delta(G) + \sqrt{cw(G^{\frac{r}{3}})})^3\right)$$

Proof. For the upper bound we generalize the construction from [9]. The lower bound comes from Theorem 1.

Observation 2. *The construction of the layout in general model from the proof of Theorem 3 is asymptotically optimal if*
$cw(G^r) = \Theta\left(n^{r-1}cw(G)\right)$ *and* $\Delta^2(G^r) << cw(G^{\frac{r}{2}}).$

2.3   Layout Volumes of Some Known Product Graphs

Table 1 contains the parameters of the factor graphs used in the framework to obtain the volumes in both models. The overview of the results for both models is in Table 2.

Table 1. Input parameters of considered factor graphs

Factor graph	n	$\Delta(G)$	$cw(G)$	$\pi(G)$
deBruijn	2^m	4	$\Theta(\frac{2^{m+1}}{m})$ [6]	$\Theta(m2^{m-1})$ [7]
Star graph	$m!$	$m-1$	$\Theta(m!)$ [1]	$\Theta(m!)$ [3]
Complete transposition graph	$m!$	$m-1$	$\Theta(mm!)$ [8]	$\Theta((m-1)!)$ [3]
Butterfly graph	$m2^m$	4	$\Theta(2^m)$	$\Theta(m^22^{m-1})$ [7]
Complete graph	m	$m-1$	$\frac{m^2}{4}$ [1]	2
CCC graph	$m2^m$	m	$\Theta(2^m)$	$\Theta(m^22^m)$ [7]
Linear array	m	2	1	$\Theta(m^2)$ [4]

Table 2. Optimal layout volumes of several product graphs

Product graph	1-AL model	General model
Complete transposition graph product	$\Theta\left((m!)^{\frac{3r}{2}}rm(m-1)\right)$	$\Theta\left((m(m!)^r)^{\frac{3}{2}}\right)$
deBruijn product	$\Theta\left(r\dfrac{n^{\frac{3r}{2}}}{\log n}\right)$	$\Theta\left(\dfrac{n^{\frac{3}{2}(r+1)}}{\log^{\frac{3}{2}}n}\right)$
Star graph product	$\Theta\left((m!)^{\frac{3r}{2}}r(m-1)\right)$	$\Theta\left((m!)^{\frac{3r}{2}}\right)$
Butterfly product	$\Theta(m^{\frac{3r}{2}-1}2^{\frac{3mr}{2}}r)$	$\Theta(m^{\frac{3r}{2}-\frac{3}{2}}2^{\frac{3mr}{2}})$
Product of complete graphs	$\Theta\left(m^{\frac{3r}{2}+1}r(m-1)\right)$	$\Theta\left(m^{\frac{3r}{2}+\frac{3}{2}}\right)$
CCC graph product	$\Theta\left(m^{\frac{3r}{2}}r2^{\frac{3mr}{2}}\right)$	$\Theta\left(m^{\frac{3r}{2}-\frac{3}{2}}2^{\frac{3mr}{2}}\right)$
Linear array product	$\Theta\left(m^{\frac{3r}{2}-1}r\right)$	$\Theta\left(m^{\frac{3r}{2}-\frac{3}{2}}\right)$
Hypercube	$\Theta\left(2^{\frac{3m}{2}}m\right)$ [9]	$\Theta\left(2^{\frac{3m}{2}}\right)$ [9]

Acknowledgment

I would like to thank to my PhD. supervisor Imrich Vrt'o for his valuable help and ideas during my work on this paper.

References

1. Chi-Hsiang Yeh, Behrooz Parhami, VLSI layouts of complete graphs and star graphs, IPL, 68 (1998) 39-45.
2. Fernandez, A., Efe, K., Efficient VLSI layouts for homogenous product networks IEEE Transactions on Computers, 46 (1997) 1070-1082.
3. Ginette, G., Edge Forwarding Index of Cayley Graphs and Star Graphs, Discrete Applied Mathematics, 80 (1997) 149-160.
4. Heydeman, M.C., Meyer, J.C., Sotteau, D., On forwarding indices of networks, Discrete Applied Mahtematics, 23 (1989) 103-123
5. Leighton, F.T., Rosenberg, A.L., Three-dimensional circuit layouts, SIAM Journal on Computing, 15 (1986) 793-813.
6. Raspaud, A., Sýkora, O., Vrt'o, I., Cutwidth of the de Bruijn graph, RAIRO, 26 (1995) 509-514.
7. Shahrokhi, F., Szekely, L.A.: An algebraic approach to the uniform concurrent multicommodity flow problem: Theory and applications, Technical report CRPDC-91-4, DCS, Uni. North texas, Denton, 1991.
8. Stacho, L., Vrt'o, I. Bisection width of transposition graphs, Discrete Applied Mathematics, 84 (1998) 221-235.
9. Torok, L., Vrt'o, I. Layout volumes of the hypercube, in: Proc. 12th Intl. Symposium on Graph Drawing, Lecture Notes in Computer Science, Springer, 2004 (to appear).

Author Index

Lecture Notes in Computer Science

For information about Vols. 1–3256

please contact your bookseller or Springer

Vol. 3303: J.A. López, E. Benfenati, W. Dubitzky (Eds.), Knowledge Exploration in Life Science Informatics. X, 249 pages. 2004. (Subseries LNAI).

Vol. 3302: W.-N. Chin (Ed.), Programming Languages and Systems. XIII, 453 pages. 2004.

Vol. 3300: L. Bertossi, A. Hunter, T. Schaub (Eds.), Inconsistency Tolerance. VII, 295 pages. 2004.

Vol. 3299: F. Wang (Ed.), Automated Technology for Verification and Analysis. XII, 506 pages. 2004.

Vol. 3298: S.A. McIlraith, D. Plexousakis, F. van Harmelen (Eds.), The Semantic Web – ISWC 2004. XXI, 841 pages. 2004.

Vol. 3296: L. Bougé, V.K. Prasanna (Eds.), High Performance Computing - HiPC 2004. XXV, 530 pages. 2004.

Vol. 3295: P. Markopoulos, B. Eggen, E. Aarts, J.L. Crowley (Eds.), Ambient Intelligence. XIII, 388 pages. 2004.

Vol. 3294: C.N. Dean, R.T. Boute (Eds.), Teaching Formal Methods. X, 249 pages. 2004.

Vol. 3293: C.-H. Chi, M. van Steen, C. Wills (Eds.), Web Content Caching and Distribution. IX, 283 pages. 2004.

Vol. 3292: R. Meersman, Z. Tari, A. Corsaro (Eds.), On the Move to Meaningful Internet Systems 2004: OTM 2004 Workshops. XXIII, 885 pages. 2004.

Vol. 3291: R. Meersman, Z. Tari (Eds.), On the Move to Meaningful Internet Systems 2004: CoopIS, DOA, and ODBASE, Part II. XXV, 824 pages. 2004.

Vol. 3290: R. Meersman, Z. Tari (Eds.), On the Move to Meaningful Internet Systems 2004: CoopIS, DOA, and ODBASE, Part I. XXV, 823 pages. 2004.

Vol. 3289: S. Wang, K. Tanaka, S. Zhou, T.W. Ling, J. Guan, D. Yang, F. Grandi, E. Mangina, I.-Y. Song, H.C. Mayr (Eds.), Conceptual Modeling for Advanced Application Domains. XXII, 692 pages. 2004.

Vol. 3288: P. Atzeni, W. Chu, H. Lu, S. Zhou, T.W. Ling (Eds.), Conceptual Modeling – ER 2004. XXI, 869 pages. 2004.

Vol. 3287: A. Sanfeliu, J.F. Martínez Trinidad, J.A. Carrasco Ochoa (Eds.), Progress in Pattern Recognition, Image Analysis and Applications. XVII, 703 pages. 2004.

Vol. 3286: G. Karsai, E. Visser (Eds.), Generative Programming and Component Engineering. XIII, 491 pages. 2004.

Vol. 3285: S. Manandhar, J. Austin, U.B. Desai, Y. Oyanagi, A. Talukder (Eds.), Applied Computing. XII, 334 pages. 2004.

Vol. 3284: A. Karmouch, L. Korba, E.R.M. Madeira (Eds.), Mobility Aware Technologies and Applications. XII, 382 pages. 2004.

Vol. 3283: F.A. Aagesen, C. Anutariya, V. Wuwongse (Eds.), Intelligence in Communication Systems. XIII, 327 pages. 2004.

Vol. 3282: V. Guruswami, List Decoding of Error-Correcting Codes. XIX, 350 pages. 2004.

Vol. 3281: T. Dingsøyr (Ed.), Software Process Improvement. X, 207 pages. 2004.

Vol. 3280: C. Aykanat, T. Dayar, İ. Körpeoğlu (Eds.), Computer and Information Sciences - ISCIS 2004. XVIII, 1009 pages. 2004.

Vol. 3279: G.M. Voelker, S. Shenker (Eds.), Peer-to-Peer Systems III. XI, 300 pages. 2004.

Vol. 3278: A. Sahai, F. Wu (Eds.), Utility Computing. XI, 272 pages. 2004.

Vol. 3275: P. Perner (Ed.), Advances in Data Mining. VIII, 173 pages. 2004. (Subseries LNAI).

Vol. 3274: R. Guerraoui (Ed.), Distributed Computing. XIII, 465 pages. 2004.

Vol. 3273: T. Baar, A. Strohmeier, A. Moreira, S.J. Mellor (Eds.), <<UML>> 2004 - The Unified Modelling Language. XIII, 454 pages. 2004.

Vol. 3272: L. Baresi, S. Dustdar, H. Gall, M. Matera (Eds.), Ubiquitous Mobile Information and Collaboration Systems. VIII, 197 pages. 2004.

Vol. 3271: J. Vicente, D. Hutchison (Eds.), Management of Multimedia Networks and Services. XIII, 335 pages. 2004.

Vol. 3270: M. Jeckle, R. Kowalczyk, P. Braun (Eds.), Grid Services Engineering and Management. X, 165 pages. 2004.

Vol. 3269: J. Lopez, S. Qing, E. Okamoto (Eds.), Information and Communications Security. XI, 564 pages. 2004.

Vol. 3268: W. Lindner, M. Mesiti, C. Türker, Y. Tzitzikas, A. Vakali (Eds.), Current Trends in Database Technology - EDBT 2004 Workshops. XVIII, 608 pages. 2004.

Vol. 3267: C. Priami, P. Quaglia (Eds.), Global Computing. VIII, 377 pages. 2004.

Vol. 3266: J. Solé-Pareta, M. Smirnov, P.V. Mieghem, J. Domingo-Pascual, E. Monteiro, P. Reichl, B. Stiller, R.J. Gibbens (Eds.), Quality of Service in the Emerging Networking Panorama. XVI, 390 pages. 2004.

Vol. 3265: R.E. Frederking, K.B. Taylor (Eds.), Machine Translation: From Real Users to Research. XI, 392 pages. 2004. (Subseries LNAI).

Vol. 3264: G. Paliouras, Y. Sakakibara (Eds.), Grammatical Inference: Algorithms and Applications. XI, 291 pages. 2004. (Subseries LNAI).

Vol. 3263: M. Weske, P. Liggesmeyer (Eds.), Object-Oriented and Internet-Based Technologies. XII, 239 pages. 2004.

Vol. 3262: M.M. Freire, P. Chemouil, P. Lorenz, A. Gravey (Eds.), Universal Multiservice Networks. XIII, 556 pages. 2004.

Vol. 3261: T. Yakhno (Ed.), Advances in Information Systems. XIV, 617 pages. 2004.

Vol. 3260: I.G.M.M. Niemegeers, S.H. de Groot (Eds.), Personal Wireless Communications. XIV, 478 pages. 2004.

Vol. 3259: J. Dix, J. Leite (Eds.), Computational Logic in Multi-Agent Systems. XII, 251 pages. 2004. (Subseries LNAI).

Vol. 3258: M. Wallace (Ed.), Principles and Practice of Constraint Programming – CP 2004. XVII, 822 pages. 2004.

Vol. 3257: E. Motta, N.R. Shadbolt, A. Stutt, N. Gibbins (Eds.), Engineering Knowledge in the Age of the Semantic Web. XVII, 517 pages. 2004. (Subseries LNAI).